Environmental Footprints and Eco-design of Products and Processes

Series editor

Subramanian Senthilkannan Muthu, SGS Hong Kong Limited,
Hong Kong, Hong Kong SAR

More information about this series at http://www.springer.com/series/13340

Kirsten Heimann · Obulisamy Parthiba Karthikeyan
Subramanian Senthilkannan Muthu
Editors

Biodegradation and Bioconversion of Hydrocarbons

Springer

Editors
Kirsten Heimann
James Cook University
Townsville, QLD
Australia

Obulisamy Parthiba Karthikeyan
James Cook University
Townsville, QLD
Australia

Subramanian Senthilkannan Muthu
Environmental Services Manager-Asia
SGS Hong Kong Limited
Hong Kong
Hong Kong SAR

ISSN 2345-7651 ISSN 2345-766X (electronic)
Environmental Footprints and Eco-design of Products and Processes
ISBN 978-981-10-0199-4 ISBN 978-981-10-0201-4 (eBook)
DOI 10.1007/978-981-10-0201-4

Library of Congress Control Number: 2016951974

Printed on acid-free paper

This Springer imprint is published by Springer Nature
The registered company is Springer Nature Singapore Pte Ltd.
The registered company address is: 152 Beach Road, #22-06/08 Gateway East, Singapore 189721, Singapore

Preface

Hydorcarbon oil contamination poses severe environmental risks and polycyclic aromatic hydrocarbons (PAHs) arising from pyrogenic origins can adversely affect human health through consumption of contaminated food. Hydrocarbon inputs into waterways, sediments and seas and oceans arise from both natural seepages and anthropogenic activities. While acute-large inputs arising from shipping and oil and gas exploration/refining accidents attract large media attention, chronic inputs from non-point sources are far less well described. In particular inputs from previously buried sediments through dredging and other sediment disturbance events can only be quantified using costly sophisticated radioisotope analyses and gas-chromatography/mass spectrometry. As oil and gas exploitation increase to service the growing energy demand of the fast rising population, so should our attempts to manage soil, water and air pollution in an economic and environmentally sustainable way. This book tackles bioremediation problems of crude oil and its recalcitrant aromatic fractions in soil, water and air. It also investigates potential of value-adding co-product development such as biosurfactants and the option to use specialist bacteria, fungi and yeast for biological upgrading of our vast heavy and extra-heavy crude oil reserves.

Chapter "The Assessment of Hydrocarbon Contamination in Contrasting Sedimentary Environments" provides a detailed summary of oil-born pollutant origins and historic trend lines, particularly for the sedimentary environment of the North Sea and reports on progress made lowering hydrocarbon-based pollution from point sources due to implementation of government and self-appointed regulations, reporting and surveillance. It highlights the importance of considering historic inputs which can be re-introduced in significant amounts based on sediment characteristics and sediment disturbance events. It concludes that, despite the expense, advanced fingerprinting techniques are required to better account for and understand hydrocarbon sources, persistence, weathering, toxicity and carcinogenicity. Specifically, a better characterisation of sediment contaminant reservoirs is required to predict release concentrations and impacts, especially PAHs, during sediment disturbance events.

Given the carcinogenic, mutagenic and cytotoxic nature of PAHs, Chapter "HC-0A-02: Analysis of Polycyclic Aromatic Hydrocarbons from Food" reviews sources of pollution and analytical techniques for the detection and quantification of pyrogenic origin PAHs in food. This chapter presents a thorough treatise of traditional, novel and automatable and miniaturised analytical approaches taken for the determination of PAHs in foods, including food matrices, parameters measured, and extraction, clean-up and determination techniques. The evaluation ends with a critical review of method validation, which highlights, that despite regulations, precision and ruggedness are often least considered.

Chapter "HC-0B-06: Biodegradation of Hydrocarbons" reviews the degradation of oil hydrocarbons in contaminated soils and groundwater. Following an introduction on the parameters that influence the speciation and bioavailability of hydrocarbon contaminants, the chapter focuses on microbial and phyto-bioremediation. On-site (in situ) and off-site (ex situ) biological systems are evaluated in the context of addition of enriched cultures of microbes (bioaugmentation) and nutrients. Based on demonstration-scale case studies, the chapter concludes that bioventing, biosparging, soil vapour extraction, enhanced bioremediation/ biorestauration, and phytoremediation are cost-efficient in situ strategies for the remediation of hydrocarbon-polluted environments.

Chapter "HC-0B-01: Biodegradation of Hydrocarbons by Extremophiles" has a focus on the biodegradation of hydrocarbon pollutants under extremophilic conditions (i.e., acidophilic (pH 1–5), alkaliphilic (pH >9) halophilic (>3 % salt), thermophilic (>50 °C), psycrophilic (<10 °C), piezophilic or barophilic (>38 MPa) and xerophilic (a_w 0.60–a_w 0.90)). It presents a detailed summary of molecular tools for identification, identified strains capable of biodegradation of PAHs, their source of isolation and conditions of maintenance. It highlights physiological and molecular adaptations required to withstand extremophilic conditions, as well as novel biodegradation pathways that are expressed by certain environmental stressors.

The biodegradation of benzene, a carcinogenic parent compound of aromatic hydrocarbons present in petroleum, under anaerobic conditions is detailed in Chapter "Biodegradation of Benzene Under Anaerobic Condition". It summarises sources of exposure, organ distribution and metabolic fate upon exposure, before summarising the environmental distribution and fate of benzene. A focal point of the chapter is devoted to anaerobic benzene metabolic pathways, as it explores the effect and effectiveness of supplementing the anaerobic degradation of benzene with various electron acceptors. Modern molecular tools and integrated metabolomic approaches are described highlighting advantages of "extremozyme" applications for benzene biodegradation under environmentally unsuitable conditions.

Renewable and sustainable hydrocarbon-based fuel supplies are at the heart of the aviation/ shipping industries and those operating heavy machinery. Following a concise introduction to catalyst-assisted hydrotreatment of plant bio-oils, Chapter "Catalytic Hydro-Cracking of Bio-oil to Bio-fuel" explores oleic acid conversion, a fatty acid present in large quantities in all plant-derived oils, to biofuels using catalytic hydro-cracking with nickel impregnated zeolites (zeolite β,

ZSM-5). Both catalysts are shown to efficiently produce Nonane, Decane, and Dodecane, the main constituents of jet fuel, but zeolite β yields were higher for the former two alkanes due to a higher surface area, while lower yields of the latter may be attributable to a greater abundance of micropores.

Environmental and health impacts of increased air pollution with volatile organic compounds (VOCs), specifically volatile hydrocarbons (VHs), have called for tighter regulations. As these compounds often show microbial toxicity and are often hydrophobic, Chapter "HC-0B-05: Two-phase Partitioning Bioreactors for Treatment of Volatile Hydrocarbons" reviews applications and progress made with using two-phase partitioning bioreactors for the bioremediation of these air pollutants. The chapter reviews design principles, mass transfer balances and selection of non-aqueous phase additives based on target VH characteristics. It identifies remediation capacities of gas mixtures with differing hydrophobicity characteristics of components and reactor design aimed to improve energy requirements as the immediate knowledge requirements for transitioning from laboratory research to implementation.

Microorganisms living in hydrocarbon-contaminated environments often produce biosurfactants for improved access to hydrophobic organic constituents. Owing to the non-toxic and biodegradable nature of biosurfactants, they have wide industrial applications, from enhancing bioremediation outcomes to healthcare and food sectors. Chapter "Biosurfactant Producing Bacteria From Hydrocarbon Contaminted Environment" provides a thorough introduction on biosurfactant types, physiological and molecular details for biosynthesis, and the regulation thereof, including effects of cultivation conditions (age, nutrient status and physico-chemical parameters) on yields. Marriage of biosurfactant environmental services and production optimization is identified as critical for economic sustainability of the biosurfactant industry.

Chapter "Biodegradation of Petroleum Hydrocarbon and its Influence on Corrosion with Special Reference to Petroleum Industry" investigates the role of the enteric bacterium *Serratia marcescens* ACE2 and *Bacillus cereus* ACE4, isolated from a diesel pipeline, on corrosion of carbon steel reservoirs and pipelines and diesel degradation. As a first tropical study, examining enteric bacterial diesel degradation, the presented data prove that both strains degrade aliphatic and aromatic diesel hydrocarbons, while formation of ferric and manganese oxides through Fenton-like reactions led to significant corrosion of API 5LX steel.

Substantial removal of high-quality crude oil resources due to increased fossil fuel-derived energy demands by the growing population requires the progressive exploitation of heavy and extra-heavy lower quality crudes. These crudes are less suitable for today's application. Chapter "HC-0C-03: Biological Treatments to Improve the Quality of Heavy Crude Oils" reviews microbial approaches for the natural upgrading of these heavy oils. Several bacterial strains and consortia, as well as filamentous fungi and yeast are identified for their preference to degrade resins and asphaltenes (present in high amounts in heavy crude oils). For the latter groups, secreted extracellular enzymes, with some belonging to the lignin degradation system, are shown to oxidise resins and asphaltenes without loss of carbon, thereby maintaining the high calorific value of heavy crude oils, rendering them and/or their enzymes particularly useful for in situ and ex situ upgrading.

The final chapter presents the findings of a study aiming to isolate local endemic bacteria useful for the bioremediation of crude oil-contaminated waste waters of the petrochemical industry in Malaysia. Of the twelve species isolated, three showed promise for n-alkane degradation in crude oil-contaminated sites, two *Acinetobacter* species and one species of the genus *Proteus*. Both *Acinetobacter* isolates preferred non-agitated cultivation conditions, with STL-(-g)-Cr8 showing the highest alkane biodegradation potential and BTL-(+g)-Cr7 a preference for long chain alkanes (C_{36} and C_{38}). In contrast, RSSF-Cr1, a Proteus species preferred agitated cultivation. Further studies using endemic strains and consortia are required for improving bioremediation of crude oil-contaminated sites in developing nations.

As detailed, this book presents a solid background and method review for students and researchers in the areas of hydrocarbon pollution mitigation and the oil and gas industries. Additionally, synthesised details on current limitations to technologies and case study data afford a great data reference for experts in these fields.

Townsville, Australia — Kirsten Heimann
Townsville, Australia — Obulisamy Parthiba Karthikeyan
Hong Kong, Hong Kong SAR — Subramanian Senthilkannan Muthu

Contents

The Assessment of Hydrocarbon Contamination in Contrasting Sedimentary Environments

David I. Little and Yakov Galperin

The past is never dead. It's not even past
William Faulkner 'Requiem for a Nun' Random House (1951).

Abstract This chapter outlines an established approach for hydrocarbon analysis in sediments, the sources of hydrocarbons in the marine environment, and the importance of understanding the sedimentary environment of deposition and the reservoir of contaminants it can represent. To supplement national monitoring programmes, government and industry have commissioned work on polycyclic aromatic hydrocarbons in several United Kingdom estuaries, including Milford Haven Waterway, Southampton Water and Sullom Voe. These areas have some of the longest-established and largest oil industry sites in the UK, and they have been monitored intensively and successfully through most of their histories as oil, gas and petrochemicals ports. However, few economically and ecologically important UK estuaries and coastal zones have a particularly long history of advanced chemical "fingerprinting" to assist monitoring the specific sources, fates and effects of sediment hydrocarbons. As such, the UK approach has not compared well with the efforts of the United States of America in terms of apportionment of contaminant sources, albeit driven in part by the latter's focus on litigation. As well as distinguishing natural and anthropogenic inputs from background conditions, hydrocarbon fingerprinting methods can directly improve the ability to discriminate between individual hydrocarbon sources using knowledge of their composition, age and relative weathering. An understanding of the manufacturing processes and their chronology can help discriminate among multiple sources of pollutant hydrocarbons and also contribute to life cycle analysis and studies of product footprints. There is a need to track the ubiquity, fate, persistence and effects of hydrocarbons and petrochemicals in the environment. However, the use of fingerprinting faces challenges that include high cost, potentially unwelcome discovery of liability,

D.I. Little (✉)
Environmental Consultancy, Swavesey, Cambridge CB24 4RL, UK
e-mail: david.i.little@btinternet.com

Y. Galperin
Environmental Geochemistry Consulting, Moorpark, CA 93021, USA
e-mail: ygalperin@geochemconsulting.com

K. Heimann et al. (eds.), *Biodegradation and Bioconversion of Hydrocarbons*,
Environmental Footprints and Eco-design of Products and Processes,
DOI 10.1007/978-981-10-0201-4_1

sediment movements and patchiness, biodegradation effects on clarity of the fingerprint, and over-printing by chronic inputs from non-point sources such as air pollution and river runoff.

Keywords Oil and PAHs inputs · Pollution chronology · Sedimentation rates · Suspended particulate matter · Milford Haven · Southampton Water · Sullom Voe

1 Introduction

1.1 Background and Aims

The provenance and mobility of fine-grained sediments and their substantial role in both introducing and scavenging water column contaminants are important processes affecting water and sediment quality in aquatic systems. Due to tidal currents, estuarine circulation and salinity changes causing flocculation and turbidity maximum zones in estuaries, the contaminants gravitate to sedimentary sinks. Sediment transport processes can effectively trap, export and episodically recycle a proportion of the pollutants in estuaries.

This chapter explores examples from selected UK estuarine, coastal and offshore sites, shown in Fig. 1, to draw out the overlapping scales and types of inputs of hydrocarbons and of PAHs in particular. It is not a comprehensive review, and does not cover biological or ecological effects. Monitoring has been performed by several groups since the late 1970s in some cases. The authors have carried out two of these projects, in the Milford Haven Waterway and Southampton Water. Chemical forensics projects and data reviews were conducted in Southampton Water (SW) from 2005 to 2006 for the Environment Agency (EA), and in the Milford Haven Waterway (MHW) between 2007 and 2014 for the MHW Environmental Surveillance Group (MHWESG) with assistance from Natural Resources Wales (NRW). The sedimentary factors affecting oil industry impacts in MHW and SW are compared with those in the less turbid waters of a more remote fjord and peat-dominated sedimentary environment in Shetland (Sullom Voe and Orka Voe), and with urbanised firths in central Scotland (Forth and Clyde).

Comparisons are made with other UK estuaries, including the Humber estuary which receives much of the non-point sources (NPS) pollution in river sediment and contaminant runoff from northern and eastern England. Finally, the role of the sedimentary environment in understanding pollution is compared in more restricted Alpine lakes (Bodensee and Mindelsee) and in an area of natural oil seepage in the southern Caspian Sea (Absheron and Nakhchivan).

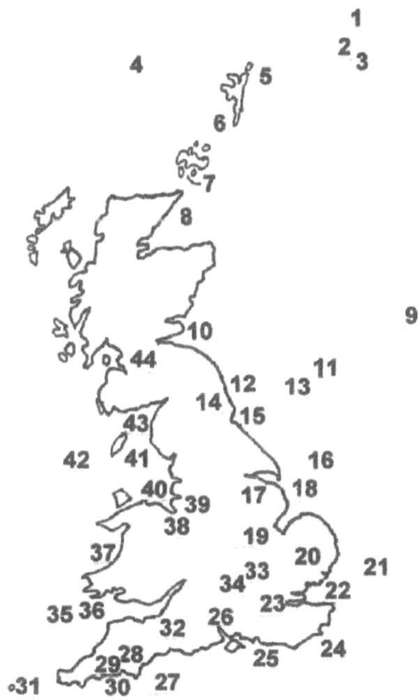

Concentrations of PAHs (also heavy metals and other contaminants) in the N. Sea, Irish Sea and some Arctic sites are unacceptable at many, mostly coastal sites:

- Some signs of decreasing offshore oil pollution inputs to the North Sea are offset by increases likely in the Arctic.
- Oil and oily waste spills are often not attributable to a responsible party in spite of aerial surveillance under the Bonn Agreement.
- Dredging has been stable at from 105-120 million t/year since 2000 but the number of dumpsites has increased.
- Regional data from UK, Belgium, the Elbe estuary (Germany) and Rotterdam harbour (The Netherlands) show that most metals concentrations in dredge spoil have declined by 60-80% since the late 1980s, but have stabilised recently.
- The outlook for inputs from land-based, sea-based and atmospheric deposition sources is uncertain

[Source: Quality Status Report (OSPAR 2010)]

Fig. 1 Locations mentioned in the text for estuaries, coastal, offshore, and sites of oil spills and onshore fires (*italics* with date): *1* Statfjord (Norway), *2* Hutton, *3* Brent, *4* West of Shetland, *5* Sullom Voe (Garths Voe, Orka Voe and *Esso Bernicia 1978*), *6 Braer 1993*, *7* Flotta (Orkney), *8* Captain, *9* Ekofisk (and *Bravo 1977* blowout, Norway), *10* Firth of Forth, *11* North Sea, *12* Clean Safe Seas Environmental Monitoring Programme (CSEMP) station 245 off Tyne, *13* CSEMP 285 off Tyne/Tees, *14* R. Wear, *15* CSEMP 295 off Tees (and R. Tees), *16* CSEMP 345 off Humber/Wash, *17* R. Humber (*Flixborough 1974*, *Sivand 1983*), *18* CSEMP 376 off Wash, *19* R. Great Ouse (CSEMP 386 Wash), *20* R. Orwell, *21* CSEMP 475 Outer Gabbard, *22* CSEMP 466 Thames, *23* R. Thames (and R. Medway), *24* CSEMP 484 Dungeness, *25* Pagham Harbour, *26* Southampton Water (Dibden Bay, R. Hamble), *27* CSEMP 536 Lyme Bay (*Napoli 2007*), *28* R. Dart, *29* R. Tamar and R. Fowey, *30* CSEMP 575 off Tamar, *31* Scilly Isles (*Torrey Canyon 1967*), *32* Bristol Channel/R. Severn, *33* Rothamsted, *34 Buncefield 2005*, *35* CSEMP 605 Celtic Deep (Celtic Sea), *36* Milford Haven Waterway (*Llanreath 1940*, *Dona Marika 1973*, *El Omar 1988*, *Sea Empress 1996*, Cosheston Pt., R. Cleddau, Pembroke Dock), *37* CSEMP 655 Cardigan Bay, *38* R. Dee, *39* R. Mersey, *40* CSEMP 715 Liverpool Bay, *41* CSEMP 805 SE. Isle of Man, *42* Irish Sea, *43* Solway Firth, *44* Firth of Clyde

1.2 Hydrocarbon Fingerprinting Methodology

The need to establish sources and associated potential liability associated with releases of hydrocarbons into the environment warrants comprehensive chemical contaminant characterization. The aim of this characterization is to identify contaminant type and sources(s), distinguish spilled product from background hydrocarbons, and assess the impact on the ecosystem. Such a characterization is commonly accomplished utilizing chemical fingerprinting, which in turn relies on evaluation of distribution patterns in different classes of hydrocarbons comprising the bulk of the hydrocarbon product (Kaplan et al. 1997).

Depending on the analytical technique(s) selected, different chemical parameters can be used for data evaluation, and are determined on a case-by-case basis. When high-resolution gas chromatography is conducted using a flame ionization detector (GC/FID), the evaluation focuses on the boiling range and distribution pattern of normal alkanes (n-alkanes) and selected isoalkanes (e.g. pristane and phytane), which are the most abundant and easily identified compounds in non-degraded hydrocarbons. As an example, high-resolution gas chromatograms for fresh and weathered reference crude oils are shown in Fig. 2.

The analysis also ascertains the boiling range and shape of the unresolved complex mixture (UCM), observed as a baseline rise over a broad retention-time window of the gas chromatogram. Depending on its origin, crude oil may contain n-alkanes with up to 40, and sometimes even 60, carbon atoms. One of the major distinctions between refined fuels is the corresponding boiling range. The hydrocarbon constituents of gasoline span a range from n-C_3 to n-C_{13}. Distillate fuels such as kerosene and diesel generally span a range from n-C_8 to n-C_{26} and exhibit a unique distribution of the product-specific homologous series of hydrocarbons. Hydrocarbon ranges for residual products, such as marine fuels, can extend to n-C_{40}. Sources of other industrial contaminants can also be evaluated based on their process-specific chemical constituents (Kaplan et al. 1997). The alkane distribution patterns in the hydrocarbon source are defined by composition of the original crude oil feedstock and refining processes. However, pattern-recognition fingerprinting of hydrocarbon products in the receiving environment is complicated by numerous physical and biochemical processes, collectively referred to as weathering, that change the chemical composition of the original product (Douglas et al. 1992; Kaplan et al. 1997).

Physical processes such as evaporation and dissolution are well understood and allow for a relatively straight forward prediction of their effect on a weathered hydrocarbon product fingerprint. These processes most noticeably affect chemical composition of light-end petroleum fractions, predominantly in the gasoline range. Their effect on the chemical fingerprint of middle-distillate (i.e. heating oil, diesel and jet fuels) and heavier (i.e. marine fuels) oil fractions is typically insignificant (Lafargue and Barker 1988; Rodgers et al. 2000). After release, the chemical composition of these crude oil fractions is mostly affected by biochemical processes resulting in selective degradation of hydrocarbons by indigenous microorganisms (i.e. bacteria, yeast and fungi) occurring naturally in the environment. Bacteria are the prevailing microorganisms involved in this degradation, because they contain a diverse set of enzymes that are involved in the metabolism of hydrocarbons (van Beilen et al. 2003). The progression of these transformations depends on the properties and composition of the hydrocarbons, parameters of the product release and environmental conditions (Atlas 1981; Leahy and Colwell 1990). Because of the complexity of biochemical processes in the hydrocarbon-contaminated

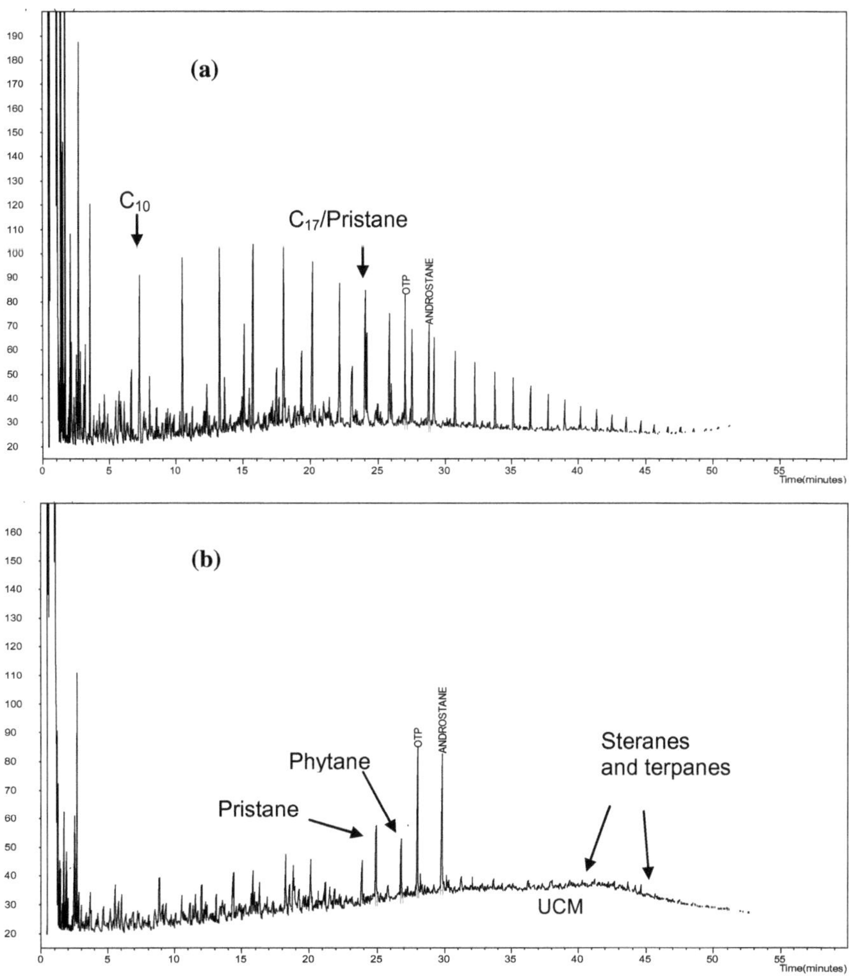

Fig. 2 Gas chromatograms of **a** fresh and **b** weathered crude oil (*Note* peaks OTP and Androstane correspond to laboratory internal standards)

environment, recognition of the fuel type(s) present is often a challenging or formidable task (Galperin and Kaplan 2011).

The n-alkanes are among the hydrocarbons most susceptible to biodegradation, and so their content and distribution change as a result of environmental exposure. As oil degrades, the resolved chromatographic peaks corresponding to n-alkanes are depleted and the UCM shifts toward the higher-molecular-weight constituents that are more resistant to degradation (Fig. 2b). The observed changes are often used to evaluate product weathering in the environment. Concentration ratios of certain recalcitrant constituents such as isoalkanes, due to their relative stability, can be used as source indicators (Kaplan et al. 1997; Wang et al. 2006).

Polycyclic aromatic hydrocarbon (PAH) compounds represent one of the homologous series of hydrocarbons used for product-type identification in environmental samples. These hydrocarbons are analyzed using gas chromatography-mass spectrometry (GC/MS) methods (Douglas et al. 1992; Kaplan et al. 1997). The target analyte list includes two-, three-, four-, five-, and six-ring PAH compounds and alkyl-substituted PAH series. Typically, 2 or 3 ring PAHs are considered low molecular weight, whereas 4 to 6 ring PAHs are considered high molecular weight. The distribution of PAH homologous series is commonly used for source identification where, in petroleum-related sources, the distribution pattern (Fig. 3a) can be generally described as:

$$C_0\text{-PAH} < C_1\text{-PAH} < C_2\text{-PAH} > C_3\text{-PAH} > C_4\text{-PAH}$$

Although fossil fuels are rich sources of saturated (aliphatic) and aromatic hydrocarbons, it is generally agreed that most of the PAH and many of the aliphatic hydrocarbons in soils and sediments are derived from incomplete combustion of organic matter at high temperatures. Combustion of any organic material, including fossil fuels, will result in the generation of a wide variety of PAH, particularly if combustion takes place at temperatures above about 400 °C in an oxygen-deficient environment. PAH assemblages produced by pyrolysis (Fig. 3a) of organic matter are complex and, unlike PAH assemblages in petroleum, usually are dominated by four-, five-, and six-ring PAH. In pyrogenic PAH assemblages, particularly those generated at temperatures above about 700 °C, the dominant compound in each homologous series is the non-alkylated parent compound (Sauer and Uhler 1994; Stout et al. 2002), as follows:

$$C_0\text{-PAH} \gg C_1\text{-PAH} > C_2\text{-PAH} > C_3\text{-PAH} > C_4\text{-PAH}$$

PAH diagnostic ratios are also commonly used to differentiate petrogenic sources in moderately weathered environmental samples. Their application is based on the finding that certain alkylated PAHs degrade following first-order kinetics. Therefore, for compounds having nearly identical rate constants, the PAH ratios retain the initial oil signature. These ratios are termed "source" ratios, in contrast to ratios that change substantially with degradation and thus are termed "weathering" ratios. Ratios of C_3-dibenzothiophenes/C_3-phenanthrenes (C3D/C3P) and C_2-dibenzothiophenes/C_2-phenanthrenes (C2D/C2P) are known to be the most stable petroleum source ratios (Douglas et al. 1996).

The most common diagnostic ratios for pyrogenic source identification are Anthracene/(Anthracene + Phenanthrene) [AN/(AN + P)], Fluoranthene/(Fluora nthene + Pyrene) [FL/(FL + PY)], Benzo(*a*)anthracene/(Benzo(*a*)anthracene + Chrysene) [BA/(BA + C)] and Indeno(*1,2,3-c,d*)pyrene/(Indeno(*1,2,3-c,d*) pyrene + Benzo(*g,h,i*)perylene) [IN/(IN + BPE)] (Yunker et al. 2002; Oros and Ross 2004; Costa and Sauer 2005). Using these ratios, it is often possible to differentiate among major sources of combustion-derived PAH such as fossil fuel and biomass.

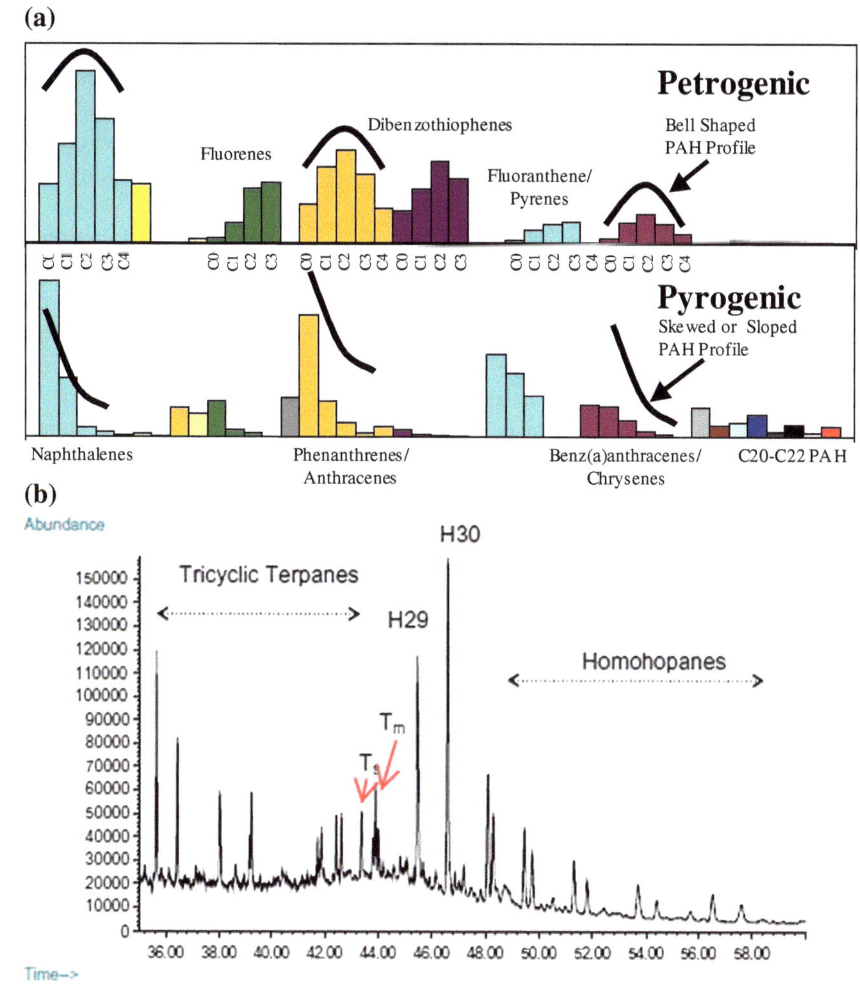

Fig. 3 **a** PAH distribution patterns in petrogenic and pyrogenic sources, and **b** Mass chromatogram of terpanes for a reference Alaska North Slope crude oil (*Note* T_s 18α-Trisnorneohopane, T_m 17α-Trisnorneohopane, *H29* Norhopane; *H30* Hopane)

Based on the PAH source ratio measurements compiled by Yunker et al. (2002), the AN/(AN + P) ratio <0.10 indicates dominance of petroleum and >0.10 indicates dominance of combustion sources; FL/(FL + PY) ratio <0.375 is typical of petroleum, 0.375–0.50 of petroleum combustion, and >0.50 of combustion of coal, grasses and wood; the BA/(BA + C) ratio <0.20 is characteristic of petroleum, 0.20–0.35 of petroleum and combustion, and >0.35 of combustion; and the IN/(IN + BPE) ratio <0.20 is indicative of petroleum, 0.20–0.50 of petroleum combustion, and >0.50 of combustion of coal, grasses and wood.

The high molecular-weight hydrocarbons steranes and terpanes represent another important fingerprinting tool for source identification: petroleum biomarkers. These are complex polycyclic alkanes originating from the polycyclic lipids found in bacteria and plants (Philp 1985; Waples and Machihara 1990). Typically, biomarkers retain all or most of the original carbon skeleton of the living organism, and this structural similarity leads to the term "molecular fossils" (Hunt 1995). Their polycyclic aliphatic structures are very stable in the environment, maintaining the relative distribution of the source crude oil and making them excellent source-specific tools (Kaplan and Galperin 1996).

Because of their unique origin, biomarkers are characteristic of petroleum sources and are not typically found in pyrogenic materials in significant concentrations. Biomarker "fingerprinting" is applied to those petroleum-derived contaminants that include abundant higher-boiling, residual range (C_{25}+) hydrocarbons (Peters and Moldowan 1992). The residual contaminants originating from petroleum sources include crude oil, marine fuels, lubricating oil, and asphalt. The commonly used biomarkers that occur within the residual range include steranes and triterpanes, which are analyzed using GC/MS. As an example, distribution of terpanes for a reference crude oil is shown in Fig. 3b.

Steranes are a class of tetracyclic saturated hydrocarbons derived from sterols. The most common steranes are in the range from C_{26} to C_{30} and are measured using m/z 217 and m/z 218 mass chromatograms. Terpanes form a broad class of complex branched cyclic alkanes mostly derived from bacterial cells, and are best monitored using m/z 191 mass chromatograms. Reviews of biomarker fingerprinting for oil source identification are presented in Wang et al. (2007) and Yang et al. (2015). Sample preparation and analytical methodology for the target compound analyses discussed in this chapter are described in Douglas et al. (1992), and Kaplan et al. (1997).

2 Inputs to the Marine and Coastal Environment

2.1 Sources of Hydrocarbons

2.1.1 General

Three major types of hydrocarbons may be present in marine and coastal sedimentary environments: (1) petrogenic, which are generated in organic-rich source rocks exposed to elevated temperatures for long periods: a category that includes crude oil and its refined products; (2) biogenic, which are generated by biological processes or in the early stages of sediment diagenesis; and (3) pyrogenic (or pyrolitic) which are generated by combustion of fossil fuels (Volkman et al. 1992).

Saturated (aliphatic) hydrocarbons comprise the majority of compounds in crude oils, typically 60–75 % by weight, followed by aromatic hydrocarbons. The former are by far the less toxic and display unique patterns associated with petroleum source, biogenic source (e.g. hydrocarbon inputs from plants), and degree of weathering. There are three types of saturated hydrocarbons in petroleum: normal alkanes (n-alkanes), branched alkanes (isoalkanes), and cyclic alkanes. The normal and branched alkanes are usually present in about equal amounts. The cyclic alkanes often are the most abundant alkanes in crude oil.

PAH is a particularly ubiquitous category of pollutant for which there are numerous sources ranging from natural-occurring to anthropogenic, all increasing the complexity of the source analysis. Both unburned and combusted fossil fuels impart their PAH signature in the marine environment. PAH can originate from a variety of anthropogenic sources in rural and urban areas by thermal combustion processes (e.g. cooking and heating oil, and burning fuel oil and coal), vehicular emissions, and biomass burning (e.g. fireplaces) including forest or grassland burning (Simoneit 1984).

PAHs have been ubiquitous pollutants across most environments thanks to their transport by atmospheric processes. Soil samples (with no local PAH sources) that were archived by Rothamsted Research since the 1840s show that total PAHs were already >300 μg/kg in concentration by the mid-nineteenth century, increasing to about 500 μg/kg between the two World Wars. From 1956 to 1980 the concentrations ranged between 1130 and 1770 μg/kg, although by the early 1990s concentrations had decreased to about 720–750 μg/kg (Jones et al. 1989). For rural areas in Wales, the median value of 253 μg/kg total PAH was suggested as a background, although the range in the whole dataset was 108–6740 μg/kg (Smith et al. 2002). Similar trends were observed in dated sediment cores sampled in marine and coastal zones such as the Firth of Clyde (Vane et al. 2011).

2.1.2 PAHs from Chronic Sources

For many industrialized areas, atmospheric deposition is a major source for PAHs entering aquatic environments. Sources of atmospheric PAHs include fuel combustion, industrial production facilities, transportation, solid waste incineration, and forest and grassland fires (Crecelius and Bloom 1989). Urban runoff is considered the most common entry pathway for chronic hydrocarbon contamination in the marine environment. In contrast to atmospheric transport and deposition, which can move contaminants a long distance from the point of origin, urban runoff tends to affect water bodies near the source. The sources of PAH in urban runoff vary but the most common sources are (1) urban dust containing combustion-related PAH, (2) street runoff containing traces of PAH from lubricating oils and vehicles' tyre wear, and (3) illegal or unintentional discharge of waste oil or petroleum products containing PAH into street and storm drains.

In spite of the presence of a petroleum component in urban runoff, the PAHs in urban runoff and, in turn, in receiving sedimentary environments are typically dominated by pyrogenic PAH (Eganhouse et al. 1982). This is for two reasons: (1) the PAHs in stormwater runoff often have a pyrogenic PAH signature to begin with, and (2) the 2- and 3-ring PAHs (predominantly petroleum-derived) are more water soluble and degradable than heavier 4-, 5- and 6-ring PAH (predominantly combustion-related). The background PAH concentrations in sediments impacted by urban runoff are very broad and highly site-dependent but are typically in the 1–50 mg/kg range (Stout et al. 2004).

2.1.3 PAHs from Acute Sources

In addition to chronic sources of sediment contamination, accidental marine oil spills occur unpredictably from tankers, offshore blowouts and acts of war (e.g. 'Torrey Canyon', 1967; Santa Barbara, 1969; 'Amoco Cadiz', 1978; 'Exxon Valdez', 1989; first Gulf War, 1991; 'Prestige', 2002; El-Jiyeh, 2006; Macondo, 2010). Spills are one of the most significant acute sources of oil and PAH contamination in the marine environment and numerous or frequent small spills in oil terminal areas may be the major local input to the coastal zone. The impact of an oil spill will depend upon the volume of the spill, duration, type of petroleum product, physical factors such as wind, wave and current conditions under which the spill occurs, and the biology of the receiving environment. Some of the lighter fractions of oil will evaporate very rapidly, others are sensitive to sunlight and oxidize to innocuous or inert compounds, and still other fractions will either dissolve, emulsify, or adsorb to sediment particles (sedimentation), depending on their physical properties. The physical fate of oil can involve several processes: spreading, emulsification, littoral deposition, physical removal, dispersion, dissolution, flushing, sedimentation, oxidation, organic uptake, etc.

Hydrocarbon degradation characteristics are influenced by a range of biological as well as physical factors. Microbial degradation (biodegradation) is one of the other main processes affecting the fate of oil, particularly the non-volatile fractions. Once oil is in the sediment, its weathering will depend not only on energy level, redox conditions, and water-washing, but also on the rate at which infiltration and further accretion takes the oil below the depth of surficial disturbance from further erosion (Little 1984; Little et al. 1981, 1986). PAH compounds may be washed differentially by water movements through sediments. Much variability in water-washing will also depend on the locus of oil contamination, whether in the sediment matrix or in macropores created by plant roots and burrowing infauna (Howard and Little 1987; Kingston et al. 1999). Some aromatic hydrocarbons may be soluble enough to be available for re-mobilisation after deposition (Rodgers et al. 2000). Benzene and toluene have water solubility at 25 °C in the range 100–2000 mg/l, whereas higher molecular-weight compounds such as Phenanthrene, Naphthalene and Pyrene range from 0.1–100 mg/l solubility. The complex aromatics such as Benzo(*a*)pyrene are practically insoluble at about 0.005 mg/l (Petrakis and Weiss 1980).

In evaluating the extent and nature of acute hydrocarbon contamination, in addition to considering potential sources of contamination, natural recovery processes rely on two primary mechanisms: burial of contaminated sediments by deposition of clean sediments, and contaminant weathering. Burial of contaminated sediments by sediment deposition results in natural capping which protects the water column (and consequently resident fish and benthos) from the vertical diffusion and advection of contaminants from surface sediments. It also reduces the risk of resuspension of contaminated surface sediments during normal or high flow events (Cardenas and Lick 1996), and reduces contaminant transport into the food chain that can occur through bioturbation and bioaccumulation in near surface sediments. Sediment burial occurs in net depositional environments where the rate of sediment deposition exceeds the rate of sediment scouring and resuspension. Most depositional environments commonly occur in slow-moving waters such as parts of estuaries, deltas, and lakes.

International conventions (and UK legislation) protecting the marine environment have stopped the disposal at sea of many acute contaminant sources including PAHs, at the following (UK) dates: open-burning of waste (1990), dumping industrial waste (1992) and sewage sludge (1998). However, dredging in navigable waters can disturb contaminants such as PAHs, polychlorinated biphenyls and metals including copper and tributyltin (TBT) that were sequestered in accreting sediments prior to their abatement or banning (Eggleton and Thomas 2004; Little et al. 2015; Nicholaus et al. 2015).

Contaminant weathering includes such mechanisms as dilution, volatilization, bio-transformation, biodegradation, and sequestration; weathering can provide permanent reduction in contaminant concentrations. For buried sediments, the primary weathering mechanisms include biotransformation, sequestration, and benthic mixing. Volatilization is prevalent for contaminants in surface sediments which can be resuspended and re-enter the water column, from which the contaminants can volatilise to the atmosphere (Jeremiason et al. 1998).

2.2 Global Estimates

Inputs of oil to the marine and coastal environment arise from a wide variety of sources, and all may make significant contributions. Releases of crude oil and refined products from a range of both non-point source (NPS) and point sources cause acute and chronic pollution in rivers, surface and ground waters, coastal and marine areas. Hydrocarbons are introduced directly into sedimentary environments from a variety of anthropogenic activities along urban waterways (e.g. petroleum terminals and refineries, power stations, aluminium smelters, gas production facilities, tar distillation plants, railway sidings, loading/unloading facilities, shipyards, breaking yards, waste management facilities, marinas, and residential areas). Petroleum and coal- or oil-derived tars are also point sources arising from natural seeps as well as from spills.

Common NPS pollution includes petroleum washed indirectly into water bodies by precipitation from urban roads and car parks, from vegetated and built surfaces. Other NPS of hydrocarbons include recreational boat traffic, ship traffic, overland flow and surface runoff, and atmospheric particulate deposition (e.g. soot from petroleum combustion, forest or grassland fires, wood-burning stoves, power stations, smelters, etc.). Most of the resulting hydrocarbon contaminants partition to sediments which are recognized as a major sink and a potential source of long-term pollution of the aquatic environment (NRC 2003). Estimates and dates of publication are given in Table 1 of the relative magnitude of local (MHW 1992), regional (North Sea 1980–2000), and global inputs (1975–2007), as percentages and total inputs from various sources.

The totals show little decline through time and wide margins of error, caused by the problems with making the estimates, which are discussed below. Although large spillages from tanker accidents and blowouts have a high media profile, most petroleum hydrocarbons enter the sea from shipping, oil and gas (O&G) exploration and production (E&P), natural oil seeps, sewage discharges, storm water outfalls, road run-off, river inputs, atmospheric fall-out due to fossil fuel combustion, and other NPS discharges.

Some hydrocarbons are also biosynthesised by phytoplankton, and specific PAHs such as Perylene are produced during early diagenesis in shallow sediments. There are considerable uncertainties in making estimates of oil inputs. For example, the NRC (1975) estimate for oil inputs to the world oceans was approximately double the figure given by NRC (1985) in Table 1 (3.2 million t). Exxon (1989) considered that there was no real reduction over the decade 1975–1985, but rather an improvement in the quality of reporting by 1985. Because of greater reductions in other sources, the relative contribution from tanker accidents actually increased over that same decade, and represented 12.5 % of inputs in addition to 22 % from tanker operations including minor spills, handling, and bilges (NRC 1985).

Almost 35 % of the global input reported in 1985 came from shipping, and this was the largest proportion attributable to the oil industry (Table 1). Although it had also been the case before 1985, the shipping input was still only <0.1 % of the annual worldwide production of oil. Focussing on sea-based activities for the period 1988–1997, estimates by GESAMP (2007) agreed well with those for 1990–1999 by NRC (2003). However, shipping inputs were exceeded by oil inputs from industrial, municipal and river runoff sources combined with other transportation. NRC (1985, 2003) considered that in spite of improvement, the accuracy in such predictions was still too poor for industrial, municipal and river runoff. This was also the case for other principal sources: atmospheric inputs and natural oil seeps. Accuracy for the latter two sources improved progressively by better knowledge of atmosphere/ocean coupling processes and sea floor/shelf geochemistry, gained by oil industry innovation in marine engineering, and advances in applied coastal, marine and shelf science (Page et al. 1997; Kvenvolden and Cooper 2003; Boehm et al. 2003).

Records of E&P performance maintained by the International Association of Oil & Gas Producers (IOGP 2014) show that in Europe 95 % of known oil production is covered by reporting companies, compared with 62 % in Africa, from 20 to 34 % in the rest of the world, and as little as 10 % in the former Soviet Union. The quality of produced water (PW) discharged during worldwide O&G operations in 2013 was between 10.5 mg/l oil (onshore) and 13.4 mg/l oil (offshore), and quantities have declined slowly, representing 8.2 t oil discharged in PW per million tonnes of oil

Table 1 Estimated annual oil inputs to Milford Haven Waterway (MHW), North Sea, and world oceans

Input source %	MHW %	North Sea %					World oceans %			
Publication date	1992[a]	1980[b]	1986[c]	1987[d]	1993[e]	2000[f]	1975[g]	1985[h]	2003[i]	2007[j]
Offshore O&G	–	2	14–21	19–41	20	15–19	1.5	1–6	2.9	2.4
Ships/accidents	2.2–4[k]	16	5–7	–	26	6–13	38.6	35	32.6	36.5
Rivers/runoff	34–65[l]	53	37–48	23–31	21	31–39	34.5	31[m]	11[m]	–
Atmosphere	9.2[n]	–	12–15	10	7	10–13	10.9	4–9	4.8	–
Coastal Effluent	–	–	5.5–8.4	7–10	7	10–13	5.4	–	–	–
Coastal Sewage	2.8–5.5[o]	8	3–8	4–10	6	9–11	5.4	–	–	–
Dredge Spoil	0.22–1.7[p]	3	–	3–7	4	1.7–3.8	–	–	–	–
Sewage Sludge	–	5	–	1–7	3	–	–	–	–	–
Refinery/terminal	16–50[q]	13	4.5–6.5	3–7	3	0.6–1.3	3.7	3.1	0.4	13.4[r]
Other (inc. seeps)	–	–	0.3–0.5	2–3	3	0.9–1.9	–	20[s]	48.3[s]	48.2
Total (t) min.	696	94,090	107,000	71,000	76,000	52,000	4.7 m	3.2 m	0.47 m	1.25 m
Total (t) max.	1338		165,000	150,000	219,000[t]	117,000[u]	6.6 m		8.3 m	

[a]Hobbs and Morgan (1992), with additional estimates for MHW by Little (2009) and Little et al. (2016) in notes below
[b]Read and Blackman (1980)
[c]Bedborough and Blackman (1986)
[d]UK Department of Environment (DoE) Quality Status Report (1987)
[e]OSPAR (1993) data 1985–1990
[f]OSPAR (2000) data 1994–1997
[g]National Research Council (1975) anthropogenic only, no seeps
[h]National Research Council (1985) including seeps
[i]National Research Council (2003) data 1990–1999, including seeps
[j]GESAMP (2007) data for 1988–1997, including seeps
[k]MHW 1960–2008 excluding 'Dona Marika' and 'Sea Empress' spills (if ~75,000 t included, proportion would be >50 %)
[l]Assuming 1 mg/l oil content in annual discharge of 450 million m^3
[m]Includes municipal and industrial input
[n]Mean substitution from North Sea and World ocean values

[o]Assuming 10 mg/l oil content
[p]No MHW disposal, but assume 2.5–38 % loss from peak annual dredging, with oil content in spoil from 300 to 500 mg/kg
[q]Assuming 10–65 mg/l oil content
[r]Includes small craft input (4.2 %)
[s]Includes other forms of transport (non-marine, e.g. pipelines)
[t]Total calibrated from runoff and seeps (mass flux and %'s given in OSPAR 1993)
[u]Total calculated from known inputs of offshore O&G, shipping, dredge spoil and refinery/terminals (OSPAR 2000), combined with OSPAR (1993) estimates for other inputs

production. Total PW is about 1.6 times the volume of produced oil (IOGP 2014). Diesel oil-based mud (OBM) dominated E&P inputs historically, but decreased since 1997. All non-aqueous drilling fluids (NADF) discharged to sea in 2013 consisted of <0.35 % PAH, and 96 % were <0.001 % PAH, equating to a probable worst case of only about 0.15 t total PAH released worldwide (IOGP 2014). This is very low, and the remaining issues are in modelling to predict NADF impacts including interactions with cuttings aggregations, particle flocculation, resuspension and other post-depositional processes (IOGP 2016).

The quantity of oil spilled during worldwide onshore and offshore E&P operations has declined steadily to 7585 t during 2013, representing 4.1 t spilled oil per 10^6 t produced (million tonnes of oil production), a much lower rate than during 2011 (15,936 t). Spills onshore are more common than offshore, and also larger in total, especially in Africa where 2667 t was spilled onshore during 2013, representing 49.6 t/10^6 t produced (IOGP 2014). This African performance deteriorated during 2014 when 3152 t was spilled, representing 54.3 t/10^6 t produced. More than 1279 t of this onshore total was spilled in two incidents in Nigeria both involving illegal third-party damage (IOGP 2015). Since 2000 in the Niger Delta, stolen oil has been traded by militant groups, security forces, politicians and businesses. The oil theft was estimated at 10,000–43,000 t/day, representing between 3 and 12 % of Nigerian oil production (Mähler 2012). In 2012, the oil theft from the three main onshore E&P consortia in the Niger Delta was an estimated 9332 t/day (NEITI 2015). The theft and illegal refining of oil causes spills and also prevents cleanup, due to lack of security in the area and ambiguous remediation targets (UNEP 2011).

The International Tanker Owners Pollution Federation (ITOPF) maintains detailed records of tanker spills worldwide and has tracked large, incremental declines in numbers of spills and amounts of oil spilled since the 1970s, especially after 'Amoco Cadiz' and 'Exxon Valdez' and the ensuing international legislation. Numbers of spills larger than 700 t declined from 24.5/year in the 1970s to 1.8/year since 2010, the oil spilled being generally <10,000 t/year since 2005. This represents less than 1 t oil spilled per million tonne-miles traded by ship, more than 20 times less than in the 1970s (http://www.itopf.com).

In monitoring programmes for some marine and near-shore areas, improved oil and chemical fingerprinting methods have been used to allocate pollutants to their various candidate sources (e.g. crude oil, fuel oil, lubricant oil, coal tar, road runoff, creosote/preservatives, etc.). Unfortunately, the analytical resolution required to do this successfully remains expensive, and is not routinely attained in inputs studies or indeed monitoring. Although use of GC/MS has increased, this has not always been across the suite of compounds and their homologues that are required for advanced chemical fingerprinting. Such methods have reached greater levels of sophistication (and cost) primarily in USA during litigation as to who spilled the oil. For chronic inputs elsewhere, such as in the UK, it has been a widespread practice to rely largely on less specific and cheaper methodologies, which also allows some responsible parties to avoid discovery of unwelcome liabilities.

2.3 North Sea Estimates

The majority of oil inputs to the UK Continental Shelf (UKCS) are into the greater North Sea, Region II of the OSPAR commission, although relative inputs more recently increased in the Celtic Sea and west of Shetland. For locations and comparability with earlier estimates in Table 1, however, the Region II data have been used in that table (OSPAR 2000). These earlier estimates were made by the E&P Forum (now IOGP) and MAFF (now Department for Environment, Food and Rural Affairs; DEFRA) for 1980 and 1986 (Read and Blackman 1980; and Bedborough and Blackman 1986).

The Conference of North Sea Ministers first started in 1984 in Bremen at the instigation of the German government after severe mortalities of oiled seabirds in 1983 in the German Bight. The 1984 conference discussed "Black and Grey Lists" substances under the Paris Commission and Rhine and Elbe Joint Commission. The Second Conference of North Sea Ministers in 1987 was the first attempt to collate data for all North Sea riparian states. However, the meeting failed to agree on estimates of illegal shipping spills and discharges that were the principal cause of mortalities of oiled birds (DoE 1987). The 1987 report was nevertheless the first Quality Status Report (QSR), and in spite of the data gaps, its estimates of inputs were generally consistent with those made by government and industry for the UKCS. Indeed, subsequent QSRs in 1993 and 2000 did not uncover large changes in the relative proportions of oil inputs, in spite of some variability in the totals. Although QSR 2010 (OSPAR 2010) did not revise a detailed overall inputs budget for the N. Sea, the main trend in offshore industry oil inputs since 2001 (when there were 9400 t of dispersed/displaced oil in PW), was a decrease of 21 % by 2006. Other positive findings since OSPAR (2000) included: continued low levels of oil spilled (between 300 to 900 t/year, except for one large spill at Statfjord in 2007); decreased oil in cuttings and organic phase fluids from a peak of 7200 t in 1997 to 1000 in 2002, and to negligible levels since 2005. Noting these reduced offshore inputs, QSR 2010 also drew prominent attention to the lack of any significant reduction in PAH concentrations in coastal UK sediments, in spite of declines in Rotterdam harbour dredge spoil from 4.6 mg/kg[1] ΣPAH_6 in 1988 to 1.2 mg/kg in 1997 and 1.4 mg/kg in 2005. There were also predicted increases above 2000 levels of inputs of SO_2 (40 %), NOx (45 %) and particulates (55 %) pollution arising from shipping (OSPAR 2010; and see also Fig. 1).

Offshore oil industry discharges consist of three separate sources: oil spills, discharges of oil from drill muds and cuttings, and discharges in association with PW. Unlike accidental oil spillage, oil in PW is permitted subject to regulatory control at national and European levels. It should be noted that discharge of OBM practically ceased from 1st January 1997 when legislative levels of oil at 10 g/kg were technically not achievable. Discharges from offshore O&G activity rose steeply in the 1980–1986 period, and then stabilised and declined, in spite of

[1] Σ represents the sum of (in this case) six individual PAH compounds (ΣPAH_6).

increasing production, mainly as a result of voluntary and legislative action to reduce and ban usage of OBM. The types of mineral oil lost from offshore operations are dominated by crude oil (43 % of volume and 54 % of incidents), diesel/fuel oils (34 % volume and 20 % incidents), and base oil (17 % volume and 1.3 % incidents). The remainder include hydraulic oil (1.8 % volume), graphite grease (1 % volume), condensate (0.7 % volume), and lube oil (0.25 % volume). Almost 7 % of incidents involved unknown material (2.7 % volume). OBM dominated the ~28,000 t/year inputs of oil from the North Sea offshore industry from 1984 until 1988. Releases from spills and flaring were also significant in 1986 and 1988. Between 1989 and 1991, oil inputs were lower (from 17,000 to 21,000 t/year) and consisted of OBM and PW in combination. Between 1992 and 1996 oil in PW became dominant with inputs ranging from 10,000 to 14,000 t/year. In 1997, oil inputs were still 17,000 t due to increasing PW volumes as fields aged, and also due to use of synthetic drilling muds.

The North Sea O&G industry for safety and environmental reasons prioritizes the prevention of oil or gas leaks. O&G operator companies report every oil spill on the UKCS, whatever its size, and carry out an investigation into its cause; measures are then put in place to prevent recurrence. Offshore spills are reported to central government and to Maritime and Coastguard Agency (MCA) and the Joint Nature Conservation Committee (JNCC). In the event of a major spill, operators have arrangements in place for oil spill countermeasures from contractors such as Oil Spill Response Limited (OSRL). Oil spills are reported in a specific format that includes the cause and source of the spill, the type and quantity of oil spilled, the area affected, weather conditions, and any actions taken to control the spill or mitigate its effects. Where appropriate to do so, more detailed investigations including environmental impact or forensics analyses may be occasionally conducted. These may or may not be funded by the spiller. Spill prevention and control policies regarding industry and government cooperation are outlined in the international Oil Pollution Preparedness and Response Convention (OPRC 1990) and promulgated in the UK Offshore Petroleum Activities (Oil Pollution Prevention and Control) Regulations 2005 (as amended; 2014). Offshore, coastal and inland surface water contaminant regulations and policies vary between states around the North Sea, but are generally based on Environmental Quality Objectives and Environmental Quality Standards (EQO/EQS) defining consents to discharge that maintain agreed water quality targets. Sediment contaminant regulations are less formalized. They are based on Dutch standards that were devised originally for soil contaminants, or increasingly, by using sediment quality guidelines derived from the USA National Oceanographic and Atmospheric Administration (NOAA) and Environment Canada models (e.g. OSPAR).

The mandatory Safety Cases prepared for each O&G installation on the UKCS help to ensure both that the prevention of oil and gas releases is considered systematically, and that control systems are in place to minimise the risk of any inadvertent loss. In 1999, Oil and Gas UK (formerly UKOOA) member companies presented their revised risk-based oil spill contingency plans in compliance with regulations introduced into UK law in 1998, arising from the OPRC (1990). Offshore installations are routinely monitored by surveillance aircraft, which over-fly offshore oil fields on behalf of MCA, using airborne scanning equipment to detect even relatively minor spills. Information from surveillance flights indicates that less than 5 % of observed spills went unreported, and when investigated, virtually all of these were from sources other than the offshore O&G industry. Nevertheless, aerial surveillance conducted under the Bonn Agreement suggests that illegal discharges of oily wastes are still occurring, in contravention of MARPOL Annex I. Data are limited on how much oil was spilt in the OSPAR area since 2000 as a result of incidental and illegal discharge. About 80 % of slicks detected using aerial surveillance do not result in identification of the polluter, making it impossible to quantify which slicks are attributable to shipping (OSPAR 2010).

Large spills such as the blowout of 30,000 t from Ekofisk Bravo in 1977 are uncommon, but when they occur they tend to dominate the record and make it difficult to discern trends. Fortunately, in most cases marine oil spills are far smaller than these, and extremely small in comparison to the UK's major tanker spills; 'Torrey Canyon' (1967), 'Braer' (1993) and 'Sea Empress' (1996). Individual smaller N. Sea incidents can be significant, for example the 1500 t crude oil released in December 1988 when the Fulmar floating storage unit broke loose from its mooring (Rowberry 1993). Other examples include the anomalous peak in 1997 that was caused by the single accidental spill of 685 t from the Captain offshore installation (Cordah 1998; Fisheries Research Services 1998), and the 450 t lost from the Hutton platform in 2000. In 1986 there were 198 reported spills in the North Sea as a whole, totalling 3800 t. By 1992, the number of reported incidents had increased to 414, and by 1995 to 621. This increase was largely due to the emergence of a better reporting culture, whereby in 1995 over 85 % of reported spills were <1 t, and the total oil mass lost had gone down to 270 t. However, releases from shipping, including accidents, showed greater inter-annual variability than all other inputs, due to the effect of large infrequent incidents. Shipping inputs in 1995 were 6750 t for the N. Sea, and in 1998 were only 291 t in the whole UKCS (OSPAR 2000).

Classification of spills by source on the UKCS, including ports, rivers and estuaries, shows that, in terms of spills (as opposed to chronic inputs), coastal areas contributed lower amounts of oil. The Advisory Committee on Protection of the Sea (ACOPS) formerly published the reported oil spill incidents in the UKCS. DEFRA (2000) also published these and other data, but they were not fully consistent with data on oil spills published prior to 2000. This reflected revisions to MCA's information management, when the remit of the survey was broadened to include additional types of discharge such as noxious liquid substances and garbage. Since the Offshore Chemical Regulations 2002, the numbers of chemical discharge reports and the volume discharged have both increased. The total number of oil spills reported has increased particularly due to aerial surveillance, but the total quantities spilled have generally fallen.

Reports include discharges attributed to vessels and offshore oil and gas installations, number of incidents and estimated volume of discharges: mineral oils were the most frequently recorded pollutants, identified in 96 % of reports. The discharges occurred most frequently in the UKCS, followed by W. Scotland and S. England. The source of pollution is reported by area: operational discharges from O&G installations on the UKCS accounted for 75 % of all confirmed discharges, excluding all reported incidents from unidentified sources in ports, near-shore and tidal waters, and land-based sources such as outfalls or industrial premises. Prosecutions and convictions for oil spills by port authorities are made public, including convictions for the illegal discharge of oil or of mixtures containing oil (e.g. ballast water) in harbours and coastal waters.

2.4 *Estuarine and Coastal Estimates*

2.4.1 Historic Inputs

In addition to contemporary inputs from coastal industry, historical inputs are probably of special importance in areas of fine-grained sediment accretion, erosion and resuspension, for example in estuaries, where older industries have been disposing of waste untreated for centuries, or partially treated in more recent times. Indeed many discharges still were untreated into the 1970s (Walker 1971; Edwards 1972). These legacy inputs need to be distinguished from contemporary pollutant sources if environmental management and regulation are to be cost–effective.

An example of why this is important is the study in Southampton Water (SW) funded by the Environment Agency (EA), regarding one of the longest-running UK ecological monitoring and restoration projects (Arthur D Little Limited 2006). Esso Petroleum Company Ltd. and ExxonMobil Chemicals Ltd. discharge effluent into Fawley saltmarsh creeks in SW, but the current inputs from Esso Fawley are negligible today compared to historical and other sources. The effluents had all but eliminated a large salt marsh in SW between 1951 and 1965 after over 10,000 t/year of oil were discharged in the effluent during the 1950s. This was reduced to 6000 t/year oil during the 1960s, decreasing to 2000 t/year during the 1970s, and to 500 t/year by 1984. These dramatic and ongoing steady improvements and the resulting ecological recovery at Esso Fawley have been well-documented since 1970 (Baker 1971, 1976, 1979; Baker et al. 1990; Dicks and Iball 1981; Dicks and Levell 1989; May 2000; Arthur D Little Limited 2006).

Radionuclide-dated sediment cores retrieved in the late 1980s for the Institute of Petroleum (now Energy Institute) from MHW, SW, and Sullom Voe showed that the inputs from oil spills and/or chronic oil industry effluent sources are superimposed upon wartime and municipal or historic industrial inputs (Little 1986; 1987a, b; Little et al. 1987, 1990; Little and McLaren 1989). This core-dating work was extended to the Humber estuary for the joint water regulators' monitoring committee, and also to the Bodensee in the Alps where it was shown that outboard 2-stroke engines made little contribution to pollutant loadings (Bartlett et al. 1989a, b). In some cases ^{137}Cs derived from the Chernobyl reactor explosion of 26th April 1986 is detected near the sediment surface, in addition to the subsurface pattern produced by the atmospheric nuclear weapons tests from the early 1950s to the peak in 1963 (Little et al. 1990). The rate of contaminant throughput in the drainage basin and estuary may thus be estimated using this man-made nuclide.

There has been an increasing understanding that although wartime inputs and impacts of mainly fuel oil during both WW1 and WW2 had been huge, a wide-scale irreversible environmental catastrophe had not apparently occurred. In fact, the data showed that previously over-exploited fisheries recovered during the war years owing to the blockades of fishing fleets which reduced benthic disturbance and allowed more fish to mature and reproduce. A similar effect is demonstrated by offshore structures that shelter sediment and epibenthos as well as juvenile fish from predators, and like shipwrecks they also represent a rare hard substrate in the UKCS. However, fishing continues to be the most widespread human impact (OSPAR 2010), and wartime wrecks have been shown to be important continuing sources of leaking oil, both in shipping lanes and elsewhere (Michel et al. 2005).

In the USA "Superfund" sites, the discrimination of historical oil/chemical sources by use of chemical forensic analysis of sediments is now an industry in its own right. An example of its success is the reduced bioavailability and impact of PAHs in Eagle Harbour, arising historically from creosote use in wood-treatment industries in Washington State, USA (Myers et al. 2008). This is being achieved partly by sediment capping in many sites in Puget Sound, but has not been attempted on a wide scale in UK waters. In contrast, when capping of legacy pollution was suggested by Associated British Ports (ABP) in its Dibden Bay container port proposals in SW, it was rejected at public enquiry. ABP creatively but not entirely altruistically had proposed to use spoil from port construction and dredging for capping nearby sediments historically contaminated by PAHs and metals (including from the Esso Fawley oil and petrochemical discharges mentioned above).

Forensics discrimination is important in other areas receiving coastal discharges such as MHW, and in E&P regions such as the North, Irish and Celtic Seas. Fuels and lubricants from historical sources such as wrecked vessels or legacy industry should be accounted for by oil companies working nearby, when interpreting their hydrocarbon data from environmental monitoring programmes, in case liability be wrongly ascribed. In this context the decommissioning of offshore O&G infrastructure may also be considered, of which the highest profile example was the controversy led by Greenpeace in 1995 surrounding the proposed (and UK government approved) deep sea disposal of Brent Spar, a redundant N. Sea oil storage facility owned by Shell and Esso.

2.4.2 Releases from Coastal Sources

In order to place oil industry releases in the context of the overall inputs to the UKCS from all other sources, as shown above, the E&P industry has contributed a relatively reliable set of input data over the years. Unfortunately, the sources and types of almost all other coastal zone inputs are not so well documented, in spite of significant and growing interest in NPS and atmospheric pollution in many countries from the 1980s (for example in UK; RCEP 1981). In addition to the inevitable data gaps, there are national, methodological, accuracy and precision issues with almost all the data that have been collected. Not least is the difficulty of data comparability through time, for which attempts were needed to inter-calibrate, validate, standardise and gain accreditation for methods (Howells et al. 1989).

Oil spills from ship transport, product transfers, and downstream facilities for example in MHW are reported to the Port of Milford Haven (formerly Milford Haven Port Authority; MHPA), and this has been thorough since MHPA began reporting in 1960. Discharges from coastal refineries and oil terminals and to a lesser extent their ecological effects have been monitored and regulated by the EA, and also monitored by operators especially in major oil ports such as MHW, SW, Thames, Humber, Mersey, Firth of Forth, and at terminals in Flotta (Orkney) and Sullom Voe (Shetland). As highlighted by RCEP (1981) and reported to Concawe, land-based O&G industry discharges were of relatively high concern because they were by definition located closer to shore, with less dilution, and thus threatened sensitive receptors more than offshore industry sources.

In 1981 for the North Sea coast as a whole 9047 t/year oil was discharged by refinery sources, and because of growing concern; this was reduced by improved process controls and wastewater treatment to 1441 t in 1993, and to only 670 t in 1997, an overall reduction in 16 years of almost 93 %. The figure for 1997 is impressive not least because there were still 58 refineries operating in OSPAR countries at that time, in spite of record low oil prices and refinery closures. Of the 670 t these refineries discharged in 1997, 556 t (83 %) of oil in effluent went into estuaries, 323 t (58 %) of which the UK refineries discharged (OSPAR 2000). The extent to which this was an improvement is shown by comparison with the early 1980s when it was estimated that the four refineries and oil-fired power station then operating in MHW discharged about 180 t/year oil to the estuary (Little et al. 1987).

Since the estimates made during the 1980s, offshore and coastal refinery inputs have been closely monitored and reported to the Department of Trade and Industry and it successors via the Environmental Emissions Monitoring System (EEMS) and to EA (and Concawe) respectively. This disclosure is in contrast to some of the diffuse sources for which there has been less focus. For example, OSPAR (2000) noted the methodology problems in making estimates of river inputs generally, but were able to quote the Dutch sector inputs, which had gone down from between 5000 and 6000 t of oil in 1980 to between 580 and 1235 t in each of the years 1990–1994. OSPAR (2000) also estimated 2000 t/year inputs of oil from harbour dredging, containing 15 t PAH. More generally, these inputs had decreased due to reductions in river inputs and siltation of ports.

In the OSPAR project on Riverine Inputs and Direct Discharges (RID) unfortunately oil inputs were not included, partly because of the above-mentioned practical and methodology problems. RID used custom software for flow-adjustment (R-Trend) and adequate flow data were particularly lacking for the Irish part of the Celtic sea. This has important implications for the EC Marine

Strategy Framework Directive and the UK Marine and Coastal Access Act 2009 in England and Wales. Nevertheless, the RID project showed that inputs of selected metals and nutrients into the North Sea and Celtic Sea have decreased significantly between 1990 and 2002, and these are all useful parameters for industrial, municipal and agricultural pollution levels (OSPAR 2005): Cd decreased by 49 % (N. Sea) and 68 % (Celtic Sea); Hg decreased 73 % (N. Sea) and 86 % (Celtic Sea); Pb decreased 33 % (N. Sea); N decreased 12 % (N. Sea); and P decreased 33 % (North and Celtic Seas).

The above decreases consolidated the earlier progress made for example in the Bristol Channel between surveys of 1974 and 1988 when metals decreased significantly in the dominant mud sediments (Zn by 25 %, Cr, Ni and Pb by 30 %, and Cu by 50 %) due to closure of metals mining and industry in South Wales and pollution abatement (Little and Smith 1994). The reduced effluent loadings are often reflected in significantly reduced sediment contamination since the estuarine peak inputs until the 1980s. This was found to be the case in MHW with mean concentrations of trace metals in the sediment samples taken for dredging control between the grouped years 1984–1992 and 1993–2006. These averages show reductions ranging from 36 % (Ni), 50 % (Zn), 62 % (Pb), and 67 % (Cu), up to 85 % (Hg).

However, the recycling of historical contaminants by sediment dynamics and spillage from dredging in MHW means that some of the recalcitrant PAHs and UCM in surficial sediments that were identified date back to oil spills in 1996 ('Sea Empress'), 1988 ('El Omar'), and during WW 2 (Little et al. 2015; and see below). The continuing NW. European decreases in inputs between 1990 and 2002 found by OSPAR (2005) were in excess of what would have been expected from industrial decline particularly in the Elbe drainage basin in former East Germany after reunification in late 1989, and from the more widespread recession at the beginning of that period. This implies that source control and environmental management processes were also driving the reductions. Although discharges from coastal refineries and oil terminals fell steadily since 1980, as a result of ongoing investment in improved process controls and wastewater treatment, as well as refinery closures, it is much less clear what the trends are in NPS oil inputs.

Dredging and dumping operations and techniques have changed little over the past 10 years (OSPAR 2010). About 90 % of all sediments dumped each year are dredged and dumped in the southern N. Sea. This is largely from maintaining navigation channels to major seaports such as Hull, Antwerp, Rotterdam, Hamburg and Esbjerg. In 2005, there were around 350 dumpsites in the OSPAR area. Between 1990 and 2007 the total annual amounts dumped at sea varied from 80 to 130 million tonnes (dry weight) with much of the variation due to capital dredging associated with port expansion and deepening of navigation channels. The level of dumping and dredging activities has been relatively stable over the past decade and is unlikely to fall. The need for dredging may be increased in coming years by a growth in ship size, requiring deeper and wider navigation channels, or a greater frequency and intensity of storm events, and thus sediment movement by waves and currents. One of the main concerns over dumping and dredging is the release of contaminants to the water column. There was a clear decrease in contaminant concentrations in dredge spoil from the southern North Sea throughout the late 1980s and 1990s, but this trend has since stabilised (OSPAR 2010).

As shown in Table 1, the diffuse NPS outweigh oil industry point sources, but this is not how the public have generally perceived the problem of marine pollution. The ‘Sea Empress’ in 1996 was perceived at least in England and Wales as the most significant tanker disaster since the ‘Torrey Canyon’. In Scotland, the 1993 *Braer* was the landmark oil spill. However, the publicity surrounding the ‘Sea Empress’ and the scale of the response that year inevitably far outweighed the gradual improvements being registered by the newly-formed EA. In fact, source control had reduced all pollution, both onshore and offshore. Nevertheless, many incidents release hydrocarbons that contribute to the background in environmental samples. Oil industry and estuarine monitoring programmes must take account of the variability and potential for recycling of inputs to the baseline in industrialised areas.

3 Inputs of Oil and PAHs to Milford Haven Waterway

3.1 Local Factors

3.1.1 Anthropogenic Background

Hydrocarbons from both natural and anthropogenic sources are common in the marine environment (Brassell and Eglinton 1980). Contaminants are typically found in areas where fine-grained sediments accumulate in estuaries with urban and industrial hinterlands, and agricultural catchments. In general, the spatial extent of estuary sediment contamination is extensive while the vertical extent is limited. Although MHW is situated in a mainly rural catchment area, its shoreline is moderately urbanised and is subject to chemical contamination from a variety of sources and pathways including agricultural, municipal and industrial. Oil inputs to MHW are estimated in Table 1. Relative to its moderate-sized population, the immediate area surrounding the estuary is highly industrialized and includes (or included in the past) metal-working, coal mining, lime-burning, whaling and fishing industries, oil refineries, power station, fuel tank farms, ship-building, and commercial and military maritime activities. These involved tar, pitch, grease, resins, anthracite, fuel and lubricant oils in the sailing and steel ship eras, and other potential contaminants (e.g. ballasts, anti-fouling paints and munitions). Numerous wrecks are present, such as the ‘Dakotian’ lying in Dale Roads anchorage whose cargo included metal ingots, and the Sunderland flying boat near Neyland. All the above hydrocarbon sources, as well as site-specific sources have been previously identified as potential causes of sediment contamination in MHW (Little 2009).

Background concentrations of PAH are ubiquitous in MHW sediments. Low concentrations of naturally occurring PAH are derived from natural fires, eroded shale/coal, and early diagenesis processes, and these pre-date modern anthropogenic activities. However, the prolific use of fossil fuels over the last 200 years has altered the loading of hydrocarbons to the environment. The resulting residues and combustion by-products have contributed a modern anthropogenic PAH background that in MHW was first observed in equal measure before and after the 'Sea Empress' oil spill in February 1996 (Levell et al. 1997). The equivalent contaminant PAH background in the Firth of Clyde sediment is characteristic of the "Anthropocene" (Vane et al. 2011).

3.1.2 Inputs of Total Hydrocarbons

Historical industrial, municipal and military activities in the near-shore zone and hinterland have contributed to the total hydrocarbon (THC) contamination of MHW surficial sediments. As would be expected in a major oil port, evaluation of the available chemical data demonstrated that general levels of hydrocarbon contamination and its composition are consistent with petroleum-related operations and spills as a major source of pollution. Relatively high levels of sediment contamination are observed in the vicinity of refinery jetties, effluent outfalls and urban areas. The spatial distribution of THC concentrations is also consistent with the fine-grained sediment transport patterns in MHW. However, chemical characteristics of the sediment samples exhibit characteristics consistent with historical rather than current hydrocarbon inputs. Chromatographic characteristics demonstrated that most of the aliphatic hydrocarbon pollution in the sediments of MHW is indicative of weathered petroleum sources of contamination. The UCM indicating weathered petroleum is ubiquitous in MHW thanks to sediment movements (Little et al. 1987; McLaren and Little 1987). The above conclusions are supported by chemical analysis of radionuclide-dated core sediments in MHW. These demonstrated a significant increase in THC concentrations coinciding with the establishment of the large scale petroleum refining industry in 1960, and a subsurface maximum at the time of peak oil throughput before the first 1970s oil crisis (Little and McLaren 1989).

Superimposed on this anthropogenic background, an order of magnitude increase in THC contamination was observed in surficial sediments in the lower reaches of the waterway shortly after the 1996 Sea Empress oil spill. Nikitik and Robinson (2003) found that a pulse of water-borne low molecular weight hydrocarbons had moved up MHW after the 'Sea Empress' spill which was thought to account for the widespread observed oil sheens, and the toxic effects on benthic amphipods lasting up to five years. Monitoring data suggest that within <5 years THC concentrations were similar to those before the spill. In fact, THC concentrations had already returned by April 1997 to pre-spill levels (1993), but in October 1997 there were renewed increases in THC almost throughout MHW. Nikitik and Robinson did not use fingerprinting and did not discuss dredging disturbance as an additional possible mechanism for release of previously-deposited contaminants, for example in major dredging campaigns of 1992, 1997 and 1998. There is reasonable evidence to suggest this connection not only for PAHs, as discussed below, but also for THC. Levell et al. (1997) found no clear fingerprinting evidence for a connection between 'Sea Empress' and subsequent sediment loadings. Analyses of PAHs by Little (2009) and Little et al. (2015) have cast doubt on the last assertion.

Nikitik and Robinson contrasted their findings with typical sediment-dominated patterns described by Rostron et al. (1986), McLaren and Little (1987) and Little and McLaren (1989). Nikitik and Robinson correctly suggested that prior to the spill the THC concentrations were higher up-estuary, whereas immediately after the spill the lower and mid-MHW reaches were more polluted. This is reasonable in that with the exception of the 'Sea Empress' and the loss of more than 2300 t of gasoline from 'Dona Marika' in 1973, no major hydrocarbon sources occurred downstream of Angle Bay.

3.1.3 Inputs of PAHs

Elevated concentrations of heavier 4–6 ring PAHs in the surficial sediments found in MHW are typical of predominantly combustion-related (pyrogenic) sources of PAH contamination. However, they are also characteristic of heavily weathered petroleum. Source-ratio evaluation suggests that most of the PAHs in surficial sediments of MHW originate from combustion sources (Little 2009). The limitations of the PAH data sets available in 1993 and 1996 did not allow differentiation among various potential pyrogenic sources of PAH contamination. However, a noticeable difference in source-ratio characteristics in five sediment samples collected in 1996 compared with 1993 suggests PAH contribution from the 'Sea Empress' oil spill. These five stations extended from the lower estuary where 'Sea Empress' grounded, past Stack Rocks, Hubberston, Newton Noyes to just east of Wear Point, and clearly following the flood tide sediment transport pathway predicted by sediment trend analysis (STA; McLaren and Little 1987).

For sediment PAHs in MHW, large differences were detected between data in 1993/1996 and three more recent surveys (2007, 2010 and 2012) employing more advanced methods. The widespread very high PAHs concentrations found in late 2007 greatly exceeded sediment quality guidelines but did not persist, although may have had transient adverse biological effects. The concentrations were reduced in late 2010 and further slightly reduced in late 2012 except where LNG-related dredging and construction were continuing. It is suggested that a pulse of suspended particulate matter (SPM) was caused by sediment remobilisation during construction activities, major dredging in many parts of the waterway in 2006 and 2010–2011, and dredging at a marina from 2004. The pulse moved up-estuary from 2010 to 2012 and coincided with PAHs bioaccumulation in mussels and ragworms near the dredging operations (Little and Bullimore 2015; Little et al. 2015, 2016).

The pyrogenic PAH source ratios and the total PAH concentration changes through time together supported the above hypothesis in terms of the forensic identification of PAH from the 'Sea Empress' HFO, refinery effluent, and possibly 'El Omar' crude and WW2 tank-farm fires as sources of both current re-working and historical pollution inputs (Little et al. 2015). The annual Clean Seas

Environment Monitoring Programme (CSEMP) time series of PAHs measured at Cosheston Point is also entirely consistent with this interpretation.

3.1.4 Sediment Behaviour

The behaviour of fine-grained sediments in MHW is an important driver of the fate and effects of hydrocarbons due to grain size, composition and surface area (Little et al. 1987). The use of STA has already been of great value in MHW in the interpretation of contaminants in terms of sediment movements (McLaren and Little 1987; Little and Bullimore 2015). Such muddy sediments are found as the highest-level sedimentation units of estuaries and deltas under regressive conditions (falling relative sea level due to infilling and prograding shorelines). Fining-upward sequences are characteristic of fluvial and estuarine meandering channels and point bars. Estuaries can nevertheless show transgressive sequences under conditions of rising relative sea level (e.g. Chesapeake Bay, USA). These produce bay-mouth shoals and sometimes sand bars that overlie muds and marshes; a coarsening-up sequence such as sand-over-mud. Such conditions may have obtained in the past in MHW as Flandrian sea level rose after the Devensian stage of the ice age. Sea level rise of almost 20 cm has also occurred in the past 100 years. Little (2009) briefly considered this in terms of erosion of small cliffs in salt marshes and the possible effect on SPM concentrations. An increased water depth relative to wave base could also mean more shelter from wave action, more sediment deposition within MHW berthing boxes and thus more frequent need for dredging. However, fining-upward sediment sequences are generally far more common in MHW and other estuaries.

Estuarine environments show bioturbation by infaunal organisms that mix sand and mud sediments, confusing the grading of sediment and the bedding patterns described above (and also the use of radionuclide dating methods). In muddy sediments, an irregular distribution of sandy pockets produced by feeding and burrowing animals gives a mottled appearance. Similarly, dark-coloured muddy flecks in light-coloured sandy sediment provide clues to the activity of biota. Sediment structures and bedding are rarely considered fully enough in pollution and biological monitoring studies.

Changes in estuarine morphology, in particular those affecting fine sediments, can have major impacts on the balance of sediment erosion, transport and deposition. In MHW the fine sediment budget has been affected as follows in the last 150 years:

- Dredging losses—5 million m^3 removed from the estuary.
- Reclamation losses—1.605 km^2 for industrial, port, marina and waste disposal.
- Historic quarrying gains—0.426 km^2 at limestone and coal quarrying sites.
- *Spartina* 'die-back' 1982–2002—1.0 km^2 E. Cleddau and Carew/Cresswell.

For the above reasons, a study of MHW sedimentology was conducted for MHWESG in 2012 using Sediment Profile Imagery (SPI) and plan views of the sediment (Germano and Associates 2013; Carey et al. 2014). SPI showed that sand-over-mud appeared in 30 stations out of 550 in total, and had not been observed during core sampling in 1982. The area affected was in mixed sediments in the centre of the dredging and/or construction activities spanning several years until 2011. This is interpreted as physical evidence of the minimum extent (i.e. in mixed sediments only) of dredging and construction disturbance (Little and Bullimore 2015; Little et al. 2015, 2016). In areas of more uniformly muddy sediments it is not surprising that the SPI data did not show any visible mottling from the physical effects of dredging or construction disturbance.

To predict the approximate amounts of oil and PAH potentially involved in any remobilisation events during dredging, and taking the average total PAH from berthing boxes from 2001 to 2006 as between 3 and 5 mg/kg, then the total PAH mass would be 600–1000 kg in 200,000 t of disposed dredge spoil (approximately the peak spoils arising in MHW in 2006, but note that not all of this was muddy). Taking the average THC concentration in 2005–2007 as 300 mg/kg then the total oil mass would be 60 t per 200,000 t of spoil. Using a higher historical THC average value, the oil mass would have been 100 t per 200,000 t of spoil. Spoil disposal can be deleterious if it is contaminated. Unfortunately, the amount of whole oil or PAH released or spilled from the spoils prior to leaving MHW for licensed offshore disposal is unknown. In addition, the contaminant content of the direct water column disposals from dredging a nearby marina is unknown.

However, assuming a minimum of 2.5 % losses during dredging (Newell et al. 1998) would mean that about 5000 t of spoil was lost from the 2006 peak of 200,000 t, which would equate to 15–25 kg PAH and 1.5 t of oil. Newell et al. (1998) reported up to 38 % losses in the worst case, and so this could mean up to 375 kg PAH and 22.8 t of oil resuspended in MHW by dredging disturbance and losses. To put the significance of the potential reservoir of PAHs in estuaries in context, this compares to about 150 kg PAHs discharged in 2013 worldwide in non-aqueous drilling fluids, assuming a worst-case of 0.001 % PAH content and adjusting upward for the approximately 71 % of oil production by non-reporting E&P companies (IOGP 2014).

3.2 Regional and National Factors

It is important to investigate whether known sedimentological and chemical processes in MHW match the temporal patterns shown by PAH and other contaminant data elsewhere. In other areas of the UK, a wide range of PAHs and other contaminants are routinely monitored. Atmospheric inputs also have contributed both direct deposition and through the tributary catchments, and from national data it is thought this was particularly true in 2005–2007. Below are some findings from a selection of these studies.

3.2.1 Estuarine Comparisons

Hutchinson et al. (2013) listed UK estuaries with high concentrations of PAH in their sediments as a result of historical contamination from industries and power stations, including MHW, the Clyde, Medway, Mersey, Tees, Tyne, and Wear. There was little evidence that PAHs concentrations were declining significantly (Woodhead et al. 1999; Vane et al. 2007a, b). Figure 4 shows the maximum sediment concentrations of ΣPAH_{22} in 12 UK estuaries ranging from those known to be contaminated to those considered "background".

Each estuary was sampled at from three up to 15 stations for DEFRA at a time when concern was rising over the persistence of PAHs in estuarine sediments (Rogers et al. 2000). The maximum PAH data for MHW shown in Fig. 4 from Little et al. (2015) illustrates that MHW is relatively low in terms of the maximum PAHs concentrations found in the DEFRA study. However, much higher PAHs were found in "hot-spots" such as Neyland Spit in 1996 and in Pembroke Dock in 2012 (Fig. 4). Patchiness in sediment contaminants is the norm.

Studies of UK nearshore sediments undertaken by CSEMP show that in the period 2000–2010, only 2 of 15 sites showed statistically significant declines in ΣPAH_{22} concentrations (Rumney et al. 2012). Ecotoxicological studies suggest that the estuaries of NE. England and parts of MHW historically had sufficiently high concentrations to be acutely toxic to some sediment infauna (Woodhead et al. 1999). PAHs concentrations are still high and likely to cause adverse effects (UKMMAS 2010). Figure 5 shows England and Wales' nearshore sediment PAHs data analysed using the same accredited methods and laboratory (Cefas) as the above-mentioned dredge licensing work in MHW. Figure 5a covers the east coast of England, which is generally higher in PAHs concentrations than the south and west coasts of England and Wales (Fig. 5b). Both regions derive their PAHs inputs

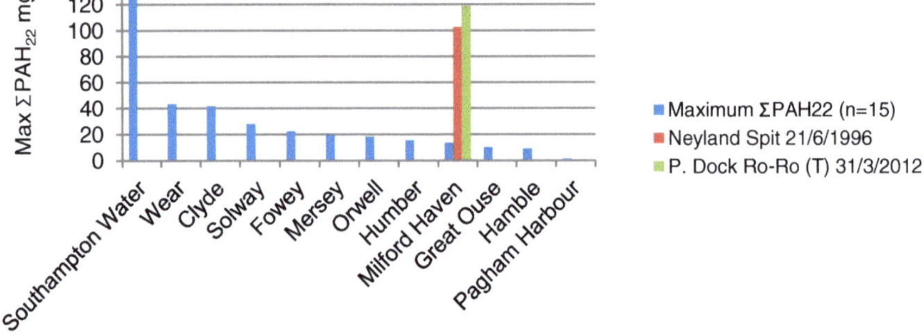

Fig. 4 Maximum ΣPAH_{22} concentrations (mg/kg) in 11 estuaries 1998–1999 (Rogers et al. 2000) plus MHW 2010 (Little et al. 2015); see Fig. 1 for locations (*Note* NMMP 646 at Neyland Spit 1996, and site 'T' Pembroke Dock Ro-Ro berth 2012 analysed by Cefas)

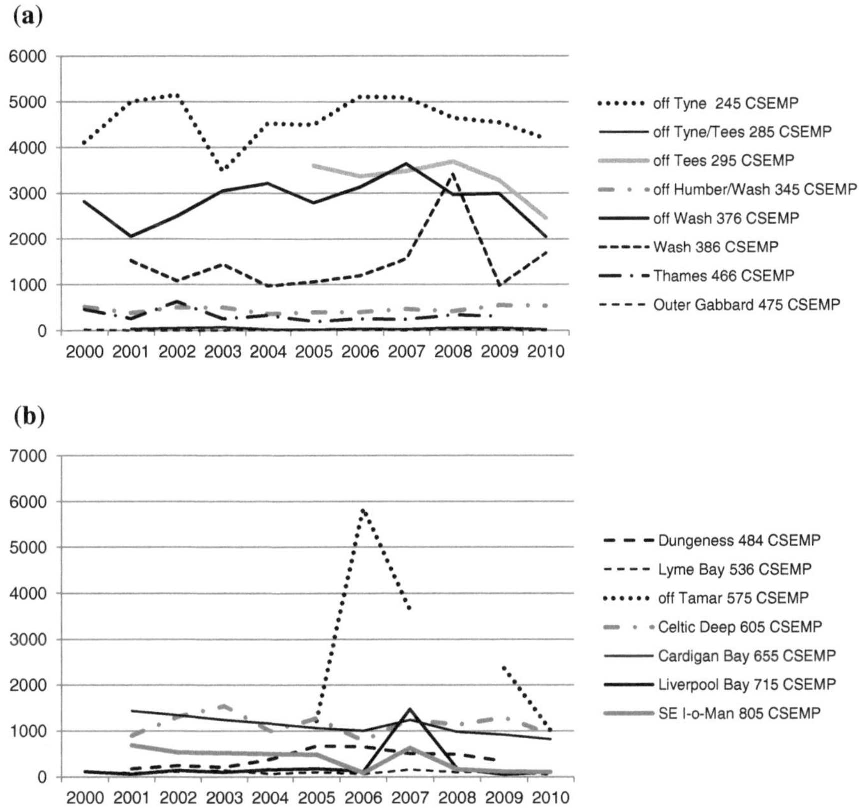

Fig. 5 Means of five replicates for ΣPAH_{22} concentrations (μg/kg) 2000–2010 at **a** eight CSEMP sites from the east coast of England, and **b** seven CSEMP sites from the south and west coasts of England and Wales (Rumney et al. 2012); see Fig. 1 for locations; (*Note* there was also an oil spill in Lyme Bay of 302 t fuel oil ('Napoli') in 2007)

from historical, industrial and mainly land-based sources. There are peaks observed in the years 2006 (offshore Tyne and offshore Tamar), 2007 (offshore Wash, Cardigan Bay, Liverpool Bay and SE. Isle of Man) and 2008 (offshore Tees and the Wash). The chi-squared test was used to evaluate how likely it was that the observed departure from homogeneity of the frequency distributions across all the CSEMP datasets arose by chance. The grouping of maximum sediment ΣPAH_{22} concentrations in 2007–2008 was not significant (i.e. could have occurred by chance). Nevertheless, the cluster of peaks between 2006 and 2008 across almost half of these nearshore sediments at the national scale shows that the timing of peak contamination in 2007 within MHW ideally should be examined more widely than only at the local and regional scales.

3.2.2 Air Quality Comparisons

It is also likely that air pollutants emitted at regional and national scales have been deposited in MHW, either directly or flushed into the tributary catchments. This additional hypothesis in particular regarding the well-wooded Eastern Cleddau was made by Morris (1985), and was explored for MHW as a whole by Little (2009) and Little et al. (2015). As with the lower and middle reaches of MHW, so the sediment fingerprints from the upper and innermost reaches in 2007 were dominated by pyrogenic PAHs. The source ratios for MHW sediments in 2010 and 2012 all show good agreement, the PAH ratios indicating pyrogenic sources dominated by biomass and coal (Little et al. 2015). In the annual CSEMP (2014) survey off Cosheston Point the fingerprint is similar, but more scattered for some years.

Pyrogenic inputs also include fallout from natural and man-made fires. The man-made pyrogenic sources likely include past and present atmospheric emissions from shipping (coal, HFO and marine diesel), and point sources from oil refineries and a former oil-fired power station, closed in 1996, that would all have contributed unknown quantities of pyrogenic material (hydrocarbons and soot) into MHW sediment. A very large pyrogenic source was tentatively identified by Little et al. (2015) using forensic comparisons with archive data on military fuel supplies used during WW2. After the Luftwaffe attack on the Admiralty dockyard fuel depot at Llanreath in August 1940, about 130,000 tons of fuel oil was burnt or lost, along with fatalities among fire-fighters. Fallout affected the apple harvest over 70 km away in N. Devon. Smaller pyrogenic inputs may also have arisen from later incidents in MHW, such as the 'Esso Portsmouth' explosion in 1960, the Amoco (now Murco) refinery tank farm fire in August 1983, and the Chevron (now Valero) refinery cracker fire in June 2011.

Total atmospheric inputs of PAHs are typically not dominated by any single source, except close to major point sources. Instead, the emissions are quickly mixed and weathered in the environment. PAH anthropogenic emissions to the UK atmosphere fell from 5604 t in 1990 to 777 t in 2013 (NPL 2015), due mainly to farming bans on open-burning (1992), and decreased inputs from metal-smelters (1999), and from waste incineration (2006). To determine sources of PAHs, molecular diagnostic ratios are often used as markers, although variations in combustion conditions and environmental degradation cause high variability in the emission and weathering of individual compounds, reducing the reliability of some ratios for source allocation (Katsoyiannis et al. 2011).

Some of the pyrogenic input to MHW is derived from road runoff into the tributaries. A proxy for this aerial/runoff input for the period 2005 to 2011 is Pembrokeshire diesel and gasoline road fuel consumption data. These data show a decline of 8.2 % since a plateau between 2006 and 2008, in line with Wales (8.3 %) and UK (7.2 %) decreases during the financial crisis (DECC 2013). Increased pollutant inputs to road runoff may have offset this decrease; tar macadam damage on roads due to budget cuts may have increased polluted runoff, and according to the Office of National Statistics, the proportion of automobiles parked overnight in garages in the UK decreased from 22 to 14 % between 2002 and 2012. It will be recalled that MHW, offshore Wash, Cardigan Bay, Liverpool Bay and SE. Isle of Man all had peak concentrations in 2007, as discussed above.

The pattern is reflected by the UK-wide continuous air sampling programme run by DEFRA. The annual average ΣPAH_{39} of up to 25 sites in 2005 was the lowest since records began at these stations and using the same methodology (modified Andersen sampler). The trough in 2005 was immediately followed by an increase through 2006 to a peak in 2007 ($n = 9$), the highest since 1996, before falling away again at those fewer stations ($n = 6$) with comparable data for 2008 (Fig. 6a). For reference, the UK air quality objective, based on the recommendations of the Expert Panel on Air Quality Standards (EPAQS), is for a mean annual air concentration of 0.25 ng/m^3 of Benzo(*a*)pyrene equivalent (B(*a*)P), and the EC target limit is 1 ng/m^3. The steep decreasing trend in measured ΣPAH_{39} air concentrations generally follows the decreases in estimated anthropogenic emissions to air of PAH (UKMMAS 2010) and of B(*a*)P (NPL 2015), but appears uncoupled in the years 2005–2008 (Fig. 6b, c). The inputs of PAHs from domestic fireplaces have increased in some parts of the UK since the financial crisis, and through the period 2001–2014 diesel vehicles have increased from 14 to 36 % of road vehicles, in turn increasing the proportion of the PAH inputs from traffic.

Notwithstanding the above warnings of Katsoyiannis et al. (2011), the sources of PAHs can sometimes be determined using double ratio cross-plots (Yunker et al. 2002; Oros and Ross 2004; Costa and Sauer 2005; Wang et al. 2015). Forensic analysis was conducted on MHW sediments sampled annually between 1999 and 2011 at Cosheston Point for CSEMP (2014). Over this same period, the cross-plots for ratios of Fluoranthene and Pyrene vs. Indeno(*1,2,3-c,d*)pyrene and Benzo(*g,h,i*) perylene shown in Fig. 7 indicate that between February of 2008 and of 2011 at Cosheston Point during these peak concentrations of PAHs the ratios in sediments

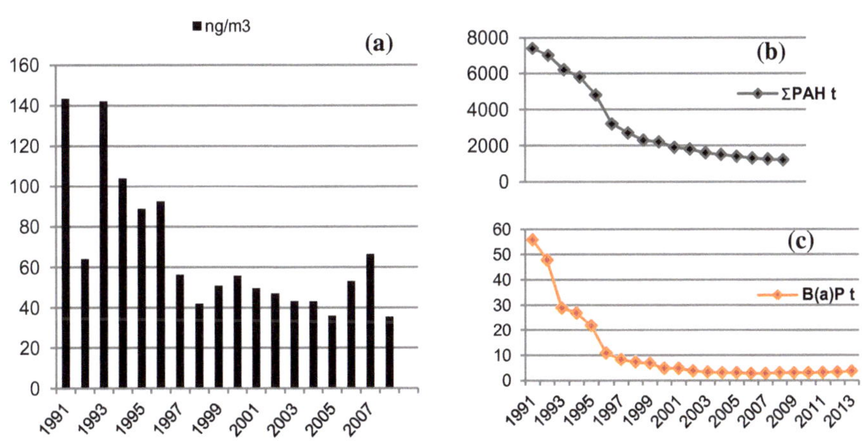

Fig. 6 **a** Mean ΣPAH_{39} concentrations (ng/m^3) in air at up to 25 UK monitoring stations 1991–2008; **b** Estimated PAH inputs (t) to air 1991–2007 (UKMMAS 2010); and **c** Estimated B(*a*)P inputs (t) to air 1991–2013 (NPL 2015)

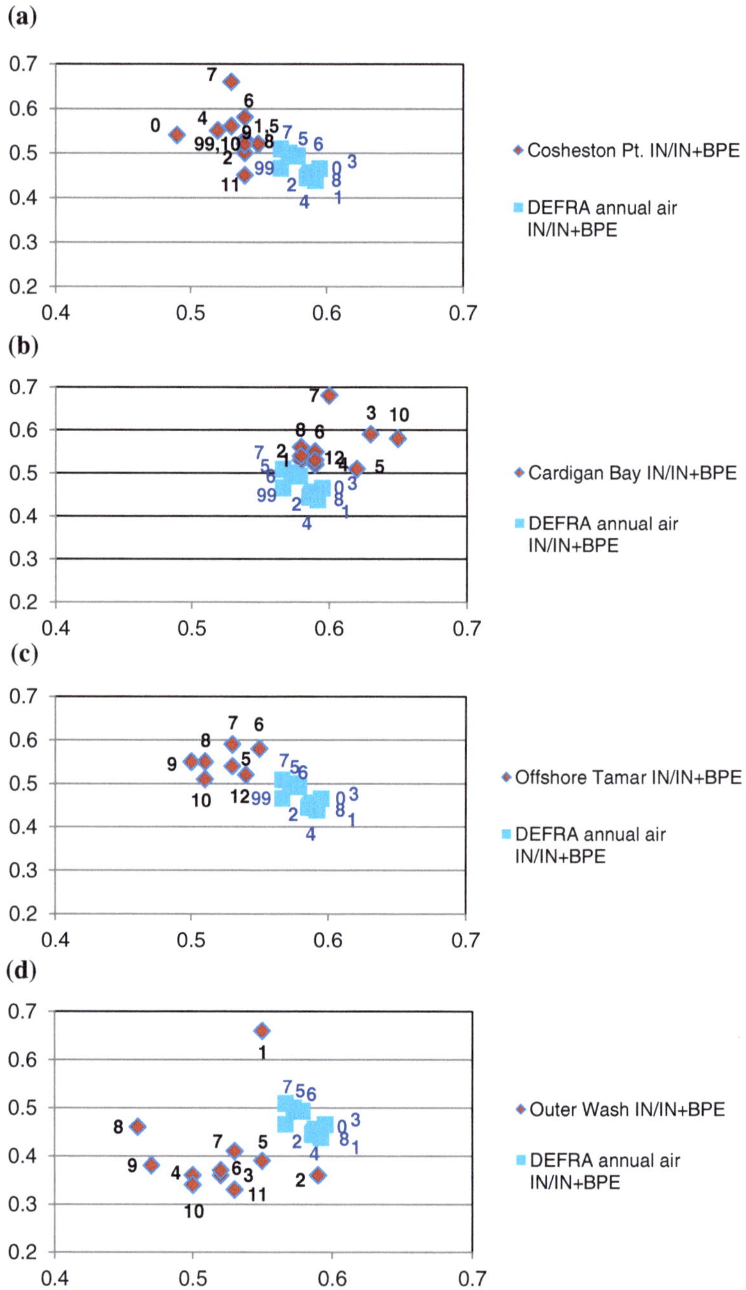

Fig. 7 FL/FL + PY (x-axis) versus IN/IN + BPE (y-axis) in 1999–2011 (*diamonds*) for sediments from **a** Cosheston Point. **b** Cardigan Bay. **c** Offshore Tamar and **d** Outer Wash CSEMP stations, and national DEFRA air sampling annual means (n = 6 – 25) in 1999–2008 (*squares*); numbers indicate year; see Fig. 1 for locations (*Note*: CSEMP 2014 via MERMAN; DEFRA 2014)

most closely resembled the composition shown in the air sample ratios between 2005 and 2007 (DEFRA 2014).

Figure 7 also shows a selection of three further CSEMP nearshore sediment sampling sites that were mentioned above as having a contaminant peak in the period 2006–2008. The sediments sampled annually between 1999 and 2011 at Cardigan Bay, offshore Tamar and offshore Wash for CSEMP (2014) suggest that the 2005 and 2007 air samples are geochemically closer than in most other years to some of the nearshore sediment samples. The timing of the inputs varies, however, and may be due to the lag in air pollution dependent on rainfall interception and runoff from the river catchments inland. The runoff into estuaries probably derives chronically from power stations, industry and domestic heating inputs. Biomass and coal combustion were dominant except at offshore Wash in all years but 2001 and at Cosheston Point in 2000, 2002 and 2011 when petroleum combustion sources dominated. Note that the average air sample ratios also indicate petroleum combustion sources every year except 2007. Although more remote from MHW, an additional possible source nationally was the oil storage depot fire at Buncefield near London 11–14th December 2005 (56,000 tons burned, the largest fire in Europe since WW2). When affected by air pollution input, entire catchments could receive atmospheric inputs directly by dry and wet deposition, indirectly after interception by vegetation, and from delayed runoff from each sub-catchment. At Cosheston Point in MHW, it was found that the years 2008 and 2011 were peaks in sediment PAHs concentrations.

The aerial pollution hypothesis is supported by the simultaneous, extremely high, but ephemeral PAHs concentrations found in late 2007 in surface sediments of rural parts of the inner MHW. The Eastern Cleddau, Carew and Cresswell Rivers are among the few muddy intertidal areas that do not show in the STA a clear domination both by fining and flood-tide pathways; nor do these areas show PAH or biomarker fingerprints that indicate a contribution from lower MHW by 'Sea Empress' HFO. The catchments of these three larger MHW tributaries are among the higher in elevation, the most easterly in location, and the more densely wooded around MHW shores. Because of somewhat higher rainfall resulting from such characteristics, these tributaries are likely to have high rates of wet deposition and effective canopy interception of widespread aerial pollutants.

The ΣPAH_{39} quarterly concentrations at 23 DEFRA air pollution monitoring sites in UK are shown in Fig. 8, where peaks are mainly caused by higher use of heating fuels in winter. Most of these stations (17/23) had high PAH concentrations in quarter 4 (Q4) of 2005, in contrast to the annual data in Fig. 6a which show that 2005 was the lowest average UK concentration. Although PAH concentration in Q4 2005 was not significantly different to the long-term Q4 average, the fact that Q4 2005 was warmer than average, perhaps reducing seasonal fuel usage, suggests that the Buncefield fire may have affected PAHs in UK air. The peak in Q4 2005 at most UK stations was not considered in reports on the Buncefield fire, which instead attributed the 'normal' ground-level concentrations to the plume reaching >3000 m

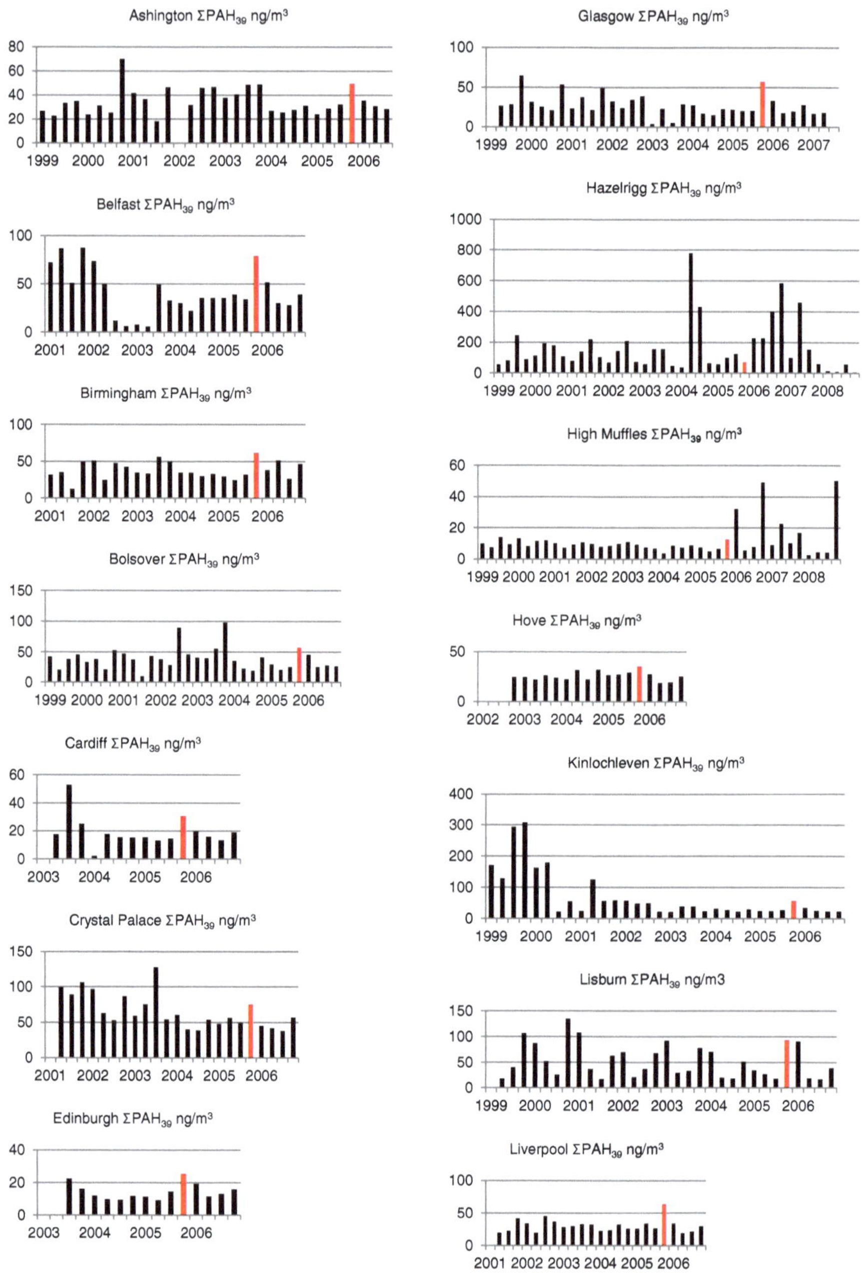

Fig. 8 Quarterly mean ΣPAH_{39} concentrations in air at 23 monitoring stations in UK 1999–2008 using Andersen samplers; note that two other stations were omitted as no Q4 2005 data were collected; also note differing axis scales (DEFRA 2014). Q4 2005 is indicated in red, contrasting with the trough in annual data for 2005 (Fig. 6a), and perhaps implicating the large fire at Buncefield oil storage depot on 11th December 2005. Monthly data that would have helped to resolve this only became available from late 2006 through 2007 at 34 stations, and used a different sampler ('Digitel' DHA-80)

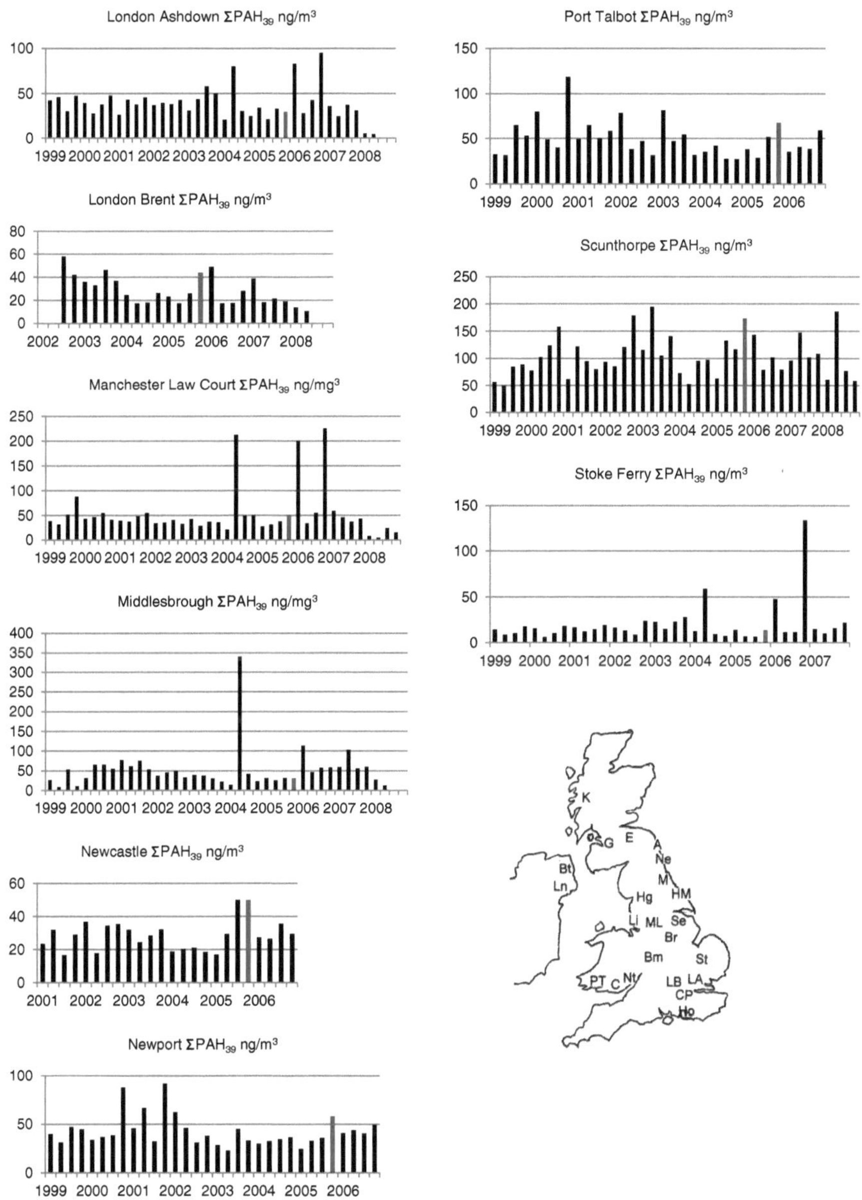

Fig. 8 (continued)

(AEA 2006; HPA 2006). Grouping the annual data in two-year bands, the chi-squared test was used to evaluate how likely it was that the observed departure from homogeneity of the frequency distributions across all the DEFRA datasets since 1999 arose by chance. The probability of the maximum PAHs concentrations in air occurring in 2005–2006 by chance was $p < 0.05$ (4 d.f.). Therefore, the hypothesis by Morris (1985) for elevated aromatic hydrocarbons in the Eastern Cleddau appears to be supported by national air quality data. In addition, the potential biological effects described by Little and Bullimore (2015) include some impacts in the relatively clear waters of the approaches and outer MHW; these could also have been partly caused by widespread atmospheric PAH inputs such as those described above.

Nam et al. (2008) determined PAHs in soil samples from background locations in the UK and Norway, to investigate environmental factors affecting their spatial distribution. Concentrations were more than four times higher in the UK soils. Proximity to sources and locations susceptible to high atmospheric deposition resulted in higher concentrations. Statistically significant relationships between PAH and total organic carbon (TOC) were observed in the Norwegian samples. High molecular weight PAHs correlated with black carbon (BC) in UK woodland soil, suggesting that TOC plays a role in the retention of PAHs in soil and that PAHs are combined with BC during combustion. PAHs with 4 and more rings comprised about 90 % of total PAHs in UK soil, but only 50 % in Norwegian soil, and Nam et al. (2008) suggested that fractionation occurred during long-range atmospheric transport and deposition. The lighter PAHs reached the more remote sites whereas the heavier PAHs remained in closer proximity to sources. In soils, however, the proportions of two- or three-ring PAHs were lower in the smaller particle sizes, indicating that the most weathered PAH mixture was located in the finest-grained material.

4 Comparisons Between MHW, Southampton Water, Sullom Voe and Other Sites

4.1 Sedimentation Rates

MHW is a drowned river valley (ria) in west Wales, SW is a coastal plain estuary in southern England, and Sullom Voe (SV) is a fjord in Shetland, northern Scotland. Using radionuclide data from the core-dating sites, Table 2 presents a summary of the estimated fine sediment accumulation rates. In addition, the table compares sedimentation rates to other aquatic systems using almost identical methods.

There are more data on sediment fluxes available for the other systems because additional bulk density measurements were made for the later work (British Standards Institution 1975). In spite of its 1980s date, the data set is a valuable contribution to knowledge of sedimentation rates in UK estuaries (Baker 1993). The chronology data for SW (Hythe area) were essentially corroborated by Croudace and Cundy (1995) and Cundy et al. (1997). No newer data have been seen for MHW and Sullom Voe.

Table 2 Mud facies sedimentation rates in MHW, SW and SV calculated using ^{137}Caesium and ^{210}Lead, and compared to other aquatic systems using similar methods

Coring site	Date	^{137}Cs peak concentration		Average sedimentation rate (mm/year)		
		dpm/g[a]	Depth (cm)	^{137}Cs < 1963[b]	^{137}Cs < 1953[c]	^{210}Pb[d]
Martinshaven, MHW	May 1985	See[e]	2–3	1.14	5.9	1.29
Daugleddau, MHW	July 1986	0.76	12–14	5.65	8.2	4.1–4.6
Jetty Area, MHW	July 1986	0.20	5–6	2.39	3.0	1.3–1.53
Pembroke River, MHW	July 1986	1.19	12–14	5.65	8.8	3.75–3.85
Cadland Creek, SW	Aug 1986	1.39	20–25	9.78	16.7	5.0–8.4
Hythe Mudflats, SW	Mar 1987	0.97	0–8[f]	8.91	8.3	7.4
Totton Marsh, SW	May 1987	1.44	5–10	3.26	4.5	2.5
Houb of Sullom, SV	June 1987	0.29	0–3[f]	5–5.5	5.1	ND[g]
Voxter Voe, SV	June 1987	0.43	0–3[f]	1.62	2.8	1.7
Pyewipe Mudflat, Humber	Aug 1987	2.19	6–8	4.1	3.8	4.5
Bodensee, Germany	Aug 1988	0.87	0–1[f]	6–18	3.9	1–6
Mindelsee, Germany	Aug 1988	0.36	3–4	5.0	4.1	1–2.5

Little (2009) and references therein

[a]dpm/g = decays per minute per gram dry sediment (dpm/g × 2.2 = pCi/g = picocuries per gram)

[b]^{137}Cs peak concentrations are usually found around 1963 due to the peak in atmospheric nuclear weapons tests worldwide at that date; this is the sedimentation rate since that time

[c]^{137}Cs first appearance is usually around 1953 due to the start of large scale atmospheric nuclear weapons tests at that date; this is the sedimentation rate since that time

[d]^{210}Pb dates determined by constant initial concentration (CIC) and constant rate of supply (CRS) models for MHW, and by the slope of the best fit line of ^{210}Pb concentration over several sections of sediment for the later studies. For the latter, the sedimentation rate in g/cm^2/year was transformed into cm/year by division by the dry in situ density (dry mass in a given volume divided by the volume) in g/cm^3

[e]94 pCi/sample (not per gram, as above)

[f]Peak ^{137}Cs was found in the surface sections of these cores, due to rain-out of nuclides from the Chernobyl accident in Ukraine, which took place in April 1986. In the surface sections of cores at Hythe and in the Bodensee the standing crop of Chernobyl ^{137}Cs was not estimated. At Houb of Sullom and Voxter Voe the standing crops of Chernobyl ^{137}Cs were 0.35 and 0.37 dpm/cm^2 respectively, whereas the bomb-produced ^{137}Cs values were 0.23 and 0.08 dpm/cm^2 respectively. To estimate sedimentation rates, therefore, the Chernobyl maxima in these surface sections were not dated at 1963, as they are in the footnotes above

[g]ND, not determined because of the flat shape of the ^{210}Pb profile throughout the surficial 24 cm of core

The core-dating work benefitted from taking place mainly after the 1986 Chernobyl incident. To assess whether catchment influx is commensurate with temporal change in contaminants, data from three sediment cores dated in July 1986 in MHW were used. It was too early (<2 months) for the Chernobyl ^{137}Cs signal to have reached the sediment of MHW via runoff from the Preseli Hills. At two sites in SW and the Humber estuary with much higher sedimentation rates, 99 and 460 days respectively were also insufficient to register Chernobyl ^{137}Cs fallout. However, two other cores in SW and one in Shetland registered Chernobyl ^{137}Cs after 365 ± 30 days (Little 1987a, 2009). This timescale for the lag of catchment sediment transport into the coastal zone may explain the temporal patterns shown in Figs. 5 and 7.

Comparing with the range of other UK estuaries, sedimentation rates are highest in SW (Table 2). MHW occupies an intermediate position in terms of sedimentation rates, and is similar to the one core taken in the Humber estuary at Pyewipe mudflats. Sedimentation rates are lower in both Sullom Voe and the Alpine lakes. Sedimentation rates are higher in the turbidity maximum zones of the contrasting estuarine systems, and in MHW the rates are thus highest in the Western Cleddau and Pembroke River, again endorsing the predicted flood-tide dominated fine-grained sediment transport paths.

The THC concentrations were also determined from the deeper pre-industrial sections of the radionuclide-dated cores. The background THC concentrations ranged from 95 mg/kg in the jetty area of MHW to 367 mg/kg in the inner basin of SV; deep core sections off Cadland Creek in SW showed only 162 mg/kg THC. The percentage mud (particles < 63 micrometres in diameter) in these core sections ranged from 13 to 98 %.

4.2 Suspended Particulate Matter

A range of estuaries was broadly characterised in terms of the hydrocarbon concentrations and major and trace element geochemistry of the SPM between 1986 and 1989 (MHW, SW and Orka Voe, Shetland). The aim was to assess similarities within and between estuaries in terms of linkages between SPM, nearby settled sediments, and STA. Sediment core-dating was performed in all three estuarine/fjord systems between 1985 and 1988. Sediment cores were sampled at multiple sites in each estuary, based on STA output. The study of oil concentrations and composition in the water column and suspended sediments showed that SPM was a major factor influencing oil (and metals) concentrations (Little et al. 1987; Little and Pritchard 1989; Little 2009). The hydrocarbon and other data are summarised in Table 3.

Table 3 Hydrocarbon concentrations in Suspended Particulate Matter (SPM) in MHW, SW, Orka (near Sullom) Voe, and adjacent settled sediments

Sampling information							SPM mg/l	LOI %	Dry sediment mg/kg			Water conc. µg/l		
Sample	Date	Location	Analyses	Type	Tidal state	Vol. (l)			AL	AR	THC	AL	AR	THC
MHW86	1/7-4/7/86	MHPA Jetty	Hydrocarbons, metals	Integrated	Neap	14.63	15.25	–	2610	870	3480	40	13	53
PR87 (MHW)	14/8-21/8/87	Pembroke R.	Hydrocarbons, metals	Integrated	Spring-Neap	84.45	16.42	15.10	295	<LOD	295	5	<LOD	5
MHW1/89	3/3-17/3/89	Popton Point	Hydrocarbons, metals	Integrated	Neap-Sp-Ne	114.68	11.26	15.40	4120	730	4850	36	6	42
SW1	5/4-7/4/87	Fawley Beacon	Hydrocarbons	Integrated	Neap	8.46	5.70	–	13,230	2160	15,390	80	10	90
SW2	7/4/1987	Fawley Beacon	Hydrocarbons	Spot	Neap	2.10	16.50	–	5810	2860	8670	90	50	140
SW3	10/5/1987	Fawley Beacon	Hydrocarbons	Spot	Neap-Spring	0.93	274.00	–	1960	<LOD	1960	540	<LOD	540
SW4	11/5/1987	Fawley Beacon	Hydrocarbons	Spot	Neap-Spring	0.27	387.00	–	2870	<LOD	2870	1110	<LOD	1110
SW5	10/5-11/5/87	Fawley Beacon	Hydrocarbons, metals	Integrated	Neap-Spring	1.12	368.00	23.80	2415	495	2910	890	180	1070
SW6	14/5/1987	Fawley Beacon	Hydrocarbons	Spot	Spring	0.53	105.00	–	8985	8985	17,970	950	950	1900
SW7	14/5-15/5/87	Fawley Beacon	Hydrocarbons, metals	Spot	Spring	10.16	135.00	85.30	948	222	1170	130	30	160
SW8	24/8-28/8/87	Fawley Beacon	Hydrocarbons, metals	Integrated	Spring	43.84	10.40	24.40	2270	2250	4520	24	23	47
OV1	29/6/1988	Orka Voe	Hydrocarbons, metals	Spot	Neap-Spring	5–10	9.5–8.4	97.10	10,500	6300	16,800	100	60	160
OV1/R	29/6/1988	Orka Voe	Metals	Spot	Neap-Spring	10.00	2.53	90.60	–	–	–	–	–	–

(continued)

Table 3 (continued)

Sampling information							SPM mg/l	LOI %	Dry sediment mg/kg			Water conc. μg/l		
Sample	Date	Location	Analyses	Type	Tidal state	Vol. (l)			AL	AR	THC	AL	AR	THC
OV2	29/6-4/7/88	Orka Voe	Hydrocarbons, metals	Integrated	Neap-Spring	45.92	0.24	33.00	1100	550	1650	30	15	45
OV3	4/7/1988	Orka Voe	Metals	Spot	Spring	10.00	3.75	90.30	–	–	–	–	–	–
OV4	4/7/1988	Orka Voe	Hydrocarbons, metals	Spot	Spring	5.40	10.53	–	7030	3510	10,540	70	30	100
MHW63	1984	Popton Point	Hydrocarbons, metals	Grab	Settled sediment				124	74	196			
PR142 (MHW)	1984	Pembroke R.	Hydrocarbons, metals	Grab	Settled sediment				311	321	632			
SW34	1988	Fawley Beacon	Hydrocarbons, metals	Grab	Settled sediment				4262	628	4890			
OV1/85	1985	Orka Voe	Hydrocarbons, metals	Grab	Settled sediment				250	23	273			

Little and Pritchard (1989)
SPM suspended particulate matter, *LOI* loss on ignition, *AL* aliphatic hydrocarbons, *AR* aromatic hydrocarbons, *THC* total hydrocarbons
Integrated sampler = WS Ocean Systems®; <LOD = below limits of detection

4.2.1 Suspended Sediment and Organic Matter Concentrations

For each SPM sample the dry mass of material retained by filtration was divided by the total sample volume, or the volume integrated by a WS Ocean Systems® sampler (WSO). This was taken from the revolution counter of the sampler, and converted to volume by calibration in the laboratory. SPM concentrations are presented in Table 3; the range in SPM concentrations was from 0.24 to 387 mg/l, with the lower part of the range (less than 10 mg/l) found in Orka Voc (OV) adjacent to the clear water of Yell Sound, Shetland. SPM concentrations were intermediate and relatively uniform in MHW (11–16 mg/l), whereas the May 1987 data from SW after antecedent stormy weather were well above this (more than 100 mg/l). The SW April and August SPM concentrations were, however, similar to those from MHW. The general SPM pattern is as would be expected from knowledge of those estuaries, ranging from clear waters in Shetland to those with well developed turbidity maximum zones.

Results of the organic content analyses by loss-on-ignition (LOI) are presented in Table 3. The data on organic content of the SPM ranged from 15 to 97 %, with lower values found in MHW than in the other estuaries; SW was intermediate (24–85 %) and OV was generally >90 %. Organic content of SPM was generally higher in May or June sampling visits than at other times of year, due to the sampling of large concentrations of plankton in spring and early summer. The data were also higher than LOI values determined in adjacent settled sediments, because of the inclusion of plankton into SPM in addition to mainly resuspended sediment, and due also to microbial degradation of organic material in the settled deposits.

4.2.2 Hydrocarbon Concentrations

Gravimetric hydrocarbon concentrations obtained from the SPM samples from MHW and the other estuaries are given in Table 3. The concentrations of THC ranged from 5 to 53 μg/kg in the MHW jetty area and Pembroke River. The SPM concentrations were from 11.3 to 16.4 mg/l. This hydrocarbon concentration range compared to a range in OV in 1988 (near the ballast water diffuser from the Sullom Voe oil terminal in Shetland) of 45–160 μg/kg, and to a range of 90–1900 μg/kg in SW in 1987. Therefore, THC concentrations in SPM were generally highest in SW, intermediate in OV and lowest in MHW. This reflected the mean sediment THC concentrations quoted by Howells et al. (1988): SW 6716 mg/kg (1986, n = 39), Sullom Voe 289 mg/kg (1987, n = 27), and MHW 144 mg/kg (1982, n = 50). The raw data are presented on a dry mass basis and then expressed as water concentrations by factoring in the SPM concentration, and so may be considered equivalent to unfiltered water samples. The general rank order in THC does not change, but it may be seen that low SPM concentrations lead to low THC concentrations, even if the THC is high on a dry mass basis. The general THC rank order is also matched reasonably closely by the data for adjacent settled sediment THC concentrations, also presented in Table 3. This should not, however, be taken as evidence that all the hydrocarbon loadings found in the SPM samples arise from resuspension of local settled sediments. The two sample matrices might simply be responding to a common input at each site. In MHW, the THC value for the SPM in Pembroke River at PR87 was low in comparison to the value for PR142 (adjacent settled sediment), suggesting that the SPM sample was not merely reflecting inputs of resuspended sediment. PR87 was probably low in THC concentration in comparison to the other SPM samples from MHW because of low turbidity at the time, and its greater distance from oil inputs from refinery effluent and jetty-related spills.

In SW, adjacent to the Fawley refinery jetties and near the mouth of Cadland Creek (which carries treated refinery effluent), the THC concentrations in the SPM were comparable to those found in the benthic sediments nearby, with a mean value of 6933 mg/kg on a dry mass basis and ranging 47–1900 μg/l as water concentrations (Table 3), compared to 5–1500 μg/l reported by Varney (2000). The average THC concentration in settled sediment was 5113 mg/kg between 1982 and 1988 at Station SW34 near Fawley Beacon between the Esso marine terminal and Cadland Creek mouth (Howells and Dodd 1985; Dicks and Levell 1989), compared with up to 3000 mg/kg reported by Varney (2000). Differences in SPM and THC concentrations from April to May 1987 were probably due to weather and tidal influences. Northerly winds predominated during May and thus the large wave fetch along the estuary probably contributed cleaner sediment suspended from other parts of SW. This appears to have increased the SPM concentration, but has reduced by dilution the overall THC concentration.

4.2.3 Hydrocarbon Composition

Analysis of aliphatic hydrocarbons in SPM for MHW using GC/FID shows considerable variability; from the medium boiling-point range UCM of MHW86 to the higher boiling-point range UCM of PR87 and MHW1/89. The former may reflect medium fuel oil inputs from port operations in MHW, but the trace from this sample also shows a homologous series of n-alkane peaks from n-C_{17} to n-C_{22} representing a likely additional input of relatively fresh crude oil. This is not unexpected because on a flood tide MHW86 is down-current of refinery effluents on up-estuary sediment transport paths (McLaren and Little 1987). The Pembroke River SPM sample (PR87), in addition to the high boiling-point UCM maximising at n-$C_{29}/_{30}$ shows relatively high concentrations of resolved peaks > n-C_{27}. The high boiling-point UCM is also a pronounced feature of the GC trace of the settled sediment at this location (PR142), maximising at n-C_{33}. However, high boiling-point UCM is very common in MHW, and so its sources within the SPM are not necessarily restricted to local sediment resuspension. The Popton Point SPM sample (MHW89) also showed some similarities to the settled sediment at that locality (MHW63), having relatively high boiling-point UCM (maximising at n-C_{30}) and a UCM shoulder at about n-C_{20}. However, resolved peaks were different between the two GC traces. Sediment transport patterns are complex in this part of

MHW, and so it is likely that the SPM would reflect inputs from multiple sources up-current, including the Texaco (now Valero) refinery effluent.

In SW in April 1987, the hydrocarbon composition of the SPM showed GC resolved peaks in the low-boiling range n-C_{12} to n-C_{24}, superimposed on UCM maximizing at n-C_{29} (spot samples) and n-C_{23} (WSO sampler). In May the composition was similar to medium fuel oil with n-alkanes between n-C_{15} to n-C_{33} and UCM maximizing at n-$C_{24.}$ In May and August the WSO sampler captured more weathered alkanes than the spot samples, and with UCM maximizing at n C_{29}. The resolved peaks and the variable positions of the UCM maxima suggest multiple sources, and most GC types were found also in settled sediment nearby, suggesting a degree of coupling of SPM with settled sediments.

In OV due to the short sampling period in June/July 1988, GC traces were more uniform than in MHW and SW, with prominent n-C_{14} and n-C_{16} peaks and homologous n-alkanes from n-C_{21} to n-C_{26} superimposed on UCM maximizing at n-C_{29}. In combination, these features were also found in nearby settled sediment in OV. However, the seabed material included a stronger odd-over-even carbon number predominance than in SPM samples, suggesting biogenic inputs in the settled sediment, probably from peat. Seabed sediment in OV also had a subordinate UCM maximum at n-C_{19} in 1987, similar to the bimodal UCM in 2012 interpreted as possibly being from weathered diesel fuel or crude oil (SGS M-Scan Limited 2012). STA results from 1983 grain size data suggest southward sediment transport past the ballast water diffuser site in Orka Voe (McLaren 1987).

4.2.4 SPM Summary

Concentrations of THC in MHW ranged from 5 to 53 µg/l (at SPM concentrations from 11.3 to 16.4 mg/l). This compared to a range in Orka Voe (near the ballast water diffuser from the Shetland oil terminal) of 45–160 µg/l (SPM range 0.2–10.5 mg/l), and to a THC range of 90–1900 µg/l (SPM range 5.7–387 mg/l) in SW in 1987. The average organic matter contents of the three contrasting SPM systems (by loss on ignition) were from 15.1 % at MHW through 44.5 % at SW, to 77.8 % at OV. Hydrocarbons in SPM when expressed as water concentrations were highest in SW, intermediate in OV and lowest in MHW. The rank orders for sedimentation rate, SPM concentration and settled sediment THC content were the same as each other (i.e. SW > MHW > OV). The rank orders were the same for LOI and the THC concentration in SPM on a dry mass basis (i.e. OV > SW > MHW). These patterns reflect the differing role in sorption of contaminants played by organic versus mineral clastic material within the SPM.

Hydrocarbons were more readily identifiable as oil industry (as opposed to background) contaminants than were the trace metals, although suggested enrichment was identified by metal concentration ratios against aluminium. Metals in SPM were higher in concentration in SW and MHW than in OV, reflecting the lesser role of organic substrates for metals sorption than for oil. The different rank order of contaminant concentration between estuaries for hydrocarbons compared to metals was also attributed to the proximity of oil industry sources, whereas the metals were more likely to be derived from local sediment resuspension. Cluster analyses of the major and trace elements confirmed that MHW SPM samples were generally of between 60 and 70 % similarity in the oil industry jetty area and Pembroke River, thus sediment transport linking the two sites is likely. This supported the settled sediment analyses of contaminants and STA (McLaren and Little 1987).

Suspended sediment concentrations are confirmed as a master variable in controlling concentrations of contaminants in estuaries over a wide range of conditions. SPM should be measured and quoted alongside measurements of water sample chemistry, whether such samples are analysed before or after filtration. The WSO integrated SPM sampler was found to give more representative integrated results than spot samples although some degradation of organic material occurred in the integrated sampler before retrieval. Studies of the fate of non-conservative contaminants and the effects of discharges or spills upon benthic ecosystems in the nearshore zone should address SPM behaviour. Further work should include measurement of fluxes of contaminants over longer time periods, preferably using bulk SPM samples (ISO 2008). Multiple sites confirmed as being SPM sinks should be coupled with further core dating and sampling and analysis using advanced chemistry fingerprinting methods. Sites in MHW could include Garron Pill, Pembroke River and the subtidal mud-bank of the main estuary on Milford Shelf.

5 Discussion

5.1 Milford Haven Waterway, Southampton Water and Sullom Voe

The sedimentary environment and the effects of operations at the complex industrial sites in MHW, SW and Sullom Voe have been interpreted at each site using STA, radionuclide-dated sediment cores (Table 2) and SPM sampling (Table 3). The similarities and differences between sites are described by Little (1987a), McLaren and Little (1987), Little and McLaren (1989), Little and Pritchard (1989), Arthur D Little Limited (2006), Little (2009), Little and Bullimore (2015) and Little et al. (2015).

5.1.1 Milford Haven Waterway and Southampton Water

The studies in MHW and SW suggest that current inputs of contaminants are now low and mostly dispersed, weathered or sequestered in fine sediment areas. Each of these cases can help to explain the currently low level of biological impacts presently observed compared to those in the recent past. This is partly due to the major improvements in pollution source control (especially SW) and closure of older facilities. The reduced industrial effluent loadings are often reflected in significantly reduced sediment hydrocarbon contamination since the peaks in both SW and

MHW before the 1980s. However, the disturbance of historical contaminants in sediment by dredging means that some of the recalcitrant petroleum contaminants, such as PAHs, have been remobilised. Some of the hydrocarbons identified in MHW using petroleum biomarkers date back to oil spill events in 1996, 1988, and during WW2 (Little et al. 2015). There is also evidence that dredging and enhanced sedimentation thereafter affected the fate of oil discharged into SW (Croudace and Cundy 1995; Arthur D Little Limited 2006).

Sedimentation rates calculated for dated sediment cores in both MHW (four sites) and SW (three sites) help to explain the differing contaminant fluxes and levels of interaction between SPM and oil sedimentation. In SW the impacts resulted from the historically oily effluents from the water-cooled Fawley refinery in the 1950s entering highly turbid waters. These early impacts of the effluent on SW particularly in the fine-grained sediments of Fawley salt marsh, intertidal and subtidal flats are well-documented, as are the subsequent Esso Fawley investments in source control and marsh restoration (Knap 1978, 1982; Dicks and Iball 1981; Armannsson et al. 1985; Dicks and Levell 1989; Arthur D Little Limited 2006). The SW impacts contrasted with the lower rates of SPM/oil sedimentation in MHW which had air-cooled refineries and greater mixing energy levels than the receiving environment in SW.

The STA confirmed the early contaminants findings at each site and has been highly consistent with scientific and engineering experience in MHW in particular, as follows: statistical residuals from multivariate regression modelling of the sediments and contaminants; pollutant chronologies in radionuclide-dated (^{210}Pb and ^{137}Cs) cores taken in potential sediment sinks predicted by the above regression residuals; knowledge of tidal streams (e.g. ebb/flood separation), seabed wave-induced orbital velocities, and subtidal and intertidal sediment facies analysis by Sediment Profile Imagery; field behaviour of SPM and contaminants, including optical backscatter data on SPM and fluorescent sediment tracer data (Little et al. 1987; Little 2009; Carey et al. 2014; Little and Bullimore 2015).

Continuously operating since the 1920s, the Esso Fawley Refinery and petrochemical plant effluent discharges into SW historically affected the saltmarsh. Ecological impacts and recovery have been monitored since the late 1960s (saltmarsh), mid-1970s (mudflat/sublittoral benthos), and early 1980s (marsh infauna). Significant progress was made in reducing the oily discharge over the last 55 years, with steady progress since 1980 ($\sim$40 mg/l down to 2.5 mg/l), and further continuous improvement. The marsh has been largely restored over the past 35 years, due to adaptive management actions (vegetation transplantation programme) linked to effluent quality improvements. The historical monitoring surveys coupled with contaminant data tracked the impacts of contaminants over time, and justified investments in effluent improvement.

Chemical analysis of radionuclide-dated core sediments (Table 2) demonstrated that a significant increase in THC concentrations coincided with the Esso Refinery expansion post-1951, and with the widespread mortality of the Fawley marsh flora (1954–1962). The date of first appearance of ^{137}Cs (1953) was soon after refinery expansion. These findings endorse similar work by Knap (1978, 1982) and are confirmed by Croudace and Cundy (1995). Peak THC was >4 % in core sediment corresponding to the mid-1960s. The THC concentrations decreased to 2 % on the mudflats in monitoring between 1982 and 1986, and this contamination extended to four stations located as far as 800 m up-estuary from the mouth of Cadland Creek. Flood-dominated net sediment transport paths showed net movement was up-estuary on mudflats and shallow subtidal bank and therefore had potential for trapping contaminants in SW (McLaren 1987). Varney (2000) reported THC concentrations of 400–500 mg/kg at Hythe and Totton.

The probable causes of historical impacts were floating oil, some or most of which had re-coalesced from effluents and possibly small spills from shipping. Historically, the effluent oil concentrations were high enough for accelerated sedimentation due to the agglomeration of oil droplets with SPM. Arthur D Little Limited (2006) showed that chemical parameters of the sediment samples exhibit characteristics consistent with historical hydrocarbon inputs rather than current ones. Because the sediments in SW are periodically dredged, and also due to the cycling processes between SPM and settled sediments, some of the historical contamination has been resuspended and distributed throughout SW. The study concluded that THC, PAHs and much of the Cu contamination is indeed historic, and now resides in fine sediment sinks of SW. The individual PAH distribution patterns as well as PAH source ratios are consistent with much more weathered material near the outfalls, and with PAH concentrations increasing farther away from the outfalls, suggesting other sources of hydrocarbons.

In the other evaluated areas of SW, the sediment PAHs exhibit characteristic "descending" patterns of major homologous series (Fig. 3a), which is typical of predominantly combustion-related (pyrogenic) sources of hydrocarbon contamination. Principal Component Analysis (PCA) of PAHs separates the sediments according to the location of known discharges in SW, and suggests some localised influence of discharges on sediment contamination (e.g. a noticeable difference between the PAH distribution patterns in samples close to the outfalls and those from more distant locations indicates that the mudflat sediments may have PAH contributions from sources other than Esso Outfalls 1 and 2). Varney (2000) reported PAH ranging 203 μg/kg (at Cadland Creek) to 6 μg/kg (at Hythe) but rising again to 143 μg/kg (at Totton).

The Esso outfall oil concentrations are now at levels where the only oil being discharged is soluble (i.e. physical removal methods cannot remove significant additional oil), and are consistently below both consented levels and the concentrations which historically enhanced agglomeration and sedimentation of oil. The cord grass *Spartina anglica* has steadily re-colonised the salt marsh in response to effluent improvement and a focused transplant programme, with positive effects of transplants being seen particularly from 1984. Within the last 15 years the dominant and many rarer marsh species have fully recovered to the pre-1951 state (before the original marsh was killed). THC concentrations in mudflat and sublittoral sediments have shown some decline, although THC and Cu are still found in high concentrations in deeper sediments in both the marsh and mudflats. However, these deposits are now substantially capped by diverse healthy vegetation that has fully resumed its habitat function and its supply of organic material to the mudflats and subtidal sediments.

Evaluation of the major pollution transfer mechanism, specifics of the hydrocarbon concentrations and composition profiles in the sediment, and temporal trends in the effluent oil contents all suggest that most of the sediment hydrocarbon contamination in the vicinity of the Esso Fawley is historic pollution from Esso operations and other SW sources. Candidates include a number of industrial point sources and also NPS such as shipping, road runoff, and atmospheric inputs. In summary, the available analytical data provide no evidence of significant current hydrocarbon contributions to the area from the Esso complex, but they do support the view that SPM and the fate of oil discharges are strongly coupled.

Fine-grained sediments act as contaminant sinks due to the high surface area and sorptive nature of clay minerals and TOC. They are an important reservoir that reduces toxicity to aquatic biota. In the case of Fawley, the oil/SPM interaction led historically to rapid agglomeration and settling processes that also scavenged oil from the water column. The SW cores showed rapid sedimentation rates of up to 16.7 mm/year near the Esso outfalls, and this result was duplicated by Croudace and Cundy (1995) whose estimate of 20 mm/year was thought to be a response to the capital dredging programme for refinery expansion in 1950–1951. In both studies, sedimentation subsequently returned to lower rates of between 5 and 10 mm/year that had prevailed before the dredging and expansion at Fawley.

5.1.2 Sullom Voe

In the clear waters of Sullom Voe, dated cores taken in Houb of Sullom and Voxter Voe saltmarshes (Table 2) show lower sedimentation rates than MHW and SW, and generally lower contaminant concentrations in settled sediments. Uniquely in this study of the three sites, the Sullom Voe data are strongly influenced by peat runoff from the sparsely-populated fjord catchment (Westwood et al. 1987; SGS M-Scan Limited 2012). Fjord environments often entrap contamination because of sills scoured by glaciers along their length, as for example in the inner basin of Sullom Voe and many Norwegian fjords (Skei 1981). However, as with MHW, THC and metals pollution took place during both World Wars in Sullom Voe, with peak THC concentrations occurring in 1919 and the early 1950s (Little et al. 1990). This did not occur in SW where WW2 contaminant signals are masked by inputs from Fawley.

In Sullom Voe, the presence of methylanthracenes and Cyclopenta(*d*,*e*,*f*) phenanthrene in the sediments and in surrounding peat suggests that the peat may be contributing to PAH levels in the sediments. A dominant input from a high-temperature source could be pyrolitic or a cracked refined petroleum product such as diesel, or HFO blended with a cracked middle distillate. The formation of peat is not a high temperature process, however, and so (in conjunction with the high percentage of 4–6 ring PAHs) this suggests that the peat may have absorbed combustion products from the atmosphere and/or undergone burning before contributing to the sediment PAHs. GC/MS analysis of the aliphatic fraction in Orka

Voe and also in Sullom Voe sediment indicates that the sterane and triterpanes most closely resemble a product of Middle Eastern crude oil origin, similar to the 1110 t HFO spilt in 1978 by 'Esso Bernicia' (SGS M-Scan Limited 2012).

These results were reconfirmed by SGS M-Scan Limited (2014), and longer-term data were also presented, updating Westwood et al. (1987, 1989). The data included sediment and hydrocarbons typically from the surficial 2 cm sampled at two year intervals, grain size since 1981, and hydrocarbons since 1992. Percentages of mud have decreased at several stations, for example, at SV1 in the inner basin from a mean of 78 % in the 1980s to 73 % in the 1990s and 71 % in the last decade. At stations SV4 and 7 in the mid-reaches the % mud decreased from a mean of 38–40 % in the 1980s to 29–34 % in the last decade. At OV1 the % mud decreased significantly from a mean of 66 % in the 1990s to only 32 % in the last decade. At most other stations there have been no major trends in texture as measured by % mud. Early monitoring from 1977/8 at SV1 and 7 showed high THC concentrations that were subject to very high variability due to the peaty nature of the sediments (Howells et al. 1989).

In the case of hydrocarbons, SV1, 7 and OV1 showed a higher carbon preference index (CPI) and higher molecular-weight aliphatics than other stations, indicative of inputs from vascular plant debris, including peat. Any fresh inputs of petrogenic hydrocarbons have been at very low levels at all stations, although at OV1 a slightly higher petrogenic component was detected, with a bimodal UCM suggesting weathered crude oil/diesel and fuel oil. Aliphatics concentrations increased slightly at SV7 near the terminal jetties and many other stations. However, in line with reduced % mud, the concentrations of aliphatics and UCM decreased by at least 20 % since the 1990s at SV1 and 4 (e.g. from 250 to 190 mg/kg aliphatics at SV1). At OV1 the decrease was more than 50 % from 109 to 51 mg/kg aliphatics and the UCM concentrations declined from 89 to 28 mg/kg. Additionally, a higher proportion of the aliphatic fraction was in the form of UCM than in the 1990s, due to weathering (SGS M-Scan Limited 2014).

The concentrations of 23 PAHs were measured at stations SV1, 4, 7, 17, 34, 6, 6f, OV1 and OV5. In general, fresh crude oils have higher concentrations of alkyl naphthalenes and alkyl phenanthrenes than do weathered oils. The alkylated PAHs are more prominent in petroleum products than they are in pyrogenic samples (Fig. 3). All samples showed a mix of petrogenic and pyrogenic patterns among the parent and alkyl group compounds. In the main axis of SV pyrogenic hydrocarbons are dominant except for phenanthrenes at SV4 and dibenzothiophenes at SV1 and 4. Conversely, in Garths Voe (presently the tug jetty and formerly Royal Air Force station, 1939–1957) and OV (ballast water diffuser) there was more evidence of petrogenic contributions. The ΣPAH_{23} data since 1992 indicated that the mean 2–6 ring PAH concentrations decreased at SV1, 4 and OV1 from the 1990s to present (e.g. from 3420 to 2283 μg/kg since the 1990s at SV1; from 597 to 518 μg/kg at SV4; from 1719 to 1227 μg/kg at OV1). There was no indication of the peaks in sediment PAHs from 2006 to 2008 described for more southerly estuaries in Sect. 3.2. The decreases were accompanied by a lower % of 4–6 ring PAH as a proportion of ΣPAH. However, at SV7 since the 1990s the ΣPAH_{23} concentrations increased from

2005 to 2150 μg/kg. OV1 had a higher proportion of Naphthalene, Phenanthrene and Dibenzothiophene (NPD) compounds than stations in the main axis of SV, although Stations SV1 and 4 had slight enrichment in the alkyl group patterns for either Phenanthrene or Dibenzothiophene. These trends mirror those in aliphatic hydrocarbons (SGS M-Scan Limited 2014). SV7 also was suggested by Westwood et al. (1989) as a sink for fine sediment and hydrocarbons resuspended from Garths Voe partly by vessel movements.

The work in Sullom Voe demonstrates the importance of monitoring sediment grain size and organic material in parallel with the contaminants in order to help control for inherent variability. In a literature study of 40 published articles on oil concentrations in marine sediments only 10 reported details of substrate grain size; of 38 articles on oil concentrations in water, only five reported SPM concentrations (Arthur D Little Limited 1997). The sequential reductions in % mud and hydrocarbons in SV for some stations since the 1980s (THC) and since the 1990s (ΣPAH_{23}) suggest that sediment type also may be a function of disturbance. As in MHW and SW, the fine sediment population of SV is subject to some resuspension by vessels, mostly tug manoeuvring in Garths Voe. McLaren (1987) suggested that the northerly transport predicted by his STA output may be due to tidal seiche effects. If so, then the effect of sea level rise due to climate change could increase the volumes of water movement during seiches, and thus be an indirect cause of sediment and contaminant changes at SV1 and SV4 in the inner and mid-Voe.

5.2 *Other UK Estuaries*

In the Firth of Forth, oil inputs since 1924 primarily originating from the BP Grangemouth refinery and petrochemical industry produced localised adverse effects. These impacts included loss of intertidal habitat or ecosystem function (e.g. reduced feeding area and quality for flatfish, shelduck and dunlin). The changes historically were due to organic enrichment and more locally to toxicity to infauna. More recently, the greatly improved discharges have had minimal current impact, but the historical contamination makes it difficult to assess ongoing effects (Elliott and Griffiths 1987). At a % mud ranging between 60 and 80 % (pers. comm. M. Elliott), mean sediment THC concentrations in 1986 were 838 mg/kg, with values ranging from 41 to 1645 mg/kg, peaking near the BP Grangemouth refinery (Howells et al. 1988).

Vane et al. (2011) analysed seven cores from the middle Clyde estuary with the aim of dating PAH inputs. The PAH concentrations in the cores are low and biogenic before the Industrial Revolution (about 1770) and also in modern relatively non-polluted sediments. The industrialisation of the Firth of Clyde increased PAH concentrations and produced PAH isomeric ratios characteristic of grass/wood/coal, petroleum, and petroleum combustion. Falling PAH levels mirrored declining coal use and increasing use of petroleum fuels in ships from 1915 and in vehicles from the 1950s. When Benzo(*g*,*h*,*i*)perylene/Benzo(*a*)pyrene ratios exceeded 3.14 in one core, it was inferred that road traffic on the A82 main road was responsible. Similar patterns of road input were found in the Tamar by Readman et al. (1984).

In the Humber estuary, sediment THC and metals were analysed in a radionuclide-dated sediment core (Table 2) taken in August 1987 from Pyewipe intertidal mudflat near Immingham (Bartlett et al. 1989a). The core profile showed generally >97 % mud except at 11 cm where uniformity was broken by a fine gravel layer dated at 1962 when a small channel possibly crossed the mudflat. The fine-grained sediment had ranged between 636 and 1126 mg/kg THC, with the higher concentrations focussed in the section dated between 1930 and 1960. Heavy metals declined from the core base dated 1910 to the surface, with some metals (Cr, V, Zn) showing similar trends to the higher oil content 1930–1960. In the surface 5 cm layer the THC concentrations were lower than at any time since 1910; the opposite of this was seen only for V, which peaked in surface sediments.

In 1987, Howells et al. (1988) established a baseline in Humber surface sediments for % mud (range 1–89 %, mean 40 %), along with THC and ΣPAH_{11}. They found a mean THC concentration of 343 mg/kg and a maximum of 885 mg/kg, rather less than the subsurface peak in the above core study, and in agreement with the more recent improvement in sediment quality suggested by Bartlett et al. (1989a). The ΣPAH_{11} in 1987 was 6 mg/kg (n = 3). Between 1987 and 1990, the ΣPAH_{11} ranged from 3.2 to 7.9 mg/kg, with an average of 5.2 mg/kg. These two very close averages were higher than in both the MHW (2.5 mg/kg in 1993) and Dee estuaries (0.49 mg/kg), very similar to the Mersey estuary (5.3 mg/kg), but lower than the Tamar (8.6 mg/kg) and Dart (9.9 mg/kg) estuaries (Howells et al. 1988; Kippen 1991; Levell et al. 1997; Little 2009; Little et al. 2015). These data show that runoff into the Humber from the UK's largest and most heavily-industrialized catchment has produced moderate PAH contamination compared to SW and the smaller estuaries in Cornwall and Devon. This pattern is generally in good agreement with PAHs data shown in Fig. 4 above (Rogers et al. 2000) and in Fig. 5 above (Rumney et al. 2012). A similar trend for THC was shown by Howells et al. (1988) except that the mean THC concentration in the Dart of 156 mg/kg (mean 45 % mud) was similar to MHW (144 mg/kg; mean of 35 % mud) but lower than many of the above-mentioned estuaries (SW, Firth of Forth and Humber which were muddier, ranging 40–92 % mud) and lower than SV with a mean of only 29 % mud.

The steranes and triterpanes analysed in Humber sediment aliphatic fractions resembled Nigerian crude similar to the 6000 t oil spilt in 1983 at the Immingham oil terminal by the 'Sivand' (Jones et al. 1985). Over the one year of study, these authors also detected fossil fuel combustion products that were less readily biodegraded in sediments than the oil-derived hydrocarbons. In another study of the fate of 'Sivand' oil over one year at two locations in the Humber (Humberston Fitties and Blacktoft Sand), it was found that the surface oil contamination persisted longer at the former, sandy site near the estuary mouth than at the latter innermost site with highly-mobile mud sediments. Sediment mobility was tracked visually and using grain size analysis (Little 1987b). Zhou et al. (1998) found variations in the ratios of the PAH compounds in SPM and surface sediment, concluding that these variations indicated different sources.

Humber estuary sediments were examined by Brettell (2013) for PAHs using isomeric diagnostic ratios, and showed inputs of power generation and vehicular emissions, cement and oil industries, smelting and incineration. River inputs were considered low relative to atmospheric inputs. The change in concentration of PAH compounds over time indicates a subtle shift in sources away from wood towards coal. The sediment PAHs are characterised by dominant 4-ring parent compounds. PAH concentrations varied between cores and ranged from 0.63 to 79.6 μg/kg. Peak concentrations were dated at 1940, some of which was due to pollution from bombing raids. Higher values were found only in older sediment. No specific input from the nearby Flixborough explosion in 1974 was detected by Brettell (2013). Between Llanreath in 1940 and Buncefield in 2005, Flixborough was the largest single seat oil industry fire on British soil (excluding the offshore Piper Alpha disaster in 1988).

5.3 *Other Aquatic Systems*

In two Alpine lakes (Bodensee and Mindelsee) in the Rhine catchment near the Switzerland/Germany border, sediment THC and metals were analysed in radionuclide-dated sediment cores (Table 2) taken in 1988 (Bartlett et al. 1989b). Two Bodensee cores had ranges of 62–99.9 % mud, and showed that contamination started before 1800 and increased sharply after 1910 to reach a peak of 407 mg/kg THC around 1960, and decreased more recently, very similar to the above-described pattern in the Humber estuary. Mindelsee was selected as an uncontaminated reference site, having no major river or small craft inputs and much lower population density. It had a slightly lower sedimentation rate and peak THC concentrations than the Bodensee system, although % mud was >94 % throughout the core depth. However, unlike the Bodensee which showed improvement in surficial sediments, the Mindelsee core showed steady increases in THC and metals concentrations up to the sediment surface. The THC value of 378 mg/kg at 0–3 cm depth was checked by analysis of four grab samples of Mindelsee surficial sediments (range 50–283 mg/kg, mean 191 mg/kg). Both lakes have become increasingly eutrophic since about 1940, and so the study enabled a comparison of aerial with urban runoff impacts, and the unexpected pattern may be explained by contaminant fallout from the atmosphere into the rural Mindelsee catchment.

In the Absheron region offshore Azerbaijan in the south Caspian Sea (water depth 220–645 m), sediments consist of fine and very fine silts of >3 % sediment TOC. Before offshore development in 1998–1999, significant petroleum inputs were evident throughout the area. Total PAH_{41} concentrations on average were 1.7 mg/kg (range 0.83–3.6 mg/kg), and showed a relatively uniform individual PAH distribution, with variations in PAH directly related to TOC and silt/clay content. However, oil seepage was occurring from an underwater "mud volcano" in the eastern part of the area. Evidence for this natural input included: significantly higher PAH levels, higher proportions of low molecular weight PAH, and higher PAH/TOC ratios compared to the surrounding sediments where TOC acts as an environmental master variable. The background levels of PAHs and aliphatics in the Absheron region are attributed predominantly to petrogenic sources with minor contributions of pyrogenic PAHs. Assessment of potential impacts from oil E&P activities should take into account the relatively high pre-exploration, petrogenic PAH levels in the region, emanating from the seabed in the Caspian Sea (Boehm et al. 2003).

Further south offshore Azerbaijan, in the Nakhchivan contract area (water depth 120–720 m), sediment was also fine- to very fine silt, with 1–3 % TOC. Before drilling between 1998 and 2000, ΣPAH_{41} concentrations on average were 0.68 mg/kg (range 0.36–1.2 mg/kg). At one station, again influenced by mud volcano inputs, the background ΣPAH_{41} concentration was 1.45 mg/kg. After drilling, by 2002, the ΣPAH_{41} increased within 50 m of the well site to 2.47 mg/kg (range 1.51–4.11 mg/kg), ranging between 0.14 and 2.24 mg/kg up to 1 km further out (Tait et al. 2004).

The Absheron and Nakhchivan hydrocarbons data may be put in the context of other regional baseline studies such as the NE. Caspian Sea in Kazakhstan waters (depth 4–9 m), where sediments consist of medium to fine sands of <0.9 % TOC. Except in an industrialised harbour where maximum ΣPAH_{26} concentrations reached 4.7 mg/kg, in the offshore zone the maximum concentration was only 0.2 mg/kg (1996–2006), similar to a maximum of 0.17 mg/kg ΣPAH_{26} determined by the Caspian Environment Programme (CEP) in the same area in 2001. The hydrocarbons in the NE. Caspian Sea sediments are generally much lower in concentration that in Russian, Iranian and Azerbaijan sediments, because the shallow shelf is winnowed of contaminated fine-grained particles by wave action and meteorological surges (Artyukhina and Little 2014).

5.4 Implications

Why do these historical sediment reservoirs of PAHs and other contaminants matter? Unless disturbed by natural processes or anthropogenic sediment movements, not much adverse impact is expected in accreting sediment systems, due to burial. Indeed, studies of the ecological services provided by the Humber estuary show that increased sediment accumulation during managed realignment of the coast would be likely to enhance storage of contaminants from SPM and reduce their inputs into the N. Sea (Andrews et al. 2006). Sediment erosion, transport and bioturbation are the causes of natural disturbance. Dredging, fishing, vessel traffic, aggregate extraction, construction of offshore installations, pipe or cable-laying activity are among the causes of anthropogenic disturbance resulting in increased SPM concentrations. This resuspension and the subsequent compaction of settled muddy sediments cause pore water expulsion. The composition of pore water differs from the overlying water column in terms of dissolved oxygen, Eh, pH, colloidal material and dissolved organic carbon (DOC). The release of pore water causes Eh and pH changes that accelerate desorption and weathering of inorganic and organic contaminants, profoundly affecting their bioavailability. For metals there is some controversy over how acutely toxic this can be, but for PAHs there is evidence that bioavailability does increase after resuspension (Eggleton and Thomas 2004).

Sediment PAH concentrations at some CSEMP sampling stations with more than 5 years' data have shown increases, exceeding NOAA sediment quality guidelines (Effects Range Medium; ERM) in parts of NE. England and the English Channel. The Effects Range Low (ERL) is exceeded in the Celtic Sea and off the Humber/Wash/Anglia area. Sediment PAH concentrations at stations with 3–4 years' data exceed ERM guidelines in most of the above areas and also in the Forth, Clyde, Humber, Thames and Severn estuaries. The ERL for sediment PAH is exceeded in parts of Liverpool Bay, Irish Sea, SE. Isle of Man and in MHW. Concentrations of PAH in filter-feeding mussels are above Environmental Assessment Criteria (EAC) in many of the above-mentioned offshore areas and estuaries (OSPAR 2010; UKMMAS 2010).

Of the 38,031 samples from up to 45 stations analysed by Cefas in the above programmes until 2011, 42.6 % exceeded ERL/EAC, and 7.7 % (2928 samples) exceeded ERM guidelines (Nicholaus et al. 2015). These data show that UK commitments to the European Commission (EC) Water Framework Directive (WFD; 2000/60/EC) to achieve Good Environmental Status (GES) were unlikely to have been met by many samples for legacy contaminants such as PAHs, particularly in the event of sediment resuspension. The GES in all water bodies may be achieved when "concentrations of contaminants are at levels not giving rise to pollution effects". For remaining "hot-spots", cost-benefit analyses will be performed on measures to improve local environmental quality. Improvements may involve dredging of ports, harbours and channels, but the costs and impacts of dredging, disposal and/or dispersal elsewhere, are likely to rule out such measures (pers. comm. DEFRA).

From the above circular logic, and in the absence of efforts to ascribe any liability and implement restoration, the "polluter pays" principle appears to have lapsed. It seems that legacy contaminants such as PAHs will continue to be sequestered in UK fine-grained sediments unless disturbed, and that in either event some contaminants will enter the food chain. Biological uptake of PAH in MHW was studied by Langston et al. (2012). The bioaccumulation data in mussel and ragworm successfully evaluated PAH uptake and potential toxicity, and showed great consistency with the monitored trends in sediment data. There is evidence that PAHs trends are coupled between sediment and biota in two key parts of MHW affected by recent dredging and construction (Little et al. 2016). However, this is not proof of biological impacts of a particular contaminant because the data are insufficient to allow forensic linkage between tissue PAHs and their sources. Additional difficulties are introduced by the different sample preparation and instrumental methods used for the analyses of biological tissue and sediments in most studies. The high-pressure liquid chromatography (HPLC) method for tissue sample analysis in MHW is significantly more prone to matrix interferences than the GC/MS method used for sediment PAH analysis, and does not separate and identify many alkylated PAH homologues that are essential for fingerprinting. Such challenges are not limited to the bioaccumulation work in MHW. For confirmation of some of the sedimentary PAHs inputs it is noted that comparable forensics work on biological material is yet to be undertaken, in spite of equal numbers of candidate sources and sediment samples being compared in the recent study of MHW sediment PAHs (Little et al. 2015).

Four years after the 'Prestige' spill of HFO off the coast of Galicia (Spain, in 2002) it was possible to distinguish the additional input of 4–5 ring PAHs for example from forest fires during 2006 from those arising from the oil spill. Vidal et al. (2011) used PAH analyses of eggs of Kentish plover, a coastal bird with its main W. European breeding grounds in Spain. Interpretation of forensic markers in tissue samples must consider not only their water-sediment partitioning and microbial degradation in the environment, but also water-tissue partitioning and the species' metabolism of the contaminants. The bioavailability of PAHs decreases with time since release into the environment and at rates varying by the individual PAH, due to their sorption to sediment organic matter and their diffusion into micropores that are inaccessible to most biota. The physiology of exposed organisms in turn affects both the kinetics and toxicology of uptake. Desorption from sediment surfaces, the concentrations and composition of SPM and DOC, all contribute to the differences in measured contaminant uptake by biota. For example, Benzo(*a*)pyrene shows three orders of magnitude variation in water partition coefficient. DOC can increase the apparent water solubility of hydrophobic compounds, yielding lower bioconcentration factors, by keeping compounds suspended in humic and fulvic SPM, thus interrupting their uptake from the truly dissolved state (Landrum et al. 2013).

The authors are not aware of reliable general mass balance models for PAHs in estuarine/fjord environments. As discussed in this chapter, the contribution of different factors to the uptake, biodegradation and removal of PAHs depends on very site-specific factors, including sediment disturbance, the forensic effect of which may be over-printed at the national scale by chronic NPS inputs from atmospheric pollution and river runoff. In atmospheric studies the role of nanoparticles in source apportionment is increasingly understood, but particulate matter (PM) air pollution is only regulated on a mass basis for particles below diameters of 10 μm (PM_{10}) and 2.5 μm ($PM_{2.5}$). In Beijing, the PAH concentrations in $PM_{2.5}$ are about 46–52 % attributable to road traffic sources depending on seasonality (Wang et al. 2015). In Australian, Kuwaiti and UK cities, the nanoparticles studied on a number basis appear to be about 75 % attributable to road traffic sources (Al-Dabbous and Kumar 2015). However, in dynamic estuarine conditions, the composition, textures, surface areas and pores within sediments, and the number and size distributions within SPM are seldom analysed in contaminant chemistry studies, much less in routine monitoring. This is perhaps because estuarine cohesive sediments are not in themselves a direct human health threat, unlike nanoparticles in breathable air. Instead, muddy sediment is a sink for contaminants which may transfer its pollutant load via the food chain, for which regulatory guidelines apply (e.g. EAC). Clearly, processes linking atmosphere, land, and sea with at-risk biological receptors cannot be studied fully in isolation.

Biological effects of contaminant mixtures are even less well understood, including onshore where declining abundance and diversity of insects since the early 1990s are not fully explained and are likely factors in the well-documented reductions in some farmland birds. Paradoxically, such declines in insects and birds correspond to a period of great progress in reducing PAH emissions and spills of hazardous materials. Hormone-disrupting chemicals (including PAHs), agricultural pesticides, alien invasive species, litter, eutrophication, acidification, climate change, flooding, and habitat simplification, severance and loss are also possible factors working in-combination in terrestrial, freshwater and marine environments (Maclean 2010).

6 Conclusions

1. Regardless of the high cost, GC/MS is the method of choice for discriminating among hydrocarbon pollutant sources in sedimentary environments with complex and long-running multiple inputs. For example, in each of Milford Haven Waterway, Sullom Voe and the Humber estuary, oil inputs from specific spills can be identified more than three, and up to eight, decades later, and this would not have been possible using more routine methods. This degree of discrimination is not an esoteric consideration relevant only to oil spills, given that a mixture of contaminants can have in-combination effects. PAHs and other NPS contaminants are preferentially associated and transferred with fine particles in water and air, and with lipids in organisms, including those in the human food chain.
2. Greater use of advanced chemical fingerprinting would add value to studies of hydrocarbon source, weathering, persistence, toxicity and carcinogenicity. This could include tracing the provenance of compounds accumulating in organisms and for analyses of environmental footprints of products and processes. Notwithstanding their interest in terms of toxicity, water-soluble hydrocarbons are variable and transient. Although able to be concentrated and monitored effectively (e.g. using lipophilic resins or by analysis of bivalve mollusc tissues), water-soluble hydrocarbons are arguably not as powerful as SPM and settled fine-grained sediments for characterization of contamination and for developing water quality management strategies. The SPM in large integrated water samples and % mud in settled sediments should be routinely determined (along with TOC, at a minimum) whenever THC or PAHs are being analysed.
3. In each study area, most of the dated cores showed sub-surface maxima in hydrocarbons concentrations. In Southampton Water capital dredging almost seven decades ago caused long-term impacts by encouraging accelerated sedimentation of refinery effluent oil/SPM flocs into the then new bathymetric lows, and also later on when the settled sediments were disturbed. In Milford Haven Waterway intensive dredging has recently exhumed PAHs from deeper sediments. In Sullom Voe there is some long-term disturbance from the terminal tug vessel operations that influences monitoring results.

4. Of the detailed study areas in this chapter, Sullom Voe has more complete time series data for abiotic parameters linked to sediment macrobenthos (also periodic studies of birds and other wildlife). Thanks to governance by Shetland Islands Council, Sullom Voe Environmental Advisory Group (SVEAG; until 1976) and Shetland Oil Terminal Environmental Advisory Group (SOTEAG; from 1977 ongoing), the approach is based on Environmental Impact Assessment (EIA) principles. The baseline studies from the 1970s are regularly followed up by monitoring, the data are reviewed, the programme is adjusted if necessary, results are made public and the findings published in the scientific literature.
5. Milford Haven Waterway developed an oil industry from 1959 before pollution monitoring in sediments began. It has since adopted a surveillance approach with monitoring less regularly and formally conducted than in Sullom Voe. Early MHW studies were funded by the universities, the Welsh government and Institute of Petroleum, and carried out largely by the Field Studies Council, a non-governmental organisation with two centres operating within MHW. Since 1992 the surveillance work is organised by MHW Environmental Surveillance Group which also maintains an open approach to publication. There are regular reviews of surveillance results by MHWESG that have facilitated and informed adaptive management. However, there was a lack of openness or indeed coordination of EIA shown by industry and government during the very large LNG projects of the 2000s. The EC is examining a complaint by Friends of the Earth against the UK government's handling of some of these developments.
6. In Southampton Water, there has been regular monitoring of Esso effluent impacts since the late 1960s in the case of saltmarsh vegetation, mid-1970s for the subtidal macrobenthos, and early 1980s for marsh infauna. Ecological recovery and restoration have been successful, and this has been publicised, but there has been very little public dissemination of the data as would be expected for example within an EIA approach. Occasional academic publications appear on wider aspects of Southampton Water and Solent, but not through a particular strategy of openness by industry and government.
7. Through periods of economic growth and slow-down, long-range atmospheric inputs of hydrocarbons appear ubiquitous even in rural areas such as Shetland, Norway, Mindelsee and N. Caspian Sea. Widespread NPS inputs give added weight to concerns regarding the increase of emissions from diesel vehicles focussed in urban areas where NOx, $PM_{2.5}$ and associated PAHs adversely affect human health. PAHs associate with airborne nanoparticles and with increasingly problematic litter, especially microscopic plastic particles in the sea. Since aerial and runoff pollution arises from the life-styles and behaviour of consumers in addition to the producers of energy and hydrocarbon products, life-cycle analysis and public awareness-raising should be used to reduce ongoing and in some cases increasing NPS inputs. Improved estimates of NPS input and studies of impact on human health are also needed.
8. Current point source inputs are low and better-managed by oil-producers and oil-handling sites in most terrestrial, coastal and offshore environments, and 90–95 % of oil production in Europe is covered by reporting E&P companies. However, major

gaps in reporting occur worldwide (surprisingly including N. America), particularly in the former Soviet Union where only 10–12 % of known oil production is covered, compared with 62–64 % in Africa. Although the reporting culture is apparently much better in Africa than everywhere else except Europe, some of the large oil spill "losses" reported onshore Nigeria probably mask the organised theft of oil that supplies a local, regional and international black market protected by some key personnel in government, law enforcement agencies, and industry.

9. Elsewhere, in those developed economies that also foster a good reporting culture, the need is not so much for more "snapshot" input budgets, but rather for robust characterization of sediment contaminant reservoirs. This should include better understanding of chemistry forensics, contaminant fluxes, particle aggregation, size and number distributions, sedimentation and resuspension. In describing the reservoirs as "hot-spots", a fuller understanding is needed to assess the sediment reservoirs' vulnerability to natural or anthropogenic resuspension events, including those due to seismic effects (e.g. tremor-triggered landslides, turbidity currents, tsunamis), and to climate change effects such as increased drought, storminess, precipitation intensity, river and coastal flooding, runoff, erosion and rising sea levels.
10. If presenting potentially high levels of risk from remobilised contaminants (e.g. from storm-water detention basins, offshore drill cuttings piles, sunken wrecks, dredging disturbance in navigation lanes, engineering works, and coastal marsh erosion as sea level rises), the principal areas sequestering legacy contaminants should be subject to cost-benefit analysis and risk assessment and remedial investigation, whereby the feasibility of remediation should be determined using the "polluter pays" principle.

Acknowledgments The work described in this chapter is the work of many; the errors are our own. Blaise Bullimore, Dave Levell and Captain Mark Andrews have encouraged us through the transparency of Milford Haven Waterway Environmental Surveillance Group, and the manuscript was reviewed by Steve Morris and colleagues. We dedicate the study to the memory of our friend and colleague Philip Wilkin Beall (1953–2014).

References

AEA (2006) Initial review of the air quality aspects of the Buncefield oil depot explosion. Report by AEA Technology, Met Office and Health Protection Agency to DEFRA, Scottish Executive, Welsh Assembly Government and Department of Environment Northern Ireland, May 2006, initial report number AEA/ENV/R/2168, p 55. http://www.defra.gov.uk/environment/airquality/buncefield/index.htm

Al-Dabbous AN, Kumar P (2015) Source apportionment of airborne nanoparticles in a Middle Eastern city using positive matrix factorization. Environ Sci Process Impacts 17:802–812. doi:10.1039/c5em00027k

Andrews JE, Burgess D, Cave RR, Coombes EG, Jickells TD, Parkes DJ, Turner RK (2006) Biogeochemical value of managed realignment, Humber estuary, UK. Sci Total Environ 371:19–30

Armannsson H, Burton J, Jones G, Knap AH (1985) Trace metals and hydrocarbons in sediments from the Southampton Water region, with particular reference to the influence of oil refining effluent. Mar Environ Res 15:31–44

Arthur D Little Limited (1997) Benchmarking of hydrocarbons in the area affected by the sea empress oil spill. Report by Arthur D. Little limited to the *Sea Empress* environmental evaluation committee, May 1997, p 101

Arthur D Little Limited (2006) Review of environmental data near Esso 1&2 discharges, Southampton Water: in support of stage 4 determination of consents. Report by Arthur D. Little limited to the Environment Agency (Southern Region), p 173

Artyukhina GV, Little DI (2014) North-east Caspian Sea bottom sediments quality. In: Ogar NP, Kriksunov YA, Little DI (eds) Environmental monitoring of the north-eastern Caspian Sea in development of oil fields: findings of Agip KCO environmental surveys, 1993–2006. North Caspian Operating Company B.V., Agip Kazakhstan North Caspian Operating Company N.V., "Terra" Centre for Remote Sensing and GIS, Ministry of Education and Science of the Republic of Kazakhstan, Almaty, pp 56–80

Atlas RM (1981) Microbial degradation of petroleum hydrocarbons: an environmental perspective. Microbiol Rev 45(1):180–209

Baker JM (1971) Refinery effluent. In: Cowell EB (ed) The Ecological effects of oil pollution on littoral communities. Applied Science Publishers, Barking, pp 33–43

Baker JM (1976) Investigations of refinery effluent impacts through field surveys. In: Baker JM (ed) Marine ecology and oil pollution. Applied Science Publishers, Barking, pp 201–225

Baker JM (1979) Ecological impact of refinery effluents in the marine and estuarine environment. In: The environmental effects of refinery effluents (Report number 5/79, pp. III-1 to III-77). Concawe, Den Haag

Baker JM (1993) Fate and effects of marine oil pollution in UK waters: with a perspective on IP-sponsored research. Institute of Petroleum Report, p 72

Baker JM, Oldham JM, Wilson CM, Dicks B, Little DI, Levell D (1990) *Spartina anglica* and oil: spill and effluent effects, clean-up and rehabilitation. In: Gray AJ, Benham PEM (eds) *Spartina anglica*—a research review. HMSO, London, pp 52–62

Bartlett PD, Little DI, Smith J, Mangini A (1989a) The sedimentary record of contaminants (hydrocarbons and metals) in Alpine lake sediments. FSC research centre report to ICOMIA Marine Environment Committee (IMEC), FSC/RC/10/89, p 19

Bartlett PD, Little DI, Smith J, Mangini A (1989b) Geochemical analysis and radiometric dating of Pyewipe core, Humber Estuary 1987. FSC Research Centre report to Institute of Petroleum FSC/RC/13/89, p 11

Bedborough D, Blackman RAA (1986) A survey of inputs to the North Sea resulting from oil and gas development. In: Environmental effects of North Sea oil and gas developments. Philos Trans R Soc B 316:495–509

Boehm PD, Maxon CL, Newton FC, Brown JS, Galperin Y (2003) Aspects of polycyclic aromatic hydrocarbons in offshore sediment in the Azeri sector of the Caspian Sea—Absheron block. Offshore oil and gas environmental effects monitoring workshop. Bedford Institute of Oceanography, Dartmouth, Nova Scotia, Canada, May 26–30, 2003

Brassell SC, Eglinton G (1980) Environmental chemistry—an interdisciplinary subject. Natural and pollutant organic compounds in contemporary aquatic sediments. In: Albaiges J (ed) Analytical Techniques in Environmental Chemistry. Pergamon Press, Oxford, pp 1–22

Brettell R (2013) Source apportionment of polycyclic aromatic hydrocarbons in sediment cores from the Humber estuary using molecular distribution. M.Sc. dissertation, University of East Anglia, p 73

British Standards Institution (1975) Methods of test for soils for civil engineering purposes. British Standards Institution, BS 1377

Cardenas M, Lick W (1996) The transport of sediments and hydrophobic contaminants in the Lower Saginaw River. J Gt Lakes Res 22:669–682

Carey DA, Hayn M, Germano JD, Little DI, Bullimore B (2014) Marine habitat mapping of the Milford Haven Waterway, Wales, UK: comparison of facies mapping and EUNIS

Classification for monitoring sediment habitats in an industrialized estuary. J Sea Res 100: 99–119. In: Proceedings of the mesh Atlantic conference, special issue. doi:10.1016/j.seares.2014.09.012

Cordah (1998) Post-spill environmental survey of sediments and fish around the captain oilfield. Report to Texaco North Sea (UK) Company, p 68

Costa HJ, Sauer TC (2005) Forensic approaches and considerations in identifying PAH background. Environ Forensics 6:9–16

Crecelius EA, Bloom N (1989) Temporal trends of contamination in Puget sound. In: Wolfe DA (ed) Processes in marine pollution series vol. 5: urban wastes in coastal marine environments, chapter 14. Robert Krieger Publishers, Malabar FLA, pp 149–155

Croudace IW, Cundy AB (1995) Heavy metal and hydrocarbon pollution in recent sediments from Southampton Water, southern England: a geochemical and isotopic study. Environ Sci Technol 29:1288–1296

CSEMP (2014) Clean safe seas environmental monitoring programme. Accessed 2014 via the MERMAN database. http://www.bodc.ac.uk/projects/uk/merman/

Cundy AB, Croudace IW, Thompson J, Lewis JT (1997) Reliability of salt marshes as 'geochemical recorders' of pollution input: a case study from contrasting estuaries in southern England. Environ Sci Technol 31:1093–1101

DECC (2013) Sub-national road transport fuel consumption. Department of Energy and Climate Change, Publication URN: 13D/107, Office of National Statistics. http://www.ons.gov.uk/. Revised June 2013

DEFRA (2000) Quality status report of the marine and coastal areas of the Irish Sea and Bristol channel. Department of Environment, Food and Rural Affairs. http://www.defra.gov.uk/science/irishbristol

DEFRA (2014) Air quality data. Department of Environment, Food and Rural Affairs. http://www.uk-air.defra.gov.uk/data/pah-data

Dicks B, Iball K (1981) Ten years of saltmarsh monitoring—the case history of a Southampton Water saltmarsh and a changing refinery effluent discharge. In: Proceedings of the 1981 oil spill conference (prevention, behaviour, control, cleanup). American Petroleum Institute, Washington, pp 361–374

Dicks B, Levell D (1989) Refinery effluent discharges into Milford Haven and Southampton Water. In: Dicks B (ed) Ecological impacts of the oil industry. Wiley, on behalf of the Institute of Petroleum, Chichester, pp 287–316

DoE (1987) Quality status of the North Sea: summary. In: Second international conference on the protection of the North Sea. Scientific and technical working group, European year of the environment, Department of the Environment. HMSO, London, p 25

Douglas GS, McCarthy KJ, Dahlen DT, Seavey JA, Steinhauer WG, Prince RC, Elmendorf DL (1992) The use of hydrocarbon analysis for environmental assessment and remediation. J Soil Contam 1(3):197–216

Douglas GS, Bence AE, Prince RC, McMillen SJ, Butler EL (1996) Environmental stability of selected petroleum hydrocarbon source and weathering ratios. Environ Sci Technol 30 (7):2332–2339

Edwards RW (1972) Pollution. In: Head JJ, Lowenstein OE (eds) Oxford biology readers. Oxford University Press, Oxford, p 16

Eganhouse RP, Blumfield DL, Kaplan IR (1982) Petroleum hydrocarbons in stormwater runoff and municipal wastes: input to coastal waters and fate in marine sediments. Thalass Jugosl 18 (1–4):411–431

Eggleton J, Thomas KV (2004) A review of factors affecting the release and bioavailability of contaminants during sediment disturbance events. Environ Int 30:973–980

Elliott M, Griffiths AH (1987) Contamination and effects of hydrocarbons on the forth ecosystem, Scotland. Proc R Soc Edinb 93B:327–342

Exxon (1989) Review of oil spill occurrences and impacts. Exxon Production Research Company, USA, p 59

Fisheries Research Services (1998) Determination and environmental assessment of fish, shellfish and sediments following an oil spill at the captain oilfield. Report FRS 9/98, Aberdeen, p 71

Galperin Y, Kaplan IR (2011) Review of microbial processes in the near-surface environment and their implications for the chemical fingerprinting of hydrocarbon fuels. Environ Forensics 12:236–252

Germano and Associates, Inc. (2013) Sediment-profile imaging survey of Milford Haven Estuary, Wales, UK—May 2012. Report to MHWESG, p 50 + Appendices

GESAMP (2007) Estimates of oil entering the marine environment from sea-based activities. Group of experts on scientific aspects of marine environmental protection. International Maritime Organization, London, p 83

Hobbs G, Morgan CI (1992) A review of the current state of environmental knowledge of the Milford Haven Waterway. Report to MHWESG by the FSC Research Centre, Pembroke, UK, p 140

Howard S, Little DI (1987) Effect of infaunal burrow structure on oil penetration into sediments. In: Proceedings of the 1987 oil spill conference (prevention, behaviour, control, clean-up). American Petroleum Institute, Washington, pp 427–431

Howells SE, Dodd NM (1985) Fawley refinery, Southampton Water. Sublittoral sediment hydrocarbon analyses, August 1984. OPRU report to Esso Petroleum Co. Ltd., p 85

Howells SE, Pritchard R, Wadsworth S, Smith J (1988) Analysis of hydrocarbons in sediments of the Humber Estuary. OPRU report to the Humber Estuary Committee, p 12

Howells SE, Dodd NM, Turner C (1989) Determination of hydrocarbons in environmental samples: sampling and analytical variability. In: Dicks B (ed) Ecological impacts of the oil industry. Wiley, Chichester, pp 135–169

HPA (2006) The public health impact of the Buncefield oil depot fire. Health Protection Agency, Dacorum and Watford and Three Rivers Primary Care Trusts, p 124

Hunt JM (1995) Petroleum geochemistry and geology. W. H. Freeman and Company, New York, p 743

Hutchinson TH, Lyons BP, Thain JE, Law RJ (2013) Evaluating legacy contaminants and emerging chemicals in marine environments using adverse outcome pathways and biological effects-directed analysis. Mar Pollut Bull. doi:10.1016/j.marpolbul.2013.06.012

IOGP (2014) Environmental performance indicators—2013 data. International association of oil and gas producers, report 2013e, p 83. http://www.iogp.org

IOGP (2015) Environmental performance indicators—2014 data. International association of oil and gas producers, report 2014e, p 84. http://www.iogp.org

IOGP (2016) Environmental fates and effects of drill cuttings and associated drilling fluids from offshore oil and gas operations. International association of oil and gas producers, report 543, p 144. http://www.iogp.org

ISO (2008) Guidance on sampling of bulk suspended solids, ISO 5667-17: 2008. International Standards Organisation, London, p 29

Jeremiason JD, Eisenreich SJ, Baker JE, Eadie BJ (1998) PCB decline in settling particles and benthic recycling of PCBs and PAHs in Lake Superior. Environ Sci Technol 32:3249–3256

Jones DM, Rowland SJ, Douglas AG, Howells S (1985) An examination of the fate of Nigerian crude oil in surface sediments of the Humber Estuary by gas chromatography and gas chromatography–mass spectrometry. Int J Environ Anal Chem 24(3):227–247

Jones KC, Stratford JA, Waterhouse KS, Furlong ET, Giger W, Hites RA, Schaffner C, Johnston AE (1989) Increases in the polynuclear aromatic hydrocarbon content of an agricultural soil over the last century. Environ Sci Technol 23:95–101

Kaplan IR, Galperin Y (1996) How to recognize a hydrocarbon fuel in the environment and estimate its age of release. In: Bois TJ, Luther BL (eds) Groundwater and soil contamination: technical preparation and litigation management. Wiley, New York, p 786

Kaplan IR, Galperin Y, Lu S-T, Lee R-P (1997) Forensic environmental geochemistry: differentiation of fuel-types, their sources and release time. Org Geochem 27:289–317

Katsoyiannis A, Sweetman AJ, Jones KC (2011) PAH Molecular diagnostic ratios applied to atmospheric sources: a critical evaluation using two decades of source inventory and air concentration data from the UK. Environ Sci Technol 45:8897–8906
Kingston P, Little DI, Harkantra S (1999) Biological impacts of oil pollution: sedimentary shores. Volume 9 in IPIECA report series. IPIECA, London, p 20
Kippen AC (1991) Analysis of hydrocarbons in sediments of the Humber Estuary, 1990. FSC Research Centre report to Humber Estuary technical committee (National Rivers Authority: Anglia, Yorkshire, Severn-Trent Regions), p 5
Knap AH (1978) The fate of non-volatile hydrocarbons in refinery effluent entering Southampton Water. Ph.D. thesis, University of Southampton
Knap AH (1982) Petroleum hydrocarbon contamination in sediments of Southampton Water estuary. Mar Environ Res 7:235–249
Kvenvolden KA, Cooper CK (2003) Natural seepage of crude oil into the marine environment. Geo-Mar Lett 23:140–146
Lafargue E, Barker C (1988) Effect of water washing on crude oil composition. Bull Am Assoc Pet Geol 72(3):263–276
Landrum PF, Lydy MJ, Eadiey BJ (2013) Dynamic of contaminant accumulation in benthos: route to understanding exposure to organic contaminants. Environ Toxicol Chem 32(6):1209–1211
Langston WJ, Pope ND, O'Hara S, Davey M, Shortridge E, Harino H, Vane CH (2012) Bioaccumulation surveillance in Milford Haven Waterway, Phase II (2010). Report by the Marine Biological Association to MHWESG, p 109
Leahy JG, Colwell RR (1990) Microbial degradation of hydrocarbons in the environment. Microbiol Rev 54(3):302–315
Levell D, Hobbs G, Smith J, Law RJ (1997) The effects of the 'Sea Empress' oil spill on the subtidal benthos of the Milford Haven Waterway: a comparison of survey data from October 1993 and October 1996. Report to the Environment Agency by OPRU/CordaH no. OPRU/22/97, p 32
Little DI (1984) Water table and drainage characteristics of intertidal sediments. OPRU report to British Petroleum and Commission of European Communities, FSC/OPRU/2/84, vol 1, p 39, vol 2, p 149 + Appendices
Little DI (1986) History of hydrocarbon contamination in sediment cores from Martins Haven saltmarsh, Milford Haven. OPRU report to the Institute of Petroleum, p 29
Little DI (1987a) Time sequence and fate of contaminant inputs in sediment cores from Milford Haven. OPRU report to the Institute of Petroleum, p 24
Little DI (1987b) Oiled sediments in the Humber Estuary following the Sivand incident. In: Proceedings of the 1987 oil spill conference. American Petroleum Institute, Washington, pp 419–426
Little DI (2009) Sediment contaminants and transport review. Report to MHWESG, p 414
Little DI, Bullimore B (2015) Discussion of McLaren, P., Sediment trend analysis (STA®): kinematic vs. dynamic modeling in Journal of Coastal Research, 2014, 30(3), 429–437. J Coast Res 31(1):224–232. doi:10.2112/JCOASTRES-D-14-00108.1
Little DI, McLaren P (1989) Sediment and contaminant transport in Milford Haven, In: Dicks B (ed) Ecological impacts of the oil industry. Wiley, on behalf of the Institute of Petroleum, Chichester, pp. 203–234
Little DI, Pritchard RW (1989) Transfer of pollutants via suspended particulate matter. OPRU report to the Institute of Petroleum, FSC/OPRU/35/89, p 26
Little DI, Smith J (1994) Appraisal of contaminants in sediments of the inner Bristol Channel and Severn Estuary. Biol J Linn Soc Lond 51:55–69
Little DI, Baker JM, Abbiss TP, Rowland SJ, Tibbetts PJC (1981) The fate and effects of dispersant-treated compared with untreated crude oil, with particular reference to sheltered intertidal sediments. In: Proceedings of the 1981 oil spill conference (prevention, behaviour, control, cleanup). American Petroleum Institute, Washington, pp 283–293

Little DI, Abbiss TP, Baker JM, Dicks B, Tibbetts PJC (1986) The effects of chemical dispersants on the flux of stranded crude oil into fine intertidal sand. Rapp P-V Reun Int Conseil Explor Mer 186:219–233

Little DI, Howells SE, Abbiss TP, Rostron D (1987) Some factors affecting the fate of estuarine sediment hydrocarbons and trace metals in Milford Haven 1978–1982. In: Coughtrey PJ, Martin MH, Unsworth MH (eds) Pollutant transport and fate in ecosystems. Blackwell, Oxford, pp 55–87

Little DI, Smith J, Howells SE, Mangini A (1990) The chronology of pollutant inputs to Sullom Voe and Southampton Water. OPRU report to the Institute of Petroleum, p 74

Little DI, Galperin Y, Bullimore B, Camplin M (2015) Environmental forensics evaluation of sources of sediment hydrocarbon contamination in Milford Haven Waterway. Environ Sci Process Impacts 17:398–420. doi:10.1039/c4em00522h

Little DI, Bullimore B, Galperin Y, Langston WJ (2016) Sediment contaminant surveillance in Milford Haven Waterway. Environ Monit Assess 188:1–30. doi:10.1007/s10661-015-5017-1

Maclean N (2010) Silent summer: the State of Wildlife in Britain and Ireland. Cambridge University Press, Cambridge, p 765

Mähler A (2012) An inescapable curse? Resource management, violent conflict, and peace-building in the Niger Delta. In: Lujala P, Rustad A (eds) High value natural resources and peace-building. Earthscan, London, pp 391–412

May SJ (2000) Saltmarsh monitoring studies adjacent to the Fawley Refinery. In: Collins M, Ansell K (eds) Solent science—a review, proceedings in marine science, 1. Elsevier, Oxford, pp 303–305

McLaren P (1987) Sediment transport and contaminant dispersal in Sullom Voe and Southampton Water. GeoSea consulting report to Oil Pollution Research Unit on behalf of the Institute of Petroleum, p 81

McLaren P, Little DI (1987) The effects of sediment transport on contaminant dispersal: an example from Milford Haven. Mar Pollut Bull 18(11):586–594

Michel J, Etkin D, Gilbert T, Urban R, Waldron J, Blocksidge C (2005) Potentially polluting wrecks in marine waters: an issue paper. In: Proceedings of the international oil spill conference. American Petroleum Institute, Washington, p 40

Morris S (1985) A benthic and physico-chemical study of the Daucleddau Estuary, Milford Haven. B.Sc. dissertation, Hatfield Polytechnic, p 121

Myers MS, Anulacion BF, French BL (2008) Improved flatfish health following remediation of a PAH-contaminated site in Eagle Harbor, Washington. Aquat Toxicol 88:277–288

Nam JJ, Thomas GO, Jaward FM, Steinnes E, Gustafsson O, Jones KC (2008) PAHs in background soils from western Europe: influence of atmospheric deposition and soil organic matter. Chemosphere 70(9):1596–1602

National Research Council (1975) Workshop on inputs, fates and the effects of petroleum in the marine environment. National Academy of Sciences, Washington, p 107

National Research Council (1985) Oil in the sea: inputs, fates and effects. The National Academies Press, Washington, p 601

National Research Council (2003) Oil in the sea III: inputs, fates and effects. The National Academies Press, Washington, p 395

NEITI (2015) Financial, physical and process audit: an independent report assessing and reconciling physical and financial flows within Nigeria's Oil and Gas Industry for 2012. Taju Adu and Company Report to Nigeria Extractive Industry Transparency Initiative, Abuja, p 76

Newell RC, Seiderer LJ, Hitchcock DR (1998) The impact of dredging works in coastal waters: a review of the sensitivity to disturbance and subsequent recovery of biological resources on the seabed. Oceanogr Mar Biol Ann Rev 36:127–178

Nicholaus EEM, Law RJ, Wright SR, Lyons BP (2015) Spatial and temporal analysis of the risks posed by polycyclic aromatic hydrocarbon, polychlorinated biphenyl and metal contaminants in sediments in UK estuaries and coastal waters. Mar Pollut Bull. doi:10.1016/j.marpolbul.2015.03.012

Nikitik CCS, Robinson AW (2003) Patterns in benthic populations in the Milford Haven waterway following the *Sea Empress* oil spill with special reference to amphipods. Mar Pollut Bull 46 (9):1125–1141

NPL (2015) Annual report for 2014 on the UK polycyclic aromatic hydrocarbons monitoring and analysis network. National Physical Laboratory Report ENV 4 to DEFRA, Department of Environment Northern Ireland, the Welsh Government and the Scottish Government, p 41

OPRC (1990) International convention on oil pollution preparedness, response and co-operation. International Maritime Organization, London. http://www.imo.org/

Oros DR, Ross JRM (2004) Polycyclic aromatic hydrocarbons in San Francisco Estuary sediments. Mar Chem 86:169–184

OSPAR (2000) Quality status report for the greater North Sea, region II. OSPAR Commission, London (including data from 1993 quality status report)

OSPAR (2005) 2005 Assessment of data collected under the OSPAR comprehensive study on riverine inputs and direct discharges for the period 1990—2002. OSPAR Commission, Assessment and Monitoring Series, Publication Number: 2005/233, London, p 41

OSPAR (2010) Quality status report 2010. OSPAR Commission, London. http://qsr2010.ospar.org/en/ch01_02.html

Page D, Boehm PD, Douglas GS, Bence AE, Burns WA, Mankiewicz PJ (1997) An estimate of the annual input of natural petroleum hydrocarbons to seafloor sediments in Prince William Sound, Alaska. Mar Pollut Bull 34:744–749

Peters KE, Moldowan JM (1992) The biomarker guide: interpreting molecular fossils in petroleum and ancient sediments. Prentice Hall, Englewood Cliffs, p 352

Petrakis L, Weiss FT (1980) Petroleum in the marine environment, advances in chemistry series, 185. American Chemical Society, Washington, p 371

Philp RP (1985) Biological markers in fossil fuel production. Mass Spectrom Rev 4:1–54

RCEP (1981) Oil pollution of the sea. Royal Commission on environmental pollution 8th report. HMSO, London, p 307

Read AD, Blackman RAA (1980) Oily water discharges from North Sea installations: a perspective. Mar Pollut Bull 11:44–47

Readman JW, Mantoura RFC, Rhead MM (1984) Distribution, composition and sources of polycyclic aromatic hydrocarbons in sediments of the River Tamar catchment and estuary. In: Parker WR, Kinsman DJJ (eds) Transfer processes in cohesive sediment systems. Plenum Press, London, pp 155–170

Rodgers R, Blumer EN, Freitas MA, Marshall AG (2000) Complete compositional monitoring of the weathering of transportation fuels based on elemental composition from Fourier transform ion cyclotron resonance mass spectrometry. Environ Sci Technol 34:1671–1678

Rogers HR, Gardner MJ, Comber SDW, Boxall ABA (2000) PAHs in harbours. WRc report to Department of the Environment, Transport and Regions, p 147

Rostron DM, Little DI, Howells SE (1986) A study of the sediments and communities in Milford Haven, Wales. Oil Chem Pollut 3:131–166

Rowberry N (1993) Oil spills in the marine environment. AURIS report to Scottish Natural Heritage, sponsored by BP, p 32

Rumney HS, Barry J, Law R (2012) Polycyclic aromatic hydrocarbons (PAH) in sediments collected within the clean seas environment monitoring programme (CSEMP) 2000–2010. Monitoring of the quality of the marine environment, 2008–2010. Sci Ser Aquat Environ Monit Rep, Cefas, Lowestoft 63:32–38

Sauer TC, Uhler AD (1994) Pollutant source identification and allocation: advances in hydrocarbon fingerprinting. Remediation, winter issue, 1994–1995, pp 25–50

SGS M-Scan Limited (2012) Chemical and macrobenthic monitoring in Sullom Voe sediments. Report to the Shetland Oil Terminal Environmental Advisory Group, The Scottish Oceans Institute, University of St Andrews, p 94

SGS M-Scan Limited (2014) Chemical and macrobenthic monitoring in Sullom Voe sediments. Report to the Shetland Oil Terminal Environmental Advisory Group, the Scottish Oceans Institute, University of St Andrews, p 360

Simoneit BRT (1984) Organic matter of the troposphere: III characterisation and sources of petroleum and pyrogenic residues in aerosols over the western United States. Atmos Environ 18:51–67
Skei J (1981) The entrapment of pollutants in Norwegian fjord sediments—a beneficial situation for the North Sea. In: Nio SD, Shuttenhelm RTE, van Weering TCE (eds) Holocene marine sedimentation in the North Sea Basin: special publications of the international association of sedimentologists, vol 5, pp 461–481
Smith B, Rawlins BG, Ferguson AJ, Fordyce F, Hutchins MG, Finnamore JR, Barr DM (2002) Information on land quality in wales: sources of information (including background contaminants). BGS Report to Environment Agency, p 87
Stout SA, Uhler AD, McCarthy KJ, Emsbo-Mattingly SD (2002) Chemical fingerprinting of hydrocarbons. In: Murphy B, Morrison R (eds) Introduction to environmental forensics. Academic Press, Cambridge, p 560
Stout SA, Uhler AD, Emsbo-Mattingly SD (2004) Comparative evaluation of background anthropogenic hydrocarbons in surficial sediments from nine urban waterways. Environ Sci Technol 38:2987–2994
Tait RD, Maxon CL, Parr TD, Newton FC, Hardin JL (2004) Impact assessment and benthic recruitment following exploration drilling in the south Caspian Sea. Soc Pet Eng 86709:12
UKMMAS (2010) Charting progress 2: an assessment of the State of UK Seas. Report by UK Marine Monitoring and Assessment Strategy, DEFRA, p 166
UNEP (2011) Environmental assessment of Ogoniland. Report by United Nations Environment Programme, Kenya, p 262
Van Beilen JB, Li Z, Duetz WA, Smits THM, Witholt B (2003) Diversity of alkane hydroxylase systems in the environment. Oil Gas Sci Technol 58(4):427–440
Vane CH, Harrison I, Kim AW (2007a) Assessment of polyaromatic hydrocarbons (PAHs) and polychlorinated biphenyls (PCBs) in surface sediments of the Inner Clyde Estuary, UK. Mar Pollut Bull 54:1301–1306
Vane CH, Harrison I, Kim AW (2007b) Polycyclic aromatic hydrocarbons (PAHs) and polychlorinated biphenyls (PCBs) in sediments from the Mersey Estuary, UK. Sci Total Environ 374:112–126
Vane CH, Chenery SR, Harrison I, Kim AW, Moss-Hayes V, Jones DG (2011) Chemical signatures of the Anthropocene in the Clyde estuary, UK: sediment-hosted Pb, $^{207/206}$Pb, total petroleum hydrocarbon, polyaromatic hydrocarbon and polychlorinated biphenyl pollution records. Philos Trans A. doi:10.1098/rsta.2010.0298
Varney M (2000) Behaviour of organic carbon in Southampton Water. In: Collins M, Ansell K (eds) Solent science—a review. Proceedings in marine science 1. Elsevier, Oxford, pp 175–185
Vidal M, Dominguez J, Luis A (2011) Spatial and temporal patterns of PAH in eggs of Kentish Plover a coastal bird from NW Iberia after a major oil spill. Sci Total Environ 409:2668–2675
Volkman JK, Holdsworth DG, Neill GP, Bavor HJ Jr (1992) Identification of natural, anthropogenic and petroleum hydrocarbons in aquatic sediments. Sci Total Environ 112:203–219
Walker CW (1971, 2nd edition 1975) Environmental pollution by chemicals, Hutchinson, p 72
Wang Z, Fingas M, Yang C, Christensen JH (2006) Crude oil and refined product fingerprinting principles. In: Morrison RD, Murphy BL (eds) Environmental forensics. Contaminant specific guide. Elsevier, Oxford, pp 339–407
Wang Z, Stout SA, Fingas M (2007) Forensic fingerprinting of biomarkers for oil spill characterization and source identification. Environ Forensics 7:105–146
Wang F, Lin T, Feng J, Fu H, Guo Z (2015) Source apportionment of polycyclic aromatic hydrocarbons in $PM_{2.5}$ using positive matrix factorization modelling in Shanghai, China. Environ Sci Process Impacts 17:197–205. doi:10.1039/c4em00570h
Waples DW, Machihara T (1990) Application of sterane and triterpane biomarkers in petroleum exploration. Bull Can Pet Geol 38:357–380
Westwood SSC, Little DI, Dodd NM, Howells SE (1987) Biological and chemical monitoring of seabed sediments at the Sullom Voe North Sea terminal: a synthesis of data for the period

1978–85. In: Kuiper J, van den Brink WJ (eds) Fate and effects of oil in marine ecosystems. Nijhoff, Amsterdam, pp 253–267

Westwood SSC, Dunnet GM, Hiscock K (1989) Monitoring the Sullom Voe terminal. In: Dicks B (ed) Ecological impacts of the oil industry. Wiley, on behalf of the Institute of Petroleum, Chichester, pp 261–285

Woodhead RJ, Law RJ, Matthiessen P (1999) Polycyclic aromatic hydrocarbons in surface sediments around England and Wales, and their possible biological significance. Mar Pollut Bull 38:773–790

Yang C, Wang ZD, Hollebone BP, Brown CE, Yang ZY, Landriault M (2015) Chromatographic fingerprinting analysis of crude oils and petroleum products. In: Fingas MF (ed) Handbook of oil spill science and technology. Wiley, New York, p 792

Yunker MB, Macdonald RW, Vingarzan R, Mitchell RH, Goyette D, Sylvestre S (2002) PAHs in the Fraser River basin: a critical appraisal of PAH ratios as indicators of PAH source and composition. Org Geochem 33:489–515

Zhou JL, Fileman TW, Evans S, Donkin P, Llewellyn C, Readman JW, Fauzi R, Mantoura C, Rowland SJ (1998) Fluoranthene and pyrene in the suspended particulate matter and surface sediments of the Humber Estuary, UK. Mar Pollut Bull 36:587–597

HC-0A-02: Analysis of Polycyclic Aromatic Hydrocarbons from Food

Rafael Pissinatti and Scheilla V.C. de Souza

Abstract Polycyclic aromatic hydrocarbons (PAHs) comprise a class of organic compounds of petrogenic origins or generated from the incomplete combustion of organic matter. Their presence in the environment can be a source of food contamination, which implies a potential risk to human health according to the International Agency for Research on Cancer. Additionally, the Food and Agriculture Organization of the United Nations (FAO/WHO) and the Scientific Committee on Food consider PAHs to be genotoxic and carcinogenic. The greatest concern has been raised regarding the possibility of cancer induction in humans exposed to PAHs from contaminated food. Therefore, many regulations and reports regarding strategies for and results of food monitoring have been published, including performance criteria for the sampling, chemical analysis and maximum permitted levels of these contaminants in food. A variety of techniques have been reported for the extraction, clean-up and determination steps of the analytical process. The details and advantages of each technique are discussed in this chapter. Additionally, because method validation is one measure that laboratories must implement to guarantee the reliability, traceability and comparability of their results, a review and critical analysis of the current practices in this area are also presented.

Keywords Polycyclic aromatic hydrocarbons · Analytical methods · Food · Method validation

---

R. Pissinatti
National Laboratory of Minas Gerais (LANAGRO-MG), Ministry of Agriculture, Livestock and Food Supply (MAPA), Av. Rômulo Joviano S/N, Centro, Pedro Leopoldo, MG 33600-000, Brazil

S.V.C. de Souza (✉)
Department of Food Science, Faculty of Pharmacy (FAFAR), Federal University of Minas Gerais (UFMG), Av. Antônio Carlos, 6627, Campus Da UFMG, Pampulha, Belo Horizonte, MG 31270-010, Brazil
e-mail: scheilla@bromatologiaufmg.com.br

K. Heimann et al. (eds.), *Biodegradation and Bioconversion of Hydrocarbons*,
Environmental Footprints and Eco-design of Products and Processes,
DOI 10.1007/978-981-10-0201-4_2

1 Introduction

Polycyclic or polynuclear aromatic hydrocarbons (PAHs) are a group of compounds composed of two or more fused aromatic rings in linear, angular, or cluster arrangements. By definition, they contain only carbon and hydrogen (Baek et al. 1991; Albers 2002).

PAHs can be formed in any incomplete combustion or high temperature pyrolytic process involving fossil fuels, plastic products, cigarette smoke, waste incineration, forest fires, combustion engines and, more generally, in processes involving materials containing C and H (Codex Alimentarius 2005; Poster et al. 2006). Furthermore, a variety of PAHs are naturally present in crude oil (0.2–7 %) and coal, with configurations ranging from two to six rings and arising from the chemical conversion of natural precursors for PAHs, such as steroids and terpenes (Harvey 1998; Albers 2002; Feng et al. 2009; Pampanin and Sydnes 2013).

PAHs were one of the first atmospheric pollutants to be identified as being carcinogenic (ATSDR 1995; WHO 2006). Their presence in food implies a potential risk to human health according to the International Agency for Research on Cancer (IARC) (IARC 2012). Additionally, the Food and Agriculture Organization of the United Nations (FAO/WHO) (WHO 2006) and the Scientific Committee on Food (SCF) (EC 2002) consider PAHs to be genotoxic and carcinogenic and recommend monitoring the presence of PAHs in food.

The main techniques used for the detection and quantification of PAHs from food matrices include gas chromatography (GC) coupled to mass spectrometry (MS) and high performance liquid chromatography (HPLC) with fluorescence detection (FL). HPLC coupled to MS or GC tandem MS (MS/MS) and time of flight (TOF) are also used. Fat extraction and clean-up steps are often required for the determination of PAHs from complex matrices such as food. The applications of both conventional and recently developed extraction and clean-up techniques, such as pressurized liquid extraction (PLE) and QuEChERS (quick, easy, cheap, effective, rugged and safe), are presented. For the clean-up step, solid-phase extraction (SPE), including conventional SPE, dispersive SPE, solid-phase micro-extraction (SPME) and molecularly imprinted polymer (MIP) SPE, as well as gel permeation chromatography (GPC) and donor–acceptor complex chromatography (DACC) are shown. The details and advantages of each technique are discussed.

The importance of validation and a review of the approaches for the in-house validation of the proposed methods for PAHs determination from food are presented and confronted with the recommendations laid in harmonized international guidelines and the regulated performance requirements.

2 Physicochemical Properties of PAHs

Knowledge of the physico-chemical properties of the PAHs (Table 1) allows for an understanding of the distribution profile of these contaminants in food matrices, as well as their interactions and stability, which will guide the choice of strategies used for their extraction from food.

In the environment, PAHs occur as compounds containing between two and seven conjugated rings and have molar masses ranging from 128 to 302 g/mol (Ming-Ho 2005). According to the United States Environmental Protection Agency

Table 1 Physico-chemical properties and classification by the International Agency for Research on Cancer (IARC) regarding the carcinogenicity of the sixteen polycyclic aromatic hydrocarbons (PAHs) prioritized by the United States Environmental Protection Agency (EPA 1986; IPCS 1998; IARC 2012)

PAHs	No of rings	Molecular weight (g/mol)	Solubility in water at 25 °C (mg/L)	Vapor pressure 25 °C (Pa)	Octanol-water partition coefficient Log KOW (−)	IARC group
Naphthalene	2	128	3.1×10^{1}	1×10^{1}	3.37	2B
Acenaphthylene	3	152	3.9	9×10^{-1}	4.00	Not classified
Acenaphthene	3	154	3.8	3×10^{-1}	3.92	3
Fluorene	3	166	1.9	8×10^{-2}	4.18	3
Phenanthrene	3	178	1.1	2×10^{-2}	4.57	3
Anthracene	3	178	4.5×10^{-2}	8×10^{-4}	4.54	3
Fluoranthene	4	202	2.6×10^{-1}	1.2×10^{-3}	5.22	3
Pyrene	4	202	1.3×10^{-1}	6×10^{-4}	5.18	3
Benzo(a) anthracene	4	228	1.1×10^{-2}	2.8×10^{-5}	5.61	2B
Chrysene	4	228	2.0×10^{-3}	8.4×10^{-5}	5.91	2B
Benzo(b) fluoranthene	5	252	1.5×10^{-3}	1.3×10^{-7}	5.80	2B
Benzo(k) fluoranthene	5	252	7.6×10^{-4}	1.3×10^{-7}	6.84	2B
Benzo(a)pyrene	5	252	3.8×10^{-3}	7.3×10^{-7}	6.50	1
Indeno(1,2,3-c, d)pyrene	6	276	6.2×10^{-2}	1.3×10^{-8}	6.58	2B
Dibenzo(a,h) anthracene	5	278	6.0×10^{-4}	1.3×10^{-8}	6.50	2A
Benzo(g,h,i) perylene	6	276	2.6×10^{-4}	1.4×10^{-8}	7.10	2B

Group 1 sufficient evidence in humans or in animals and strong mechanistic data in humans, *Group 2A* limited evidence in humans and sufficient evidence in animals, *Group 2B* limited evidence in humans and less than sufficient evidence in animals, *Group 3* inadequate evidence in humans and inadequate or limited evidence in animals

(USEPA), sixteen PAHs are considered particularly important regarding the environmental monitoring of organic pollutants (EPA 1986).

The physico-chemical characteristics of these PAHs, such as solubility and vapor pressure, are important factors that drive the distribution of these contaminants in soluble and particulate phases throughout atmospheric, aqueous and biotic media. In general, the solubility of PAHs in water decreases with increasing molecular weight and ranges from highly insoluble, such as for benzo(g,h,i)perylene (solubility 0.0003 mg/L), to slightly soluble, such as for naphthalene (solubility 31 mg/L). PAHs can also be classified as moderately to highly soluble, with octanol-water partition coefficients (log KOW) ranging between 3.37 and 7.10 (IPCS 1998; Meire et al. 2007).

Most PAHs are of low volatility and have a high tendency to adsorb onto organic particulate matter. In the atmosphere, PAHs containing five or more aromatic rings are found predominantly in association with particulates, usually on small (<2.5 μm) particles, such as fly ash and soot. PAHs with two or three rings are almost entirely found in the vapor phase, whereas those with four rings exhibit an intermediate behavior (EC 2002). In aqueous environments, PAHs are generally found adsorbed on particulates and humic matter or dissolved in any oily contaminant that may be present in the water, sediment or soil (EFSA 2008b).

PAHs are chemically stable and resistant to degradation by hydrolysis. In the presence of light, they are susceptible to oxidation and photo-degradation. Depending on various parameters (such as the type of adsorption onto particles and molecular mass), the half-lives of PAHs in air range from a few hours to days. In soil, PAHs may also be degraded by microbial activity. The estimated half-lives of PAHS in soils vary from several months to several years (Harvey 1998; EC 2002).

3 Toxicity of PAHs

The toxicological properties of PAHs highlight the importance of determining these compounds in food, even at very low levels.

PAHs are the most extensively studied compounds with regard to their cancer-inducing properties in laboratory animals. Researchers have shown that certain PAHs can cause cancer when inhaled, ingested and even after skin contact. Immunosuppression, hepatic hypertrophy and changes in the growth of other tissues have also been reported in laboratory animals (ATSDR 1995). The evidence of carcinogenicity in humans comes from occupational studies in workers exposed to known sources of PAHs, such as oil refineries, charcoal ovens and chimneys (ATSDR 1995; WHO 2006).

Many PAHs are considered toxic even at low concentrations. Lower molecular weight compounds consisting of two or three rings (such as naphthalene, anthracene, and phenanthrene) have high acute toxicity but little or no carcinogenic potential. Conversely, compounds with higher molecular masses that consist of four, five or six rings (such as benzo(a)pyrene) have low acute toxicity but greater carcinogenic potential (Meire et al. 2007). Acute effects affect mainly the liver and kidneys, leading to cutaneous inflammation, ulcerations and hyperkeratosis. Changes in the lymph nodes and immunosuppressive induction have also been reported (WHO 2006).

In 2012, the IARC updated the classification of benzo(a)pyrene to Group 1-Carcinogenic to humans (IARC 2012). This group includes 116 agents (or mixtures) with sufficient evidence of carcinogenicity in humans. In Table 1, the classification of the sixteen PAHs prioritized by the EPA is presented.

PAHs are considered to be pro-carcinogenic, i.e., they must be altered by metabolic activation to generate the active carcinogen capable of reacting with DNA and other molecules (Ming-Ho 2005). The carcinogenicity of PAHs is associated with the complexity of the molecule, i.e., the number of benzene rings present. The mechanism of toxicity involves the oxidation of the aromatic rings by the enzymes of the cytochrome P450 family, which generates epoxide intermediates that covalently bind to critical sites in the DNA, causing replication errors and allowing mutation to occur (Boström et al. 2002; EC 2002). The main sites are the amino groups of adenine and guanine (WHO 2006).

Some PAHs and their metabolites also bind the aryl hydrocarbon receptor, resulting in the up-regulation of several enzymes involved in the metabolism of PAHs and exhibiting complex and non-linear dose-response curves when mixtures of PAHs are present (WHO 2006). This phenomenon hinders the establishment of safe limits for the intake of these contaminants.

Thus, a minimum safe threshold for chronic exposure has not been established by international agencies and health authorities, such as the Agency for Toxic Substances and Disease Registry (ATSDR), the SCF and the EFSA. Instead, these agencies recommend that exposure to PAHs be as low as is reasonably achievable (Bulder et al. 2006).

Other international organizations have calculated a so-called "virtually safe dose" based on extrapolating the data obtained from experiments with laboratory animals. The EPA associates an intake of 0.14 ng/kg body weight/day of benzo(a) pyrene with a cancer risk of 1×10^{-6}. The Dutch National Institute of Public Health and Environment (RIVM) calculated the virtually safe dose for the same cancer risk to be 0.50 ng/kg body weight/day (Bulder et al. 2006).

4 PAH in Food

4.1 Contamination and Occurrence

The presence of PAHs in food has been extensively described in the literature (Diletti et al. 2005; Houessou et al. 2005; Ballesteros et al. 2006; Liguori et al. 2006; Ciecierska and Obiedzinsk 2007; Rose et al. 2007; Stanciu et al. 2008; Windal et al. 2008; Veyrand et al. 2007; Ziegenhals et al. 2009; Belo et al. 2012;

Pissinatti et al. 2014). There are several routes of food contamination by PAHs, and contamination can arise from both environmental pollution (fruits, vegetables and grains grown in industrial regions) and food processing (such as smoking and roasting) (Camargo and Toledo 2002; Tfouni et al. 2012). PAHs from environmental pollution have both natural sources (bitumens, coal, forest and prairie fires, oil seeps, plant debris) and anthropogenic sources (fossil fuels and combustion). Since the PAH compositions of the two sources overlap, the significance of anthropogenic PAH in the environment have been evaluated against a dynamic background of natural PAH (Yunker et al. 2002). Some authors have proposed the estimation of isomeric ratios to predict the source of the contamination (Budzinski et al. 1997; Yunker et al. 2002; Mannino and Orecchio 2008; Orecchio 2010).

The values of ratios between anthracene (An) and anthracene plus phenanthrene (An + Phe) when lower than 0.10 indicate low temperature sources (petroleum) while ratios bigger than 0.10 indicate a dominance of combustion (Yunker et al. 2002), because phenanthrene and anthracene are isomers and phenanthrene is more thermodynamically stable than anthracene (Budzinski et al. 1997). BaA/(BaA + Chr) ratios lower than 0.20 involve petroleum, from 0.20 to 0.35 indicate either petroleum or combustion and upper than 0.35 combustion. Also, IP/(IP + BghiP) ratios between 0.20 and 0.50 imply liquid fossil fuel combustion (Yunker et al. 2002). The ratio between fluoranthene (Fl) and fluoranthene plus pyrene (Fl + Py) have also been applied for PAHs source typing (Budzinski et al. 1997).

In vegetables and fruits, PAH contamination mainly occurs via the deposition of particles from air pollution on their surfaces. The concentrations depend on the location of cultivation and production, and samples cultivated in highly industrialized areas or near roads and express highways generally have higher levels of PAHs than those cultivated in rural areas. The association between PAHs with higher molecular masses and suspended particulate matter in the atmosphere is a major source of contamination (Nielsen et al. 1999). The concentrations of PAHs are generally higher at the surfaces of plants, such as fruits shells and edible leaves, than in the internal tissues. The higher levels of PAHs found in plants grown in areas with polluted air are especially evident in plants with a large exposed surface area, for example, lettuce, cabbage and spinach (EC 2002).

During the process of smoking, PAHs may be deposited on the surface and migrate into the food that is being smoked. Numerous factors in the smoking process influence the composition of the smoke and the absorption of PAHs by smoked food, with the combustion temperature during the generation of the smoke being a particularly critical parameter (Wretling et al. 2010).

In general, due to the lipophilic properties of PAHs, oils and fats are more susceptible to concentrated contamination from environmental sources or during the process of seed drying prior to oil extraction. Smoked foods, such as meat and fish, can also have high levels of PAH contamination. However, the contribution of these foods to the total ingested PAHs is not significant because the levels of these products in the population's diet are usually low (Codex Alimentarius 2009).

A survey published in 2008 by EFSA, which involved data from sixteen European countries, indicated that fish and seafood, vegetable oils, meat, coffee and tea, food supplements and spices as the main food groups susceptible to PAH contamination. Those types of food presented levels exceeding 10 μg/kg for the sum of the following eight PAHs: benzo(a)anthracene, chrysene, benzo(b)fluoranthene, benzo(k)fluoranthene, benzo(a)pyrene, indeno(1,2,3-c,d)pyrene, dibenzo (a,h)anthracene and benzo(g,h,i)perylene ($\sum$PAH8). These results suggest that attention should be paid to coffee and tea, food supplements and spices because there are no current regulations for the PAHs levels in these food matrices (EFSA 2008a, b).

Studies conducted in Italy (Lodovici et al. 1995) highlighted the fact that exposure to PAHs from contaminated food intake is significantly higher than environmental exposure or exposure through inhalation or skin absorption. According to the EFSA Report, the diet is the largest non-occupational source of PAH exposure for non-smokers. In this European study, based on the dietary profiles of the 16 countries involved, an average dietary exposure for the whole population (including the non-consumers) would provide 235 ng/day of benzo(a) pyrene and 1729 ng/day of all eight PAHs, whereas a diet high in PAHs would provide 389 ng/day of benzo(a)pyrene and 3078 ng/day of all eight PAHs. The products with the largest contributions to PAH exposure, considering the median consumption estimated for consumers only, were seafood, cereals, vegetables, meat, oils and fats, fish and coffee in terms of both benzo(a)pyrene alone and all eight PAHs (Table 2). It must be emphasized that, in this study, the number of cereal and coffee samples was significantly lower than the other matrices studied (EFSA 2008a, b).

In this context, the use of benzo(a)pyrene as a marker for the presence of PAHs in food, as proposed by the SCF (EC 2002), is controversial. The survey data collected by the EFSA (EFSA 2008a, b) also show that the concentrations of benzo (a)pyrene and other PAHs, such as pyrene and benzo(a)anthracene, which are also considered toxic, presented a low correlation in products such as fish, crustaceans, tea and coffee. Among 9714 samples comprising 95 groups of Codex Alimentarius food matrices, 33 % had values exceeding the detection limit for one or more priority PAH, and benzo(a)pyrene was not detected.

According to the report published by the Agence Française de Sécurité Sanitaire des Aliments (AFSSA 2004), in which PAHs occurrence data from more than eight thousand samples in 44 different food groups were compiled, the highest levels of benzo(a)pyrene were found in dried fruits (48.10 μg/kg wet weight), olive pomace oil (17.7 μg/kg), smoked fish (5.28 μg/kg), grape seed oil (4.2 μg/kg), smoked meat products (3.27 μg/kg), fresh mollusks (3.09 μg/kg), and spices/sauces and condiments (2.16 μg/kg).

Several studies have reported evidence of increased PAH concentrations in seafood after oil spills compared with baseline levels. Law et al. (2002) examined PAH levels based on datasets covering background levels and seafood after oil spills from 19 studies and found average total PAH concentrations between 20 and 1600 μg/kg in baseline monitoring studies and between 104 and 27,400 μg/kg after

Table 2 Consumer exposure estimated by European Food Safey Authority (EFSA) to benzo(a) pyrene (BaP) alone, ∑PAH2 (BAp and chrysene), ∑PAH4 (PAH2, benz(*a*)anthracene, and benzo(*b*)fluoranthene), and ∑PAH8 (PAH4, benzo(*k*)fluoranthene, benzo(*ghi*)perylene, dibenz(*a, h*)anthracene, and indeno(*1,2,3-cd*)pyrene) for each food category for which occurrence data were available. (EFSA 2008a, b)

Food category	Consumption median (g/day)	Exposure (ng/day)			
		∑BaP	∑PAH2	∑PAH4	∑PAH8
Alcoholic beverages	413	4	12	25	74
Cereals and cereal products	257	67	129	257	393
Cheese	42	6	12	20	30
Coffee, tea and cocoa (expressed as liquid)	601	21	55	106	156
Fats (vegetal and animal)	38	26	112	177	239
Fish and fishery products	41	21	84	170	210
Fruits	153	5	40	75	87
Meat and meat products (including substitutes)	132	42	107	195	279
Seafood and seafood products	27	36	140	289	421
Sugar and sugar products (including chocolate)	43	5	13	25	39
Vegetables, nuts and pulses	194	50	124	221	378

Median consumption: estimated by the Member States for consumers only

different oil spills. Studies of oils spills suggest that several factors may play a role in determining the duration of PAH contamination, including the amount of sedimentation and the likelihood of subsequent resuspension of the oil, the composition of the oil, the rate of biodegradation (which tends to be higher in warmer climates), and the particular species of interest (Gohlke et al. 2011). Among fishery products, finfish species (i.e., tuna, mackerel and salmon) generally present lower contamination levels than shellfish (i.e., bivalve mollusks), even if they originated from polluted areas. Indeed, unlike bivalves, fish oxidize and metabolize PAHs to water-soluble compounds, which are eventually excreted (EU 2011b; Gohlke et al. 2011; Purcaro et al. 2013).

4.2 Regulation

Until 2011, the maximum permitted level described in European legislation was applied only for benzo(a)pyrene. The report published by the EFSA (2008a, b) concluded that benzo(a)pyrene was not a suitable marker for the occurrence of PAHs in food and that a system of four specific substances would be appropriate to determine the carcinogenic potency of PAHs in food. Thus, the European regulation 1881/2006 was amended by the regulation 835/2011 (EU 2011b) with regard to the permitted levels for benzo(a)pyrene only, as well as for the sum of benzo(a)pyrene, chrysene, benz(a)anthracene, and benzo(b)fluoranthene ($\sum$PAH4).

The regulation EU/835/2011 (EU 2011b) also updated the matrices to be monitored based on new occurrence data. Therefore, it was concluded that maintaining a maximum level for PAHs in fresh fish (muscle) was no longer appropriate, as PAHs are quickly metabolized in fresh fish. Other limits were added, namely, for cocoa beans and derivatives and coconut oil (limits described in terms of fat). Other matrices affected by this legislation are as follows: oils and fats, smoked products, bivalve mollusks and baby food for infants and young children. For the latter, the maximum permitted levels were set at 1.0 μg/kg (benzo(a)pyrene and $\sum$PAH4) and ranged up to 6 μg/kg (benzo(a)pyrene) or 35 μg/kg ($\sum$PAH4) for bivalve mollusks.

5 Analytical Methods for PAHs from Food

In Table 3, a review of the published analytical methods for the determination of PAHs from food is shown in chronological order. The matrices and measured parameters are listed, in addition to the analytical steps and parameters, such as sample size, extraction and clean-up techniques, and detection and quantification techniques, including details about chromatographic columns, detectors and the use of internal standards.

A discussion about the characteristics of the different techniques presented in Table 3, with notes concerning limitations and advantages of some of them, is made in the following Sect. 5.1–5.3.

5.1 Extraction and Clean-up

Traditional methods for the extraction and clean-up of PAHs from food matrices are laborious and require large amounts of solvent. Usually, sample extraction involves saponification with potassium hydroxide solution, followed by liquid–liquid partition with organic solvents and column purification using silica gel, alumina or Florisil as the stationary phase (Badolato et al. 2006; Stanciu et al. 2008). These methods have significant drawbacks, including the need for high purity solvents and the formation of emulsions that are difficult to break, which may affect the recovery of some measured parameters (Orecchio et al. 2009).

The need for more reliable analytical methods, as well as consideration of environmental and sustainability issues, has driven the development of new technologies for the extraction and purification of samples and the miniaturization of analytical systems.

Table 3 Review of analytical approaches for the determination of PAHs in foods: food matrices, parameters measured, and extraction, clean-up and determination techniques

References	Food matrix	Parameter measured (PAHs and respective monitored ions)	Analytical method						
			Sample size	Extraction	Clean-up	Detection and quantification	Column	Detector details	Internal standard
Diletti et al. (2005)	Olive pomace oil	BaA (228 226), BbF (252 250), BkF (252 250), BaP (252 250), BeP (252 250), IP (276 277), DBahA (278 279) and BghiP (276 277)	10 g	Liquid–liquid partition (10 mL n-pentane followed by 15 + 10 + 10 mL dimethyl sulfoxide and 50 mL cyclohexane three times)	TLC (silica gel)	GC–MS	DB-5MS (30 m × 0.25 mm i.d., 0.25 μm)	Ion Trap (Polaris Q, Thermo)	Isotope labeled (^2H)
Houessou et al. (2005)	Coffee brew	Fl, BbF, BaP	50 mL	SPE (0.5 g polystyrene-divinylbenzene cartridge) Extraction (methanol:tetrahydrofuran 10:90, v/v)	No further clean-up	HPLC–FL	Supelcosil LC-PAH (150 mm × 3.0 mm i.d., 5 μm)	–	Not used
Badolato et al. (2006)	Coffee	BaP	20 g	Soxhlet (250 mL acetone) Saponification (50 mL methanol: water 9:1, v/v and 1.49 g potassium hydroxide) Liquid–liquid partition (40 mL cyclohexane three times)	Column (5 g silica gel)	HPLC–FL	Lichrosphere 100 RP18 (150 mm × 4.6 mm i.d., 5 μm)	–	Not used
Ballesteros et al. (2006)	Olive and olive-pomace oil	BkF (252 > 226 + 250 + 251), BeP (252 > 226 + 250 + 251), BaP (252 > 226 + 250 + 251) and BghiP (276 > 274 + 275)	2 g	Extraction (2 mL n-hexane, 10 mL acetonitrile and 3 mg of anhydrous sodium sulfate)	GPC (CL_2CH_2)	GC–MS/MS	HP-5 (30 m × 0.25 mm i.d., 0.25 μm)	Ion Trap (Saturn 2200, Varian)	p-terphenyl
Liguori et al. (2006)	Blue mussel, salmon fillet, fish oil and fish feed	Na (128), 2-MeNa (142), 2,6-DiMeNa (156), Ap (152), Ac (154), 2,3,6-TriMeNa (170), F (166), Phe (178), A (178), 2-MePhe (192), 3,6-DiMePhe (206), Fl (202), Py (202), 1-MePy (216), Tph (228), Ch (228), BaA (228), BkF (252), BbF (252), BaP (252), 6-MeBaP (266), IP (276), DBahA (278) and BghiP (276)	5 g (mussel), 1.3 g (salmon), 1.5 g (fish feed) and 0.25 g (fish oil)	PLE in cell clean-up (extraction solvent: dichloromethane and adsorbents: 22 g aluminum oxide, 16 g silica gel and 3 g magnesium sulfate)	GPC (CL_2CH_2)	GC–MS	HP-5MS (30 m × 0.25 mm i.d., 0.25 μm)	Single quadrupole (5972, HP)	Isotope labeled (^2H)

(continued)

Table 3 (continued)

References	Food matrix	Parameter measured (PAHs and respective monitored ions)	Analytical method						
			Sample size	Extraction	Clean-up	Detection and quantification	Column	Detector details	Internal standard
Aguinaga et al. (2007)	Milk	Na (128, 127), Ap (152, 151), Ac (154, 153), F (166 165), Phe (178, 176), A (178, 176), Fl (202, 101), Py (202, 101), BaA (228, 114), Ch (228, 114), BbF (252, 126), BkF (252, 126), BaP (252, 126), IP (278, 276), DBahA (278 276), BghiP (276, 278)	5 mL (liquid sample) 5 g (solid sample)	Immersion SPME (100 μm polydimethylsiloxane, 65 μm olydimethylsiloxane-divinylbenzene and 85 μm polyacrylate)	No further clean-up	GC–MS	HP-5MS (30 m × 0.25 mm i.d., 0.25 μm)	Single quadrupole (5973, Agilent)	Not used
Purcaro et al. (2007b)	Vegetable oils	15 (+1) EU priority PAHs	200 μL	Immersion SPME (15 μm Carbopack Z/polydimethyl-siloxane)	No further clean-up	GC × GC–MS	BPX5 (30 m × 0.25 mm i.d., 0.25 μm) BPX50 (1 m × 0.1 mm i. d., 0.1 μm)	Pegasus III (LECO Corporation)	Not used
Purcaro et al. (2007a)	Vegetable oils	BaP (252)	200 μL	Immersion SPME (15 μm Carbopack Z/polydimethyl-siloxane)	No further clean-up	GC–MS	SPB-5 (30 m × 0.25 mm i.d., 0.25 μm)	Quadrupole (QP 2010 Ultra, Shimadzu)	BaP-D12
Rose et al. (2007)	Olive pomace, sunflower and coconut oil	Ap (152 151), Ac (154 153), F (166 165), A (178 176), Phe (178 176), Fl (202 200), Py (202 200), BaA(226 224), BghiF (226 224), benzo (b)naphtha(2,1-d)thiophene (234 232), Ch (228 226), CPP (226 224), 5-MeCh (242 241), BbF (252 250), BjF (252 250), BkF (252 250), BeP (252 250), BaP (252 250), IP	5 g	Saponification (200 mL 2 mol/L methanolic potassium hydroxide) Liquid–liquid partition (100 mL cyclohexane and 50 mL dimethylformamide)	Column (2.5 g silica gel)	GC–MS	DB-5MS (60 m × 0.25 mm i.d., 0.25 μm)	Single quadrupole (TRACE, Thermo Finnigan)	Isotope labeled (^{13}C)

(continued)

Table 3 (continued)

References	Food matrix	Parameter measured (PAHs and respective monitored ions)	Analytical method						
			Sample size	Extraction	Clean-up	Detection and quantification	Column	Detector details	Internal standard
		(276 277), BghiP (276 277), An (276 277), DBahA (278 279), DBalP (302 303), DBaeP (302 303), DBaiP (302 303), DBahP (302 303) and COR (300 301)							
Veyrand et al. (2007)	Food and oil	Phe (178 > 176, 152), A (178 > 176, 152), Fl (202 > 200, 152), Py (202 > 200, 152), CPP (226 > 224, 200), BaA (228 > 226, 202), Ch (228 > 226, 202), 5-MeCh (242 > 240, 226), BbF (252 > 250, 226), BjF (252 > 250, 226), BkF (252 > 250, 226), BaP (252 > 250, 226), IP (276 > 274, 272), DBahA (278 > 276, 252), BghiP (276 > 274, 272), DBalP (302 > 300, 298), DBaeP (302 > 300, 298), DBaiP (302 > 300, 298) and DBahP (302 > 300, 298)	10 g (food) and 1 g (oil)	Freeze-drying and PLE in cell clean-up (extraction solvent: hexane: acetone 50:50, v/v and adsorbents: 15 g Florisil and 1 g Celite)	SPE (polystyrene divinylbenzene cartridge)	GC–MS/MS	Zebron ZB-5MS (30 m × 0.25 mm i.d., 0.25 μm)	Triple quadrupole (Quattro micro, Waters, Micromass)	Isotope labeled (^{13}C)
Stanciu et al. (2008)	Coffee	BaA, BkF, BbF, BaP, BghiP, DBahA and IP	10 g	Soxhlet (hexane) Liquid–liquid partition (petroleum ether and acetonitrile)	Column (5 g aluminum oxide, 5 g silica-gel and 1 g anhydrous sodium sulfate)	HPLC–FL	Column SS (100 mm × 4.6 mm i.d.)	–	Not used

(continued)

Table 3 (continued)

References	Food matrix	Parameter measured (PAHs and respective monitored ions)	Analytical method						
			Sample size	Extraction	Clean-up	Detection and quantification	Column	Detector details	Internal standard
Windal et al. (2008)	Fish oil and dried plants	BcL, CPP, BaA, Ch, 5MeCh, BjF, BbF, BkF, BaP, DBahP, DBahA, BghiP, IP, DBaeP, DBaiP and DBahP	2 g	Dilution (10 mL dichloromethane:cyclohexane 50:50, v/v)	Automatic clean-up with DACC (oil) or SPE (0.5 g silica gel cartridges) followed by DACC (dirty matrices)	HPLC–UV/FL	Varian Pursuit 3 PAH (100 mm × 4.6 mm i.d., ¼")	Fluorescence (2475, Waters) UV (487, Waters)	Not used
Danyi et al. (2009)	Food supplements (dried plants and plant extracts)	15 (+1) EU priority PAHs	1 g	Extraction (15 mL dichloromethane:cyclohexane 50:50, v/v, followed by centrifugation)	SPE (silica gel column)	HPLC/UV–FL	C18 Pursuit 3 PAH (100 mm × 4.6 mm i.d., 3 μm)	Fluorescence (996 PDA and 2475, Waters)	^{2}H-DBaiP
Lund et al. (2009)	Smoked fish	16 EPA priority PAH + BcL, DBalP, DBaeP, DBaiP, DBahP, Per, 5-MeCh and CPP	10 g	PLE in cell clean-up (extraction solvent: hexane:acetone 1:1, v/v and adsorbents: Ottawa sand and polyacrylic acid)	No further clean-up	GC–MS	ZB-5 (40 m × 0.18 mm i.d., 0.25 μm)	Single Quadrupole (5975B, Agilent)	Ap-D8, BaP-D12, BaA-D12, IP-D12 and DBaiP-D14
Orecchio et al. (2009)	Coffee (brew)	Ap (152, 151, 76), Ac (154, 152, 76), F (166, 165, 164), A (178, 188, 89), Phe (178 > 152, 176), Fl (202, 200, 101), Py (202, 200, 101), BaA (228, 114, 226), Ch (228, 114, 226), BaP (252, 126, 250), BbF (252, 126, 250), BkF (252, 126, 250), DBahA (278, 279, 139), BeP (252, 126, 250), Per (252, 126, 250), DBalP (302, 151), DBaeP (302, 151), DBaiP (302, 151), DBahP (302, 151)	15 mL	Digestion under reflux (50 mL methanolic solution of potassium hydroxide 2 mol/L, 3 h) Liquid–liquid partition (10 mL hexane, three times)	No further clean-up	GC–MS	Equity-5 (30 m × 0.25 mm id, 0.5 μm)	Quadrupole (QP5000, Shimadzu)	Ac-D10, Phe-D10, Ch-D12 and Per-D12

(continued)

Table 3 (continued)

References	Food matrix	Parameter measured (PAHs and respective monitored ions)	Analytical method						
			Sample size	Extraction	Clean-up	Detection and quantification	Column	Detector details	Internal standard
Smoker et al. (2010)	Shrimp	Na (128 > 102, 78), Ap (152 > 126, 102), Ac (154 > 152, 127), F (166 > 115, 139), A (178 > 152, 176), Phe (178 > 152, 176), Fl (202 > 200, 150), Py (202 > 200, 151), BaA (228 > 226, 200), Ch (228 > 226, 200), BaP (252 > 250, 226), BbF (252 > 224, 250), BkF (252 > 250, 226), BghiP (276 > 274, 250), IP (276 > 274, 250) and DBahA (278 > 276, 250)	10 g	QuEChERS (10 mL acetonitrile, 6 g magnesium sulfate, 1.5 g sodium chloride, followed by centrifugation)	Dispersive SPE (PSA and magnesium sulfate)	LC–MS/MS	PAH C18 (150 mm × 4.6 mm i.d., 5 μm)	Triple quadrupole (PhotoSpray APPI source and 4000 QTrap, Applied BioSystems)	Not used
Belo et al. (2012)	Soy, sunflower and olive oil	BaA (226, 228), BbF (250, 252), BkF (250, 252), BaP (250, 252), Ch(226, 228), IP (277, 276), DBahA (279, 278) and BghiP(277, 276)	2 g	Low volume liquid–liquid partition (hexane and dimethylformamide)	SPE (octadecyl and silica gel cartridges)	GC–MS	DB-5MS (30 m × 0.25 mm i.d., 0.25 μm)	Single quadrupole (DSQ, Thermo)	Isotope labeled (^{13}C)
Cai et al. (2012)	Oysters	Na (128.1), Ap (152.1), Ac (154.1), F (166.1), Phe (178.1), A (178.1), Fl (202.1), Py (202.1), BaA (228.1), Ch (228.1), BbF (252.1), BkF (252.1), BaP (252.1), DBahA (278.1), BghiP (276.1) and IP (276.1)	10 g	QuEChERS (15 mL acetonitrile, 6 g magnesium sulfate, 1.5 g sodium chloride, followed by centrifugation)	Dispersive SPE (50 mg PSA and 150 mg magnesium sulfate)	UHPLC–MS	Zorbax Eclipse PAH column Rapid Resolution HT (50 mm × 2.1 mm i.d., 1.8 μm)	Single quadrupole (G6140A, with Syagen's PhotoMate® APPI® source, Agilent)	Not used

(continued)

Table 3 (continued)

References	Food matrix	Parameter measured (PAHs and respective monitored ions)	Analytical method						
			Sample size	Extraction	Clean-up	Detection and quantification	Column	Detector details	Internal standard
Drabova et al. (2012)	Vegetable oils and tea	15 (+1) EU priority PAHs	2 g	QuEChERS (10 mL ethyl acetate, 4 g magnesium sulfate, 2 g sodium chloride, followed by centrifugation)	MIP SPE-PAH (50 mg/3 mL)	GC × GC–TOFMS	BPX-50 (30 m × 0.25 mm i.d., 0.25 μm) BPX-5 (1 m × 0.1 mm i. d., 0.1 μm) SGE	Pegasus III (LECO Corporation)	Isotope labeled (^{13}C)
Londoño et al. (2013)	Milk	Ap, Ac, F, A, Phe, Fl, Py, BaA, Ch, BaP, BbF, BkF, BghiP, IP and DBahA	1 g	Low volume liquid–liquid partition (10 mL hexane three times)	SPE (silica gel cartridge, 690 mg, 55–105 μm particle size)	HPLC–UV/FL	PAH C18 (250 mm × 4.6 mm i.d., 5 μm)	UV–Vis (2698, Waters) Fluorescence (2475, Waters)	Not used
Payanan et al. (2013)	Edible oils	16 EPA priority PAHs	1 g	Low-temperature liquid–liquid partition (two extractions with 8 mL of acetonitrile: acetone 4:1, v/v and freezing for 24 h)	SPE (alumina cartridge)	HPLC–FL	PAH C18 (250 mm × 4.6 mm i.d., 5 μm)	–	Not used
So-Young et al. (2013)	Edible oils	15 EU priority PAHs	2 g	Extraction (8 mL isooctane:cyclohexae 1:1, v/v)	SPE (polystyrene divinylbenzene cartridges)	GC–MS/MS	VF-5 ms 5 % phenyl-95 % polydimethyl-siloxane (30 m × 0.25 mm i.d., 0.25 μm)	320-MS Bruker system (Bremen, Germany)	IP-D12 and Ch-D12
Taylor et al. (2013)	Fish and mussel	Na (127,128), Ap (151, 152), Ac (153, 154), F (165, 166), A (178, 176), Phe (178, 176), Fl (202, 201) Py (202, 201), BaA (228, 226), Ch (228, 226), BaP (252, 250), BbF (252, 250), BkF (252, 250), BghiP (276, 138), IP (276, 138), DBahA (278, 139) and 2-MeNa (142, 141)	10 g	QuEChERS (20 mL methylene chloride; 2 g magnesium sulfate)	No further clean-up	GC × GC–MS	BPX50 50 % phenyl polysilphenylene siloxane (30 m × 0.25 mm i.d., 0.50 μm). BPX1 100 % polydimethyl-siloxane (2 m × 0.1 mm i.d., 0.1 μm)	QP 2010 Ultra Shimadzu	F-D10 and Fl-D10

(continued)

Table 3 (continued)

References	Food matrix	Parameter measured (PAHs and respective monitored ions)	Analytical method						
			Sample size	Extraction	Clean-up	Detection and quantification	Column	Detector details	Internal standard
Pincemaille et al. (2014)	Tea	BaA (228, 228), Ch (228, 228), BbF (252, 126) and BaP (252, 126) (PAH4)	5 g	QuEChERS (acetonitrile:acetone 60:40, v/v, magnesium sulfate, sodium chloride, sodium citrate, followed by centrifugation)	SPE (octadecyl and Florisil® cartridges)	GC–MS/MS	HP-5MS (30 m × 0.25 mm i.d., 0.25 μm)	Triple Quadrupole (7000 A, Agilent)	BaP-D12
Pissinatti et al. (2014)	Coffee	Fl (202, 200) Py (202, 200), BaA (228, 226), Ch (228, 226), BaP (252, 250), BbF (252, 250), BkF (252, 250), BghiP (277, 276), IP (277, 276), DBahA (278, 279) and BghiP (277, 276)	10 g	PLE in cell clean-up (extraction solvent: hexane:dichloromethane 85:15, v/v and adsorbent: silica gel)	SPE (silica gel column)	GC–MS	DB-5MS (30 m × 0.25 mm i.d., 0.25 μm)	Ion Trap (Polaris Q, Thermo)	Isotope labeled (^{13}C)

1-MePy 1-mepyrene, *15 (+1) EU priority PAHs* cyclopenta(c,d)pyrene, benzo(a)anthracene, chrysene, 5-methylchrysene, benzo(b)fluoranthene, benzo(j)fluoranthene, benzo(k)fluoranthene, benzo(a)pyrene, indeno(1,2,3-c,d) pyrene, dibenzo(a,h)anthracene, benzo(g,h)perylene, dibenzo(a,l)pyrene, dibenzo(a,e)pyrene, dibenzo(a,i)pyrene, dibenzo(a,h)pyrene. *(+1)* benzo(c)fluorine; *16 EPA priority PAH* naphthalene, acenaphthylene, acenaphthene, fluorene, phenanthrene, anthracene, fluoranthene, pyrene, benz(a)anthracene, chrysene, benzo(b)fluoranthene, benzo(k)fluoranthene, benzo(a)pyrene, indeno(1,2,3-c,d)pyrene, dibenz(a,h)anthracene, benzo(g,h,i) perylene); *2,3,6-TriMeNa* 2,3,6-trimenaphthalene; *2,6-DiMeNa* 2,6-dimenaphthalene; *2-MeNa* 2-menaphtahlene; *2-MePhe* 2-mephenanthrene; *3,6-DiMePhe* 3,6-dimephenanthrene; *5-MeCh* 5-methylchrysene; *6-MeBaP* 6-mebenzo(a) pyrene; *A* anthracene; *Ac* acenaphthene; *An* anthanthrene; *Ap* acenaphthylene; *BaA* benzo(a)anthracene; *BaP* benzo(a)pyrene; *BbF* benzo(b)fluoranthene; *BcL* benzo(c)fluorine; *BeP* benzo(e)pyrene; *BghiF*: benzo(g,h,i) fluoranthene; BghiP: benzo(g,h,i)perylene; BjF: benzo(j)fluoranthene; BkF: benzo(k)fluoranthene; Ch: chrysene; Cl_2CH_2: dichloromethane; *COR* coronene; *CPP* cyclopenta(c,d)pyrene; *DACC* donor–acceptor complex chromatography; *DBaeP* dibenzo(a,e)pyrene; *DBahA* dibenzo(a,h)anthracene; *DBahP* dibenzo(a,h)pyrene; *DBaiP* dibenzo(a,i)pyrene; *DBalP* dibenzo(a,l)pyrene; *DeP* benzo(e)pireno; *F* fluorine; *Fl* fluoranthene; *GC–MS/MS* gas chromatography coupled to tandem mass spectrometry; *GC–MS* gas chromatography coupled to mass spectrometry; *GC × GC–MS* two-dimensional gas chromatography coupled to mass spectrometry; *GC × GC–TOFMS* two-dimensional gas chromatography coupled to time of flight mass spectrometry; *GPC* gel permeation chromatography; *HPLC–FLD* high-performance liquid chromatography coupled to fluorescence detector; *HPLC–UV/FLD* high-performance liquid chromatography coupled to ultraviolet/fluorescence detector; *IP* indene(1,2,3-c,d)pyrene; *MIP* molecularly imprinted polymer; *Na* naphthalene; *Py* pyrene; *Per* perylene; *Phe*: phenanthrene; *PLE* pressurized liquid extraction; *PSA* primary secondary amine; *QuEChERS* quick, easy, cheap, effective, rugged and safe; *SPE* solid-phase extraction; *TLC* thin-layer chromatography; *Tph* triphenylene

Some authors have adopted GPC in the clean-up step to remove lipids and high molecular weight compounds from food extracts (Ballesteros et al. 2006; Liguori et al. 2006). However, the prevailing technique is SPE, in which the solvent volume and the time spent are considerably lower than those needed for liquid–liquid extraction and which is amenable to automation (Tobiszewski et al. 2009). Houessou et al. (2005) evaluated the extraction efficiency of octadecyl (C18, 0.5 g) and polystyrene-divinylbenzene copolymer (PS-DVB, 0.2–0.5 g) cartridges for the determination of PAHs from coffee brew samples. The PS-DVB phase was also used by Veyrand et al. (2007) and So-Young et al. (2013) for food and edible oil matrices, respectively. This phase is selective for PAHs due to its lipophilicity and ability to form π-π bonds with the aromatic rings in PAHs. Although the results found by Houessou et al. (2005) indicated similar recovery values for both solid phases used, C18 and PS-DVB, better reproducibility was observed when the PS-DVB phase (0.5 g) was employed.

Silica gel have been the most widely used phase in the clean-up step by SPE to remove interfering molecules during the determination of PAHs from various food matrices, both in the form of packaged cartridges containing 0.5–2.0 g of silica gel (Windal et al. 2008; Belo et al. 2012; Londoño et al. 2013) and as columns prepared with up to 5 g of silica gel (Badolato et al. 2006; Rose et al. 2007; Stanciu et al. 2008; Danyi et al. 2009; Pissinatti et al. 2014). The alumina-N SPE cartridge showed satisfactory PAH recovery and highly efficient elimination of the lipid-interferences from the extracts of edible oils (Payanan et al. 2013).

The use of MIP SPE for the clean-up of PAH extracts from food samples has also been reported, including for vegetable oils and tea samples. SPE cartridges containing molecular recognition elements engineered to bind structurally related PAHs with high selectivity are commercially available. It is assumed that multiple interactions (such as hydrogen bonding, ionic, van der Waals and hydrophobic forces) take place between the measurands and MIP cavities and are responsible for their binding. Operatively, the MIP SPE technique is very similar to the traditional SPE performed on non-specific supports, and the usual steps of column conditioning, sample loading, column washing and measurands elution are carried out (Drabova et al. 2012).

Another variation of SPE, known as dispersive SPE (which is commonly used as part of QuEChERS extraction), has been reported for the determination of PAHs from matrices such as shrimp, oysters, fish and mussel (Smoker et al. 2010; Cai et al. 2012; Taylor et al. 2013). The procedure of dispersive SPE clean-up is simple: the sample extract and the sorbent are stirred together and centrifuged, and the supernatant is collected. This procedure results in better recoveries and more effective clean-up that removes the interferences and generates less waste due to the lower volume of organic solvent employed in the sample preparation when compared to traditional cartridge SPE (Oshita and Jardim 2015).

SPME is a rapid, simple and low solvent consuming technique widely used for environmental pollutants, including PAHs. Extraction is performed into a silica fiber coated with an appropriate stationary phase and can be carried out immersing the fiber in the liquid sample (direct immersion) or from the sample headspace

(Aguinaga et al. 2007; Purcaro et al. 2007a, b). For vegetable oils, Purcaro et al. (2007b) concluded that the use of GC $\times$ GC was necessary in order to achieve a better resolution between interferents and PAHs of interest, while GC–MS, using the deuterated congerene as internal standard, provided enough selectivity when only benzo(a)pyrene was considered (Purcaro et al. 2007a). SPME followed by a GC–MS analysis was successfully applied for quantification of the 16 EPA priority PAHs in milk products (Aguinaga et al. 2007).

Donor–acceptor complex chromatography (DACC) was first developed for the analysis of PAHs in oil and fat samples (Windal et al. 2008). The samples are eluted using a column with a modified stationary phase that acts as an electron acceptor. This column retains electron donors, including PAHs, by π–π interactions and is used for sample clean-up (ISO 2012). Commercial instruments have been designed for the on-line coupling of a DACC column to an HPLC reversed-phase column and fluorescence detector, allowing for the automated clean-up, separation and quantification of PAHs (Windal et al. 2008). This technique can be used for fats and oils without any other clean-up steps (ISO 2012) or may be applied to dirty matrices, such as dried plants, if an additional clean-up is performed (Windal et al. 2008).

PLE is a type of extraction that uses conventional organic solvents at high temperature (100–180°C) and pressure (6.9×10^6 to 1.0×10^7 Pa) to extract organic components from solid or semi-solid samples. It was first used in 1995, and in 1996, this technique was approved by the EPA as a standard method for the analysis of organic contaminants, such as pesticides, chlorinated herbicides, polychlorinated biphenyls (PCBs) and dioxins, in environmental matrices (Mitra 2003). The elevated temperature and pressure affect the solvent, the sample and the interactions between them. For example, the boiling point of the solvent is increased by the high pressure, which enables the extraction to occur at higher temperatures, resulting in lower viscosity and lower surface tension. Additionally, high temperatures increase the solubility of the compounds by weakening the interactions, such as the Van der Waals forces and hydrogen bonds, thus allowing faster mass transfer. High pressure also allows the solvent to penetrate deeper into the matrix. Thus, using high pressure and temperature assures faster extraction and better recoveries (Mitra 2003). On average, one to three extraction cycles of up to 15 min each is sufficient. Using this technique, the solvent volume and extraction time are reduced compared to traditional extraction procedures.

Many authors have demonstrated the efficiency of the use of PLE for PAH analysis. Wang et al. (2007) compared PLE, Soxhlet and solvent extraction assisted by microwave (MAE) for PAHs extraction from soil samples. According to these authors, PLE showed higher extraction efficiency, in addition to the advantages described above.

A further advantage of PLE is that some instruments allow a simultaneous purification step to be performed by adding adsorbents to the extraction cell. The most commonly used adsorbents are fat retainers, such as alumina, silica or Florisil® (Liguori et al. 2006; Veyrand et al. 2007; Lund et al. 2009; Pissinatti et al. 2014). For low complexity matrices, it is possible to carry out a one-step extraction/purification method (Kim et al. 2003), but because food matrices are complex, most authors have used pre-purification within the PLE cell to simplify later clean-up steps.

Recently, QuEChERS extraction, which has been widely applied for the analysis of pesticides, has been used to extract PAHs from food matrix (Smoker et al. 2010; Cai et al. 2012; Taylor et al. 2013; Pincemaille et al. 2014). Briefly, in this technique, samples are extracted with acetonitrile and inorganic salts (e.g., magnesium sulfate and sodium chloride). After centrifugation, an aliquot of the acetonitrile layer is filtered or cleaned up in dispersive mode. An acetonitrile QuEChERS extraction with PSA (clean-up sorbent that contains primary and secondary amines) and magnesium sulfate clean-up was used by Smoker et al. (2010) and Cai et al. (2012) for shrimp and seafood samples, respectively. The authors observed that spiked samples showed ion suppression and lower recovery rates without PSA clean-up. Pincemaille et al. (2014) used two different SPE purification steps after the initial QuEChERS extraction from tea samples: a first step on a C18-support to extract non-polar compounds and a second step on a polar support (Florisil®) to trap polar impurities. This technique represents an advance in determining PAHs mainly due to environmental friendly and high throughput characteristics.

5.2 *Determination and Quantification*

Many authors have reported the use of HPLC with a UV/FL detector as the detection and quantification technique for the analysis of PAHs from food (Houessou et al. 2005; Badolato et al. 2006; Stanciu et al. 2008; Windal et al. 2008; Danyi et al. 2009; Londoño et al. 2013; Payanan et al. 2013). Liquid chromatography coupled to MS has also been applied (Smoker et al. 2010; Cai et al. 2012). However, CG is the most widely used technique for these analytical scopes (Diletti et al. 2005; Ballesteros et al. 2006; Liguori et al. 2006; Aguinaga et al. 2007; Purcaro et al. 2007a, b; Rose et al. 2007; Veyrand et al. 2007; Lund et al. 2009; Orecchio et al. 2009; Belo et al. 2012; Drabova et al. 2012; So-Young et al. 2013; Taylor et al. 2013; Pincemaille et al. 2014; Pissinatti et al. 2014). According to Poster et al. (2006), for these compounds, GC has better selectivity, resolution and sensitivity compared to liquid chromatography. These authors also report the ease of coupling GC with MS, allowing for the confirmation and quantification of the compounds of interest and indicating that GC is preferable to LC. The thermal properties of PAHs are an important factor: PAHs are readily volatile but are not degraded by the higher temperatures used in GC. Typical chromatograms obtained for PAHs in food matrices by GC coupled to MS are illustrated in Fig. 1.

A study of the mass spectra of PAHs was carried out by Veyrand et al. (2007), who aimed to develop a method for the analysis of contaminants in food. According to these authors, the minimal fragmentation observed when using electron ionization under conventional conditions is attributable to the stability of PAH molecules, which results in a highly intense molecular ion signal and the presence of ions with

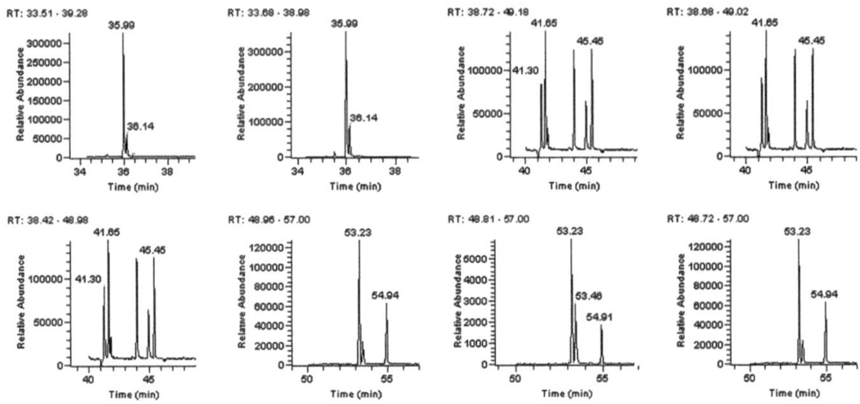

Fig. 1 Typical chromatograms obtained for eight polycyclic aromatic hydrocarbons from food matrices (roasted coffee and fish samples) by gas chromatography coupled to mass spectrometry. Monitored ions (m/z): BaA and Ch (226–228); BbF, BkF and BaP (250–252); IP and BghiP (276–277); and DBahA (278–279). RT: retention time. Roasted coffee: BaA 19.66 min; Ch 19.78 min; BbF 23.58 min; BkF 23.67 min; BaP 24.65 min; IP 28.29 min; DBahA 28.47 min; and BghiP 29.22 min. Fish: BaA 35.99 min; Ch 36.14 min; BbF 41.30 min; BkF 41.65 min; BaP 45.45 min; IP 53.23 min; DBahA 53.46 min and BghiP 54.94 min. BaA: benzo(a)anthracene; BaP: benzo(a)pyrene; BbF: benzo(b)fluoranthene; BghiP: benzo(g,h,i)perylene; BkF: benzo(k)fluoranthene; Ch: chrysene; DBahA: dibenzo(a,h)anthracene; and IP: indene(1,2,3-c,d)pyrene

the loss of two hydrogens. The same observation was reported by Poster et al. (2006), who concluded that this stability makes PAHs amenable to GC. However, this characteristic makes the use of tandem MS, in which the precursor ion is fragmented into specific product ions, problematic. As a result, very few works have described the use of MS/MS for PAH analysis (Ballesteros et al. 2006; Veyrand et al. 2007; Smoker et al. 2010).

HPLC coupled to MS is more commonly used for environmental samples than food samples. The higher detection limits needed for environmental samples can be easily achieved using this technique. The most commonly used ion sources in LC–MS analysis are electro spray ionization (ESI) and atmospheric-pressure chemical ionization (APCI). However, such sources are inefficient for non-polar compound ionization. Post-run chemical derivatization was proposed by several authors to overcome this problem, but this technique may result in deposits in the instrument and create the need for more frequent maintenance. The atmospheric-pressure photoionization (APPI) ion source extends the range of ionizable compounds to many non-polar substances. Furthermore, this type of interface shows less ion suppression than APCI or ESI. Additionally, to further increase the ionization efficiency, a dopant (e.g., acetone or toluene) can be used (Purcaro et al. 2013). Thus, the low sensitivity of HPLC–MS has only recently been overcome (Cai et al. 2012).

5.3 Use of Internal Standards

The isotope dilution mass spectrometry (IDMS) technique was first developed during the 1950s for analyzing inorganic elements. In 1970, it was extended to the field of organic chemistry, with applications in trace analysis to determine persistent organic pollutants and in medical tests (Sargent et al. 2002; Mechlinska et al. 2010). ID consists of modifying the natural isotopic composition of a target measurand present in the sample by adding a known amount of an isotopically labeled analog (internal standard). In MS, unlike spectrophotometric techniques, there is a fixed relationship between the quantity or concentration of a particular substance and the instrument response. The sensitivity for a given compound may vary over time or in accordance with the calibration of the equipment. These variations are added to variations caused by, for example, losses during extraction or the introduction of the analytical sample into the chromatographic system. Adding the internal standard at the beginning of the analytical procedure allows for compensation of losses and errors throughout the analytical process (Sargent et al. 2002).

The most important criterion for selecting a standard substance is that it mimics as closely as possible the physico-chemical properties of the target analyte. This is achieved by using similar molecules that isotopically labeled, especially with ^{13}C, ^{37}Cl or ^{2}H. Because the amount of internal standard added to the sample is known, the recovery percentage can be calculated and used as an indirect measurement of the target compound recovery (Mechlinska et al. 2010).

IDMS is often used for the analysis of PAHs to overcome systematic errors in analysis of PAH. In fact, almost all of the work published using MS have used this technique (Diletti et al. 2005; Liguori et al. 2006; Rose et al. 2007; Veyrand et al. 2007; Danyi et al. 2009; Lund et al. 2009; Orecchio et al. 2009; Belo et al. 2012 Drabova et al. 2012; So-Young et al. 2013; Taylor et al. 2013, Pincemaille et al. 2014; Pissinatti et al. 2014). Thus, because extraction methods are usually time consuming and laborious and because the volatility of the compounds can result in loss during the extraction procedure, the use of isotope dilution is almost a necessity for this type of analysis, allowing for acceptable accuracy, even at low concentrations (μg/kg).

^{13}C-labeled PAHs are preferred in MS methods due to their relatively high stability compared to deuterated species (Rose et al. 2007; Veyrand et al. 2007; Belo et al. 2012; Drabova et al. 2012; Pissinatti et al. 2014), but the high cost of these PAHS limits their use. Although a number of ^{13}C-labeled PAH standards are commercially available, there is still a need for ^{13}C-labeled analogues for various PAHs, such as benzo(j)fluoranthene, dibenzo(a,h)pyrene, dibenzo(a,l)pyrene cyclopenta(c,d)pyrene, 5-methylchrysene and benzo(c)fluorine (EFSA 2008a, b). The lack of ^{13}C-labeled standards may partly explain the high variation in the results reported for some PAHs in foods (Rose et al. 2007).

5.4 Validation Practices

Despite the significant number of works dedicated to the development and application of new technologies and equipment for the determination of PAHs in food samples, contemporary analytical techniques remain limited in their ability to obtain reliable information regarding the levels and forms of PAHs in the environment. This has caused researchers to alter their approaches to the determination of these compounds, which are present at low concentrations in highly variable and complex matrices, with regard to control issues and quality assurance of the results (Konieczka et al. 2010). In the context of laboratory quality management systems, specifically regarding the quality assurance of the analytical results, method validation is an extremely important tool.

Method validation is defined as the confirmation, through the provision of objective evidence that the requirements for a specific intended use or application have been fulfilled (ISO 2005). Inherent in this definition is the need to evaluate the performance characteristics or parameters of the method, although judgement of the suitability of the method is also important (Magnusson and Ornemark 2014).

The performance parameters proposed by EURACHEM (Magnusson and Ornemark 2014) and harmonized by the Association of Official Analytical Chemists (AOAC International), International Organization for Standardization (ISO) and International Union of Pure and Applied Chemistry (IUPAC) (Thompson et al. 2002) in their respective guidelines for single-laboratory validation include the following: selectivity, linearity (tests for general matrix effects), trueness, precision (under repeatability and intermediate precision conditions), recovery, range, detection limit, limit of quantification, sensitivity, ruggedness and measurement uncertainty.

In this context, the European Commission established analytical performance criteria for the determination of the regulated PAHs (benzo(a)pyrene, benz(a)anthracene, benzo(b)fluoranthene and chrysene) in food. Fully validated methods (i.e., methods validated by collaborative trials) were recommended when appropriate and available. Alternatively, in-house validated methods were indicated, provided that they fulfill some performance criteria: applicability (demonstrated using the regulated food matrices), selectivity (free from matrix or spectral interferences, verification of positive detection), recovery (between 50 and 120 %), relative standard deviation under repeatability and reproducibility conditions (HORRAT less than 2, HORRAT being the observed relative standard deviation divided by the value estimated from the modified Horwitz equation), limit of detection (≤ 0.30 µg/kg for each of the four substances) and limit of quantification (≤ 0.90 µg/kg for each of the four substances). Additionally, the in-house validated methods must produce results with a combined standard measurement uncertainty less than the established maximum standard measurement uncertainty (EU 2011a).

In Table 4, the validation procedures reported in the literature for methods used for the determination of PAHs in food are presented. This review showed that the current practices for validation do not address all of the parameters needed to assess

Table 4 Review of current validation procedures practiced for the determination of PAHs in foods

References	Method validation								
	Linearity/range/sensitivity	Selectivity/matrix effects	Recovery	Trueness	Precision	Detection limit	Limit of quantification	Measurement uncertainty	Ruggedness
Diletti et al. (2005)	Estimation of regression parameters, the respective deviations and R^2 from 0.9942–0.9999 (0.1–10 μg/mL, n = 8)	Ion relative abundances (spiked samples 2ug/kg and calibration solutions) No significant peaks interfering with the compound (blank samples and CRM blank of coconut oil, n = 5)	69–97.5 % (spiked samples at 2, 5 and 20 μg/kg, n = 6)	97.0–110.3 % (CRM BCR 458, coconut oil, for 4 PAHs, n = 5)	RSD repeatability from 3.6 to 15.5 % (spiked samples at 2, 5 and 20 μg/kg, n = 6) Intermediate precision not studied	0.2–0.4 μg/kg (signal/noise ≥ 3)	Not studied	8.5–11.4 %	Not studied
Houessou et al. (2005)	Estimation of regression parameters and R^2 from 0.991 to 0.9993(0.4–50 μg/mL, n = 12, 12 concentration levels)	Analysis of blank samples Influence of coffee brew matrix in recovery by comparison with water	81.5–105.2 % (spiked samples at 2 μg/kg, n = 3)	Not studied	RSD repeatability from 5.1 to 17.3 % (spiked samples at 2 μg/kg, n = 3) Intermediate precision not studied	0.19–2.49 μg/kg (analytical) and from 0.76 to 9.96 μg/kg (method) (signal/noise = 3 based on the analysis of blanks, n = 7)	0.63–1.11 μg/kg (analytical) and from 2.52 to 33.2 μg/kg (method) signal/noise = 3 based on the analysis of blanks, n = 7)	Not studied	Not studied
Badolato et al. (2006)	Estimation of regression parameters and R^2 of 0.9993 (0.1×10^{-3} to 0.1 μg/mL)	Not studied	76–116 % (spiked samples at 1.01, 2.02 and 3.03 μg/kg, n = 3)	Not studied	RSD repeatability from 12 to 18 % (spiked samples at 1.01, 2.02 and 3.03 μg/kg, n = 3) Intermediate precision not studied	0.03 μg/kg (signal/noise = 3 based on the analysis of blanks, n = 6)	0.10 μg/kg (signal/noise = 10 based on the analysis of blanks, n = 6)	Not studied	Not studied
Ballesteros et al. (2006)	R^2 from 0.9922 to 0.9980 (0.3–200 μg/kg, matrix-matched calibration curve)	Ion analysis	84–110 % (spiked samples at 1, 2 and 10 μg/kg, n = 3).	Not studied	RSD repeatability from 5.11 to 7.76 % (1 μg/kg, n = 11)	0.05–0.07 μg/kg (signal/noise ≥ 3)	0.10–0.20 μg/kg (signal/noise ≥ 3)	Not studied	Not studied

(continued)

Table 4 (continued)

References	Method validation								
	Linearity/range/sensitivity	Selectivity/matrix effects	Recovery	Trueness	Precision	Detection limit	Limit of quantification	Measurement uncertainty	Ruggedness
					Intermediate precision not studied				
Liguori et al. (2006)	R^2 of 0.999 (six concentration points between 10 and 600 ng/mL)	Compound signals in the presence of potential components of the sample were compared with the responses of standard solution	75–121 % (salmon fillet, fish feed and fish oil spiked at 60 and 100 ng, n = 3 and blue mussels spiked at 100 ng n = 5)	CRM NIST SRM 2977 mussel tissue and T0621 olive oil, for four PAHs, n = 3	RSD repeatability from 0.2 to 19 % (salmon fillet, fish feed and fish oil spiked at 60 and 100 ng, n = 3 and blue mussels spiked at 100 ng, n = 5) Intermediate precision not studied	0.42–3.4 pg/g (wet weight) for salmon fillet, 0.36–2.9 pg/g for fish feed, 2.2–18 pg/g for fish oil and 0.11–0.88 pg/g (dry mass) for blue mussels (signal/noise = 3)	1.4–11 pg/g (wet weight) for salmon fillet, 1.2–9.8 pg/g for fish feed, 7.2–30 pg/g for fish oil and 0.36–2.9 pg/g (dry mass) for blue mussels (signal/noise = 10)	Not studied	Not studied
Aguinaga et al. (2007)	Estimation of regression parameters and R^2 from 0.9956 and 0.9999 (0.01–100 μg/L, 5 concentration levels, standard addition calibration curves)	Significant matrix effects (comparison of the slope of the calibration curve in solvent with the slope of the standard addition calibration curves by t test)	87.6–112 % (spiked samples of 4 matrices at 0.6–3.2 mg/L)	Not studied	RSD repeatability from 2 to 19.6 % (spiked sample at five times the quantitation limit of each compound, n = 8) Intermediate precision not studied	0.003–1.56 μg/L (signal/noise = 3)	Not studied	Not studied	Not studied

(continued)

Table 4 (continued)

References	Method validation								
	Linearity/range/sensitivity	Selectivity/matrix effects	Recovery	Trueness	Precision	Detection limit	Limit of quantification	Measurement uncertainty	Ruggedness
Purcaro et al. (2007a)	Estimation of regression parameters and R^2 of 0.999 Linearity tested by Mandel's fitting test and evaluation of regression assumptions, including homoscedasticity and lack of fit test (0.46–15.79 μg/kg, 7 concentration levels, n = 3)	Not studied	+6 to −2 % (spiked samples at 0.4 and 2.0 ng)	Not studied	RSD repeatability from 1.6 to 5.3 % RSD intermediary precision from 1.93 to 5.76 % (spiked samples at 0.4 and 2.0 ng)	0.03 ng corresponding to 0.17 μg/kg (average signal of the blank sample plus 2 times the blank standard deviation and the constant of the t Student distribution, n = 8)	0.09 ng corresponding to 0.46 μg/kg (average signal of the blank sample plus 10 times the blank standard deviation and the constant of the t-Student distribution, n = 8)	Not studied	Not studied
Purcaro et al. (2007b)	Estimation of regression parameters and R^2 from 0.957 to 0.998 (1–36.8 μg/kg	Not studied	+20.9 to −69.4 (spiked sample at 2 levels, n = 6)	Not studied	RSD repeatability from 12.8 to 34.5 % (one sample, n = 6)	0.1–1.4 μg/kg (signal/noise = 3)	0.4–4.6 μg/kg (signal/noise = 10)	Not studied	Not studied
Rose et al. (2007)	Weighted least-squares linear regression	Use of ^{13}C-isotopes in every sample aiming to correct for any variation due to matrix effects	53 a 128 % (spiked samples at 1, 2, 10 and 50 μg/kg, measured in four analytical runs)	CRM BCR 458, coconut oil, for 6 PAHs, n = 120	Repeatability not studied RSD intermediate precision from 0.02 to 0.48 % (unfortified, 1, 2, 10 and 50 μg/kg, measured in 4 analytical runs)	Less than 2 μg/kg (standard uncertainty at zero concentration multiplied by 3)	Lowest concentration that could be measured with a sufficiently low relative uncertainty	Less than 0.2 for 24 PAHs	Not studied
Veyrand et al. (2007)	Estimation of regression parameters R^2 from 0.9986 to 1 (0–4 μg/kg of dry matter)	Ion analysis Matrix effects not studied	12–70 % (same sample analyzed 20 times by two operators)	CRM NIST SRM 2977 mussel tissue for 4 PAHs Participation in an	Repeatability not studied RSD intermediate precision from 2.9 to 20.5 % (same sample	0.08–0.15 μg/kg (signal observed at the lowest point of the calibration curve, extrapolation at signal/noise = 3)	0.25–0.915 μg/kg (signal observed at the lowest point of the calibration curve, extrapolation at signal/noise = 10)	Not studied	Analysis of different types of food

(continued)

Table 4 (continued)

References	Method validation								
	Linearity/range/sensitivity	Selectivity/matrix effects	Recovery	Trueness	Precision	Detection limit	Limit of quantification	Measurement uncertainty	Ruggedness
				inter-laboratory assay	analyzed 20 times by two operators)				
Stanciu et al. (2008)	R^2 0.9946–0.9993 (0.2–10 μg/L)	Not studied	83–105 % (spiked sample at 2 μg/kg)	Not studied	Not studied	0.02–0.04 ng/kg (signal/noise = 3, based on analysis of blank samples, n = 7)	0.2 ng/kg (signal/noise = 3, based on analysis of blank samples, n = 7)	Not studied	Not studied
Windal et al. (2008)	Linearity tested by Mandel's fitting test (5.2, 32, 66, 90 pg/L for the CPP and 0.2, 2, 6, 10 pg/L for the other PAH except for BjF for which the first level was 0.5 pg/L)	No significant matrix effects (comparison of the slope of the calibration curve in acetonitrile with the slope of the matrix-matched calibration curve by t test)	60 and 115 % (spiked samples at 2 μg/kg for all PAHs except BjF and IcP 10 μg/kg and CPP 200 μg/kg, n = 6)	Participation in an inter-laboratory assay	RSD repeatability from 2.3 to 19 % (spiked samples at 2 μg/kg for all PAHs except BjF and IcP 10 μg/kg and CPP 200 μg/kg, n = 6 on the same day) RSD intermediate precision from 1.3 to 13 % (spiked samples at 2 μg/kg for all PAHs except BjF and IcP 10 μg/kg and CPP 200 μg/kg, n = 6, analysis of one sample on different days)	0.3–5.8 pg (54 pg for CPP) (instrumental—standard solutions in acetonitrile., signal/noise = 3) 0.07–0.75 μg/kg (7.8 pg for CPP) (method—spiked matrices	0.6–12 pg (104 pg for CPP) (instrumental—standard solutions in acetonitrile., signal/noise = 3) 0.13–1. 5 μg/kg (16 pg for CPP) (method—spiked matrices	Not studied	Not studied

(continued)

Table 4 (continued)

References	Method validation								
	Linearity/range/sensitivity	Selectivity/matrix effects	Recovery	Trueness	Precision	Detection limit	Limit of quantification	Measurement uncertainty	Ruggedness
Danyi et al. (2009)	R^2 from 0.98 to 1.00	Not studied	63–118 % (spiked samples at three levels, corresponding to the LOQ, 2 and 4 μg kg, except for BjF, IP and CPP)	Not studied	Repeatability HORRAT from 0 to 1.2 (spiked samples at 2 μg kg except for BjF, IP and CPP, n = 6, on the same day) Intermediate precision HORRAT from 0 to 0.4 (spiked samples at 2 μgkg except for BjF, IP and CPP, n = 6, one time during 6 independent days)	<0.1 μg/kg (extrapolation at signal/noise = 3 measured in the chromatogram of unspiked sample)	0.2–0.3 μg/kg except for CPP, IP, BjF and BcL (extrapolation at signal/noise = 6 measured in the chromatogram of unspiked sample)	Not studied	Not studied
Lund et al. (2009)	R^2 from 0.995 to 1.000	Not studied	53–108 %	Not studied	RSD repeatability from 3 to 87 % Intermediate precision not studied	0.2–4.4 ng/g	0.7–14.6 ng/g	Not studied	Fat-retention capacity (5 g fish homogenate n = 6, between 5 and 30 % fat, analyzed in random order in two runs during 2 days)

(continued)

Table 4 (continued)

References	Method validation								
	Linearity/range/sensitivity	Selectivity/matrix effects	Recovery	Trueness	Precision	Detection limit	Limit of quantification	Measurement uncertainty	Ruggedness
Orecchio et al. 2009	Five standard solutions	Not studied	≥ 85 %	Not studied	RSD repeatability from 4 to 16 % (3 replicates of 13 sample) Intermediate precision not studied	0.67–18 ng/L	<1 µg/kg	Not studied	Not studied
Smoker et al. (2010)	R^2 from 0.996 to 1.000 (0.2, 0.5, 1, 2, and 10 µg/mL)	Ion analysis Matrix effects not studied	49–129 % without PSA and 74–122 % with PSA (spiked samples at 0.2 and 1.0 µg/g)	Not studied	Not studied	0.02–0.42 µg/g without PSA and from 0.02 to 0.51 µg/g with PSA (signal/noise = script of Analyst software)	Not studied	Not studied	Not studied
Belo et al. (2012)	Weighted least-squares linear regression and R^2 from 0.9845 to 0.9896 (0, 0.75, 1, 2, 3 and 4 µg kg, matrix matched calibration curve, n = 6)	Ion analysis Significant matrix effects (slopes of matrix-matched calibration curves were compared with slopes of pure solvent curves by t test)	54.01–114.69 % (spiked samples at 0.75, 2.00 and 4.0 µg/kg, n = 18)	Participation in inter-laboratory assay	RSD repeatability from 4.58 to 54.60 % and HORRAT from 0.31 to 3.72 RSD intermediate precision from 4.04 to 44.48 % and HORRAT from 0.18 to 2.02 (spiked samples at 0.75, 2.00 and 4.0 µg/kg, n = 18, analyzed	0.04–0.48 µg/kg (three times the standard deviation of the mean of blank oil sample, n > 20)	0.12–1.34 µg/kg (theoretical—ten times the standard deviation of the mean of blank oil sample, n > 20) The lowest concentration level of the calibration curve (practical)	14.5–92.3 % (top-down approach, considering the uncertainty of the calibration curve and intermediate precision)	Not studied

(continued)

Table 4 (continued)

References	Method validation								
	Linearity/range/sensitivity	Selectivity/matrix effects	Recovery	Trueness	Precision	Detection limit	Limit of quantification	Measurement uncertainty	Ruggedness
					on 3 different days, by 2 different analysts)				
Cai et al. (2012)	Estimation of regression parameters and R^2 from 0.995 to 1.000 (1.5–400 pg, 18 concentration levels, n = 3)	Ion analysis Definition of a gradient elution program in order to separate isobaric mass matrix interference peak responses from target measurands	77–110 % (spiked samples at 0.1–2 ppm level 1 and at 0.5–10 ppm level 2, n = 3)	Not studied	RSD repeatability from 0.3 to 6.7 % (spiked samples at 0.1–2 ppm level 1 and at 0.5–10 ppm level 2, n = 3) Intermediate precision not studied	8–105.6 pg (instrument) and from 0.013 to 0.129 ppm (method) (six times the standard deviation of manually selected background signal/noise = 3 based on the analysis of low level calibration standards, n = 3)	Not studied	Not studied	Not studied
Drabova et al. (2012)	Estimation of regression parameters and R^2 from 0.9963 to 0.9997 (0.05–100 ng/mL)	Ion analysis (use of GC × GC for separation of otherwise well-known critical groups of PAHs) Matrix effects not studied	73–103 % (spiked samples at 0.5, 2.5 and 5 μg/kg, n = 6)	Comparison with the routinely used method on the naturally contaminated sample	RSD repeatability from 2 to 12 % (spiked samples at 0.5, 2.5 and 5 μg/kg, n = 6) Intermediate precision not studied	Not studied	0.05–0.2 μg/kg	Not studied	Not studied
Londoño et al. (2013)	A least three orders of magnitude had $R^2 > 0.991$	No significant peaks interfering with the measurands (blank samples)	79–105 % (spiked samples at 0.1–53 μg/kg, n = 3)	Verified through a reference solution	RSD repeatability from 1.2 to 9.4 % (spiked samples at 0.1–53 μg/kg, n = 3)	0.0004–4.9 μg/kg (signal/noise = 3 based on the analysis of blanks, n = 6)	0.0014–16.4 μg/kg (signal/noise = 10 based on the analysis of blanks, n = 6)	Not studied	Not studied

(continued)

Table 4 (continued)

References	Method validation								
	Linearity/range/sensitivity	Selectivity/matrix effects	Recovery	Trueness	Precision	Detection limit	Limit of quantification	Measurement uncertainty	Ruggedness
					Intermediate precision not studied				
Payanan et al. (2013)	R^2 from 0.9739 to 0.9989 (spiked samples at 0.25–150.0 μg/kg)	Not studied	45.9–118.5 % (spiked samples at 0.5, 1.0, 2.0 and 6.0 μg/kg, n = 10)	RM FAPAS T0631, olive oil, for 5 PAHs	RSD repeatability from 0.1 to 6.2 % (spiked samples at 0.5, 1.0, 2.0 and 6.0 μg/kg, n = 10) Intermediate precision not studied	0.13–3.13 μg/kg (n = 10)	0.25–6.25 μg/kg (n = 10)	Not studied	Not studied
So-Young et al. (2013)	R^2 from 0.9763 to 0.9999 (0.5, 1, 5, 10, and 20 μg/kg)	Ion analysis Matrix effects not studied	55.1–105 % (spiked samples at 100 μg/kg, n = 3)	96–97.4 % (CRM BCR 458, coconut oil, for 4 PAHs, n = 5)	RSD repeatability from 0.8 to 7.5 % (spiked samples at 100 μg/kg, n = 3) Intermediate precision not studied	0.01–0.06 μg/kg (ratio between the standard deviation of the response and the slope of the calibration curve, multiplied by 3)	0.03–0.17 μg/kg (ratio between the standard deviation of the response and the slope of the calibration curve, multiplied by 10)	Not studied	Not studied
Taylor et al. (2013)	Not studied	Not studied	Not studied	CRM NIST SRM 2947, mussel tissue, for 8 PAHs Comparison with a standard	Not studied	Not studied	Not studied	Not studied	Not studied

(continued)

Table 4 (continued)

References	Method validation								
	Linearity/range/sensitivity	Selectivity/matrix effects	Recovery	Trueness	Precision	Detection limit	Limit of quantification	Measurement uncertainty	Ruggedness
				HPLC–FLD method					
Pincemaille et al. (2014)	R^2 from 0.9811 to 0.9991 (0, 0.1, 0.5, 1, 5 and 10 μg/kg for tea leaves and μg/L for infusions, matrix-matched calibration curve)	No significant peaks interfering with the measurands (blank samples, n = 5) Significant matrix effects (slopes of matrix-matched calibration curves were compared with slopes of non-matrix-matched calibration curves)	53–88 % (spiked samples with 1 μg/kg, n = 5)	Not studied	Repeatability HORRAT from 0.2 to 1.0 Reproducibility HORRAT from 0.2 to 0.6	0.1–0.3 μg/kg or L (lowest concentration where both the quantifying and the qualifying transition presented a signal/noise = 3)	0.2–0.6 μg/kg or L (lowest concentration where both the quantifying and the qualifying transition presented a signal/noise = 10)	Not studied	Not studied
Pissinatti et al. (2014)	Estimation of regression parameters, the respective deviations, R^2 from 0.996 to 1.000 and evaluation of regression assumptions, including lack of fit test (50–800 pg/μL corresponding to 0.25–4.00 μg/kg, 6 concentration levels, n = 3)	Ions analysis No significant peaks interfering with the measurands (blank samples and CRM blank of coconut oil, n = 18) No significant matrix effects (comparison of the slope of the calibration curve in solvent with the slope of the matrix-matched calibration curve by t test)	87.08–111.28 % (samples spiked at 0.25, 1.00, and 3.00 μg/kg, n = 18)	Not studied	RSD repeatability from 3.26 to 23.75 % and HORRAT 0.22–1.62 RSD intermediate precision 3.29–33.34 % and HORRAT 0.15–1.52 (samples spiked at 0.25, 1.00, and 3.00 μg/kg, n = 18, analyzed on 3 different days, by	0.03–0.18 μg/kg (instrument) (signal/noise = 3 based on the analysis of reagent blanks, n = 21)	0.11–0.59 μg/kg (instrumental) (signal/noise = 10 based on the analysis of reagent blanks, n = 21) 0.22–4.5 μg/kg (theoretical) (10 times the standard deviation of the response obtained for unspiked samples, n = 18) 0.26–3.09 (method) (lower concentration levels for which acceptable	11.85–66.06 % (top-down approach, considering the uncertainty of the calibration curve and intermediate precision)	Not studied

(continued)

Table 4 (continued)

References	Method validation								
	Linearity/range/sensitivity	Selectivity/matrix effects	Recovery	Trueness	Precision	Detection limit	Limit of quantification	Measurement uncertainty	Ruggedness
					different analysts, employing different equipment and brands/batches of reagents)		recovery and precision were observed, n = 18)		

BcL benzo(c)fluorine; *BjF* benzo(j)fluoranthene; *CPP* cyclopenta(c,d)pyrene; *CRM* certified reference material; *GC × GC* two-dimensional gas chromatography; *HORRAT* the observed relative standard deviation divided by the value estimated from the modified Horwitz equation; *HPLC–FLD* high-performance liquid chromatography coupled to fluorescence detector; *IP* indene(1,2,3-c,d)pyrene; *n* number of observations; *PSA* cleanup sorbent that contains primary and secondary amines, used here in dispersive mode; *RM* reference material; *RSD* relative standard deviation, R^2 coefficient of determination the fitness of the method for the stated purpose. Even for the evaluated parameters, experimental design and data analysis often did not follow the guidelines recommendations, as well as the regulated minimum performance criteria.

Although the use of the coefficient of determination as a test for linearity has been strongly discouraged (Thompson et al. 2002), this approach prevails in validation methods for PAHs in food (Diletti et al. 2005; Houessou et al. 2005; Badolato et al. 2006; Ballesteros et al. 2006; Liguori et al. 2006; Aguinaga et al. 2007; Purcaro et al. 2007b; Veyrand et al. 2007; Stanciu et al. 2008; Danyi et al. 2009; Lund et al. 2009; Orecchio et al. 2009; Cai et al. 2012; Drabova et al. 2012; Londoño et al. 2013; Payanan et al. 2013; So-Young et al. 2013; Pincemaille et al. 2014). A proper assessment of lack of fit and the verification of the regression assumptions (Souza and Junqueira 2005) have not been performed by most authors. Information about the experimental design revealed that several important aspects, such as the number of concentrations levels, evenly spaced distributions over the concentration range of interest and truly random replicates, have not been considered.

One relevant parameter that is frequently neglected is the matrix effects (Badolato et al. 2006; Ballesteros et al. 2006; Purcaro et al. 2007a, b; Veyrand et al. 2007; Stanciu et al. 2008; Danyi et al. 2009; Lund et al. 2009; Orecchio et al. 2009; Drabova et al. 2012; Payanan et al. 2013; So-Young et al. 2013; Taylor et al. 2013). According to the harmonized guideline (Thompson et al. 2002), the effects of a possible general matrix mismatch must be assessed during validation if the calibration standards were prepared as simple solutions of the measurands. Somes strategies that were used to compensate for matrix effects were the matrix-matched calibration curves (Ballesteros et al. 2006; Belo et al. 2012; Pincemaille et al. 2014) and the calibration by standard addition (Aguinaga et al. 2007). Rose et al. (2007) argued that the use of ^{13}C-isotopes corrected for any variation due to matrix effects. In fact, IDMS is discussed as an approach to overcome matrix effects. The use of matrix-matched curves is regarded as unnecessary to compensate matrix effects in this case, considering that the relative responses between the target measurand and the labeled analogue remain constant (Sargent et al. 2002; Hewavitharana 2011).

Trueness was not frequently investigated in the current practices for method validation because commonly certified reference materials were not used in the studies (Houessou et al. 2005; Badolato et al. 2006; Ballesteros et al. 2006; Aguinaga et al. 2007; Purcaro et al. 2007a, b; Stanciu et al. 2008; Danyi et al. 2009; Lund et al. 2009; Orecchio et al. 2009; Cai et al. 2012; Pincemaille et al. 2014; Pissinatti et al. 2014). Often, when these materials were employed, they did not represent the matrices defined in the scopes of the validation processes. This fact reflected another problem in this specific area that is the lack of materials for a significant range of food matrices. Comparison with a reference method (Drabova et al. 2012; Taylor et al. 2013) and participation in interlaboratorial assays (such as proficiency testing schemes) (Veyrand et al. 2007; Windal et al. 2008; Belo et al. 2012) were adopted by the researchers as alternatives in the absence of certified reference materials.

The validation processes were restricted to the evaluation of precision under repeatability conditions in most cases (Diletti et al. 2005; Houessou et al. 2005; Badolato et al. 2006; Ballesteros et al. 2006; Liguori et al. 2006; Aguinaga et al. 2007; Purcaro et al. 2007b; Lund et al. 2009; Orecchio et al. 2009; Cai et al. 2012; Drabova et al. 2012; Londoño et al. 2013; Payanan et al. 2013; So-Young et al. 2013). For precision assessment in a single-laboratory validation, two sets of conditions are important: repeatability conditions, related to the variations observed during a single run, and precision under run-to-run or intermediate precision conditions, related to variations in run bias. The variation in conditions among the runs must represent what happen in the routine use of the method, including representative variations in reagent batches, analysts, and instruments (Thompson et al. 2002). These findings indicated that an important aspect of precision had not been considered in the literature related to validation of methods for PAHs from food.

This fact may explain the small number of studies that considered the measure of uncertainty in the validation process (Diletti et al. 2005; Rose et al. 2007; Belo et al. 2012; Pissinatti et al. 2014), since the estimates of standard deviations obtained from the precision experiments represents the uncertainties associated with the random error (repeatability) and run (intermediate precision) terms, which are combined to estimate measurement uncertainty (Thompson et al. 2002).

Ruggedness is highlighted as one of the least studied parameters. In any method there are certain stages which, if not carried out sufficiently carefully, will produce significant effect on method performance. These stages should be identified by ruggedness tests. However, usually, the ruggedness is understood by the authors as part of the method development instead of the method validation (Magnusson and Ornemark 2014).

6 Conclusions

Techniques for the determination of PAHs in food have advanced, especially with regard to automation and miniaturization, as well as increased sensitivity and selectivity. However, the reported validation practices show that, in general, the performance evaluation of the methods is not conducted properly, which may compromise the reliability and comparability of the reported measurements and, consequently, the correctness of the actions taken.

References

AFSSA (Association for French Studies in Southern Africa) (2004) Collection of occurrence data on polycyclic aromatic hydrocarbons in food. Final report Scoop Task 3.2.12 occurrence data on PAH in food. Maisons-Alfort, France

Aguinaga N, Campillo N, Viñas P, Hernández-Córdoba M (2007) Determination of 16 polycyclic aromatic hydrocarbons in milk and related products using solid-phase microextraction coupled to gas chromatography–mass spectrometry. Anal Chim Acta 596:285–290

Albers PH (2002) Petroleum and individual polycyclic aromatic hydrocarbons. In: Hoffman DJ et al (eds) Handbook of ecotoxicology, 2nd edn. CRC Press, Boca Raton, pp 341–371

ATSDR (Agency for Toxic Substances and Disease Registry) (1995) Toxicological profile for polycyclic aromatic hydrocarbons. ATSDR, Atlanta

Badolato ESG, Martins MS, Pimentel SA, Alaburda J, Kumagai EE, Baptista GG, Rosnethal A (2006) Sistematic study of benzo(a)pyrene in coffee samples. J Braz Chem Soc 17(5):989–993

Baek SO, Field RA, Goldstone ME, Kirk PW, Lester JN, Perry R (1991) A review of atmospheric polycyclic aromatic hydrocarbons: sources, fate, and behavior. Water Air Soil Pollut 60: 279–300

Ballesteros E, Sánchez AG, Martos NR (2006) Simultaneous multidetermination of residues of pesticides and polycyclic aromatic hydrocarbons in olive and olive-pomace oils by gas chromatography/tandem mass spectrometry. J Chromatogr A 1111:89–96

Belo RFC, Nunes CM, Santos EV, Augusti DV, Pissinatti R (2012) Single laboratory validation of a SPE method for the determination of PAHs in edible oils by GC–MS. Anal Methods 4: 4068–4076

Boström C, Gerde P, Hanberg A, Jernström B, Johansson C, Kyrklund T, Rannug A, Törnqvist M, Victorin K, Westerholm R (2002) Cancer risk assessment, indicators and guidelines for polycyclic aromatic hydrocarbons in the ambient air. Environ Health Perspect 110:451–488

Bulder AS, Hoogenboom LAP, Kan CA, Raamsdonk LWD, Traag WA, Bouwmeester H (2006) Initial risk assessment of polycyclic aromatic hydrocarbons (PAHs) in feed (materials). Report 2006.001. RIKILT—Institute of Food Safety, Wageningen

Budzinski H, Jones I, Bellocq J, Piérard C, Garrigue P (1997) Evaluation of sediment contamination by polycyclic aromatic hydrocarbons in the Gironde estuary. Mar Chem 58: 85–97

Cai S, Stevensb J, Syagea JA (2012) Ultra high performance liquid chromatography–atmospheric pressure photoionization-mass spectrometry for high-sensitivity analysis of US Environmental Protection Agency sixteen priority pollutant polynuclear aromatic hydrocarbons in oysters. J Chromatogr A 1227:138–144

Camargo MCR, Toledo MCF (2002) Polycyclic aromatic hydrocarbon contamination in different commodity groups. Braz J Food Technol 5:19–26

Ciecierska M, Obiedzinsk M (2007) Influence of smoking process on the polycyclic aromatic hydrocarbons's content in meat products. Acta Sci Pol Technol Aliment 6(4):17–20

Codex Alimentarius Commission (2005) Discussion paper on polycyclic aromatic hydrocarbons (PAH) contamination. JOINT FAO/WHO Food Standards Programme Codex Committee on Food Additives and Contaminants. Thirty-seventh Session The Hague, The Netherlands, 25–29 April, 2005

Codex Alimentarius Commission (2009) Code of practice for the reduction of contamination of food with polycyclic aromatic hydrocarbons (PAH) from smoking and direct drying processes. CAC/RCP 68

Danyi S, Brose F, Brasseur C, Schneider YJ, Larondelle Y, Pussemier L, Robbens J, De Saeger S, Maghuin-Rogister G, Scippo ML (2009) Analysis of EU priority polycyclic aromatic hydrocarbons in food supplements using high performance liquid chromatography coupled to an ultraviolet, diode array or fluorescence detector. Anal Chim Acta 633:293–299

Diletti G, Scortichini G, Scarpone R, Gatti G, Torreti L, Migliorati G (2005) Isotope dilution determination of polycyclic aromatic hydrocarbons in olive pomace oil by gas chromatography–mass spectrometry. J Chromatogr A 1062:247–254

Drabova L, Pulkrabova J, Kalachova K, Tomaniova M, Kocourek V, Hajslova J (2012) Rapid determination of polycyclic aromatic hydrocarbons (PAHs) in tea using two-dimensional gas chromatography coupled with time of flight mass spectrometry. Talanta 100:207–216

EC (European Commission), SCF (Scientific Committee on Food) (2002) Opinion of the Scientific Committee on Food on the risks to human health of polycyclic aromatic hydrocarbons in food (expressed on 4 December 2002)

EFSA (European Food Standard Agency) (2008a) Findings of the EFSA data collection on polycyclic aromatic hydrocarbons in food—a report from the unit of data collection and exposure on a request from the European Commission. EFSA/DATEX/002 (revision 1)

EFSA (European Food Standard Agency) (2008b) Polycyclic aromatic hydrocarbons in food—scientific opinion of the panel on contaminants in the food chain. EFSA J 724:1–114

EPA (Environment Protection Agency) (1986) Quality criteria for water, 1986. EPA 440/5-86-001. United States EPA, Washington

EU (2011a) Commission regulation 2011/836/EU of 19 August 2011 amending Regulation (EC) No 333/2007 laying down the methods of sampling and analysis for the official control of the levels of lead, cadmium, mercury, inorganic tin, 3-MCPD and benzo(a)pyrene in foodstuffs. Off J Eur Union L215:9–16

EU (2011b) Commission regulation 2011/836/EU of 19 August 2011 amending Regulation (EC) No 1881/2006 as regards maximum levels for polycyclic aromatic hydrocarbons in foodstuffs. Off J Eur Union L215:4–8

Feng X, Pisula W, Müllen K (2009) Large polycyclic aromatic hydrocarbons: synthesis and discotic organization. Pure Appl Chem 81(12):2203–2224

Gohlke JM, Doke D, Tipre M, Leader M, Fitzgerald T (2011) A review of seafood safety after the deepwater horizon blowout. Environ Health Perspect 119:1062–1069

Harvey RG (1998) Environmental chemistry of PAHs. In: Hutzinger O (ed) The handbook of environmental chemistry, volume 3, 1st edn. Springer, Berlin, pp 1–55

Hewavitharana AK (2011) Matrix matching in liquid chromatography–mass spectrometry with stable isotope labelled internal standards—is it necessary? J Chromatogr A 1218:359–361

Houessou JK, Benac C, Delteil C, Camel VC (2005) Determination of polycyclic aromatic hydrocarbons in coffee brew using solid-phase extraction. J Agric Food Chem 53:871–879

IARC (International Agency for Research on Cancer) (2012) Chemical agents and related occupations: a review of human carcinogens. IARC monographs on the evaluation of carcinogenic risks to humans, vol 100F. IARC, Lyon, p 599

IPCS (International Programme on Chemical Safety) (1998) Environmental criteria 202. Selected non-heterocyclic PAHs. World Health Organization, Geneva

ISO (International standard Organization) (2005) ISO 9000:2005 quality management systems—fundamentals and vocabulary. ISO, Geneva

ISO (International Standard Organization) (2012) ISO 22959 animal and vegetable fats and oils—determination of polycyclic aromatic hydrocarbons by on-line donor–acceptor complex chromatography and HPLC with fluorescence detection, 1st edn. ISO, Geneva

Kim JH, Moon JK, Li QX, Cho JY (2003) One-step pressurized liquid extraction method for the analysis of polycyclic aromatic hydrocarbons. Anal Chim Acta 498:55–60

Konieczka P, Wolska L, Namiesnik J (2010) Quality problems in determination of organic compounds in environmental samples, such as PHAs and PCBs. Trend Anal Chem 29(7): 706–717

Law RJ, Kelly C, Baker K, Jones J, McIntosh AD, Moffat CF (2002) Toxic equivalency factors for PAH and their applicability in shellfish pollution monitoring studies. J Environ Monit 4: 383–388

Liguori L, Heggstad K, Hove HT, Julshamn K (2006) An automated extraction approach for isolation of 24 polyaromatic hydrocarbons (PAHs) from various marine matrixes. Anal Chim Acta 573–574:181–188

Lodovici M, Dolara P, Casalini C, Ciappellano S, Testolin G (1995) Polyciclic aromatic hydrocarbon contamination in the Italian diet. Food Addit Contam 12:703–713

Londoño VAG, Garcia LP, Scussel VM, Resnik S (2013) Polycyclic aromatic hydrocarbons in milk powders marketed in Argentina and Brazil. Food Addit Contam 30(9):1573–1580

Lund M, Duedahl-Olesen L, Christensen JH (2009) Extraction of polycyclic aromatic hydrocarbons from smoked fish using pressurized liquid extraction with integrated fat removal. Talanta 79:10–15

Magnusson B, Ornemark U (2014) Eurachem guide: the fitness for purpose of analytical methods—a laboratory guide to method validation and related topics, 2nd edn. EURACHEM, p 62

Mannino MR, Orecchio S (2008) Polycyclic aromatic hydrocarbons (PAHs) in indoor dust matter of Palermo (Italy) area: extraction, GC–MS analysis, distribution and sources. Atmos Environ 42:1801–1817

Mechlinska A, Wolska L, Namiesnik J (2010) Isotope-labeled substances in analysis of persistent organic pollutants in environmental samples. Trend Anal Chem 29(8):820–831

Meire RO, Azererdo A, Machado JP (2007) Aspectos ecotoxicológicos de hidrocarbonetos policíclicos aromáticos. Oecol Bras 11(2):188–201

Ming-Ho Y (2005) Volatile organic compounds. In: Ming-Ho Y (ed) Environmental toxicology: biological and health effects of pollutants, 2nd edn. CRC Press, Florida, pp 171–184

Mitra S (2003) Sample preparation techniques in analytical chemistry. Wiley, New Jersey

Nielsen T, Feilberg A, Binderup ML (1999) The variation of street air levels of PAH and other mutagenic PAC in relation to regulations of traffic emissions and the impact of atmospheric processes. Environ Sci Pollut Res 6(3):133–137

Orecchio S, Ciotti VP, Culotta L (2009) Polycyclic aromatic hydrocarbons (PAHs) in coffee brew samples: analytical method by GC–MS, profile, levels and sources. Food Chem Toxicol 47:819–826

Orecchio S (2010) Assessment of polycyclic aromatic hydrocarbons (PAHs) in soil of a Natural Reserve (Isola delle Femmine) (Italy) located in front of a plant for the production of cement. J Hazard Mater 173:358–368

Oshita D, Jardim ICSF (2015) Evaluation of dispersive and cartridge SPE clean-up procedures using the modified QuEChERS method for the analysis of pesticides in strawberries. Anal Methods 7:982–989

Pampanin DM, Sydnes MO (2013) Polycyclic aromatic hydrocarbons a constituent of petroleum: presence and influence in the aquatic environment. In: Kutcherov V (ed) Hydrocarbon. InTech, ISBN: 978-953-51-0927-3. doi:10.5772/48176

Payanan T, Leepipatpiboon N, Varanusupakul P (2013) Low-temperature cleanup with solid-phase extraction for the determination of polycyclic aromatic hydrocarbons in edible oils by reversed phase liquid chromatography with fluorescence detection. Food Chem 141:2720–2726

Pincemaille J, Schummer C, Heinen E, Moris G (2014) Determination of polycyclic aromatic hydrocarbons in smoked and non-smoked black teas and tea infusions. Food Chem 145: 807–813

Pissinatti R, Nunes CM, Souza AG, Junqueira RG, Souza SVC (2014) Simultaneous analysis of 10 polycyclic aromatic hydrocarbons inroasted coffee by isotope dilution gas chromatography–mass spectrometry: optimization, in-house method validation and application to an exploratory study. Food Control 51:140–148

Poster DL, Schantz MM, Sander LC, Wise SA (2006) Analysis of polycyclic aromatic hydrocarbons (PAHs) in environmental samples: a critical review of gas chromatographic (GC) methods. Anal Bioanal Chem 386:859–881

Purcaro G, Moret S, Conte LS (2013) Overview on polycyclic aromatic hydrocarbons: occurrence, legislation and innovative determination in foods. Talanta 105:292–305

Purcaro G, Moret S, Conte LS (2007a) Rapid validated method for the analysis of benzo(a)pyrene in vegetable oils by using solid–phase microextraction–gas chromatography-mass spectrometry. J Chromatogr A 1176:231–235

Purcaro G, Morrison P, Moret S, Conte LS, Marriott PJ (2007b) Determination of polycyclic aromatic hydrocarbons in vegetable oils using solid-phase microextraction-comprehensive two-dimensional gas chromatography coupled with time-of-flight mass spectrometry. J Chromatogr A 1161:284–291

Rose M, White S, Macarthur R, Petch RG, Holland J, Damant AP (2007) Single-laboratory validation of a GC/MS method for the determination of 27 polycyclic aromatic hydrocarbons (PAHs) in oils and fats. Food Addit Contam 24:635–651

Sargent M, Harte R, Harrington C (2002) Guidelines for achieving high accuracy in isotope dilution mass spectrometry (IDMS), 1st edn. Royal Society of Chemistry, Cambridge 34 p

Smoker M, Tran K, Smith RE (2010) Determination of polycyclic aromatic hydrocarbons (PAHs) in shrimp. J Agric Food Chem 58:12101–12104

Souza SVC, Junqueira RG (2005) A procedure to assess linearity by ordinary least squares method. Anal Chim Acta 552:25–35

So-Young J, Ju-Sung P, Min-Su C, Mi-Sun K, Sang-Mi L, Jung-Hun K, Young-Zoo C (2013) A Simple method for the determination of polycyclic aromatic hydrocarbons (PAH) in edible oil employing solid phase extraction (SPE) cartridge purification. Food Sci Biotechnol 22(1):241–248

Stanciu G, Dobrinas S, Birghila S, Popescu M (2008) Determination of organic compounds from different types of coffee by HPLC and GC–ECD analysis. J Environ Eng 7(6):661–666

Taylor CM, Raines JM, Rodriguez JM (2013) Determination of polycyclic aromatic hydrocarbons in seafood. Food Anal Methods 6:1330–1336

Tfouni SAV, Serrate CS, Carreiro LB, Camargo MCR, Teles CRA, Cipolli KMVAB, Furlani RPZ (2012) Effect of roasting on chlorogenic acids, caffeine and polycyclic aromatic hydrocarbons levels in two *Coffea* cultivars: *Coffea arabica* cv. Catuaí Amarelo IAC-62 and *Coffea canephora* cv. Apoatã IAC-2258. Int J Food Sci Technol 47:406–415

Thompson M, Ellison SLR, Wood R (2002) Harmonized guidelines for single-laboratory validation of methods of analysis. Pure Appl Chem 74:835–855

Tobiszewski M, Mechlińska A, Zygmunt B, Namieśnik J (2009) Green analytical chemistry in sample preparation for determination of trace organic pollutants. TrAC Trends Anal Chem 28:943–951

Veyrand B, Brosseaud A, Sarcher L, Varlet V, Monteau F, Marchand P, Andre F, Le Bizec B (2007) Innovative method for determination of 19 polycyclic aromatic hydrocarbons in food and oil samples using gas chromatography coupled to tandem mass spectrometry based on an isotope dilution approach. J Chromatogr A 1149:333–344

Wang W, Meng B, Lu X, Liu Y, Tao S (2007) Extraction of polycyclic aromatic hydrocarbons and organochlorine pesticides from soils: a comparison between Soxhlet extraction, microwave-assisted extraction and accelerated solvent extraction techniques. Anal Chim Acta 602:211–222

WHO (World Health Organization) (2006) Evaluation of certain food contaminants—report of the Joint FAO/WHO Expert Committee on Food Additives, 64th meeting, Rome, 8–17 February, 2005, p 1 and 61. WHO Technical Report Series, No. 930

Windal I, Boxus L, Hanot V (2008) Validation of the analysis of the 15 + 1 European-priority polycyclic aromatic hydrocarbons by donnor–acceptor complex chromatography and high-performance liquid chromatography–ultraviolet/fluorescence detection. J Chromatogr A 1212:16–22

Wretling S, Eriksson A, Eskhult GA, Larsson B (2010) Polycyclic aromatic hydrocarbons (PAHs) in Swedish smoked meat and fish. J Food Compos Anal 23:264–272

Yunker MB, Macdonald RW, Vingarzanc R, Mitchelld RH, Goyettee D, Sylvestre S (2002) PAHs in the Fraser River basin: a critical appraisal of PAH ratios as indicators of PAH source and composition. Org Geochem 33:489–515

Ziegenhals K, Speer K, Jira W (2009) Polycyclic aromatic hydrocarbons (PAH) in chocolate on the German market. J Verbrauch Lebensm 4:128–135

HC-0B-06: Biodegradation of Hydrocarbons

Emenike Chijioke Uche and Arezoo Dadrasnia

Abstract Hydrocarbons and associated derivatives are unavoidably common in the environment. The use of crude oils, petrol, diesel, heavy oils, lubricants etc. has continued to cause widespread aquatic, terrestrial and even atmospheric pollution. Considering the importance of hydrocarbons in various industries, its occurrence in natural and blended forms is inevitable. However, possible recovery of systems contaminated with hydrocarbons has given rise to the interest on the degradation of hydrocarbons, especially in a biologically-driven process known as biodegradation. Therefore, the Chapter will provide an overview of the basic behaviour of hydrocarbons in the environment, chances of biological degradation and factors pivotal for biodegradability. Similarly, among other issues related to biodegradation of hydrocarbons, the core remediation techniques (bio-, chemico-, and physico-treatments) adopted in most hydrocarbon biodegradability ventures are evaluated alongside the evaluation of the mechanisms of degradation. The use of plant technology in the remediation of polluted sites often referred to as phytoremediation is one of the options employed for the bio-recovery of hydrocarbon-impacted systems due to the various mechanisms involved. Therefore, regardless of the fact that some phyisco-chemically driven processes of hydrocarbon removal are abundantly used at some contaminated sites, none has been as green, cost effective and sustainable like the adoption of a biodegradation process.

Keywords Bioremediation · Bioaugmentation · Hydrocarbon · Phytoremediation · Removal

E.C. Uche · A. Dadrasnia (✉)
Institute of Research Management and Monitoring, Deputy Vice Chancellor (Research and Innovation), University of Malaya, 50603 Kuala Lumpur, Malaysia
e-mail: are.dadrasnia@gmail.com

E.C. Uche
e-mail: emenike@um.edu.my

K. Heimann et al. (eds.), *Biodegradation and Bioconversion of Hydrocarbons*, Environmental Footprints and Eco-design of Products and Processes, DOI 10.1007/978-981-10-0201-4_3

1 Introduction

Modern industrial society is built upon and ruled by petroleum hydrocarbons. At a global level, crude oil productions are estimated to be more than 12 million metric tons annually and about 1.7–8.8 million metric tons of oil are released into the aquatic environment and soil respectively per annum (EIA 2011). About 90 % of this emission is directly related to human activities including deliberate illegal waste disposal. Petroleum is essential to the current global networked economy, without it, our economic order would cease to function, bringing disaster to many populations. Yet the blessings of hydrocarbons are mixed, there is a growing awareness that imperfect petroleum technologies are changing ecosystems in ways that decrease the ability of these systems to support human populations. Effects of crude oil pollution in the environment will change from one source to another because of crude oil and its derivatives or mixture of organic compounds (Onwurah et al. 2007). Fuel oil may enter the water or soil environment as a result of spillages during transportation and by leakages from the storage facilities or pipelines. The more volatile components of fuel oils (low molecular weight alkanes) can be degraded in both water and soil and could volatise to enter into the atmosphere where they will form contaminants (Dadrasnia and Agamuthu 2013). The removal of contaminants from the environment is a crucial approach towards returning any environmental medium to its natural/original state; hence the term "environmental restoration". While the introduction of contaminants or pollution of air, water and soil can be easy, rapid and persistent, the removal is often a daunting task. Remediation of petroleum-contaminated systems can be achieved by physical, chemical or biological methods. However, the unintended negative consequences of physical and chemical approaches are currently directing greater attention to the use of the biological alternatives. Therefore, the potential of remediation techniques will depend on the area where the spill has accrued among other factors. Oil spills in the water environment may affect microorganisms physically or induce direct toxicity. Hence, this chapter is designed to elucidate biological treatments, especially phytoremediation potential, as a green clean up technique for the restoration of the environment.

1.1 Hydrocarbons and the Environment

Hydrocarbons, especially polycyclic aromatic hydrocarbons (PAHs) are commonly found as organic pollutants in the environment (Dadrasnia and Agamuthu 2014). The presence of a huge number of commercial trucks, private automobiles, ships and boats, locomotives, industrial engines (Roy 1997), tractors and heavy vehicles has resulted in an increase in the use of petroleum products. For instance, in the UK, approximately 120,000 contaminated petrol station sites exist with an associated remediation cost of two billion dollars (Collins 2007). The US petroleum industry has spent about one billion dollars in 2001 on remediation. It has been estimated that in 2040, contaminated site treatment may cost approximately two trillion dollars (Yu and Hill 2006).

PAHs are toxic to aquatic organisms due to oxidation when it is exposed to solar UV radiation, whereby toxic compounds are released (Onwurah et al. 2007). There are so many individual constituents of petroleum hydrocarbons; hence, it is difficult to determine the effect of each constituent within the context of a hydrocarbon mixture. However, aromatic compounds tend to be more toxic than aliphatic compounds.

Crude oil, as a result of PAH content, interrupts the survival, reproduction, development and growth of organisms. This may increase the risk of mortality from infectious diseases (Onwurah et al. 2007). Soil pollution with petroleum spills has resulted in negative effects on the food cycle. For instance, the number of oil polluted soil increased significantly between 1978 and 1979 in Nigeria, affecting farmlands used to grow some of the crops, especially cassava, maize, rice and plantain (Onwurah et al. 2007).

Crude oil can affect soil chemistry, germination, growth and fertility of plants, but the effects depend on the type and degree of the oil spill. Petroleum and its derivatives in contaminated land can affect some soil properties such as the cation exchange capacity and organic matter composition. There is an undeniable connection between the environment and human health (Onwurah et al. 2007). Toxic materials in oil may affect human health via inhibition of nerve synapse function, protein synthesis, damage to the plasma membrane with consequences for membrane transport systems (Afuwale and Modi 2012; Onwurah et al. 2007). Light oils contain a high ratio of saturated hydrocarbons; hence, these can be more hazardous than heavy oils (Kauppi et al. 2011). A National Priorities list was compiled in 1999 on prioritized chemicals based on the frequency of their occurrence and their risk for the environment and human health. On this list, PAHs were collectively ranked ninth, and benzo(a)pyrene was ranked eighth (Olson et al. 2003).

A number of PAHs have been identified as probable human carcinogens. Benzene is also of concern because it is a known human carcinogen (EPA 2006). Chronic effects of naphthalene are seen by changes in the nervous system, liver, kidneys, blood and heart, due to relative insolubility and carcinogenic potential (Roy 1997).

1.2 Biodegradability of Hydrocarbon: An Overview

Environmental oil spills elicit a wide variety of biological, physical and chemical processes. The capability and interaction of animals, plants, and microorganisms are identified as limiting degradation factors. Most hydrocarbons are insoluble in water, resulting in poor mobility, hence their bioavailability is limited in the degradation process (Phlip et al. 2005). Biodegradation is defined as the breakdown of organic compounds by actinomycetes, fungi and bacteria. Microorganisms are provided with metabolic systems to use petroleum as a source of energy and carbon (Van Hamme et al. 2003). The growth of microorganisms on hydrocarbons depends on the emulsification of the undissolved carbon source in the culture medium (Chrzanowski et al. 2006). The ability of microbes to break down hydrocarbons depends on the enzymatic make up of the microorganisms (Mancera-López et al. 2008). Such property enable growth on petroleum contaminated sites. *n*-alkanes are completely biodegradable, with their break down occurring at the beginning [early] of the biodegradation process (Van Hamme et al. 2003). Pelletier et al. (2004) reported that over 90 % of *n*-alkanes degrade in the first six months and most light aromatics (2–3 rings) disappeared during the first year of observation. Pristane (C_{19}) and phytane (C_{20}) are the most easily recognized compounds in the remaining diesel fuel mixture at this point, which are isoprenoids.

2 Agents, Mechanisms and Factors for Hydrocarbon Degradation

There are a number of mechanisms known for degradation of organic compounds. During degradation, bacteria are selected for their ability to degrade and this involves several mechanisms for the breakdown of the substance molecules (Das and Chandran 2011). The bacteria have to be an electron acceptor for degradation of organic molecules. The mechanism is based on two processes: (i) growth and (ii) co-metabolism (Fritsche and Hofrichter 2005). With regards to growth, bacteria use organic carbon as a source of energy and carbon (mineralization of organic pollutants) (Angelidaki and Sanders 2004). Co-metabolism is the metabolism of an organic compound in the presence of a growth substrate that is used as the primary carbon and energy source (Fritsche and Hofrichter 2005). Therefore, the degradation mechanism depends on the hydrocarbon enzyme system. Other mechanisms are (1) attachment of microbial cells to the substrates and (2) production of biosurfactants (Das and Chandran 2011). Enzymes, such as dioxygenases, monooxygenases and hydroxylases, play important roles in the microbial degradation of oil. An enormous number of bacteria and fungi have the ability to degrade organic pollutants. However, a single bacterial species does not have the enzymatic capability to metabolize all of the organic compounds in contaminated soils (Fritsche and Hofrichter 2005). Thus a mixed microbial community has a more powerful biodegradation potential compared to a single species, because the genetic information of more than one organism is necessary to degrade the complex mixtures of organic compounds present in polluted areas (Fritsche and Hofrichter 2005). In summary, the essential specifications of aerobic microorganisms degrading organic pollutants are as follows:

- Metabolic processes for optimizing the contact between the organic pollutants and the microbial cells must be efficient, making the chemicals accessible to the organisms carrying out biodegradation activities (Fritsche and Hofrichter 2005).

- Next, the degradation path transforms organic pollutants systematically into intermediates of the central mediator metabolism (Fritsche and Hofrichter 2005).
- The initial intracellular attack on organic pollutants is an oxidative process; the activation and incorporation of oxygen is the enzymatic key reaction catalyzed by oxygenases and peroxidases (Fritsche and Hofrichter 2005).

2.1 Degradation of Hydrocarbon Fractions

Microorganisms easily utilize gaseous hydrocarbons. Dry gases are dominated by methane relative to the higher homologs with a dryness coefficient $[C_1/\Sigma\ (C_1 - C_5)]$ of >95 %, while wet gases are rich in C_{2+} components with a dryness coefficient of <90 % (Pallasser 2000). Whiticar (1994) indicated that microorganisms can metabolize methane, and methanotrophs have been identified in heavily degraded oil reservoirs. Biodegraded gases are often compositionally dry with very few exceptions. The C_{6-15}-*n*-alkane groups are among the most rapidly biodegraded components of oil, although they are also susceptible to removal by extensive water washing. Benzene or toluene are less affected by biodegradation than *n*-heptane, 3-methylhexane, cyclohexane and methylcyclohexane (Masterson et al. 2001). Cyclic and branched-chain alkanes are more stable to biodegradation than linear alkanes (Rojo 2009). There is a tendency of reduced capability to biodegradation with greater alkyl replacement for alkylcyclohexanes, isoalkanes, alkycyclopentanes and alkylbenzenes (George et al. 2002). The location of methylation has a strong effect on susceptibility to biodegradation. Among the branched alkanes, 2-methylalkanes are more susceptible to degradation than 4-methylalkanes, which in turn are more susceptible than 3-methylalkanes (George et al. 2002). Adjacent methyl group reduce the susceptibility of an isomer to biodegradation; for example, 1,1-dimethylcyclopentane and 1,1-dimethylcyclohexane are the most resistant than alkylcyclohexanes and alkylcyclopentanes; 1,2,3-trimethylbenzene and 1,2,3,4-tetramethylbenzene are more resistant to biodegradation than other C_3- and C_4-alkylbenzenes (George et al. 2002). Hence, concentrations of 3-methylpentane/*n*-hexane, 1,1-dimethylcyclo pentane/*n*-heptane, and 3-methylpentane/2-methylpentane increase with increasing degree of biodegradation.

2.2 Biodegradation of Aromatic Compounds

Polycyclic aromatic hydrocarbons (PAHs) which consist of two or more benzene rings are typically more resistant than aliphatic hydrocarbons. The most important aromatic petroleum hydrocarbons are toluene, benzene, xylene, ethyl benzene, and xylene. Benzene and its properties are important because all aromatic hydrocarbons are derivatives of benzene. Benzene has three double bonds with a six membered ring and it is flat (C_6H_6). It is unsaturated structurally; similar to the cyclic alkenes. However, it is stable and does not participate in reactions that are specific of alkenes.

Understanding and manipulating the microbial ecology of contaminated land are increasingly critical steps towards meeting the challenge of remediating the considerable legacy of industrial pollution worldwide. Microbial ecology links community structure, phenotypic potential, community function and interactions of microorganisms within communities and with their environments. The ability of microbial communities to degrade contaminants in the environment is dependent on the complex appropriation of nutrients, electron acceptors and other conditions including pH, redox potential, moisture and temperature. Metabolic cooperation is often an important feature of microbial communities.

The main differences between aromatic and aliphatic hydrocarbons provide a useful method for classifying these compounds. For example, the anaerobic degradation process of aromatic hydrocarbons is slow and uncommon compared to aerobic biodegradation (Foght 2002; Chakraborty and Coates 2004). Alkanes are chemically not very reactive due to the lack of functional groups, the presence of sigma bonds only determines the non-polar nature and the very low solubility in water. Although *n*-alkanes are degraded faster than the other components of the petroleum hydrocarbons, they are quantitatively the major fraction of crude oil (Head et al. 2006). The exact mechanisms involved in the first steps of the activation of alkanes in anoxic conditions are not yet completely understood. In addition, degradation of aliphatic hydrocarbons in aerobic conditions requires oxygen as a terminal electron acceptor. Essentially, aromatic hydrocarbons can support the growth of bacteria when they are present as the sole source of carbon and energy (Boonchan et al. 2000; Field and Alvarez 2007). Therefore, aromatic hydrocarbons are not as easy to biodegrade compared to branched alkanes which are slightly more degradable than the alicyclic hydrocarbons (Cao et al. 2009). Some recent reports on the application of microbial strains for the remediation of crude oil are presented in Table 1.

2.3 *Biodegradation of* n-*Alkanes*

Aliphatic hydrocarbons are more biodegradable than PAHs, as they are less toxic and have a higher bioavailability. The *n*-alkanes (C_6–C_{15}) are among the most rapidly biodegraded components of oil, although they are also susceptible to removal by extensive water washing. A number of microbial have been reported capable of degrading alkanes (Colombo et al. 1996; Sonia et al. 2002; Hidayat and Tachibbana 2012).

Margesin and Schinner (2001a, b) identified *Rhodococcus fascians* as one of the hydrocarbon degrading bacteria which are able to produce bio-emulsifiers when

Table 1 Specific bioaugmentation researches

Type of contaminant	Initial TPH concentration	Microorganisms added	TPH reduction (%)	Duration of research (days)	References
TPH (diesel oil)	16,300 ppm	*Candida tropicalis* SK 21	>80	120	Fan et al. (2014)
TPH (crude oil)	2 % (v/v)	*Bacillus salmalaya* 139SI	>70	60	Dadrasnia et al. (2016)
TPH (diesel oil)	10,000 ppm	*Bacillus cereus, Gordoni rubripertincta, Kociria rosea, Bacillus subtilis, Aspergillus erreus, Aspergillus arneus*	>50	90	Díaz-Ramírez et al. (2013)
TPH (diesel oil)	3 % (v/v)	*Bacillus amyloliquefaciens An6*	>70	18	Ben Ayed et al. (2014)
TPH (diesel oil)	5000 ppm	*Pseudomonas fluorescens*	>60	270	Pirnay et al. (2003)
TPH (diesel oil)	2000 ppm	*Rhodococcus* sp. EH831	>90	46	Lee et al. (2011)
TPH (biodiesel)	Unknown	*Pseudomonas aeruginosa, Arthrobacter xylosoxidans*	>30	32	Colla et al. (2014)

grown with *n*-alkanes as the sole carbon source. Bacteria can uptake and transport soluble alkanes that are dissolved in the liquid phase (Ahmed 2004). Indeed, bacteria can only utilize solubilized hydrocarbons. There is a different mechanism for biodegradation of aliphatics. Ron and Rosenberg (2002) reported that microbial degradation of aliphatic and aromatic compounds depends on the microorganisms which grow in oil polluted sites and have an important role in the biological treatment of the contaminants. One of the limiting factors in this process, especially at low temperature, is the bioavailability of the many fractions of the oil (Ahmed 2004). These surface-active materials increase the surface area of hydrophobic substrates and increase their bioavailability, thereby enhancing the growth of bacteria and the rate of bioremediation (Ahmed 2004). *n*-alkane in the C_{10}–C_{13} range are typically metabolized first, which probably reflects an optimal carbon number with increasing enthalpy of reaction and decreasing water solubility as the alkane carbon number increases. Aromatic hydrocarbons are typically more resistant than aliphatic hydrocarbons. In general, branched-chain alkanes and cyclics are more resistant to biodegradation than linear alkanes.

2.4 *Factors Influencing Hydrocarbon Biodegradation*

Numerous parameters influence the biodegradation process of hydrocarbons. Some environmental conditions such as pH, temperature, oxygen, nutrients, and soil moisture can influence biodegradation results (Gavrilescu 2010; MSMG and MSMD 2012). In addition, microbial population and availability of pollutants to the microbial population play key role in the degradation process. Air availability, pH, nutrient levels, and moisture contents are the initial controlling factors (Boopathy 2000; Chaillan et al. 2006). Generally, the greatest degradation occurs in a range of pH between 6.5 and 7.5 and temperatures between 20 and 30 °C (depending on the microbial species) because microbial growth and activities are readily affected by pH, temperature, and moisture (Harekrushna and Kumar 2012). The efficiency of the biodegradation process depends on the provision of optimal conditions to improve the degradation process (Fig. 1).

2.4.1 Oxygen

The amount of oxygen determines whether the system is anaerobic or aerobic. In some cases, it is possible to introduce hydrogen peroxide or magnesium peroxide so as to increase the amount of reactive oxygen in the soil. A primary step in the catabolism of hydrocarbons by bacteria and fungi involves the oxidation of the substrate by oxygenase. The availability of oxygen in the soil is dependent on the rate of microbial oxygen consumption in the vicinity of utilizable substrate (Bartha and Bossert 1984). Aerobic bacteria, such as *Pseudomonas, Mycobacterium, Sphingomonas* and *Rhodococcus*, are found to rapidly degrade hydrocarbons (Juwarkar et al. 2010). These microbes have often been reported to degrade pesticides and hydrocarbons, both alkanes and polyaromatic compounds (MSMG and MSMD 2012). The bacteria in an aerobic experiment were able to degrade 20–25 % of the organic material and 90–95 % of the alkanes (Van 2011). In a 50-day anaerobic experiment, 15–18 % of the organic material and only 20–25 % of the alkenes were degraded (Van 2011). Delivering air or oxygen to contaminated soils may be difficult for a number of reasons: the soil porosity may not be favorable and therefore mass transfer from the gas phase to the aqueous phase will be limited.

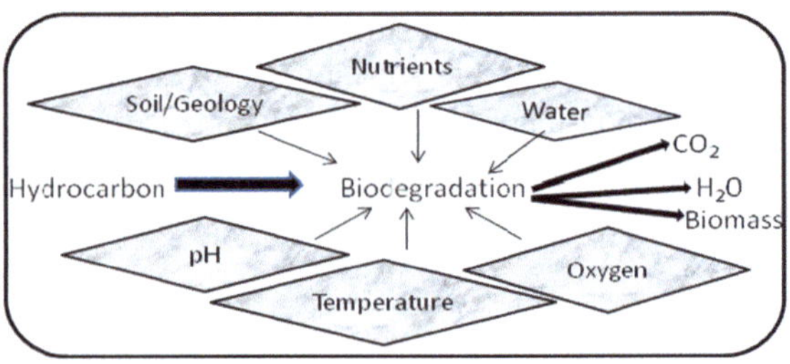

Fig. 1 Physico-chemical factors affecting biodegradation

Also relatively low solubility of oxygen in water is a primary limiting factor (Phlip et al. 2005). The complete oxidation of aromatic compounds and hydrocarbons to carbon dioxide is difficult in the absence of molecular oxygen due to the great stability of C–H and C–C bonds. Although anaerobic microorganisms have the potential to metabolize organic contaminants, and do so in many field situations, yet oxygen is often an integral part in the oxidation of many organic pollutants, including hydrocarbons, because molecular oxygen is required to oxidize the carbon moiety.

2.4.2 Soil Properties and Nutrient Availability

Physico-chemical properties of soil are another important factor in increasing the speed of the degradation process. Some materials such as organic carbon can be applied in the reclamation of soil structure in order to improve delivery of air, water and nutrients. In environments, evaporation of volatile organic compound (VOCs) was observed from all soil types, especially at freshly polluted sites. The decrease in the concentration of oil will be a result of microbial degradation rather than volatilization. Binding to soil particles strongly affects biodegradation rates, yet this parameter is difficult to measure (Kauppi 2011).

Nutrient status has direct impacts on biodegradation and microbial activity (Jain et al. 2011). Nutrients serve as sole sources of electron donors, carbon and energy. Basically, fungi and bacteria require inorganic nutrients, vitamins and amino acids for growth. Nitrogen and phosphorus, and in some cases iron, are important elements needed for cellular metabolism which could become a limiting factor and thus affect the biodegradation processes (Das and Chandran 2011). Atlas and Bragg (2009) reported that availability of nutrients, especially N and P were important in the degradation process of oil spills irrespective of whether they occurred in freshwater and marine environments. Therefore, addition of nutrients is carried out for reclamation and enhancement of biodegradation (Zhu et al. 2004). The effects of organic fertilizers, composts, poultry manure, banana skins, melon shell, wood chips, rice husk mixtures, soy cake, sewage sludge, mushroom and animal droppings on the degradation process has been studied by several researchers (Hickman and Reid 2008; Park et al. 2001; Abioye et al. 2009; Adesodun and Mbagwu 2008). Bento et al. (2005) reported that the addition of N and P increased the degradation rate from 16 % to more than 90 % in contaminated soil.

2.4.3 Moisture and Temperature

Water availability is a requirement of life and affects microbial activity and growth. Irrigation is often required to achieve optimal moisture levels (Harekrushna and Kumar 2012). Water holding capacities recommended bioremediation range from 25 to 28 % (Pandey and Fulekar 2012). Increasing soil moisture positively affected the removal of polycyclic aromatic hydrocarbon from soil. Gong et al. (2005)

demonstrated the significant effect of moisture on extraction of PAH from sunflowers grown in soil at a manufactured gas plant. Previous work illustrated the great effect of moisture on reduced organic adsorption capacities (Thibaud et al. 1993). Temperature can affect biodegradation due to changes in metabolic activities of microbes (Eriksson et al. 2001). This factor has an important role in the diversity of the microbial flora, its physiology, and metabolism. Temperature can affect the rate of biodegradation as well as the chemical and physical properties of the oil (Margesin and Schinner 2001a, b). Generally, microbial activity will reduce at low temperatures (Delille 2000; Gibb et al. 2001; Baraniecki et al. 2002; Eckford et al. 2002). Also, the bioavailability of some substances, such as polyaromatic hydrocarbons and aliphatics, is dependent on the temperature. Diffusion rates of organic compounds increase with rising temperature (Northcott and Jones 2000). The highest degree of degradation in soil environments occurs between 30–40 and 20–30 °C in water (Das and Chandran 2011). Although, many microorganisms in cold climates can survive and grow at temperatures below 5 °C, but it is essential that contaminated sites be at the optimum temperature for bioremediation to progress successfully. In addition, the solubility and bioavailability of a contaminant will increase as temperature increases, but oxygen solubility will be reduced (Margesin and Schinner 1999). Optimal temperatures are, however, also species-specific. For example, a selection of *Rhodococcus* species isolated from an Antarctic soil were able to successfully degrade a number of *n*-alkane at −2°C but was severely inhibited at a higher temperature (Bej et al. 2001).

2.4.4 Bioavailability of Hydrocarbon

Bioavailability has been defined as the accessibility of a chemical for assimilation and possible toxicity (Puglisi et al. 2007). This has also been defined as the level to which a substance is free to move onto or into an organism and it is known that bioavailability differs between species and organisms. The bioavailability of organic contaminants is the main factor determining their fate, ecological risk, toxicity, and losses in the environment (Oleszczuk 2009). It provides information on the actual risk relating to the presence of contaminants. Two main factors that determine the amount of a chemical that is bioavailable are the rate of transfer of the compound from the soil to the living cell and the rate of uptake and metabolism (Semple et al. 2003). The bioavailability of a chemical is determined by the rate of mass transfer relative to the intrinsic activity of the soil biota (Semple et al. 2003). Bioavailability controls biodegradation because microbial cells must consume energy to induce catabolic gene systems used in biodegradation (Madsen 2002). The bioavailability of hydrocarbons is also dependent on their physical state, hydrophobicity; sorption onto soil particles, volatilization and solubility of hydrocarbons greatly affects the extent of biodegradation.

Even under optimal conditions for hydrocarbon biodegradation, it has been shown that a residual fraction of hydrocarbon remains undegraded. Mainly, after its arrival in the soil, an organic contaminant may be lost by biodegradation, leaching or volatilization, or it may accumulate within the soil biota or be sequestered and complexed within the soil's mineral and organic matter fractions (Sihag et al. 2014). The rate of transfer to the cell and the rate of uptake and metabolism by the microorganisms is controlled by a number of physical–chemical processes such as sorption/desorption, diffusion, and dissolution.

As the fate of hydrocarbons in the subsurface depends on the processes of transport, multiphase flow, volatilization, dissolution, geochemical reactions, biodegradation, and sorption, an interdisciplinary investigation of these processes is critical to successfully evaluate the potential for migration of hydrocarbons into the subsurface. To oxidize contaminants into less-harmful metabolites, bacteria extract and utilize oxygen from its surrounding environment, producing CO_2 and H_2O. Algae use photosynthesis to assimilate inorganic carbon, which requires CO_2 and H_2O. Photosynthesis, in turn, releases oxygen that bacteria can utilise for further contaminant oxidation, thus repeating the cycle (Fig. 2).

One of the key components that affects microbial degradation of hydrocarbon is the hydrophobicity of oil, which limits its transfer to the cell surfaces of microorganisms (Lee et al. 2006; Schein et al. 2009). This limitation may be overcome either by growing surfactant producing microorganisms or by an addition of surface-active agents. This increases the bioavailability of diesel oil to microorganisms. Several constraints can limit the bioavailability of organic compounds in the environment. These are low aqueous solubility, sorption, micropore exclusion, and content of organic carbon (Huesemann et al. 2004; Froehner et al. 2012). Puglisi et al. (2007) reported that phenanthrene degradation is usually lower in soils with high organic matter due to higher adsorption and lower diffusion to the water

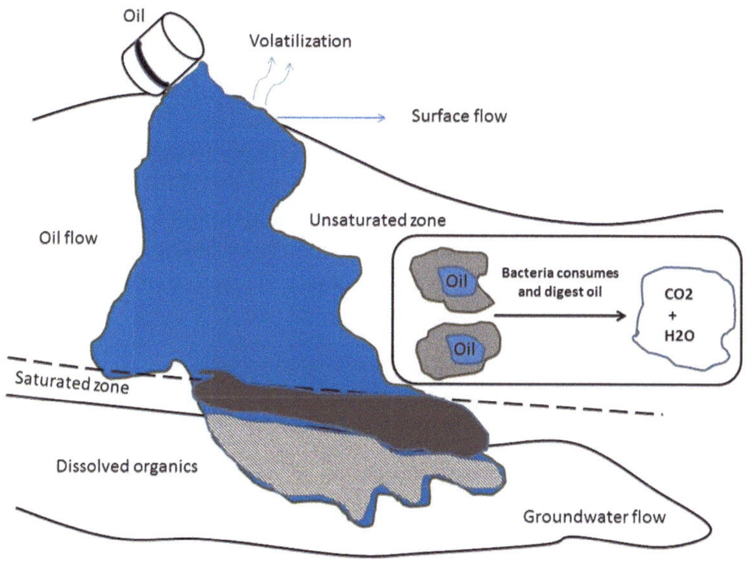

Fig. 2 Fate of hydrocarbons in soil

phase, while in soils with low levels of organic matter, variations in biodegradation may instead, be related to the amount of clay in the soils. The rate of transfer is determined by the equilibrium and the actual concentration in the bulk phase and aqueous phase. This is central to the concept of bioavailability as it relates to biodegradation.

2.4.5 Concentration of Petroleum Hydrocarbon

Concentration of petroleum hydrocarbon determines to a greater extent the rate of breakdown of the hydrocarbons from the environment (Abioye et al. 2012). Hydrocarbon concentration has an important role in its biodegradability and level of toxicity. High concentration of hydrocarbon can affect the activities of microorganisms. Concentrations of 1–100 μg/ml of water or 1–100 μg/g of soil or sediment are not generally considered to be toxic to common heterotrophic bacteria and fungi. Ijah and Antai (2003) indicated a high rate of degradation of hydrocarbons in soil contaminated with low concentration (10 and 20 %) compared to those contaminated with the high concentration (30 and 40 %) of crude oil within a period of one year. In contrast, Rahman et al. (2002) reported that an increase in the concentration of oil from 1 to 10 % decreased biodegradation from 78 to 52 %. Indeed, high concentrations of hydrocarbons can be associated with increased amounts of heavy oil in polluted sites which lead to inhibition of biodegradation due to limitation of nutrients. Table 2 shows the major factors that affect biodegradation processes.

3 Hydrocarbon Remediation Strategies

Conventional methods such as (1) soil vapour extraction, (2) pump and treat systems, (3) containment, and (4) incineration are available for the removal, reduction, or mitigation of the impacts of hydrocarbons in nature. However, the mentioned methods are characterized by drawbacks which include some level of risk during the treatment of contaminated media. Hence, the use of biologically-driven options can bridge the risk while enhancing sustainability. Therefore treatment options can

Table 2 Factors affecting the hydrocarbon biodegradation process

Parameters	Condition required	Optimum value for oil degradation
Temperature (°C)	14–45	Oil degradation
Nutrient content	N and P for microbial growth	C:N:P = 100:1:1
Oxygen content	Minimum 10 %	10–40 %
Soil pH	5.5–8.5	6.5–8
Soil moisture	25–28 %	30–90 %
Type of soil	–	Silt or low clay

be in situ or ex situ dependent on the type, site and level of contamination. Significantly, the ability of microbes and plants to breakdown (degrade) and transform contaminants enhance the cleanup of hydrocarbons from spills and storage sites. Such knowledge informed the basis for most in situ remediation options in practice and the majority of ex situ treatment activities (Speight and Arjoon 2012). Therefore, any treatment option can target specific or multiple needs (wastewater, groundwater, soil/sludge, gas) with the aim of either immobilizing contaminants or mineralization of organic compounds which includes hydrocarbons.

3.1 In Situ Remediation of Hydrocarbons

Hydrocarbon contaminated media can be recovered by the adoption of an in situ bioremediation (ISB) process. This process is basically the application of bioremediation at the subsurface. The application can be undertaken within an unsaturated/vadoze zone or saturated soils and groundwater. Originally, the ISB technology was developed as a cheaper and significantly more effective alternative to the usual pump-and-treat methods often employed in the cleanup of aquifers and organo-chemically contaminated soils. However, this technology has since gone beyond organics to incorporate treatment options in other situations like toxic metals, explosives and inorganics. The advantages of ISB stretch from allowing for complete destruction of the contaminant (hydrocarbon), lowering risk to site workers, to utilization of simpler equipment reducing operating costs.

There are two broad ways to categorize ISB, namely (1) by the type of metabolism involved, and (2) by the degree of human intervention. Aerobic and anaerobic conditions are the higher level categories of metabolism considered. Specificity to contaminants of interest determines the target metabolism for an ISB system. Hydrocarbons are degraded through an aerobic pathway whereas some other contaminants can be degraded anaerobically or either way (anaerobic or aerobic) as in the case of carbon tetrachloride and trichloroethene, respectively.

In the case of the degree of human intervention, accelerated ISB is a typical situation at one end of the scale. Under accelerated ISB, a contaminated system is amended with nutrient or substrate typically aimed to enhance the growth of specific bacteria consortia. The specific bacteria are often seen as being "resident"/ "indigenous" to the system. However, the degradation of the contaminants is enhanced when enriched cultures of bacteria (isolated from other sites) are introduced into the contaminated media via a condition known as bioaugmentation. The process is often adopted when there are limited nutrients, electron acceptor and/or donor which slow down the rate of contaminant biotransformation. Hence the amendments are expected to increase biotransformation of contaminant (e.g., hydrocarbon), while every amendment is determined by the type of target metabolism for the targeted contaminant. Most times, ISB under aerobic condition may need the addition of electron acceptor only (usually oxygen), whereas anaerobic ISB often measure the provision of both electron donor (e.g., benzoate, lactate) and electron acceptor (e.g., sulphate, nitrate).

However, the other end of the scale has a method of implementing in situ bioremediation without human intervention; monitored natural attenuation. This approach is multi-faceted, wherein one aspect of it is the transformation/degradation of contaminants by the resident/indigenous microbes in the absence of human intervention. It is a situation where whatever nutrient available in the media (e.g., soil, aquifer) is utilized by the resident microbes without any amendment or enrichment.

Therefore, it is pertinent to note that in situ remediation of contaminated systems/media includes chemical and biological methods (Doelman and Breedveld 1999). Intrinsically, in situ bioremediation is also a natural attenuation process which depends on natural, bio-, physic-, and chemical subsurface processes to remove, degrade or biotransform contaminants (EPA 1999). Still, the potentials of natural attenuation under certain environments need to be thoroughly investigated because of the finite capacity of the process (Rittmann 2000, 2004). A number of the commonly and widely used in situ bioremediation options for hydrocarbon contaminated soils are listed in Table 3.

More so, the adoption of accelerated ISB or MNA for a hydrocarbon-contaminated site must be determined based on the properties of the contaminated media, chemical/contaminant concentrations, objective of the remediation project and the financial implications of the anticipated approach. Usually, degradation

Table 3 In-situ bioremediation methods for hydrocarbon contaminated soil

Methods	Description	Cost	Hydrocarbon treated	References
Intrinsic remediation/monitored natural attenuation (MNA)	Depends on natural subsurface processes.	According to the duration of monitoring	Benzene, toluene, ethyl benzene and xylene, petroleum hydrocarbons.	Renner (1998), Mulligan and Yong (2004) and Salminen et al. (2004)
Biosparging	Oxygen is introduced below groundwater surface	50–110 €/t	Organic contaminants	Doelman and Breedveld (1999)
Bioventing	Oxygen is induced to soil vapour phase	25–120 €/t	Petroleum hydrocarbons	FRTR (2005) and EPA (2005)
Phytoremediation	Plants are utilized	Method specific	Diesel, lubricating oil, petroleum hydrocarbons	Adams et al. (2000) and Dadrasnia and Agamuthu (2013)

under natural attenuation is slow as opposed to enhanced bioremediation (microbes-oriented or phyto-based). In situ bioremediation strategies are discussed as follows;

3.1.1 Bioventing, Biosparging and Soil Vapour Extraction

The bioventing technology supplies additional oxygen into hydrocarbon-contaminated media, which stimulates natural aerobic biodegradation. It uses low airflow rates to induce sufficient oxygen for sustaining microbial activity. Direct air injection is used to introduce oxygen into the residual contamination in soil. As the vapours move slowly across the biologically active soil, hydrocarbon residues are absorbed.

Biosparging is another option used to enhance the soil's activities (biological). Most times, pure oxygen replaces air injections as the former increases degradation rates. Unfortunately, such treatments are considered expensive, alongside with the limited degree of dissolved oxygen available for microbes. Hence, hydrogen peroxide (H_2O_2) can be an alternative to supply additional oxygen to most polluted sites (Schlegel 1977). This is considered more efficient in terms of enhancing microbial activity when adopting bioremediation approaches for contaminated groundwater and soil (Brown et al 1996), though toxicity is most likely to occur (Brown and Norris 1994).

Similarly, soil vapour extraction aids in the direct removal of vapour (usually from the soil above the water table). The system usually involves the boring of extraction wells within the contaminated zone. A vacuum is applied to the well for the purpose of pulling air and vapours from the soil depth to the surface. In most cases, special equipment for air pollution is used to collect the extracted air for the purpose of separating the harmful vapours from the clean air. Basically, air used in soil vapour extraction helps microbial growth within the polluted system. Hence, bioventing, biosparging and soil vapour extraction work on the principle of oxygen or air utilization to enhance microbial activities within a contaminated system as a means of enhancing removal of contaminants. These systems have been successfully utilized for the recovery of petroleum hydrocarbons from contaminated soils.

To ensure successful bioremediation of hydrocarbon-contaminated sites when adopting the techniques, the following criteria need to be satisfied;

1. As a way of ensuring aerobic environment, air in sufficient levels must be allowed through the system.
2. Microbes which naturally degrade hydrocarbons are required and have to exist in substantial levels necessary to initiate higher biodegradation rate. Hence, it is necessary to initially determine the degree of air penetrability and in situ respiration rates of the soil before actual implementation of remediation plan.

Limitations:

The effective application of the techniques could be mitigated by certain issues which include:

1. Performance of bioventing can be reduced due to distance of the water table from the surface, saturated soil lenses or low soil permeability.
2. Sometimes, vapours may accumulate in pockets within the radius influenced by the air injection wells.
3. Biodegradation may be limited due to extremely low moisture content of the soil.
4. Processes may be slowed due to low temperatures yet success has been recorded on remediation within extremely cold weather climates.

3.1.2 Enhanced Bioremediation

This is a process which involves the use of indigenous/resident or inoculated microbes to degrade or metabolize hydrocarbons or other organic pollutants identified in soil and/or groundwater, and convert them to safe end products. Different amendments, nutrients or oxygen are used to boost bioremediation and desorption of contaminants from soil or materials at the subsurface. The microbes function when oxygen is relatively abundant (aerobic condition) alongside other nutrient elements, to significantly convert the hydrocarbon pollutants into carbon dioxide (CO_2), water and microbial cell mass.

Therefore, the core process is that the nutrients percolate the contaminated area which is additionally saturated with dissolved oxygen. Sometimes the percolation or injection may be supplemented with acclimatised microbes (known as bioaugmentation) and/or hydrogen peroxide as a source of oxygen. Basically, injection wells are utilized when contamination is deep in the soil, while spray irrigation or an infiltration gallery is used on shallow-contaminated soils. Temperature level impacts the extent of enhanced bioremediation. Such in situ bioremediation has been a success in cold regions of the globe, yet the bioremediation process can be delayed by low temperature as found in the case of bioventing, biosparging or soil vapour extraction. Therefore, heat blankets are used in situations of low temperature-contaminated soil to enhance rates of degradation.

Success of enhanced bioremediation has been recorded for the remediation of soils, sludge, and groundwater contaminated with petroleum hydrocarbons, solvents, pesticides, wood preservatives, and other organic chemicals. This treatment was also effective for anaerobic microbial degradation of nitrotoluenes in soils contaminated with munition wastes (FRTR 2005). Contamination with PAHs, non-halogenated SVOCs (excluding PAHs) and BTEX are commonly treated with this method especially when contaminations are due to wastes that emanate from the preservation of wood, refining and reuse of petroleum. To ensure successful bioremediation of hydrocarbon-contaminated sites when adopting enhanced bioremediation, the following conditions need to be evaluated;

1. Identification of distinct characteristics of contaminants which include leaching potential (water solubility and soil sorption coefficients), chemical reactivity and biodegradability.
2. Soil characteristics which include soil depth and extent of the contaminated zone, contaminants concentrations, soil type and properties (pH, permeability, moisture and nutrient levels, water-holding capacity and organic content), oxygen competition, availability or absence of potentially toxic compounds to microbes, and contaminant degradation ability of the soil microbes.

Limitations:

The effective application of the techniques could be mitigated by certain issues which include:

1. Where the soil matrix does not allow contact between contaminant and microbes, then the goals of the cleanup may not be reached.
2. Mobility of contaminant may be undesirably enhanced due to the spread of solutions (water-based) through the soil, as this may require the treatment of core groundwater.
3. Water injection into the wells and nutrient clogging may lead to preferential colonization of microbes.
4. This option is not suitable for clay, heterogeneous subsurface or highly layered areas due to oxygen transfer limitations. Absolute reduction in contact between injected fluids and contaminants may occur throughout the contaminated zone due to preferential flow paths.
5. Low temperatures reduce bioremediation process.
6. Activity of microbes will be inhibited if more than 100–200 ppm of hydrogen peroxide is introduced into groundwater.

3.1.3 Phytoremediation

Phytoremediation is a plant-based technology used for the remediation (removal, stabilization, transfer and/or destruction of hydrocarbon contaminants in soils and sediments. The system uses different mechanisms that include enhanced rhizosphere biodegradation, phytoaccumulation (phytoextraction), phytodegradation and phytostabilization. This implies that the hydrocarbons may be degraded, extracted from the soil and deposited in the plant.

For hydrocarbon decontamination, the core phytoremediation mechanisms actively involved are plant uptake and phytodegradation that initiate volatilization and rhizodegradation. In rhizodegradation, anaerobic and aerobic degradation pathways have to be considered because microhabitats exist for microbes under both conditions (Walton et al. 1994). Hydrocarbon uptake in plants can be at the root level (Trapp et al. 1990) and in the above-ground biomass (Chaineau et al. 2000). Enhanced rhizosphere biodegradation occurs in the soil immediately surrounding the plant roots. The existing microbes in turn obtain nutrients via natural substances released by the plants to encourage their biological activities. As plant roots loosen soil compartments, routes for water and aeration are created. Such activity draws water to the surface and dries the lower saturated zone. Poplar trees are most commonly used due to the fast-growth and ability to thrive in a wide range of climates. While drawing water to the surface, the plants also draw the hydrocarbons (pollutants) from the contaminated system which possibly reduces the level of contaminant flushed through or out of the soil or aquifer.

Phytodegradation is actually the metabolism of pollutants within plant tissues. Such degradation can be enhanced by enzymes like dehydrogenases and oxygenases produced by plants. On the other hand, phytoaccumulation is the uptake of pollutants (e.g., hydrocarbons) by the plant roots prior to possible translocation to shoots and leaves. However, chemical compounds can be produced by plants and used to immobilize the pollutant. This often takes place at the interface of roots and soil.

Remediation via the phyto-approach is conveniently applied to crude oil-contaminated sites as well as in other situations like pesticides, landfill leachates, metals, solvents, and PAH pollution. Plants with known hydrocarbon remediation potential include but are not limited to *Vallisneria spiralis* (Liu et al. 2014), *Echinacea purpurea* (Purple cornflower), *Callistephus chinensis* (Aster Callistephus), *Festuca arundinacea Schreb* (Fawn), and *Medicago sativa Linn* (Alfalfa) (Xiao et al. 2015). To ensure successful phytoremediation of hydrocarbon-contaminated media, especially soil, it is important to evaluate the following; the kinds of soil used in the project, Concentrations of reductive oxygen, water movement, growth and structure of the plant root.

Limitations:

The effective application of the techniques could be mitigated by certain conditions:

1. High concentrations of pollutant can initiate toxicity to the plant.
2. Depth of treatment zone can only be determined by the utilized plant. In fact, it only works with shallow soils in most cases.
3. The process may be seasonal and depends on the geographical location of contaminated media.
4. The process may immobilize the pollutant in the groundwater or it might bioaccumulate in animals.
5. The process is often unfamiliar to most regulators.

3.2 *Ex-Situ Remediation of Hydrocarbons*

The similarity of this system of technology to ISB ends with being a biological process, because ex situ bioremediation (ESB) involves the excavation of the solid media (e.g., soil) or extraction of aqueous media (e.g., wastewater, groundwater), with treatment occurring in a structured above-ground treatment area. It can involve aeration to enhance the degradation impact of resident/indigenous microbial populations. Similarly, anaerobic conditions can be applied to stimulate specific and non-oxygen dependent microbes that metabolize organics like petroleum hydrocarbon mixtures, PAHs, phenols, cresols and other hydrocarbons, before degrading them to carbon dioxide and water. ESB is useful approach for bioremediation of polar soils (Aislabie et al. 2006; Nugroho et al. 2010).

Ex-situ bioremediation can be used to tackle a variety of hydrocarbon-polluted environments such as oil fields, petroleum storage sites, and oil spills in marine waters. The remediation situations may apply to general hydrocarbons, phenols and cresols, kerosene, PAHs and semi-volatile organic compounds, lubricating oils and diesel, straight chain oliphatics, non-chlorinated hydrocarbons with carbon chain lengths C_6–C_{14}, chlorinated hydrocarbons with carbon chain lengths C_{15}–C_{32} (but it is time consuming). Basically chlorinated hydrocarbons take a longer time to degrade.

Various ESB methods are available for the remediation of hydrocarbon-impacted media ranging from land-farming, turned windrows (composting), forced vented bioplies, to slurry phase biological treatment.

3.2.1 Land-Farming

Sometimes considered the simplest form of bioremediation, this technology involves the excavation of the contaminated media before spreading out in layers (approx. 0.3 m thick) on a lined treatment area. The remediation activity is boosted when the bed is turned periodically along with nutrients addition. Land-farming options are optimized through control of certain soil conditions that eventually enhances the rate of hydrocarbon (contaminant) degradation. Such controlled conditions are:

- Moisture content—using spraying or irrigation.
- pH—using crushed limestone or agricultural lime to buffer to near neutrality.
- Aeration—using tilling at a predetermined frequency to ensure proper soil mixing.

This technique is applied to remediation of petroleum hydrocarbon situations. It can be used to treat diesel fuel, Nos. 2 and 6 fuel oils, oily sludge and certain pesticides.

Limitations:

Issues associated to the technique are;

1. It requires a large operational space.
2. It is difficult to control certain conditions (e.g., rainfall, temperature) which influence biodegradation of contaminants, and as such increases the duration of remediation.
3. Pre-treatment is required for remediation of volatile contaminants to avoid air pollution.

4. Runoff collection systems need to be constructed.
5. To determine the best design for the facility, the topography, erosion and climatic issues need to be evaluated.

3.2.2 Turned Windrow

Considered to be a composting approach, this is a controlled biological process where microorganisms are employed to convert hydrocarbons or other organic contaminants into safe, stabilized by-products. The process involves excavation and mixing of soil with bulking agents and organic amendments like manure, hay, wood chips and even food wastes. When the amendment is properly done, it allows for sufficient porosity and an appropriate carbon to nitrogen balance that enhance microbial activity (usually thermophilic).

The excavated soil is piled up to 1.5–2 m high with a 6 m width within a lined area. Proper aeration and periodic turning is carried out on the pile using specialized machinery. During the turning of windrows, it can be augmented with organic amendments and nutrients. This approach is often cost-effective. The system can be applied on areas contaminated with PAH, explosives (TNT, RDX and HMX) and ammonium picrate.

Limitations:

Issues associated to the technique are;

1. It requires large operational space.
2. During excavation of contaminated soil, there is high risk of uncontrolled VOC release.
3. The system leads to volumetric increase in material due to additional amendment materials.

3.2.3 Forced Vented Biopiles

Biopile treatment is a full-scale technology where soil that was excavated is blended with soil conditioners and kept at the site of treatment. The treatment system incorporates leachate collection and aeration (vacuum pump or air injection blower system). This process is used to reduce concentrations of petroleum constituents from the contaminated media via biodegradation. The biopile is up to 3 m in height to occupy less space than the windrow. Furthermore, the vapour that emanates from the biopile can be collected and treated on-site via the use of air bio-filters or granular activated carbon (GAC) to mitigate air pollution.

The system is viewed to be a short-term technology, and can be operated and maintained for a few weeks to some months. It is seen as a composting system that operates on the principle of an aerated static pile. The process can be readily applied to contaminations involving non-halogenated VOCs and fuel hydrocarbons.

Limitations:
Issues associated to the technique are:

1. It requires complete excavation of contaminated soil.
2. Treatability tests must investigate the biodegradability of contaminants, oxygenation and nutrient loading rates.
3. Some doubt still exists on the effectiveness for the treatment of halogenated compounds especially when a solid phase process is used.
4. Clean-up times are longer at similar batch sizes compared to windrow
5. The static nature of the treatment can easily lead to non-uniform treatment as opposed to systems that utilize periodic mixing.

3.3 Advanced Remediation Technologies

A typical system that represents advanced bioremediation technology is the slurry phase biological treatment. This system is ex situ based and ideally involves the precise remediation of excavated contaminated soil using a bioreactor. In this process, physical separation of bulk materials like stones and rubble is carried out before mixing the soil with water. The extent of mixing depends on contaminant concentration, biodegradation rate and physical properties of the soil. Sometimes pre-washing is required in order to concentrate the contaminant. Subsequently, clean sand can be removed while only contaminated fines and wash water (to be bio-treated) are left behind. The slurry composition is often 10–30 % solids by weight.

Furthermore, the solids generated from the slurry are maintained in suspension in a bioreactor vessel where they undergo mixing with nutrients and oxygen. The pH level may be adjusted when necessary by the addition of acid or alkali. Also, the desired microbes can be introduced, especially when the resident population is found insufficient in enhancing biodegradation. Upon completion of the biodegradation process, de-watering of the slurry must be carried out. Various dewatering devices such as clarifiers, vacuum filters, centrifuges, pressure filters or sand drying beds can be installed.

This system is considered as a short- to medium-term technology which has been successfully employed for the remediation of petroleum hydrocarbons, petrochemicals, sediments impacted with pesticides and other organic chemicals. Its cutting edge over ISB is that it takes care of heterogeneous soils, soils with low permeability, situations where groundwater is not easy to capture. The bioreactor or system is also a faster treatment option.

Limitations:
The effectiveness of the technique could be mitigated by certain conditions;

1. Except for its implementation on lagoons, it is compulsory to excavate any contaminated media.

2. The cost of resizing materials before depositing them into bioreactor vessels can be high and even difficult. This is typical of heterogeneous soil.
3. Cost of de-watering soil fines after treatment can be high.
4. Non-recyclable wastewaters generated in the process need to be disposed in an acceptable manner.

It is also important to note that the residence time in the bioreactor varies with contaminants of interest based on the nature, concentration and intended level of remediation.

4 Case Studies

The principles of bioremediation of hydrocarbon-associated contamination have been applied in different places. Applications have shown varying degrees of effectiveness often due to differences in climates, cost, concentrations of contaminants of interest, extent of contamination etc. Here, some case studies are outlined to reflect the applicability and effectiveness of some in situ bioremediation options as discussed in the previous chapter. The case studies presented show specific remediation technologies adopted in various clean-ups. Such technologies could be extrapolated to any other contaminated site while putting into considerations the associated costs and complexities of hydrocarbon in combination with other pollutants.

4.1 Treatment of Petroleum, Oil and Lubricant (POL) Contaminated Soils in Alaska Using Rhizosphere-Enhanced Bioremediation

A field-scale remediation of three (3) sites in Alaska (Annette Island, Galena/Campion and Barrow) was carried out using rhizosphere-enhanced bioremediation. The sites were characterized for petroleum hydrocarbons especially gasoline range organics (GRO) and diesel range organics (DRO), and even minute levels of tetrachloroethane (PCE). Cold-tolerant grass/clover species were used to seed the contaminated sites which were also amended with fertilizer. The fertilizer addition and seeding were installed in the summer of 1998, except for the installation at Barrow site which was completed in the summer of 1999. Remediation at all sites was monitored until the summer of 2000. Results showed that significant plant growth was observed at the contaminated sites and the contaminant levels were reduced. While the cleanup goals were not quantitatively shown, the capital cost of the activities was \$7,250 with an additional \$1,400 per year for operation and maintenance. General information on the remediation activity is stated as follows.

Site Names and locations: Pollution due to POL was recorded at three (3) different locations in Alaska, namely:

- Former U.S. army air force landing field (Annetter Island) in Metlakatla, Alaska
- Former air force station (Galena/Campion) in Galena
- Former tank farm, and former dry cleaning facility, both at Barrow

Type of Cleanup: Classified as field demonstration

Adopted technology: rhizosphere-enhanced bioremediation

- Cold-tolerant grasses were used (a seed mixture of three species) namely Arctared red fescue (60–70 %), annual ryegrass (10–15 %), and white clover (20–25 %).
- Minimal soil preparation was carried out prior to seeding.
- Seeds were applied to the surface by directly pressing the seed into the soil with the aid of hand-held seeders.
- <2,000 mg nitrogen/kg of soil (permissible level) was introduced to serve as standard agricultural fertilizer
- Controls were fertilized conditions without seeds and seeding without fertilization.
- ***Contaminants:*** The major pollutants were organic especially DRO and GRO
  - there was no specific information on the organic pollutant in Annette Island but due to land use, soil-fuel-related contaminants were present.
  - the assessment of 1995 data from Galena/Campion showed that about 36–75,000 and 59–7,500 mg/kg range of organic contaminants were DRO and GRO, respectively.
  - at Barrow, the soil of the dry cleaning facility contained 230–810 mg/kg of DRO whereas the GRO concentration ranged from below detection limit to 85 mg/kg. Similarly, soil from the tank farm contained 47–9,400 mg/kg of TPH; 8.1–365 mg/kg of Pb while other compounds detected included PAHs, phenolics, BTEX, halogenated aliphatics, solvents and inorganics.
- ***Media treatment***: The core medium involved was soil.
  - predominately surface soils (coarse sand and gravel marine beach deposits), however in vegetated areas the soil was silty.
  - the area of petroleum-contaminated soil was about 5351.884 m^3 (7,000 cubic yards).
  - soils are frozen at most times of the year, except between August or September with a maximal depth of 55 in. which refreeze by late October.
  - there is no significant groundwater flow, since groundwater is found just within the thawed zone above the permafrost
- ***Purpose/Significance of Application:***
  - To use cold-tolerant plants to assess the bioremediation of POL-contaminated surface soils in a cold-climate region.

- To comparatively evaluate the degree of effectiveness of bioremediation on unfertilzed or unplanted areas against fertilized and planted sections.
- To understand the interaction between root-zone microbes and degradation of the contaminant
- To understand the associated population dynamics of the microbes and the associated composition

- ***Regulatory Requirements/Cleanup Goals:*** expected to reduce the concentration of contaminants (short-term goal); to recover land (Native American) contaminated by hydrocarbons (long-term goal).
- ***Results:***
  - although quantitative results were not provided but there was reduction in the concentration of contaminants.
  - pronounced plant growth was recorded for fertilized areas
  - it was concluded that continued remediation especially during August and September will help to achieve the long-term cleanup goals.
- ***Cost Factors:***
  - \$7,250 was used as capital cost and yearly operation and maintenance costs were \$1,400.
  - the cost of other activities especially excavation and disposal of residues, regulatory oversight and compliance, and even prolonged monitoring, amounted to \$6,000 per year.
  - the total cost incurred every 10 years of operation is estimated to be \$27,250. This is based on treatment space [929.03 m^2 (10,000-ft^2)] and depth [0.62 m^2 (2 ft)].

4.2 Enhanced in Situ Anaerobic Bioremediation of Groundwater Contaminated with Fuel

Due to replacement of a steel fuel tank with a fiberglass type, there was leakage at the Naval Weapons Station (NWS) Seal Beach located in southern California (between Long Beach and Huntington Beach) in 1984. The leaked fuel (approximately 5800 gallons) penetrated the soil profile to groundwater. This pollution was a big threat to existing wildlife.

Of the 20 acres of contaminated area, only a portion of the site was used for the remediation test. The project was designed to compare the effect of various nitrate and sulfate concentrations added to the three zones in relations to the three injection wells and one extraction well. The assessment indicated an overall reduction in BTEX concentrations. However, preferential degradation was observed on toluene.

Degradation of ethylbenzene and xylenes were similar, while flushing was used to remove benzene. Projected costs for full-scale implementation of this technology were 30 % less than for the usual pump and treat approach.

- ***Site Name and location:*** Naval Weapons Station Seal Beach, Southern CA.
- ***Duration:*** activities were conducted from September, 1997 to October, 1998.
- ***Cleanup type:*** classified as field demonstration.
- ***Adopted technology:*** in situ bioremediation.
- the system involved the use of injections wells (3 zones) at 180 m^3 each and an extraction well.
- rate of injection was 1.5 L/min/well while the rate of extraction was 4.5 L/min.
- each zone was augmented with different electron acceptors – sulfate, one with sulfate and nitrate, and one without augmentation.
- Automated sampling systems were used.
- ***Contaminants:*** Fuel hydrocarbons and BTEX
- Benzene, ethlybenzene and m + p-xylenes were present in the groundwater at 4,000 μg/L, 250 μg/L, 500 μg/L, respectively.
- ***Waste source***: Leaks from USTs
- ***Type/quantity of media treated***: Groundwater (in situ), Soil (in situ), light non-aqueous phase liquids
- 20 acres of contaminated area, a portion was used to conduct the demonstration
- depth to groundwater low; transmissivity of groundwater was more than 2 ft^2/day and groundwater velocity was 0.7 cm/s.
- groundwater had been anaerobic for more than 10 years
- ***Purpose/significance of application***: Aimed at demonstrating anaerobic bioremediation as a treatment option for fuel hydrocarbons
- ***Regulatory requirements/cleanup goals:***
  - to assess the technical viability of treating petroleum hydrocarbons with the adopted technology, and the use of nitrate and sulfate to stimulate biodegradation of BTEX.
  - cleanup goal was not specified.
- ***Results:***
  - reduction in BTEX compounds concentrations, preferential degradation of toluene was noticed.
  - nitrate addition stimulated the degradation of ethylbenzene and m + p-xylene to less than 10 and 20 μg/L, respectively.
  - sulfate addition stimulated the degradation of O-xylene to less than 10 μg/L.
  - flushing was the main cause of benzene removal rather than biodegradation.
- ***Cost Factors:***
  - More than 9000 samples were collected and the activities cost $875,000, which included equipment, labor, laboratory supplies, travel, and overheads.

5 Conclusion

The abundance of hydrocarbons and associated derivatives in the environment, especially in soil and groundwater is often due to activities ranging from storage of fuels, transportation to maintenance and repairs. There has been serious interest in the cleanup, recovery and restoration of media (soil, water, air) contaminated with hydrocarbons. As discussed in this chapter, the ability of most hydrocarbon compounds to undergo biological breakdown has formed the basis for the adoption of bioremediation as not only a green approach towards cleanup of contaminated media, but also a cost-effective and sustainable option.

The chapter has elucidated the on-site (in situ) and off-site (ex situ) biological systems that can be adopted for the remediation of hydrocarbon contaminated media. Such systems consider the type of metabolism suitable for degrading a particular contaminant or even across functional groups, and the degree of human intervention. Prior to the adoption of ISB and ESB, pump-and-treat methods were prevalent but with demerits that include high cost of operation and time consuming. The addition of enriched cultures of microbes (bioaugmentation) has proven to be a formidable approach towards the degradation of hydrocarbon contaminants. Therefore, methods like bioventing, biosparging, soil vapour extraction, enhanced bioremediation/biorestauration, and phytoremediation are in situ strategies adopted for the remediation of hydrocarbon-polluted environments.

Similarly, the adoption of land-farming, turned windrow and forced vented biopiles are ex situ methods that provide suitable conditions for the biodegradation of hydrocarbons. Though the options discussed have significant potentials in remediating hydrocarbon-polluted environments, yet they are not devoid of issues which include, but not limited to, the need for large operational space, controlled environmental conditions (temperature, moisture content), tedious excavation, and pre-treatment etc. Case studies have demonstrated the biodegradation potentials for both in situ and ex situ strategies as short- and long-term options that are cost-effective. Therefore, it is imperative to conclude that bioremediation is a sound technology for the degradation and removal of hydrocarbons in the environment. However, the suitability of discussed options is harnessed when any bioremediation project considers the properties of contaminated media, chemical/contaminant concentrations, and objective of the remediation project and the financial implications of the anticipated approach.

References

Abioye OP, Agamuthu P, Abdul Aziz AR (2012) Biodegradation of Used Motor Oil in Soil Using Organic Waste Amendments. Biotechnol Res Int 2012:8. doi:10.1155/2012/587041

Abioye OP, Alonge OA, Ijah UJJ (2009) Biodegradation of Crude Oil in Soil Amended with Melon Shell. AU J Technol 13(1):35–38

Adams N, Carroll D, Madalinski, K, Rock, S, Wilson T, Pivetz B (2000) Introduction to phytoremediation. National Risk Management Research Laboratory. Office of Research and Development. US EPA Cincinnati, Ohio. EPA/600/R-99/107

Adesodun JK, Mbagwu JSC (2008) Biodegradation of waste-lubricating petroleum oil in a tropical alfisol as mediated by animal droppings. Bioresour Technol 99(13):5659–5665

Afuwale CD, Modi HA (2012) Preparation of bacterial consortium for enhancing degradation of crude oil. J Adv Dev Res 3:63–69

Ahmed AM (2004) Psychrophilic degradation of long chain alkanes. der Technischen Universität Hamburg-Harburg

Aislabie J, Saul DJ, Foght JM (2006) Bioremediation of hydrocarbon-contaminated polar soils. Extremophiles 10:171–179

Angelidaki I, Sanders W (2004) Assessment of the anaerobic biodegradability of macropollutants. Rev Environ Sci Bio/Technol 3:117–129

Atlas R, Bragg J (2009) Bioremediation of marine oil spills: when and when not—the Exxon Valdez experience. Microb Biotechnol 2(2):213–221. doi:10.1111/j.1751-7915.2008.00079.x

Baraniecki CA, Aislabie J, Foght JM (2002) Characterization of Sphingomonas sp. Ant 17, an aromatic hydrocarbon-degrading bacterium isolated from Antarctic soil. Microb Ecol 43:44–54

Bartha R, Bossert I (1984) Petroleum microbiology. In: Atlas RM (ed) The treatment and disposal of petroleum refinery wastes. Macmillan, New York, pp 553–557

Bej AK, Saul D, Aislabie J (2001) Cold-tolerant alkane-degrading Rhodococcus species from Antarctica. Polar Biol 23:100–105

Ben Ayed H, Jridi M, Maalej H, Nasri M, Hmidet N (2014) Characterization and stability of biosurfactant produced by Bacillus mojavensisA21 and its application in enhancing solubility of hydrocarbon. J Chem Tech Biotech. 89:1007–1014

Bento FM, Beech IB, Gaylarde CC, Englert GE, Muller IL (2005) Degradation and corrosive activities of fungi in a diesel–mild steel–aqueous system. World J Microb Biotech 21(2):135–142

Boonchan S, Britz ML, Stanley GA (2000) Degradation and Mineraliza-tion of high molecular weight polycyclic aromatic hydrocarbons by defined fungal bacterial cocultures. Appl Environ Microb 66(3):1007–1019

Boopathy R (2000) Factors limiting bioremediation technologies. Bioresour Technol 74:63–67

Brown RA, Norris RD (1994) The evolution of a technology: hydrogen peroxide in In-situ bioremediation. In: Hinchee RE, Alleman BC, Hoeppel RE, Miller RN (eds) Hydrocarbon bioremediation. CRC Press, Boca Raton, pp 148–162

Brown ED, Norcross BL, Short JW (1996) An introduction to studies on the effects of Exxon valdez oil spill in early life history stages of Pacific herring,(clupae pallasi), in Prince Williian Sound, Alaska. Can J Fish Aquat Sci 53:2337–2342

Cao B, Nagarajan K, Loh K (2009) Biodegradation of aromatic compound: current status and opportunities for biomolecular approaches. Appl Microb Biotechnol Adv 85:207–228

Chaillan F, Chaıneau CH, Point V, Saliot A, Oudot J (2006) Factors inhibiting bioremediation of soil contaminated with weathered oils and drill cuttings. Environ Pollut 144:228–237

Chaineau CH, Morel JL, Oudot J (2000) Biodegradation of fuel oil hydrocarbons in the rhizosphere of maize. J Environ Qual 29:569–578

Chakraborty R, Coates JD (2004) Anaerobic degradation of monoaromatic hydrocarbons. Appl Microb Biotechnol 64:437–446

Colla T, Andreazza R, Bücker F, Souza M, Tramontini L, Prado G, Frazzon A, Camargo F, Bento F (2014) Bioremediation assessment of diesel–biodiesel-contaminated soil using an alternative bioaugmentation strategy. Environ Sci Pollut Res 21:2592–2602

Chrzanowski Ł, Kaczorek E, Olszanowski A (2006) The ability of Candida Maltosa for hydrocarbon and emulsified hydrocarbon degradation. Polish J Environ Stud 15(1):47–51

Collins CD (2007) Implementing phytoremediation of petroleum hydrocarbons. In: Willey N (ed) phytoremediation: methods and reviews, vol 23. Wiley-Blackwell, Totowa, p 512

Colombo JC, Cabello M, Arambarri AM (1996) Biodegradation of aliphatic and aromatic hydrocarbons by natural soil microflora and pure cultures of imperfect and lignolitic fungi. Environ Pollut 94(3):355–362

Dadrasnia A, Agamuthu P (2014) Diesel toxicity and soil remediation—technical report. Malays J Sci 33(2):92–97

Dadrasnia A, Agamuthu P (2013) Diesel fuel degradation from contaminated soil by *Dracaena reflexa* using organic waste supplementation. Int J Jpn Petrol Inst 56(4):236–243

Dadrasnia A, Usman MM, Wei KSC, Velappan RD, Jamali H, Mohebi N, Salmah I (2016) Native soil bacterial isolate in Malaysia exhibit promising supplements on degrading organic pollutants. Proc Safet Environ Prot 100:264–271

Das N, Chandran P (2011) Microbial degradation of petroleum hydrocarbon contaminants: an overview. Biotechnol Res Int. doi:10.4061/2011/941810

Delille D (2000) Response of Antarctic soil assemblages to contamination by diesel fuel and crude oil. Microb Ecol 40:159–168

Díaz-Ramírez I, Escalante-Espinosa E, Schroeder RA, FócilMonterrubio R, Ramírez-Saad Hugo (2013) Hydrocarbon biodegradation potential of native and exogenous microbial inocula in Mexican tropical soils. Biodegrad Hazard Spec Prod. doi:10.5772/56233

Doelman P, Breedveld G (1999) In situ versus on site practices. In: Andriano DC, Bollag JM, Frankenberger WT, Sims RC (Eds) Bioremediation of contaminated soils. Agronomy No. 37, American Society of Agronomy, Inc. Soil Science Society of America, Inc. Madison, Wisconsin, USA, pp 539–558

Eckford R, Cook FD, Saul D, Aislabie J, Foght J (2002) Free-living heterotrophic nitrogen-fixing bacteria isolated from fuel-contaminated Antarctic soils. Appl Environ Microb 68:5181–5185

EIA (2011) U.S. Environmental Infirmation Agency, China electricity. http://www.eia.gov/

EPA (2006) U.S. Environmental Protection Agency. Integrated Risk Information System website. http://epa.gov/iris/

EPA (2005) EPA REACH IT database. Remediation and characterization innovative technologies. Cited http://www.epareachit.org/ April 23rd, 2015

EPA (1999) Use of monitored natural attenuation at superfund, RCRA corrective action and underground storage tank sites. OSWER Directive Number 9200.4-17P. United States Environmental Protection Agency, Office of Solid Waste and Emergency Response, Washington

Eriksson M, Ka JO, Mohn WW (2001) Effects of low temperature and freeze-thaw cycles on hydrocarbon biodegradation in arctic tundra soil. Appl Environ Microbiol 67(11):5107–5112

Fan MY, Xie RJ, Qin G (2014) Bioremediation of petroleum-contaminated soil by a combined system of biostimulation-bioaugmentation with yeast. Environ Tech. 35(4)

Field JA, Alvarez SR (2007) Biodegradability of chlorinated aromatic compounds. University of Arizona. http://www.eurochlor.org/media/14960/8-5-12_sd_biodegradability_aromatic.pdf

Foght J (2002) Anaerobic biodegradation of aromatic hydrocarbons: pathways and prospect. J Mol Microbiol Biotechnol 15:93–120

Fritsche W, Hofrichter M (2005) Aerobic degradation of recalcitrant organic compounds by microorganisms. In: Jördening H-J, Winter J (Eds) Environmental biotechnology. Wiley-VCH Verlag GmbH & Co. KGaA, Weinheim, pp 203–227

Froehner S, Dombroski L, Machado K, Scapulatempo Fernandes C, Bessa M (2012) Estimation of bioavailability of polycyclic aromatic hydrocarbons in river sediments. Int J Environ Sci Technol 9(3):409–416

FRTR (2005) Federal remediation technologies roundtable remediation screening matrix and reference guide http://www.frtr.gov/matrix2/top_page.html

Gavrilescu M (2010) Environmental biotechnology: achievements, opportunities and challenges. In: Dynamic biochemistry, process biotechnology and molecular biology 4(1):1–36

George SC, Boreham CJ, Minifie SA, Teerman SC (2002) The effects of minor to moderate biodegradation on C5 to C9 hydrocarbons in crude oils. Org Geochem 33(12):1293–1317 http://www.sciencedirect.com/science/article/pii/S0146638002001171

Gibb A, Chu A, Wong RCK, Goodman RH (2001) Bioremediation kinetics of crude oil at 5 degrees. Canadian Journal of Environmental Engineering 127:818–824

Gong Z, Wilke BM, Alef K, Li P (2005) Influence of soil moisture on sunflower oil extraction of polycyclic aromatic hydrocarbons from a manufactured gas plant soil. Sci Total Environ 343 (1–3):51–59

Harekrushna S, Kumar DC (2012) A review on: bioremediation. Int J Res Chem Environ 2(1):13–21

Head IM, Jones DM, Roling WFM (2006) Marine microorganisms make a meal of oil. Nat Rev Microbiol 4:173–182

Hickman ZA, Reid BJ (2008) The co-application of earthworms (*Dendrobaena veneta*) and compost to increase hydrocarbon losses from diesel contaminated soils. Environ Int 34 (7):1016–1022

Hidayat A, Tachibbana S (2012) Biodegradation of Aliphatic Hydrocarbon in Three Types of Crude Oil by Fusarium sp. F092 under Stress with Artificial Sea Water. J Environ Sci Technol 5(1):64–73

Huesemann MH, Hausmann TS, Fortman TJ (2004) Does bioavailability limit biodegradation? A comparison of hydrocarbon biodegradation and desorption rates in aged soils. Biodegradation 15(4):261–274

Ijah UJJ, Antai SP (2003) The potential use of chicken-drop micro-organisms for oil spill remediation. Environmentalist 23(1):89–95

Jain PK, Gupta VK, Gaur RK, Lowry M, Jaroli DP, Chauhan UK (2011) Bioremediation of petroleum oil contaminated soil and water. Res J Environl Toxic 5:1–26

Juwarkar A, Singh S, Mudhoo A (2010) A comprehensive overview of elements in bioremediation. Rev Environ Sci Biotechnol 9(3):215–288

Kauppi S (2011) Bioremediation of diesel oil contaminated soil and water. University of Helsinki, Lahti

Kauppi S, Sinkkonen A, Romantschuk M (2011) Enhancing bioremediation of diesel-fuel-contaminated soil in a boreal climate: comparison of biostimulation and bioaugmentation. Int Biodeterior Biodegrad 65(2):359–368

Lee EH, Kang YS, Cho KS (2011) Bioremediation of diesel-contaminated soils by natural attenuation, biostimulation and bioaugmentation employing Rhodococcus sp EH831. Korean J Microb Biotech 39(1):86–92

Lee M, Kim MK, Singleton I, Goodfellow M, Lee ST (2006) Enhanced biodegradation of diesel oil by a newly identified Rhodococcus baikonurensis EN3 in the presence of mycolic acid. J Appl Microb 100:325–333

Liu H, Meng F, Tong Y, Chi J (2014) Effect of plant density on phytoremediation of polycyclic aromatic hydrocarbons contaminated sediments with *Vallisneria spiralis*. Ecol Eng 73:380–385

Madsen EL (2002) Report on bioavailability of chemical wastes with respect to the potential for soil bioremediation. Department of Microbiology, Cornell University, Ithaca

Mancera-López ME, Esparza-García F, Chávez-Gómez B, Rodríguez-Vázquez R, Saucedo-Castañeda G, Barrera-Cortés J (2008) Bioremediation of an aged hydrocarbon-contaminated soil by a combined system of biostimulation–bioaugmentation with filamentous fungi. Int Biodeterior Biodegrad 61(2):151–160

Margesin R, Schinner F (1999) Biodegradation of diesel oil by cold-adapted microorganisms in presence of sodium dodecyl sulfate. Chemosphere 38(15):3463–3472

Margesin R, Schinner F (2001a) Biodegradation and bioremediation of hydrocarbons in extreme environments. Appl Microbiol Biotechnol 56:650–663

Margesin R, Schinner F (2001b) Bioremediation (natural attenuation and biostimulation) of diesel-oil-contaminated soil in an alpine glacier skiing area. Appl Environ Microb 67:3127–3133

Masterson WD, Dzou LIP, Holba AG, Fincannon AL, Ellis L (2001) Evidence for biodegradation and evaporative fractionation in West Sak, Kuparuk and Prudhoe Bay field areas, North Slope, Alaska. Org Geochem 32:411–441

Msmg N, Msmd D (2012) Review paper on—parameters affecting bioremedaition. Int J Life Sci Pharma Res 2(3):77–80

Mulligan CN, Yong RN (2004) Natural attenuation of contaminated soils. Environ Int 30:587–601
Northcott GL, Jones KC (2000) Experimental approaches and analytical techniques for determining organic compound bound residues in soil and sediment. Environ Pollut 108:19–43
Nugroho A, Effendi E, Karonta Y (2010) Petroleum degradation in soil by thermophilic bacteria with bopile reactor. Makara Teknol 14(1):43–46
Oleszczuk P (2009) Application of three methods used for the evaluation of polycyclic aromatic hydrocarbons (PAHs) bioaccessibility for sewage sludge composting. Bioresour Technol 100 (1):413–420
Olson PE, Reardon KF, Pilon-Smits EAH (2003) Ecology of rhizosphere bioremediation. In: Mccutcheon S, Schnoor J (eds) Phytoremediation: transformation and control of contaminants. Wiley, Hoboken, pp 317–353
Onwurah INE, Ogugua VN, Onyike NB, Ochonogor AE, Otitoju OF (2007) Crude oil spills in the environment, effects and some innovative clean-up biotechnologies. Int J Environ Res 1 (4):307–320
Pallasser RJ (2000) Recognizing biodegradation in gas/oil accumulations through the data C-13 compositions of gas components. Org Geochem 31:1363–1373
Pandey B, Fulekar M (2012) Bioremediation technology: a new horizon for environmental cleanup. Biol Med 4(1):51–59
Park JA, Hur JM, Jang BK, Son BS (2001) Evaluation of compost addition and its effect on biodegradation of diesel oil in contaminated soil composting. J Ind Eng Chem 7(3):127–136
Pelletier E, Delille D, Delille B (2004) Crude oil bioremediation in sub-Antarctic intertidal sediments: chemistry and toxicity of oiled residues. Mar Environ Res 57(4):311–327
Philp J, Bamforth S, Singleton I, Atlas R (2005) Environmental pollution and restoration: a role for bioremediation. In: Atlas R, Philip J (Eds) Bioremediation. ASM Press, Washington, DC, pp 1–48. doi:10.1128/9781555817596.ch1
Pirnay JP, De Vos D, Cochez C, Bilocq F, Pirson J, Struelens M, Duinslaeger L, Cornelis P, Zizi M, Vanderkelen A (2003) Molecular epidemiology of *Pseudomonas aeruginosa* colonization in a burn unit: persistence of a multidrug-resistant clone and a silver sulfadiazine-resistant clone. J Clin Microbiol 41(3):1192–1202
Puglisi E, Cappa F, Fragoulis G, Trevisan M, Del Re A (2007) Bioavailability and degradation of phenanthrene in compost amended soils. Chemosphere 67:548–556
Rahman KSM, Thahira RJ, Laksh P, Banat IM (2002) Towards efficient crude oil degradation by a mixed bacterial consortium. Bioresour Technol 85(3):257–261
Renner R (1998) Intrinsic remediation under the microscope. Environ Sci Technol 32(7):180A–182A
Rittmann B (2000) Natural attenuation's promise and application. Water 21:20–21 https://asu.pure.elsevier.com/en/publications/natural-attenuations-promise-and-application
Rittmann BE (2004) Definition, objectives and evaluation of natural attenuation. Biodegradation 15:349–357
Rojo F (2009) Degradation of alkanes by bacteria. Environ Microbiol 11(10):2477–2490
Ron EZ, Rosenberg E (2002) Biosurfactants and oil bioremediation. Curr Opin Biotechnol 3:249–252
Roy JI (1997) Environmental contaminants encyclopedia diesel oil entry. National Park Service, Colorado
Salminen JM, Tuomi PM, Suortti AM, Jorgensen KS (2004) Potential for aerobic and anaerobic biodegradation of hydrocarbons in boreal subsurface. Biodegradation 15:29–39
Schein A, Scott JB, Mos L, Hodson PV (2009) Oil dispersion increases the apparent bioavailability and toxicity of diesel to rainbow trout (*Oncorhynchus mykiss*). Environ Toxic Chem 28(3):595–602
Schlegel HG (1977) Aeration without air: oxygen supply by hydrogen peroxide. Biotechnol Bioeng 19:413
Semple KT, Morriss AWJ, Paton GI (2003) Bioavailability of hydrophobic organic contaminants in soils: fundamental concepts and techniques for analysis. Eur J Soil Sci 54:809–818

Sihag S, Pathak H, Jaroli DP (2014) Factors affecting the rate of biodegradation of polyaromatic hydrocarbons. Int J Pure Appl Biosci 2(3):185–202

Sonia V, Häggblom M, Young LY (2002) Biodegradation of aromatic and aliphatic compounds by rhizobial species. Soil Sci 167(12):802–810

Speight JG, Arjoon KK (2012) Bioremediation of petroleum and petroleum products. Scrivener Publishing LLC and Wiley, New York

Thibaud C, Erkey C, Akgerman A (1993) Investigation of the effect of moisture on the sorption and desorption of chlorobenzene and toluene from soil. Environ Sci Technol 27(12):2373–2380

Trapp S, Matthies M, Scheunert I, Topp EM (1990) Modelling the bioconcentration of organic chemicals in plants. Environ Sci Technol 24:1246–1253

Van Hamme JD, Singh A, Ward OP (2003) Recent advances in petroleum microbiology. Microb Mol Biol Rev 67(4):503–549

Van RM (2011) Environmental degradation of petroleum hydrocarbons. Utrecht University, Binnenstad

Walton BT, Guthrie EA, Hoylman AM (1994) Toxicant degradation in the rhizosphere. In: Anderson TA, Coats JR (eds) Bioremediation through rhizosphere technology, ACS symposium series, vol 563. American Chemical Society, Washington, pp 11–26

Whiticar MJ (1994) Correlation of natural gases with their sources. In: LB Magoon, WG Dow (ed) The petroleum system—from source to trap. AAPG memoir 60. American Association of Petroleum Geologist, Tulsa, pp 261–283

Xiao N, Liu R, Jin C, Dai Y (2015) Efficiency of five ornamental plant species in the phytoremediation of polycyclic aromatic hydrocarbon (PAH)—contaminated soil. Ecol Eng 75:384–391

Yu R, Hill GA (2006) Bioremediation of polycyclic aromatic hydrocarbon (PAH)-contaminated soils in a roller baffled bioreactor. University of Saskatchewan. Dept Chem Eng. Master thesis, pp 1–149 http://www.collectionscanada.gc.ca/obj/s4/f2/dsk3/SSU/TC-SSU-07182006114607.pdf

Zhu X, Venosa AD, Suidan MT (2004) Literature review on the use of commercial bioremediation agents for cleanup of oil-contaminated estuarine environments. U.S. Environmental Protection Agency, Cincinnati 9:1–61 https://www.epa.gov/sites/production/files/2014-10/documents/lit_review_on_the_use_of_commercial_bioremediation.pdf

HC-0B-01: Biodegradation of Hydrocarbons by Extremophiles

P. Arulazhagan, Sami Mnif, J. Rajesh Banu, Q. Huda and M.A.B. Jalal

Abstract Polycyclic aromatic hydrocarbons (PAHs) present in the petroleum wastewater are ubiquitous environmental pollutants. They are generated from both natural and anthropogenic processes, and pose a serious concern on the health of aquatic life and human beings through bioaccumulation. PAHs are known to be cytotoxic, mutagenic and carcinogenic. They persist in the environment due to their hydrophobic nature and are difficult to treat using chemical methods. This chapter details the biodegradation of PAHs under extremophilic condition. Bacteria, fungi and algae are reported to be used in the treatment of PAHs. Extremophilic conditions such as acidophilic (pH 1–5), alkaliphilic (pH > 9) halophilic (>3 % salt), thermophilic (temperature > 50 °C), psycrophilic (temperature < 10 °C), piezophilic or barophilic (pressure > 38 MPa) and xerophilic (a_w 0.60–a_w 0.90) conditions. Collection of samples, isolation of extremophilic micro-organism, mineralization of hydrocarbons and identification of microbes using molecular techniques are detailed. Despite the microbial ability to degrade petroleum hydrocarbons, there are factors such as temperature, pH and nutrients that influence the degradation of hydrocarbons. Information on mechanisms and pathways for

P. Arulazhagan (✉) · Q. Huda · M.A.B. Jalal
Centre of Excellence in Environmental Studies, King Abdulaziz University,
Jeddah 21589, Saudi Arabia
e-mail: arulazhagan_p@yahoo.co.in

Q. Huda
e-mail: hagri@kau.edu.sa

M.A.B. Jalal
e-mail: jbasahi@kau.edu.sa

S. Mnif
Laboratory of Molecular and Cellular Screening Processes,
Centre of Biotechnology of Sfax, Road Sidi Mansour Km 6, Sfax, Tunisia
e-mail: sami.mnif@gmail.com

J. Rajesh Banu
Department of Civil Engineering, Regional Centre of Anna University,
Tirunelveli, Tamil Nadu 627007, India
e-mail: rajeshces@gmail.com

K. Heimann et al. (eds.), *Biodegradation and Bioconversion of Hydrocarbons*,
Environmental Footprints and Eco-design of Products and Processes,
DOI 10.1007/978-981-10-0201-4_4

petroleum hydrocarbon degradation under extreme conditions is scarcely known and recently few studies reported on enzymes, genes and metabolism of hydrocarbons. Microbial cell interaction with petroleum hydrocarbons are also detailed in this chapter. Extremophiles play a vital role in the degradation of petroleum hydrocarbons and in the treatment of refinery wastewater.

Keywords Biodegradation · Petroleum hydrocarbons · Extremophiles · Crude oil · Refinery wastewater

1 Introduction

Hydrocarbons are the major constituent of crude oil pollution which persist in the environment for long periods due to their hydrophobic nature. In recent decades, environmental pollution due to petroleum and petrochemical products attracted more researchers to work in this area of research. Polycyclic aromatic hydrocarbons (PAHs) are ubiquitous with toxic, mutagenic and carcinogenic nature. The fate of PAHs in the environment is associated with both abiotic and biotic factors including volatilization, adsorption and microbial transformation (Wick et al. 2003; Pathak et al. 2009). They are generated from both natural and anthropogenic processes, and pose a serious concern on the health of aquatic life and human beings through bioaccumulation.

Marine ecosystems are often highly contaminated due to frequent oil spills during transportation and cleaning up is a highly difficult task. Once the oil spilled into the marine environment, several process of degradation take place; spreading, evaporation, sedimentation, biodegradation, oxidation and uptake by biota (Fig. 1). Several methods are deployed for cleanup such as physical, chemical and biological processes. Among the methods used, bioremediation appears to be the best process for solving the problem as it does not adversely affect the ecological system; i.e. it is environmental friendly. Microorganisms act as biocatalysts for the biodegradation of hydrocarbons (Fuentes et al. 2014). Biodegradation is limited due to hydrocarbon hydrophobicity, bioavailability and environmental conditions. Several reports are available on the biodegradation of hydrocarbons under normal conditions. A limited number of studies were, however, performed on extremophilic conditions. Extremophiles are defined as microbes that sustain and grow under extreme conditions such as high temperature, salinity, pH and pressure.

Microbes are assigned to different groups based on favoured growth conditions such as thermophiles (higher temperature), psycrophiles (low temperature), halophiles (high salinity), acidophiles (acidic condition), alkaliphiles (alkaline condition) and barophiles (high pressure). Among the extremophiles, research has focused preferentially on halophiles, thermophiles and acidophiles. Biodegradation of a marine oil spill is strongly influenced by salinity. In order to overcome degradation problems, research was recently attracted to examining the capabilities of extremophilic strains, which resulted in the discovery of many unidentified and

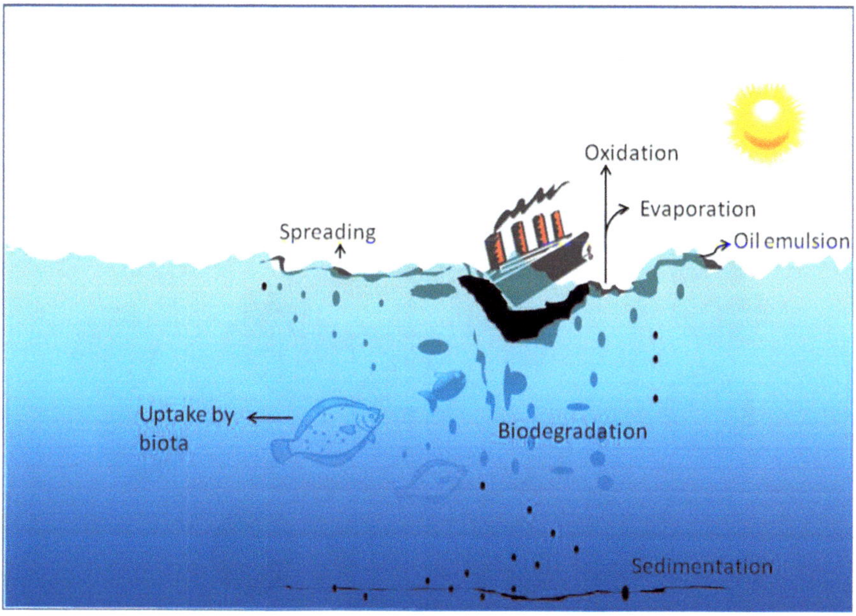

Fig. 1 Schematic om the fate of a crude oil spill in a marine environment

highly effective bacterial, fungal and algal strains. This chapter details extremophilic microbial communities involved in the biodegradation of petroleum hydrocarbons under extreme conditions. Isolation, enrichment and molecular analyses of extremophilic microbial communities from different contaminated sites will be covered.

2 Extremophiles

Extemophiles are defined as the microorganisms that grow under stressful conditions. Based on the stress factors in the environment, extremophiles are divided into different groups: halophiles (high salt concentration), thermophiles (high temperature), psychrophiles (low temperature), alkaliphiles (high pH), acidophiles (low pH) and barophiles (high pressure) (Fig. 2; Table 1).

The low solubility and high hydrophobicity of PAHs limit transport into microbial cells and thus biodegradability. Direct interfacial uptake of aliphatic hydrocarbons facilitated by hydrophobic cell surfaces has been reported for several bacteria belonging to the genus *Acinetobacter, Rhodococcus* and *Pseudomonas* (Van Hamme and Ward 2001; Whyte et al. 1999; Goswami and Singh 1991).

Microorganisms capable of changing the structure of their outer membrane and enhancing cell surface hydrophobicity to facilitate direct uptake of hydrocarbons

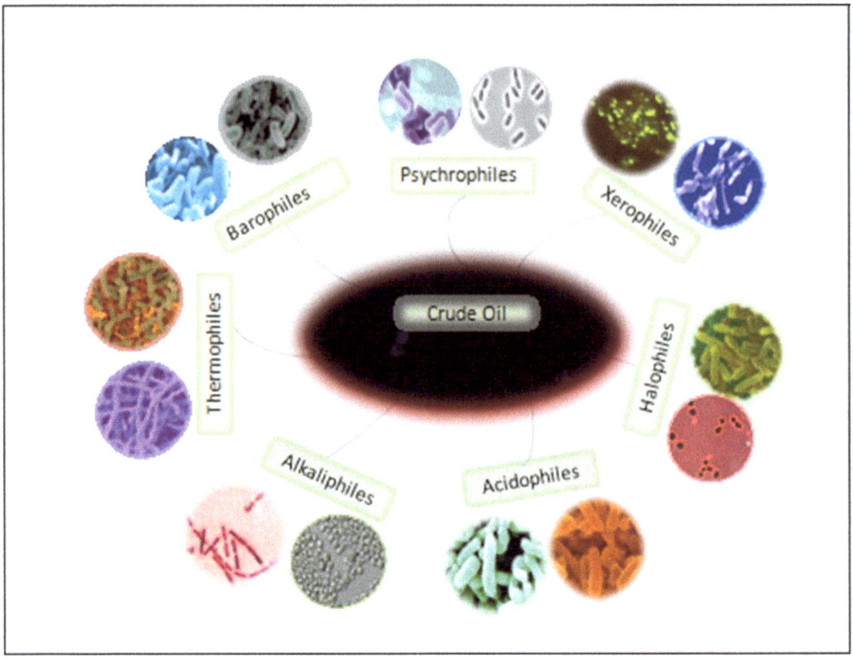

Fig. 2 Types of extremophiles for biodegradation of crude oil

Table 1 Different types of extremophiles Adopted and revised from Stan-Lotter and Fendrihan (2012)

S. no	Extremophiles	Sources/habitat	Lineage	Tolerance stress limit		
				Min	Opt	Max
1	Halophiles	Salt lakes and flats, saline industrial effluent, saline sediments	Archaea/Bacteria	4 % NaCl	9–15 % Nacl	35 % NaCl
2	Thermophiles	Oil drilling sites, hot undersea hydrothermal vents, black smoker fluids	Archaea/Bacteria	60 °C	80 °C	121 °C
3	Psychrophiles	Sea ice, sea mug (Antartic region)	Bacteria	0 °C or below	5 °C	10 °C
4	Acidophiles	Mining regions, acidic hot springs	Archaea	pH 0	pH 1-2	pH 4
5	Alkaliphiles	Soda lakes	Bacteria	pH 8.5	pH 10	pH 12.5
6	Barophiles	Deep sea water (ocean) sediments	Bacteria	50 MPa	70 MPa	100 MPa
7	Xerophiles	Mouldy fruit	Fungi	a_w 0.60	a_w 0.80	a_w 0.90

may employ various mechanisms such as: changes in extracellular polymeric substances (EPS) (Iwabuchi et al. 2000), accumulation of lipopolysaccharides (LPS)/nonionic biosurfactants on the cell wall (Moormann et al. 1997; Ortalo-Magne et al. 1996) and release/structural modification in LPS present in the outer membranes (Norman et al. 2002; Al-Tahhan et al. 2000). Under high saline conditions, solubility of PAHs is very low (0.00026–31 mg/L) based on the number of benzene rings present (Means 1995; Lu et al. 2011) resulting in low biodegradation rates. So there is a need for halotolerant or halophilic bacterial strains or consortia for biodegradation of PAHs. Halophilic bacteria producing biosurfactants result in efficient biodegradation of petroleum hydrocarbons (Kumar et al. 2007). Raising the temperature provides an alternative method to increase bioavailability and biodegradation of PAH compounds (Annweiler et al. 2000; Margesin and Schinner 2001; Feitkenhauer et al. 2001, 2003). Lefebvre et al. (2006) explored the microbial community under hypersaline conditions of wastewater treatment. Thus, the extremophiles have wide applications in the field of wastewater treatment.

2.1 *Mineralization of PAHs*

Complete mineralization of PAHs is required to be achieved since the by-products formed during degradation may be more toxic than the parent compounds. Penet et al. (2004) reported the ratio between the amount of carbon dioxide produced during mineralization of PAHs and residual PAHs gives the exact detail on hydrocarbon degradation. Respirometers (Bioscience, USA) can be used to analyse the amount of carbon dioxide evolved and monitor oxygen uptake during PAH degradation. The mineralisation of PAH can be performed using the following experimental set up: mineral salt medium with PAH (100 mg/L) + bacterial consortia used as test sample, while PAH + MSM (Mineral salt medium) and HSM + bacterial consortia served as control. The amount of CO_2 evolved was calculated based on the amount of CO_2 trapped in 5 mL of potassium hydroxide (KOH) solution (45 %) in the CO_2 trap present along with the electrolyte cell. The electrolyte cell was filled with 1 N H_2SO_4 solution. Five mL of 2 N barium chloride was added to the KOH solution and titrated against 0.25 N HCl (hydrochloric acid) with phenolphthalein as an indicator, where the change from pink to colourless indicated the endpoint. The amount of HCl consumed showed the amount of CO_2 trapped in the vial. The amount of CO_2 evolved was analysed every 12 or 24 h (Arulazhagan et al. 2015). Thus the PAHs complete degradation correlated with the amount of carbon dioxide evolved.

2.2 Identification of Extremophiles by Molecular Techniques

Molecular techniques used to explore extremophile microbial classification are cloning and sequencing of rDNA, nucleic probes and hybridization techniques, ARDRA (Amplified Ribosomal DNA Restriction Analysis), T-RFLP (Terminal Restriction Fragment Length Polymorphism), Real time PCR, etc. Ribosomal RNAs such as 5S, 16S and 23S for prokaryotice microorganisms (Bacteria and Archeaa), 5.8S, 18S and 28S for eukaryotic microorganisms (Eucarya) are used in molecular techniques to identify extremophiles (Tailliez et al. 2002).

2.2.1 Phylogenetic Analysis

Bacterial DNA can be extracted using DNA extraction kits. Amplification of DNA was performed in a thermal cycler (Applied Biosystems). PCR (Polymerase Chain Reaction) was performed with 27F and 1492 as primers (Frank et al. 2008). PCR amplification was performed using the PCR Supermix from Invitrogen (USA) following the protocol detailed by Arulazhagan and Vasudevan (2009). The PCR amplified products obtained from extracted DNA were used for DGGE-PCR (Arulazhagan et al. 2010). DGGE PCR was performed with 968 F (with GC clamp) and 1492 primers (Szekely et al. 2009).

Denaturing agents such as urea and formamide (continuous gradient) were used in the DGGE. The gel contained 40 % of polyacrylamide (Nikolausz et al. 2008). The samples were mixed with an equal volume of laoding dye and loaded in the wells. The electrophoresis tank was filled with 7 L of 1× Tris-Acetate-Ethylenediamine tetraaceticacid (TAE) running buffer. The DGGE was run at 60 °C in a DCode Universal Mutation Detection System (Bio-Rad, USA). DGGE was performed at 60 °C with 130 V for 8 h. The gel was stained with 250 mL of running buffer and 25 μL of 10 mg/mL of ethidium bromide for 5–15 min. After staining, the gel was transferred into a dish containing 250 mL of 1× running buffer and destain for 5–20 min. The gels were scanned in a GelDoc system (Bio-Rad).

Identified bands were excised from the DGGE gel using sterile blades. The bands were kept in sterile water overnight and a part was used as template for PCR. Gel extraction kit was used to purify the PCR product and further cloning was performed using pGEM-T Easy Vector systems II (Promega, USA). The ABI DNA analyser was used to record the nucleotide sequence of the bacterial DNA. BLASTN search confirmed the genus of the bacterial strain and further neighbourhood relationship was obtained by evolutionary analyses using Molecular Evolutionary Genetic Analysis (MEGA v5).

3 Biodegradation of Hydrocarbons by Extremophiles

3.1 *Halophiles*

Halophiles were classified as halotolerant or slight halophiles (2–5 % NaCl), moderate halophiles (5–20 % NaCl) and extreme halophiles (20–35 % NaCl). The main origin of the halophiles are from hypersaline environments such as salt flats, lakes, solar salterns, saline river beds, industrial effluent with high salinity and the dead sea (Veil et al.2004; Arulazhagan and vasudevan 2009, 2011a; Sei and Fathepure 2009; Al-Mailem et al. 2010; Bonfa et al. 2011; Erdogmus et al. 2013). The carbon-free halophilic mineral salts medium normally contained NH_4Cl-2.5 g, KH_2PO_4-5.46 g, Na_2HPO_4-4.76 g, $MgSO_4$-0.20 g, NaCl-40.0 g, pH-7.4 ± 0.2, and distilled water-1L.

3.1.1 Growth Conditions

Halophiles grow under saline conditions; decrease in salinity may retard the growth of halophiles. Most of the studies reported aerobic conditions. Under aerobic conditions, petroleum hydrocarbon degradation is activated by mono or di-oxygenase enzymes resulting in the dihydroxylation of aromatic rings. Degradation of hydrocarbons under anaerobic condition is activated by: (1) Fumarate addition by glacyl-radical enzymes, (2) Hydroxylation by molybdenum cofactor containing enzymes of alkyl substituent through dehydrogenase, (3) methylation of unsubstituted aromatics and (4) carboxylation initiated by uncharacterized enzymes leads to methylation and fumarate addition (Boll and Heider 2010). Lack of nutrients such as phosphate and nitrogen strongly inhibit the growth of halophiles during biodegradation of hydrocarbons. Addition of nutrients accelerates the biodegradation of hydrocarbons under extreme saline condition.

3.1.2 Physiological Adaptation of Halophiles and Petroleum Wastewater Treatment

Halophilic or halotolerant bacterial strains play key roles in the degradation of PAHs present in produced water (petroleum wastewater) from petroleum refineries. The produced water is the water which has intimate contact with the hydrocarbons at various processes of the petroleum refinery. The sources of hydrocarbons are mainly from desalter effluent, sour water, tank bottom draws and spent caustic. Desalter effluent is defined as the inorganic salts (predominantly sodium chloride) present along with crude oil removed by chemical and electrical methods. The level of contamination in desalter water depends on operating pH, effectiveness of interface control device and frequency of mud wash. The steam used for distillation and catalytic cracking condensed into an aqueous phase is removed as sour water.

Water drawn from crude tanks, gasoline tanks and slop tanks result in tank bottom draws. Acidic components such as organic acids, phenols, H_2S, CO_2 and hydrogen cyanide are extracted using spent caustic solution. Cooling water used at various stages of the refining process also constitute contamination for the wastewater.

Halophiles adopt two osmoregulatory mechanisms to withstand ionic strength and considerable water stress which are the salt-in-cytoplasm and organic-osmolyte mechanism (Fig. 3). The salt-in-cytoplasm is a mechanism by which the thermodynamic adjustment of the cell can be achieved by increasing the salt concentration in the cytoplasm similar to salt concentrations similar to the surrounding environment. In the organic osmolytic mechanism, the cytoplasm is kept free of NaCl and the chemical potential of cell water is reduced by accumulation of uncharged, highly water-soluble, organic solutes. Compared to normal strains, halophiles make a slight adjustment in their interior proteins which are more acidic than cytoplasmic proteins in *E. coli*.

Halophiles such as *Cycloclasticus* sp. (Geiselbrecht et al. 1998), *Paenibacillus* sp. (Daane et al. 2001, 2002), *Bacillus napthovorans*. (Zhuang et al. 2002), *Novosphingobium pentaromativorans* (Sohn et al. 2004), *Pseudoalteromonas* sp. (Hedlund and Staley 2006), *Ochrobactrum* sp. (Arulazhagan and Vasudevan 2011a), survive under high saline conditions and are capable of degrading PAHs in crude oil. Current research (KACST 2015) with halophilic bacterial consortia enriched from brine water determined degradation of 86 % of phenanthrene (1500 ppm) at 200 g/L of NaCl concentration. The consortia potentially degraded low and high molecular weight PAHs as sole carbon source under saline condition. The organisms present in the consortia were *Ochrobactrum* and *Pseudomonas* sp. The study of halophilic bacterial strains capable of degrading PAHs earned more

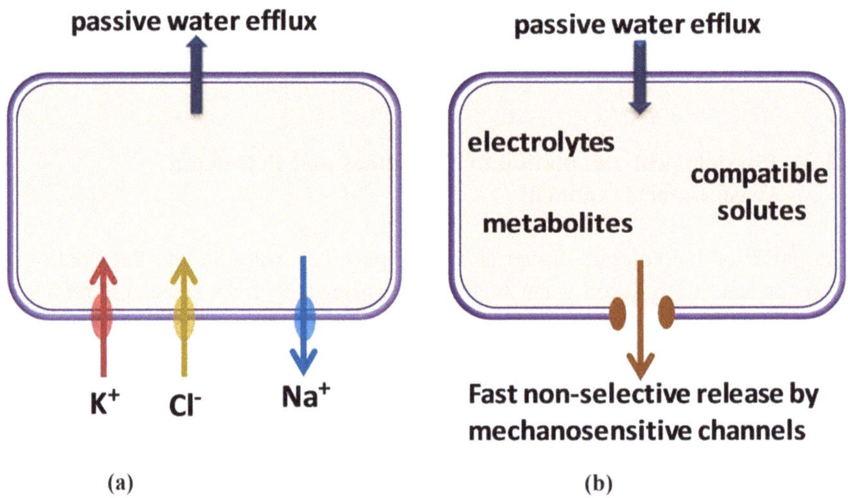

Fig. 3 Mechanism of osmoregulation in halophilic bacterial strains. **a** Salt in cytoplasm and **b** organic osmolyte

focus, since the produced water from petroleum refineries contains more PAHs in high salinity. All the above studies either required additional carbon sources or the time taken for degradation was comparatively longer. Thus, additional carbon sources play a key role in PAHs degradation under saline condition (Arulazhagan and Vasudevan 2011b). Table 2 details different types of halotolerant and halophilic microbial strain or consortia involved in the biodegradation of petroleum hydrocarbons under different saline conditions. Few reports on extremely halophilic bacterial strains are available. Main sources of extremely halophilic bacteria are derived from salt marshes (Bertrand et al. 1990) with optimum growth at 22 % salinity, oil deposits (Kulichevskaya et al. 1991) at 15–32 % salinity degraded mineral oil hydrocarbons, Mangrove sediments (Diaz et al. 2000) at 7.5 % degraded crude oil, harbour sediment (Yang et al. 2000), hypersaline pond in salt producing area (Tapilatu et al. 2010) at 22.5 % degraded heptadecane. Cui et al. (2014) reported the degradation of anthracene by the halophilic bacterium *Mortelella* sp. AD-3. The metabolism of anthracene proceeds via two routes In route I, the anthracene metabolism proceeds via benzocoumarin, 3-hydroxy-2-naphthoic acid, salicylic acid and gentisic acid. In route II, anthracene is metabolized to 9,10-anthraquinone (Fig. 4). In fact, the biodegradation of anthracene by *Martelella* sp. strain AD-3 suggests that both dioxygenases and monooxygenases catalyse the initial attack on the aromatic ring. Since positional isomers of cis dihydrodiols are formed, it may also be hypothesized that several dioxygenase are present in *Martelella* sp. strain AD-3. The broad range of PAHs that are degraded by *Martelella* sp. strain AD-3 may also be due to a relaxed specificity of the same dioxygenase for initial attack on PAHs (Cui et al. 2014). Kumar et al. (2007) reported *Bacillus* sp. DHT is capable of degrading naphthalene and pyrene at 10 % salinity with biosurfactant production. The strain grows both in halophilic and thermophilic conditions and the biosurfactant facilitates the degradation of PAHs under these conditions. DNA samples of the extremophile bacterial strains were extracted and bands were separated by using Denaturation Gradient Gel Electrophoresis (DGGE) or Terminal Restriction Fragment Length Polymorphism (TRFLP). 16S rRNA sequencing was used to identify the extremophile bacterial strains. 18S rRNA was used in the case of fungi. Molecular analyses supported the phylogenetic analysis of the extremophiles enriched from extreme environments. A limited number of studies are performed on the degradation of PAHs by algae. Abed et al. (2006) reported cyanobacterial mats with organisms belonging to beta-, gamma- and delta proteobacteria, the Cytophaga–Flavobacterium–Bacteroides (CFB) group and Spirochetes. The strains potentially degraded phenanthrene and dibenzothiophene at 3.5 and 8 % salinity. The halotolerant or halophilic bacterial strains were capable of degrading LMW (low molecular weight) PAHs almost completely under saline condition, a very limited number of strains were capable of degrading HMW (high molecular weight) PAHs under saline condition. The stress due to salt and the benzene rings influenced the degradation process.

Table 2 Halophilic microorganisms degrading petroleum hydrocarbons under saline conditions

Microorganisms	NaCl concentration (%)	Compounds	References
Bacteria			
Ochrobactrum, Pseudomonas aeruginosa and *Stenotrophomonas maltophila*	3 %	Crude oil	Varjani et al. (2015)
Acinetobacter	3.5 %	Fuel oil	Fatajeva et al. (2014)
Georgenia, Microbactrerium, Bacillus, Cellulomonas, Stappia and *Isolptericola*	2.5–5 %	n-alkanes and aromatic compounds	Al-Awadhi et al. (2007)
Marinobacter, Pseudomonas and *Cycloclasticus*	4–5 %	Heptadecane, naphthalene, crude oil	Jurelevicius et al. (2013)
Ochrobactrum, Enterobacter cloceae and Stenotrophomonas maltophila	3–6 %	PHN,AN,FL, PY, B(e)PY	Arulazhagan and Vasudevan (2009), Arulazhagan et al. (2010)
Dietzia natronolimnaea	0–6 %	Aniline	Jin et al. (2012)
Novosphingobium pentaromativorans	0–7 %	PY, B(a)AN, CHR, B(b) FLN and B(a) PY	Sohn et al. (2004)
Alcanivorax dieselolei	3–7.5 %	Alkanes	Liu and Shao (2005)
Achromobacter, Marinobacter and *Rhodanobacter*	3–9 %	B(e)PY	Arulazhagan et al. (2014)
Halomonas sp.	10 %	Crude oil	Mnif et al. (2009)
Cellulomonas sp., *Bacillus marisflavi, Dietzia maris, Halomonas eurihalina*	10 %	Diesel oil	Riis et al. (2003)
Marinobacter	14.5	BTEX	Nicholson and Fathepure (2004), Al-Awadhi et al. (2007)
Alcanivorax	14.5 %	Benzene, toleune	Hassan et al. (2012)
Rhodococcus, Micrococcus, Arthrobacter	15 %	Hydrocarbon	Kulichevskaya et al. (1991)
Matelella	0.1–15 %	PHN, AN	Cui et al. (2014)
Planococcus	20 %	BTEX	Li et al. (2006)
Halobacterium piscisalsi, Halorubrum ezzemoulense, Halobacterium salinarium, Haloarcula hispanica, Haloferax, Halorubrum, and *Haloarcula*	20 %	NAP, PHN, PY	Erdogmus et al. (2013)

(continued)

Table 2 (continued)

Microorganisms	NaCl concentration (%)	Compounds	References
Marinobacter, Erwinia ananas, Bacillus	22 %	Petroleum hydrocarbons	Diaz et al. (2000, 2002)
Acidovorax delafieldi, Pseuodomonas, Halobacillus salinus, Bacillus simplex	0–23 %	BTEX	Nicholson and Fathepure (2005)
Arhodomonas	3–23 %	Benzene, toluene	Dalvi et al. (2012)
Halobacterium salinarum	15–32 %	Diesel oil	Grant et al. (2001), Lee et al. (2004), Erdogmus et al. (2013)
Fungi			
Fusarium	3.5 %	Crude oil	Hidayat and Tachibana (2012)
Phanerochaete chrysosporium	0–4 %	Crude oil	Behnood et al. (2014)
Papulaspora	5 %	Crude oil	Obuekwe et al. (2005)
Fusarium lateritium and Drechslera	10 %	Crude oil	Obuekwe et al. (2005)
Rhodotorula mucilaginosa	15 %	PHN, AN	Lahav et al. (2002)
Algae			
Oscillatoria salina, Plectonema terebrans and *Aphanocapsa*	3 %	Crude oil, AN, PY	Raghukumar et al. (2001)
Cyanobacteria	21 %	Crude oil	Abed et al. (2006), Cohen (2002)

AN Anthracene, *B(e)PY* Benzo(e)Pyrene, *NAP* Naphthalene, *CHR* Chrysene, *B(a)AN* Benzo(a) anthracene, *PHN* Phenanthrene, *B(b)FLN* Benzo(b)fluoranthene, *FL* Fluorene, *PY* Pyrene, *BTEX* Benzene, Toluene, Xylene

3.2 *Thermophiles*

3.2.1 PAH Degradation Under Thermophilic Conditions

Hydrocarbon degradation using hydrocarbonoclastic bacteria was first described by Zobell (1946). Procaryotes and eucaryotes have the enzymatic capacities which allows for the oxidation of aromatic and polyaromatic compounds. Polycyclic aromatic hydrocarbons are commonly classified into two groups based on their molecular structures. Low molecular weight (LMW) PAHs have three or fewer aromatic rings, while high molecular weight (HMW) PAHs have four or more rings. Differences in the structure and size of individual PAHs result in substantial

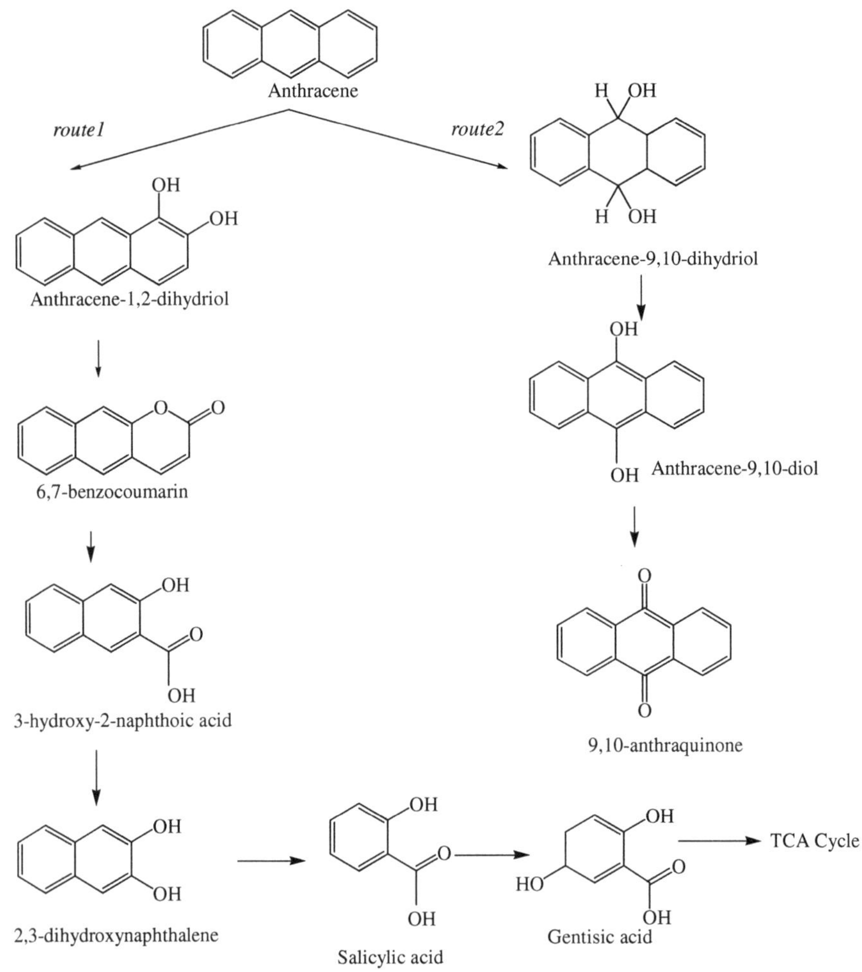

Fig. 4 Metabolism of anthracene by the halophilic strain *Martelella* sp. AD-3 on the basis of identified metabolites (Cui et al. 2014)

variability in the physical and chemical properties of these substances (CCME 1999). The need for thermophilic PAH-degrading strains is justified by the fact that increased temperature has a significant influence on the bioavailability of PAH. This is due to the decrease in viscosity and increase in diffusion coefficient of PAH compounds with elevated temperature. However, studies on hydrocarbon degradation under thermophilic conditions are as yet very few. For example, in 1998, Muller et al. described the isolation of bacteria converting two rings (naphthalene) or three rings (phenanthrene and anthracene) aromatic hydrocarbons. The described metabolic pathways differ from those reported under mesophilic conditions.

3.2.2 Aerobic Condition

PAH degradation under thermophilic conditions had been reported previously. Many publications described the degradation of (LMW) PAH such as naphthalene, phenanthrene and anthracene. For example, Shimura et al. (1999) reported the isolation of a naphthalene and biphenyl utilizing strain at 60 °C. This *Bacillus* sp. strain JF8 was isolated from compost obtained from a field in Okayama, Japan. One year later, Annweiler et al. (2000) reported the degradation of naphthalene by the bacterium *Bacillus thermoleovorans* at 60 °C. The naphthalene degradation pathway differed from the known pathways found for mesophilic strains. It has been reported that thermophilic microorganisms contain enzymes that could exclusively function at elevated temperature and are more stable (Bubinas et al. 2008). In fact, apart from typical metabolites known from mesophilic bacteria (like 1-Naphthol, 2-Naphthol and salicylic acid), novel intermediates such as 2,3 dihydroxynaphthalene, 2 carboxycinnamic acid, benzoic acid and phthalic acid were exclusively identified as metabolites of naphthalene degradation. A moderately thermophilic bacterium, *Nocardia otitidiscaviarum* strain TSH1, isolated from petroindustrial wastewater-contaminated soil in Iran, could transform naphthalene at 50 °C. The elucidated metabolic pathway suggests that strain TSH1 initiated its naphthalene degradation by the dioxygenation at its C-1 and C-2 positions to give 1,2 dihydro-1,2-dihydroxynaphthalene. The intermediate 2-hydroxycinnamic acid, characteristic of the meta-cleavage of the resulting diol was identified. A part from typical naphthalene degradation reported for mesophiles, benzoic acid was identified as an intermediate for the naphthalene pathway of this *Nocardia* strain. Neither phthalic acid nor salicylic acids were detected (Zeinali et al. 2008) (Table 3).

In another study, the degradation of naphthalene using *Geobacillus* sp. G27 proceeded through protocatechuic acid via the ortho-cleavage pathway (Bubinas et al. 2008). It was also demonstrated, that this ortho-cleaving pathway is assured by the protocatechuate 3,4-dioxygenase purified from this thermophilic bacterium (Bubinas et al. 2007) (Table 3).

For the case of anthracene and phenanthrene, the strain *Nocardia otitidiscaviarum* TSH1 could degrade the two or three benzene rings aromatic hydrocarbons at 50 °C. In fact, four metabolites were detected in the case of anthracene: 1,2-dihydroxy-1,2-dihydroanthracene, 3-(2-carboxyvinyl)naphthalene-2carboxylic acid, 2,3-dihydroxynaphthalene and benzoic acid. However, degradation studies of phenanthrene showed the presence of 2,2′-diphenic acid, phthalic acid, 4-hydroxyphenylacetic acid, o-hydroxyphenyl-acetic acid, benzoic acid, phenanthrene dihydriol, 4-[1-hydroxy(2-naphthyl)]-2-oxobut-3-enoic acid and dihydroxy-2-naphthoic acid (1H2NA). The proposed degradation pathway closely matches with the phenanthrene pathways already described in the case of mesophilic bacteria. Apart from typical metabolites found with the mesophilic bacteria, it was noticed that intermediates such as benzoic acid, 4-hydroxyphenylacetic acid and *o*-hydroxyphenylacetic acid were identified for a first time with strain TSH1 (Zeinali et al. 2007).

Table 3 Characteristics of metabolites produced by thermophilic bacteria growing with naphthalene as carbon source compared with mesophilic conditions

Product	Presence (+) of			
	Intermediates produced by thermophilic *B. thermoleovorans* (Annweiler et al. 2000)	Intermediates produced by thermophilic *N. otitidiscaviarum* (Zeinali et al. 2008)	Intermediates produced by thermophilic G. sp. G27 (Bubinas et al. 2008)	Intermediates produced by mesophilic organisms (Cerniglia 1984, 1992; Cerniglia et al. 1989; Seo et al. 2006)
1-Naphthol	+		+	+
2-Naphthol	+		+	+
Cis-1,2-Dihydro-1,2-dihydroxynaphthalene		+		+
1,2-Dihydroxynaphthalene (I)		+	+	+
2,3-Dihydroxynaphthalene (VI)	+			
4-(2-Hydroxyphenyl)-2-oxo-but-3-enoic acid		+		+
2-Hydroxycinnamic acid (III)		+		+
Coumarin (chromen-2-one)	+			+
3-(2-Hydroxyphenyl)-propanoic acid	+			
2,3-Dihydroxycoumarin (chroman-2-one)	+			
2-Hydroxybenzoic acid (salicylic acid)	+			+
Benzene-1,2-diol (catechol)		+		+
3-(2-Carboxyphenyl)-2-propenoic acid (2-carboxycinnamic acid)	+			
Protocatechuic acid			+	
Phthalic acid	+		+	
Benzoic acid	+	+		

On the other hand, the degradation of (HMW) PAH under thermophilic condition is often thought to occur through co-metabolic processes (Kanaly et al. 2000). Reports on the roles of co-substrates such as (LMW) PAH are contradictory. On the one hand, the presence of a co-substrate could impact the biodegradation of (HMW) PAH by co-substrate toxicity to microbes and/or competitive inhibition. For example, the degradation of benzo[a]pyrene by *Mycobacterium* sp. strain RGJII-135 was inhibited in the presence of three or four ring aromatic hydrocarbons such as anthracene, benzo[a]anthracene and pyrene (McLellan et al. 2002). On the other hand, the addition of (LMW) PAH could facilitate and enhance the degradation of (HMW) PAH. Boonchan et al. (2000) reported that in the presence of pyrene (at 250 mg/L), a mixed culture composed of a bacterium and fungus could degrade 29.05 mg/L of benzo[a] pyrene during 56 days of continuous incubation. The degradation of benzo[a]pyrene under mesophilic condition was largely restricted by its low bioavailability due to its low solubility and mass transfer rate. Elevated temperatures can increase the solubility and mass transfer rates of PAHs (Viamajala et al. 2007). Zhao and Wong (2010) demonstrated the potential of two isolated thermophilic bacterial strains, *Bacillus subtilis* BUM and *Mycobacterium vanbaalenii* BU42 in the degradation of benzo[a]pyrene under thermophilic conditions using different co-substrates. Results obtained showed statistically significant correlations between the degradation rates of benzopyrene and phenanthrene under these thermophilic conditions. The study by the authors recorded complete degradation of (LMW)—and (HMW) PAHs under thermophilic condition (60 °C) by a bacterial consortium enriched from an oil drilling site of Saudi Aramco (KACST 2015). The consortia possessed bacterial strains of *Pseudomonas* sp. with biosurfactant production. The intermediates formed during biodegradation were further degraded to carbon dioxide and water, but in some cases toxic intermediates formed during mineralization of PAHs resulting in inhibition of the bioremediation process.

3.2.3 Anaerobic Conditions

The anaerobic biodegradation of PAHs is less studied. Long considered unfeasible, recent results showed that there is real potential anaerobic flora to degrade these compounds. For example, Trably et al. (2003) measured the removal of 13 unsubstituted PAHs from municipal sewage sludge in anaerobic stirred tank bioreactors under thermophilic methanogenic conditions and found that degradation was enhanced at 55 °C compared with 35 and 45 °C, particularly for the larger PAHs. However, in that study, abiotic losses accounted for a significant portion of the total loss of the smaller PAHs at high temperatures, and no attempts were made to unequivocally demonstrate complete mineralization rather than partial co-metabolic oxidation. Trably et al. (2003) also observed PAH losses from municipal sludge incubated as mesophilic and thermophilic enrichments and found that bioaugmentation with adapted inoculums enhanced PAH degradation. This limited number of studies indicates the potential for waste stream processing although more research in this area is required, especially demonstrating mass balances to document complete oxidation of the hydrocarbons. Recent biogeochemical evidence also suggests that PAH can be methanogenically biodegraded in situ. Aitken et al. (2004) detected several anaerobic PAH metabolites in biodegraded crude oils sampled from reservoirs across the globe, offering evidence that PAH can serve as substrates in deep anaerobic subsurface environments. Such a finding, coupled with isotopic analyses indicating that a substantial portion of the CH_4 and CO_2 present in oil-bearing formations is biogenic, also suggests that biodegradation of oil components in reservoirs, including PAHs, occurs under methanogenic conditions and can lead to the formation of heavy oil (Gray et al. 2010). Zhang et al. (2012) described methanogenic anthracene-amended cultures, and Gieg et al. (2010) showed that a thermophilic methanogenic consortium consumed specific PAHs and alkanes in crude oil-amended incubations compared to controls, with overall hydrocarbon loss correlating with methane production.

3.3 Psychrophiles

Psychrophiles are organisms which grow at low temperatures ranging from 0 to 15 °C. The psychrophiles adapt to the extreme cold condition by liquid water phenomena in which the freezing point of the intracellular water determines the lowest temperature for growth. In the presence of high concentrations of solutes (antifreeze solutions), the freezing point is lowered to below 0 °C. Surprisingly, a large number of bacterial strains was reported to occur in cold conditions (Cowan et al. 2002). Irgens et al. (1996) reported γ-proteobacteria *Polaromonas vacuolata* isolated from antartic marine water was able to grow at 4 °C. Cold-adapted petroleum hydrocarbon degrading bacterial strains are (i) antarctic region: *Pseuodomonas syringae*, *P. fluorescens*, *Sphingomonas* (Aislabie et al. 2000, 2006), *Oleispira* sp., *Rhodococcus* (Yakimov et al. 2003, 2004; Saul et al. 2005), *Halomonas* (Pepi et al. 2005) (ii) Artic region: *Shewanella* sp., *Marinobacter* sp., *Pseudomonas* sp. (Gentile et al. 2003; Gerdes et al. 2005), *Psychrobacter*, *Pseudoalteromonas*, *Ageria* (Deppe et al. 2005). Petroleum hydrocarbon degradation at low temperature (7 °C) under anaerobic and nitrate-reducing condition by cultures enriched from arctic soils has been described (Eriksson et al. 2003). Margesin (2007) and Brakstad (2008) established studies on cold-adapted hydrocarbon degrading cultures from alpine soils and marine environments. Recently, Okere et al. (2012) reported the degradation of phenanthrene by indigenous microorganisms in soils collected in Livingstone Island (South Shetlands Islands, Antarctica) with no history of pollution at 4, 12 and 22 °C. Results obtained showed that the degradation occurred even at 4 °C, where it was weak compared to the degradation at 22 °C. Borresen et al. (2003) and Brakstad et al. (2008) observed arctic hydrocarbon degraders at 5 °C and also recorded increased total and heterotrophic microorganisms counts in crude oil-contaminated sea ice compared to an unpolluted site. Several reports on biodegradation of hydrocarbons in soil where evaporation and photo-oxidation are completely absent under cold showed slow biodegradation rates (Rike et al. 2003; Aislabie et al. 2001, 2004; Saul et al. 2005). Cui et al. (2008) identified *Cycloclasticus* sp. capable of degrading phenanthrene and *Alteromonas* dominated a naphthalene degrading consortium. *Cycloclasticus* sp. was found to be the dominant strain in a pyrene degrading bacterial consortium isolated from deep sea sediments of the west Pacific Ocean (Wang et al. 2008). Cold-adapted hydrocarbon degrading fungi such as *Phialophora* sp. and *Hormoconis resinae* were isolated from contaminated antarctic soils (Kerry 1990; Aislabie et al. 2001). Hughes et al. (2007) reported hydrocarbon degrading *Mortierella* sp. HC8D isolated from soils near Rhotera Research Station (Antarctic Peninsula). The ability of petroleum hydrocarbon degradation by the most dominant filamentous fungi in antarctic microbial communities is not well explored and more research needs to occur.

3.4 Acidophiles

Acidophiles are microorganisms capable of growing under strongly acidic condition. A limited number of acidophiles are capable of degrading petroleum hydrocarbons at pH below 4. Acidophiles have been isolated from air biofilters amended to treat volatile hydrocarbons (Kennes and Veiga 2004; Estevez et al. 2005), coal tar pits and mine runoff (Stapleton et al. 1998; Uyttebroek et al. 2007) and oil seepages (Hamamura et al. 2005; Roling et al. 2006). Most of the acidophilic hydrocarbon degrading fungi were isolated from air biofilters and most of them belong to the ascomycetes. Stapleton et al. (1998) reported the presence of the yeast *Pichia* along with the bacterial strains *Acidocella* and *Acidiphilium* capable of degrading naphthalene and toluene under acidic condition (pH 2).

Gemmell and Knowles (2000) reported 23 hydrocarbon degrading isolates but the sequencing results proved two of the strains to belong to *Acidocella* sp. and *Mycobacterium montefiorense* showed 73 % mineralization of pyrene to carbon dioxide. Acidophiles adapt to extreme acidic condition by producing organic acids during hydrocarbon degradation which avoid the uncoupling of the respiratory chain through cell diffusion of protonated forms followed by proton dissociation (Baker-Austin and Dopson 2007). In 2003, Dore et al. reported the isolation of five bacterial strains using naphthalene as a sole carbon source at acidic condition (pH 3 and pH 5). These naphthalene degrading strains were identified as *Clavibacter*, *Arthrobacter* and *Acidocella*. Extensive care should be taken in isolation and enrichment of acidophiles since change in pH can lead to failure of culture growth even on phenanthrene and pyrene after transfer to fresh medium (Uyttebroek et al. 2007). The current study (KACST 2015) obtained a fungal and bacterial mixed consortium showing potential for degrading (LMW) PAHs (1000 ppm) and (HMW) PAHs (100 ppm). It is recommended to maintain acidic conditions throughout the cultivation phase. The physiology, biochemistry and genetics of the hydrocarbon degrading acidophiles remains largely unexplored.

3.5 Alkaliphiles

Alkaliphiles adapt to the alkaline conditions by using sodium ions instead of protons for bioenergetic transformations and balance the inverted pH gradient by strong membrane potentials. *Mycobacterium* sp. isolated from soil is capable of degrading anthracene, phenanthrene, pyrene and fluoranthene at pH 9 (Habe et al. 2004). Gerbeth et al. (2004) reported pyrene degrading *Micrococcus* strain KB-K5 at pH 10.3 isolated from PAH-contaminated building rubble. They also described *Dietzia* strain SK3 as degrading fluorene and phenanthrene at pH 9.5. Alkalohilic bacterial strains involved in petroleum hydrocarbon degradation are *Bacillus licheniformis* (Swaathy et al. 2014), *Marinobacter*, *Micrococcus*, *Dietzia*, *Bacillus* (Al-Awadhi et al. 2007). The metabolism of anthracene was also reported at alkaline condition (pH 9) with *Bacillus badius*. The detection of 1,2-dihydoxyanthracene suggests a metabolism pathway involving a dioxygenase. Phthalic acid was also detected, which was further degraded to simple aliphatic compounds via the phthalic acid pathway to TCA (Tricarboxylic acid) cycle intermediates (Ahmed et al. 2012).

3.6 Barophiles

Barophiles are present in deep ocean and lithospheric sites. Life at high hydrostatic pressure requires special adaptations of the enzymes, as increased pressure decreases the binding capacity of enzymes for their substrates. Protein synthesis and membrane transport are often pressure-sensitive in non-barophilic microorganisms. Relatively few proteins appear to be pressure-controlled in barophiles. In the barophile *Photobacterium* sp. strain SS9, the expression of *ompH* and *ompL* genes encoding outer membrane porin proteins depends on the pressure. The OmpH protein is maximally abundant at a pressure optimum of 28 MPa, while the OmpL protein is produced in greatest quantity at 0.1 MPa. Several additional genes were found whose expression was regulated by pressure or appear to be critical for baroadaptation (Kato and Bartlett 1997). Recently Schedler et al. (2014) reported naphthalene degrading *Sphingobium yanoikuyae* B1 capable of degrading naphthalene up to 8.8 MPa, but the growth of the bacterial strain was completely inhibited by pressure at 12 MPa.

3.7 Xerophiles

Xerophiles are organisms which grow under low water content (a_w 0.8). Most of the xerophiles or xerotolerant strains are fungi. Xerophiles sense low a_w by osmosensors and balance the internal and external osmotic pressure by accumulating glycerol as a compatible solute which is retained in the cell by membrane modification (Pettersson and Leong 2011). Recently Boldu et al. (2012) reported a complex of fungal and bacterial strains capable of degrading toluene, ethylbenzene and p-xylene under xerophilic condition (water content-23 %) enriched from biofilter possessed *Exophiala oligosperma* (Chaetotyriales), chaetothyrialean fungi, dominated by bacterial strains of the Burkholderiales and Xanthomonadales family. *Acremonium kiliense* was also detected in the toluene, ethylbenzene and p-xylene degrading microbial consortium under xerophilic condition.

4 Summary and Conclusions

Petroleum hydrocarbons poses serious threats to the environment, treatment of the hydrocarbons under extreme conditions plays a vital role to save the ecosystem which leads to sustainable development. Mostly marine and terrestrial ecosystem are highly prone to PAH contamination. The present chapter describes different extremophiles and their role in degradation of PAHs under stressed conditions. Cell interactions under stressed conditions requires more research. Current research performed on extremophiles proved that PAH contamination can be bio-treated with efficient microorganisms under stressed conditions. Limited studies are available on extremophiles apart from the treatment of wastewater, the bioproducts from extremophiles are employed in various fields. Further research is required to develop these cultures for deployment in the treatment of PAH-contaminated wastewater by technology transfer.

Acknowledgments This project was funded by the National Plan for Science, Technology and Innovation (MAARIFAH)—King Abdulaziz City for Science and Technology—the Kingdom of Saudi Arabia—award number (11-ENV2012-03). The authors also, acknowledge with thanks Science and Technology Unit, King Abdulaziz University for technical support.

References

Abed RMM, Al-Thukair A, de Beer D (2006) Bacterial diversityofa cyanobacterialmat degrading petroleum compounds at elevated salinities and temperatures. FEMS Microbiol Ecol 57:290–301

Al-Awadhi H, Sulaiman RHD, Mahmoud HM, Radwan SS (2007) Alkaliphilic and halophilic hydrocarbon-utilizing bacteria from Kuwaiti coasts of the Arabian Gulf. Appl Microbiol Biotechnol 77:183–186

Ahmed AT, Othman MA, Sarwade VD, Gawai KR (2012) Degradation of Anthracene by Alkaliphilic Bacteria *Bacillus badius*. Environ Pollut 1:97–104

Aitken CM, Jones DM, Larter SR (2004) Anaerobic hydrocarbon biodegradation in deep subsurface oil reservoirs. Nature 431:291–294

Aislabie J, Fraser R, Duncan S, Farrell RE (2001) Effects of oil spills on microbial heterotrophs in antarctic soils. Polar Biol 24:308–313

Aislabie J, Foght J, Saul D (2000) Aromatic hydrocarbon degrading bacteria from soil near Scott Base, Antarctica. Pol Biol 23:183–188

Aislabie J, Balks MR, Foght JM, Waterhouse EJ (2004) Hydrocarbon spills on Antarctic soils: effects and management. Environ Sci Technol 38:1265–1274

Aislabie J, Saul DJ, Foght JM (2006) Bioremediation of hydrocarbon-contaminated Polar soils. Extremophiles 10:171–179

Al-Mailem DM, Sorkhoh NA, Al-Awadhi H, Eliyas M, Radwan SS (2010) Biodegradation of crude oil and pure hydrocarbons by extreme halophilic archaea from hypersaline coasts of the Arabian Gulf. Extremophiles 14:321–328

Al-Tahhan RA, Sandrin RT, Bodour AA, Maier RM (2000) Rhamnolipid-induced removal of polysaccharide from *Pseudomonas aeruginosa*: Effect on cell surface properties and interaction with hydrophobic substrates. Appl Environ Microbiol 66:3262–3268

Annweiler E, Richnow HH, Antranikian G, Hebenbrock S, Garms C, Franke S, Francke W, Michaelis W (2000) Naphthalene degradation and incorporation of naphthalene derived carbon into biomass by the thermophile *Bacillus thermoleovorans*. Appl Environ Microbiol 66:518–523

Arulazhagan P, Vasudevan N (2009) Role of moderately halophilic bacterial consortium in biodegradation of polyaromatic hydrocarbons. Mar Pollut Bull 58:256–262

Arulazhagan P, Vasudevan N (2011a) Role of nutrients in the utilization of PAHs by halotolerant bacterial strain. J Environ Sci 23:282–287

Arulazhagan P, Vasudevan N (2011b) Biodegradation of polycyclic aromatic hydrocarbons by a halotolerant bacterial strain *Ochrobatrum* sp.VA1. Mar Pollut Bull 62:388–394

Arulazhagan P, Vasudevan N, Yeom IT (2010) Biodegradation of polycyclic aromatic hydrocarbons by a halotolerant bacterial consortium from marine environment. Int J Environ Sci Technol 7:639–652

Arulazhagan P, Sivaraman C, Adish Kumar S, Aslam M, Rajesh Banu J (2014) Co-metabolic degradation of benzo(e)pyrene by halophilic bacterial consortium at different saline conditions. J Environ Biol 35(3):445–452

Arulazhagan P, Rajesh J, Jeyakumar D, Yeom IT (2015) Biodegradation of 1,4 dioxane by *Rhodanobacter* sp. AYS5 and the role of additional substrates. Ann Microbiol. doi:10.1007/s13213-015-1060-y

Baker-Austin C, Dopson M (2007) Life in acid: pH homeostasis in acidophiles. Trends Microbiol 15:165–171

Bubinas A, Giedraitytė G, Kalėdienė L, Nivinskiene N, Butkiene R (2008) Degradation of naphthalene by thermophilic bacteria *via* a pathway, through protocatechuic acid. Cent Eur J Biol 3:61–68

Bubinas A, Giedraityte G, Kalediene L (2007) Protocatechuate 3,4-dioxygenase from thermophilic *Geobacillus* sp. strain. Biologija 18:31–34

Boldu FXP, Guivernau M, Gallastegui G, Vinas M, Hoog GSD, Elias A (2012) Fungal/bacterial interactions during the biodegradation of TEX hydrocarbons (toluene, ethylbenzene and p-xylene) in gas biofilters operated under xerophilic conditions. FEMS Microbiol Ecol 80:722–734

Boonchan S, Britz ML, Stanley GA (2000) Degradation and mineralization of high-molecular-weight polycyclic aromatic hydrocarbons by defined fungal–bacterial co-cultures. Appl Environ Microbiol 66:1007–1019

Bonfa MRL, Grossman MJ, Mellado E, Durrant LR (2011) Biodegradation of aromatic hydrocarbons by Haloarchaea and their use for the reduction of the chemical oxygen demand of hypersaline petroleum produced water. Chemosphere 84:1671–1676

Bertrand JC, Almallah M, Acquaviva M, Mille G (1990) Biodegradation of hydrocarbons by an extremely halophilic archaebacterium. Lett Appl Microbiol 11:260–263

Behnood M, Nasernejad B, Nikazar M (2014) Biodegradation of crude oil from saline wastewater using white rot fungus *Phanerochaete chrysosporium*. J Indus Eng Chem 20:1879–1885

Brakstad OG (2008) Natural and stimulated biodegradation of petroleum in cold marine environments. In: Margesin R, Schinner F, Marx J-C, Gerday C (eds) Psychrophiles: from biodiversity to biotechnology. Springer, Berlin, pp 389–407

Brakstad OG, Nonstad I, Faksness LG, Brandvik PJ (2008) Responses of microbial communities in Arctic sea ice after contamination by crude petroleum oil. Microb Ecol 55:540–552

Borresen M, Breedveld GD, Rike AG (2003) Assessment of the biodegradation potential of hydrocarbons in contaminated soil from a permafrost site. Cold Reg Sci Technol 37:137–149

Boll M, Heider J (2010) Anaerobic degradation of hydrocarbons: mechanisms of C–H-bond activation in the absence of oxygen. In: Timmis KN (ed) Handbook of hydrocarbon and lipid microbiology. Springer, Berlin

CCME (Canadian Council of Ministers of the Environment) (1999) Canadian sediment quality guidelines for the protection of aquatic life: polycyclic aromatic hydrocarbons (PAHs). In: Canadian environmental quality guidelines, Canadian Council of Ministers of the Environment, Winnipeg

Cerniglia CE (1984) Microbial metabolism of polycyclic aromatic hydrocarbons. Adv Appl Microbiol 30:31–71

Cerniglia CE (1992) Biodegradation of polycyclic aromatic hydrocarbons. Biodegradation 3:351–368

Cerniglia CE, Campbell WL, Freeman JP, Evans FE (1989) Identification of a novel metabolite in phenanthrene metabolism by the fungus *Cunninghamella elegans*. Appl Environ Microbiol 55:2275–2279

Cohen Y (2002) Bioremediation of oil by marine microbial mats. Int Microbiol 5:189–193

Cui Z, Lai Q, Dong C, Shao Z (2008) Biodiversity of polycyclic aromatic hydrocarbon degrading bacteria from deep sea sediments of the Middle Atlantic Ridge. Environ Microbiol 10(8):2138–2149

Cui C, Ma L, Shi J, Lin K, Luo Q, Liu Y (2014) Metabolic pathway for degradation of anthracene by halophilic *Martelella* sp. AD-3. Int Biodeter Biodeg 89:67–73

Cowan DA, Russell NJ, Mamais A, Sheppard DM (2002) Antarctic Dry Valley mineral soils contain unexpectedly high levels of microbial biomass. Extremophiles 6:431–436

Dore SY, Clancy QE, Rylee SM, Kulpa CF (2003) Naphthalene-utilizing and mercury-resistant bacteria isolated from an acidic environment. Appl Microbiol Biotechnol 63:194–199

Daane LL, Harjono I, Zylstra GJ, Haggblom MM (2001) Isolation and characterization of polycyclic aromatic hydrocarbon-degrading bacteria associated with the rhizosphere of salt marsh plants. Appl Environ Microbiol 67(6):2683–2691

Daane LL, Harjono I, Barns SM, Launen LA, Palleroni NJ, Haggblom MM (2002) PAH-degradation by *Paenibacillus* spp. and description of *Paenibacillus naphthalenovorans* sp. nov., a naphthalene-degrading bacterium from the rhizosphere of salt marsh plants. Inter J Syst Evol Microbiol 52:131–139

Dalvi S, Azetsu S, Patrauchan MA, Aktas DF, Fathepure BZ (2012) Proteogenomic elucidation of the initial steps in the benzene degradation pathway of a novel halophile, *Arhodomonas* sp. StrainRozel, isolated from a hypersaline environment. Appl Environ Microbiol 78:7309–7316

Diaz MP, Boyd KG, Grigson SJW, Burgess JG (2002) Biodegradation of crude oil across a wide range of salinities by an extremely halotolerant bacterial consortium MPD-M, immobilized onto polypropylene fibers. Biotechnol Bioeng 79:145–153

Diaz MP, Grigson SJW, Peppiatt CJ, Burgess JG (2000) Isolation and characterization of novel hydrocarbon degrading euryhaline consortia from crude oil and mangrove sediments. Mar Biotechnol 2:522–532

Deppe U, Richnow H-H, Michaelis W, Antranikian G (2005) Degradation of crude oil by an Arctic microbial consortium. Extremophiles 9:461–470

Erdogmus SF, Mutlu B, Korcan SE, Guven K, Konuk M (2013) Aromatic hydrocarbon degradation by halophilic archaea isolated from Çamalti Saltern, Turkey. Water Air Soil Pollut 224:1449

Eriksson M, Sodersten E, Yu Z, Dalhammar G, Mohn WW (2003) Degradation of polycyclic aromatic hydrocarbons at low temperature under aerobic and nitrate-reducing conditions in enrichment cultures from Northern soils. Appl Environ Microbiol 69:275–284

Estevez E, Veiga MC, Kennes C (2005) Biodegradation of toluene by the new fungal isolates *Paecilomyces variotii* and *Erophiala oligosperma*. J Ind Microbiol Biotechnol 32:33–37

Fatajeva E, Gailiūtė I, Paliulis D, Grigiškis S (2014) The use of *Acinetobacter* sp. for oil hydrocarbon degradation in saline waters. Biologija 60(3):126–133

Feitkenhauer H, Schnicke S, Muller R, Markl H (2001) Determination of the kinetic parameters of the phenol-degrading thermophile *Bacillus themoleovorans* sp. A2. Appl Microbiol Biotechnol 57:744–750

Feitkenhauer H, Muller R, Markl H (2003) Degradation of polycyclic aromatic hydrocarbons and long chain alkanes at 60–70 °C by *Thermus* and *Bacillus* spp. Biodegradation 14:367–372

Frank JA, Claudia IR, Sharma S, Weisbaum JS, Wilson BA, Olsen GJ (2008) Critical evaluation of two primers commonly used for amplification of bacterial 16S rRNA genes. Appl Environ Microbiol 74(8): 2461–2470

Fuentes S, Méndez V, Aguila P, Seeger M (2014) Bioremediation of petroleum hydrocarbons: catabolic genes, microbial communities and applications. Appl Microbiol Biotechnol 98 (11):4781–4794

Gemmell RT, Knowles CJ (2000) Utilisation of aliphatic compounds by acidophilic heterotrophic bacteria. The potential for bioremediation of acidic wastewaters contaminated with toxic organic compounds and heavy metals. FEMS Microbiol Lett 192:185–190

Gerbeth A, Krausse S, Gemende B, Muller RH (2004) Search of microorganisms that degrade PAHs under alkaline conditions. Eng Life Sci 4:311–318

Gray ND, Sherry A, Hubert C, Dolfing J, Head IM (2010) Methanogenic degradation of petroleum hydrocarbons in subsurface environments: remediation, heavy oil formation, and energy recovery. Adv Appl Microbiol (Allen SS, Laskin I and Geoffrey MG, eds) Academic Press, San Diego, pp 137–161

Gieg LM, Davidova IA, Duncan KE, Suflita JM (2010) Methanogenesis, sulfate reduction and crude oil biodegradation in hot Alaskan oilfields. Environ Microbiol 12:3755–3762

Goswami P, Singh DH (1991) Different modes of hydrocarbon uptake by two *Pseudomonas* species. Biotechnol Bioeng 37:1–11

Gerdes B, Brinkmeyer R, Dieckmann G, Helmke E (2005) Influence of crude oil on changes of bacterial community in Arctic sea-ice. FEMS Microbiol Ecol 53:129–139

Geiselbrecht AD, Hedlund BP, Tichi MA, Staley JT (1998) Isolation of marine polycyclic aromatic hydrocarbon (PAH)-degrading *Cycloclasticus* strains from the Gulf of Mexico and comparison of their PAH degradation ability with that of Puget Sound *Cycloclasticus* strains. Appl Environ Microbiol 64:4703–4710

Gentile G, Bonasera V, D'Amico C, Giuliano L, Yakimov MM (2003) *Shewanella* sp. GA-22, a psychrophilic hydrocarbonoclastic Antarctic bacterium producing polyunsaturated fatty acids. J Appl Microbiol 95:1124–1133

Grant WD, Kamekura M, McGenity TJ, Ventosa A (2001) Class III. Halobacteria class. nov. In: Boone DR, Castenholz RW, Garrity GM (eds) Bergey's manual of systematic bacteriology. Springer, New York, pp 294–334

Habe H, Kanemitsu M, Nomura M, Takemura T, Iwata K, Nojiri H et al (2004) Isolation and characterization of an alkaliphilic bacterium utilizing pyrene as a carbon source. J Biosci Bioeng 98:306–308

Hamamura N, Olson SH, Ward DM, Inskeep WP (2005) Diversity and functional analysis of bacterial communities associated with natural hydrocarbon seeps in acidic soils at Rainbow Springs, Yellowstone National Park. Appl Environ Microbiol 71:5943–5950

Hassan HA, Rizk NMH, Hefnawy MA, Awad AM (2012) Isolation and characterization of halophilic aromatic and chloroaromatic degrader from wadi el-natrun soda lakes. Life Sci J 9 (3):1565–1570

Hedlund BP, Staley JT (2006) Isolation and characterization of *Pseudoalteromonas* strains with divergent polycyclic aromatic hydrocarbon catabolic properties. Environ Microbiol 8:178–182

Hidayat A, Tachibana S (2012) Biodegradation of aliphatic hydrocarbon in three types of crude oil by *Fusarium* sp. F092 under stress with artificial sea water. J Environ Sci Technol 5:64–73

Hughes KA, Bridge P, Clark MS (2007) Tolerance of Antarctic fungi to hydrocarbons. Sci Tot Environ 372:539–548

Irgens RL, Gosink JJ, Staley JT (1996) *Polaromonas vacuolata* gen. nov., sp. nov., a psychrophilic, marine, gas vacuolated bacterium from Antarctica. Int J Syst Bacteriol 46:822–826

Iwabuchi N, Sunairi M, Anzi H, Nakajima M, Harayama S (2000) Relationship between colony morphotypes and oil tolerance in *Rhodococcus rhodochrous*. Appl Environ Microbiol 66:5073–5077

Jin Q, Hu Z, Jin Z, Qiu L, Zhong W, Pan Z (2012) Biodegradation of aniline in an alkaline environment by a novel strain of the halophilic bacterium, *Dietzia natronolimnaea* JQ-AN. Biores Technol 117:148–154

Jurelevicius D, Alvarez VM, Marques JM, deSousa Lima LR, DiasFde A, Seldin L (2013) Bacterial community response to petroleum hydrocarbon amendments in freshwater, marine and hypersaline water-containing microcosms. Appl Environ Microbiol 79:5927–5935

KACST (King Abdulaziz City of Science Technology) funded research project (2015) Biodegradation of petroleum hydrocarbons using extremophilic bacterial consortia from contaminated sites (11-ENV-2012-03) at Centre of Excellence in Environmental Studies. King Abdulaziz University, Jeddah

Kanaly RA, Bartha R, Watanabe K, Harayama S (2000) Rapid mineralization of benzo[a]pyrene by a microbial consortium growing on diesel fuel. Appl Env Microbiol 66:4205–4211

Kato C, Bartlett DH (1997) The molecular biology of barophilic bacteria. Extremophiles 1(3):111–116

Kumar M, Vladimir L, de Sistro Materano A, Ilzins OA (2007) A halotolerant and thermotolerant *Bacillus* sp. degrades hydrocarbons and produces tension active emulsifying agent. World J Microbiol Biotechnol 23:211–220

Kulichevskaya IS, Milekhina EI, Borzenkov IA, Zvyagintseva IS, Belyaev SS (1991) Oxidation of petroleum hydrocarbons by extremely halophilic archaebacteria. Microbiology 60:596–601

Kennes C, Veiga MC (2004) Fungal biocatalysts in the biofiltration of VOC-polluted air. J Biotechnol 113:305–319

Kerry E (1990) Effects of temperature on growth rates of fungi from subantarctic Macquarie Island and Casey, Antarctica. Polar Biol 10:293–299

Lahav R, Fareleira P, Nejidat A, Abelovich A (2002) The identification and characterization of osmotolerant yeast isolates from chemical wastewater evaporation ponds. Microb Ecol 43:388–396

Lee MS, Joo WA, Kim CW (2004) Identification of a novel protein D3UPCA from *Halobacterium salinarum* and prediction of its function. Proteomics 4(11):3622–3631

Lefebvre O, Vasudevan N, Thanasekaran K, Moletta R, Godon JJ (2006) Microbial diversity in hypersaline wastewater: the example of tanneries. Extremophiles 10:505–513

Li H, Liu YH, Luo N, Zhang XY, Luan TG, Hu JM et al (2006) Biodegradation of benzene and its derivatives by a psychrotolerant and moderately haloalkaliphilic *Planococcus* sp. strain ZD22. Res Microbiol 157:629–636

Liu C, Shao Z (2005) *Alcanivorax dieselolei* sp. nov., a novel alkane-degrading bacterium isolated from sea water and deep-sea sediment. Int J Syst Evol Microbiol 55:1181–1186

Lu X, Zhang T, Fang HH (2011) Bacteria-mediated PAH degradation in soil and sediment Appl Microbiol Biotechnol 89:1357–1371

Margesin R (2007) Alpine microorganism: useful tools for low-temperature bioremediation. J Microbiol 45:281–285

Margesin R, Schinner F (2001) Potential of halotolerant and halophilic microorganisms for biotechnology. Extremophiles 5:73–83

McLellan SL, Warshawsky D, Shann JR (2002) The effect of polycyclic aromatic hydrocarbons on the degradation of benzo[a]pyrene by *Mycobacterium* sp. strain RJGII-135. Environ Toxicol Chem 21:253–259

Means JC (1995) Influence of salinity upon sediment-water partitioning of aromatic hydrocarbons. Mar Chem 51:3–16

Moormann M, Zahringer U, Moll H, Kaufmann R, Schmid R et al (1997) A new glycosylated lipopeptide incorporated into the cell wall of a smooth variant of *Gordona hydrophobica*. J Biol Chem 272:10729–10738

Mnif S, Chamkha M, Sayadi S (2009) Isolation and characterization of *Halomonas* sp.strain C2SS100, a hydrocarbon degrading bacterium under hypersaline conditions. J Appl Microbiol 107:785–794

Muller R, Antranikian G, Maloney S, Sharp R (1998) Thermophilic degradation of environmental pollutants. Adv Biochem Eng Biotechnol 61:155–169

Nicholson CA, Fathepure BZ (2004) Biodegradation of benzene by halophilic and halotolerant bacteria under aerobic conditions. Appl Environ Microbiol 70(2):1222–1225

Nicholson CA, Fathepure BZ (2005) Aerobic biodegradation of benzene and toluene under hypersaline conditions at the Great Salt Plains, Oklahoma. FEMS Microbiol Lett 245:257–262

Nikolausz M, Kappelmeyer U, Székely A, Rusznyák A, Márialigeti K, Kästner M (2008) Diurnal redox fluctuation and microbial activity in the rhizosphere of wetland plants. Eur J Soil Biol 44:324–333

Norman RS, Frontera-suau R, Morris PJ (2002) Variability in *Pesudomonas aeruginosa* lipopolysaccharide expression during crude oil degradation. Appl Environ Microbiol 68:5096–5103

Okere UV, Cabrerizo A, Dachs J, Jones KC, Semple KT (2012) Biodegradation of phenanthrene by indigenous microorganisms in soils from Livingstone Island, Antarctica. FEMS Microbiol Lett 329:69–77

Ortalo-Magne A, Lemassu A, Marie-Antoinette L, Bardou F, Silve G et al (1996) Identification of the surface-exposed lipids on the cell envelops of *Mycobacterium tuberculosis* and other Mycobacterial species. J Bacteriol 178:456–461

Obuekwe CO, Badrudeen AM, Al-Saleh E, Mulder JL (2005) Growth and hydrocarbon degradation by three desert fungi under conditions of simultaneous temperature and salt stress. Int Biodeterior Biodegradation 56:197–205

Pathak H, Kantharia D, Malpani A, Madamwar D (2009) Naphthalene degradation by *Pseudomonas* sp. HOB1: in vitro studies and assessment of naphthalene degradation efficiency in simulated microcosms. J Hazard Mater 166:1466–1473

Penet S, Marchal R, Sghir A, Monot F (2004) Biodegradation of hydrocarbon cuts used for diesel oil formulation. Appl Microbiol Biotechnol 66:40–47

Pepi M, Cesaro A, Liut G, Baldi F (2005) An Antarctic psychrotrophic bacterium *Halomonas* sp. ANT-3b, growing on n-hexadecane, produces a new emulsifying glycolipid. FEMS Microbiol Ecol 53:157–166

Pettersson VO, Leong SIL (2011) Fungal xerophiles (Osmophiles). John Wiley & Sons Ltd, New Jersey. doi:10.1002/9780470015902.a0000376.pub2

Raghukumar C, Vipparty V, David J, Chandramohan D (2001) Degradation of crude oil by marine cyanobacteria. Appl Microbiol Biotechnol 57(3):433–436

Riis V, Kleinsteuber S, Babel W (2003) Influence of high salinities on the degradation of diesel fuel by bacteria consortia. Can J Microbiol 49:713–721

Rike AG, Haugen KB, Børresen M, Engene B, Kolstad P (2003) In situ biodegradation of petroleum hydrocarbons in frozen Arctic soils. Cold Reg Sci Technol 37:97–120

Roling WFM, Ortega-Lucach S, Larter SR, Head IM (2006) Acidophilic microbial communities associated with a natural, biodegraded hydrocarbon seepage. J Appl Microbiol 101:290–299

Schedler M, Hiessl R, Juarez AGV, Gust G, Muller R (2014) Effect of high pressure on hydrocarbon-degrading bacteria. AMB Express 4:77

Saul DJ, Aislabie JM, Brown CE, Harris L, Foght JM (2005) Hydrocarbon contamination changes the bacterial diversity of soil around Scott Base, Antarctica. FEMS Microbiol Ecol 53:141–155

Sei A, Fathepure BZ (2009) Biodegradation of BTEX at high salinity by an enrichment culture from hypersaline sediments of Rozel Point at Great Salt Lake. J Appl Microbiol 107:2001–2008

Seo JS, Keum YS, Hu Y, Lee SE, Li QX (2006) Phenanthrene degradation in *Arthrobacter* sp. P1-1: Initial 1,2-, 3,4- and 9,10-dioxygenation, and meta- and ortho-cleavages of naphthalene-1,2-diol after its formation from naphthalene-1,2-dicarboxylic acid and hydroxyl naphthoic acids. Chemosphere 65:2388–2394

Shimura M, Mukerjee-Dhar G, Kimbara K, Nagato H, Kiyohara H, Hatta T (1999) Isolation and characterization of a thermophilic *Bacillus* sp. JF8 capable of degrading polychlorinated biphenyls and naphthalene. FEMS Microbiol Lett 178:87–93

Sohn JH, Kwon KK, Kang JH, Jung HB, Kim SJ (2004) *Novosphingobium pentaromativorans* sp. nov., a high-molecular-mass polycyclic aromatic hydrocarbon-degrading bacterium isolated from estuarine sediment. Int J Syst Evol Microbiol 54:1483–1487

Stan-Lotter H, Fendrihan S (2012) Adaption of microbial life to environmental extremes: novel research results and applications. Wien, Newyork 2

Stapleton RD, Savage DC, Sayler GS, Stacey G (1998) Biodegradation of aromatic hydrocarbons in an extremely acidic environment. Appl Environ Microbiol 64:4180–4184

Swaathy S, Kavitha V, Pravin AS, Mandal AB, Gnanamani A (2014) Microbial surfactant mediated degradation of anthracene in aqueous phase by marine *Bacillus licheniformis* MTCC 5514. Biotechnol Rep 4:161–170

Szekely AJ, Sipos R, Berta B, Vajna B, Hajdú C, Márialigeti K (2009) DGGE and T-RFLP analysis of bacterial succession during mushroom compost production and sequence-aided T-RFLP profile of mature compost. Microb Ecol 57:522–533

Trably E, Patureau D, Delgenes JP (2003) Enhancement of polycyclic aromatic hydrocarbons removal during anaerobic treatment of urban sludge. Water Sci Technol 48:53–60

Tapilatu YH, Grossi V, Acquaviva M, Militon C, Bertrand JC, Cuny P (2010) Isolation of hydrocarbon-degrading extremely halophilic archaea from an uncontaminated hypersaline pond (Camargue, France). Extremophiles 14:225–231

Tailliez OM, Brochier C, Forterre P, Philippe H (2002) Archaeal phylogeny based on ribosomal proteins. Mol Biol Evol 19:631–639

Uyttebroek M, Vermeir S, Wattiau P, Ryngaert A, Springael D (2007) Characterization of cultures enriched from acidic polycyclic aromatic hydrocarbon-contaminated soil for growth on pyrene at low pH. Appl Environ Microbiol 73:3159–3164

Van Hamme JD, Ward OP (2001) Physical and metabolic interactions of *Pseudomonas* sp. Strain JA5-B45 and *Rhodococcus* sp. Strain F9-D79 during growth on crude oil and effect of a chemical surfactant on them. Appl Environ Microbiol 67:4874–4879

Varjani SJ, Rana DP, Jain AK, Bateja S, Upasani VN (2015) Synergistic ex-situ biodegradation of crude oil by halotolerant bacterial consortium of indigenous strains isolated from on shore sites of Gujarat, India. Int Biodeter Biodeg 103:116–124

Veil JH, Puder MG, Elcock D, Redweik RJ (2004) A white paper describing produced water from production of crude oil, natural gas and coal bed methane in U.S. Department of energy, National Energy Technology Laboratory. Prepared by Argonne National Laboratory (Washington, DC) pp 1–87

Viamajala S, Peyton RM, Richards LA, Petersen JN (2007) Solubilization, solution equilibria, and biodegradation of PAH's under thermophilic conditions. Chemosphere 66:1094–1106

Wang BJ, Lai QL, Cui ZS, Tan TF, Shao ZZ (2008) A pyrene-degrading consortium from deep-sea sediment of the West Pacific and its key member *Cycloclasticus* sp. P1. Environ Microbiol 10:1948–1963

Wick LY, Pasche N, Bernasconi SM, Pelz O, Harms H (2003) Characterization of multiple-substrate utilization by anthracene-degrading *Mycobacterium frederiksbergense* LB501T. Appl Environ Microbiol 69(10):6133–6142

Whyte LG, Slagman SJ, Pietrantonio F, Bourbonniere L, Koval SF et al (1999) Physiological adaptations involved in alkane assimilation at a low temperature by *Rhodococcus* sp. Strain Q15. Appl Environ Microbiol 65:2961–2968

Zeinali M, Vossoughi M, Ardestani SK (2007) Degradation of phenanthrene and anthracene by *Nocardia otitidiscaviarum* strain TSH1, a moderately thermophilic bacterium. J Appl Microbiol 105:398–406

Zeinali M, Vossoughi M, Ardestani SK (2008) Naphthalene metabolism in *Nocardia otitidiscaviarum* strain TSH1, a moderately thermophilic microorganism. Chemosphere 72:905–909

Zhao Z, Wong WC (2010) Rapid biodegradation of Benzo[a]pyrene by *Bacillus subtilis* BUM Under thermophilic condition. Environ Eng Sci 27(11):939–945

Zhang S, Wang Q, Xie S (2012) Stable isotope probing identifies anthracene degraders under methanogenic conditions. Biodegradation 23:221–230

Zhuang WQ, Tay JH, Maszenan AM, Tay STL (2002) *Bacillus naphthovorans* sp. nov. from oil-contaminated tropical marine sediments and its role in naphthalene biodegradation. Appl Microbiol Biotechnol 58:547–553

Zobell CE (1946) Action of microorganisms on hydrocarbons. Microbiol Rev 10:1–49

Yang L, Lai C, Shieh WK (2000) Biodegradation of dispersed diesel fuel under high salinity conditions. Water Res 34:3303–3314

Yakimov MM, Giuliano L, Gentile G, Crisafi E, Chernikova TN, Abraham W-R, Lünsdorf H, Timmis KN, Golyshin PN (2003) *Oleispira antarctica* gen. nov., sp. nov., a novel hydrocarbonoclastic marine bacterium isolated from Antarctic coastal sea water. Int J Syst Evol Microbiol 53:779–785

Yakimov MM, Gentile G, Bruni V, Cappello S, D'Auria G, Golyshin PN, Giuliano L (2004) Crude oil induced structural shift of coastal bacterial communities of Rod Bay (Terra Nova Bay, Ross Sea) and characterization of cultured cold-adapted hydrocarbonoclastic bacteria. FEMS Microbiol Ecol 49:419–432

Biodegradation of Benzene Under Anaerobic Condition

Bhawana Pathak, Shalini Gupta and M.H. Fulekar

Abstract Global industrialization has largely expanded the edges of petroleum hydrocarbon (PHC) exploration. A large amount of various hydrocarbons are introduced into the environment during the stages of oil extraction, refinement, storage, transportation and disposal. Benzene is the parent hydrocarbon among the aromatic organic compounds which naturally occurs in petroleum products. It is a well-known carcinogenic organic compound. Its contamination is a widespread problem in soil as well as groundwater due to lack of oxygen in subsurface soils. Various physical and chemical methods are known to clean up aromatic hydrocarbons but they are too expensive and lead to adverse effects. Bioremediation technology has gained a great attention for the cleanup of hazardous aromatic compounds. There are advantages to rely on indigenous microorganisms rather than adding microbes to degrade waste. Emerging technologies have been developed in the field of environmental biotechnology for enhance degradation and complete removal of organic contaminant. This chapter reviews on recent progress in anaerobic degradation of benzene along with its sources, environmental fate and anaerobic mineralization pathways in the presence of different electron acceptors and also focuses on enhanced benzene degradation by enrichment and immobilization-based culture technique, factors affecting the rate of anaerobic degradation, role of enzymes and molecular tools to assess bioremediation.

Keywords Aromatic hydrocarbon · Benzene · Anaerobic condition · Biodegradation · Electron acceptors · Molecular tools

B. Pathak (✉) · S. Gupta · M.H. Fulekar
School of Environment and Sustainable Development, Central University of Gujarat, Sector-30, Gandhinagar, Gujarat 382030, India
e-mail: bhawanasp@hotmail.com

K. Heimann et al. (eds.), *Biodegradation and Bioconversion of Hydrocarbons*, Environmental Footprints and Eco-design of Products and Processes,
DOI 10.1007/978-981-10-0201-4_5

1 Introduction

The wide consumption and demand of petroleum based products like fuel oil and gasoline has led to expansion of the petrochemical industry. The petrochemical industry produces approximately 70 million tons of aromatics including benzene, toluene, ethylbenzene, xylene (BTEX). A petrochemical industry includes the global processes of exploration, extraction, refining, and transportation of the raw material which give rise to great variety of pollutants into the environment. Among the organic pollutants, aromatic organic compounds are a major pollution problem in terrestrial and aquatic ecosystem. The transportation sector uses a significant amount of petrochemicals and gasoline. Commercial gasoline is comprised of almost 30 % (wt/wt) benzene, toluene, ethylbenzene, and xylenes (BTEX) (Seagren and Becker 2002; Villatoro-Monzón et al. 2008). The mandatory decrease of lead alkyls in gasoline in 2004, has led to an increase in aromatic hydrocarbon content of gasoline to maintain high octane levels and anti-knocking properties. It is one of the natural constituent of crude petroleum recovered from refinery streams during catalytic reformation. It is used as an important raw material for the manufacture of synthetic rubbers, gums, lubricants, dyes, pharmaceuticals, and agricultural chemicals. When it released into the environment, it acts as groundwater contaminant and also tends to adsorb on soil matrices (Cole 1994; Bennett 1999; Senthil Kumar et al. 2013; Coates et al. 2002). Residence time of benzene vapor in air varies between a few hours and days depending on certain other factors such as climate and the concentration of other chemicals (i.e. nitrogen and sulfur dioxides) (USEPA 2005; Senthil Kumar et al. 2013). Benzene is a frequent groundwater contaminant as a result of gasoline and diesel fuel spills into the environment. Benzene is defined as a toxic compound by the Canadian Environmental Protection act and is classified as a group A: human carcinogen by the US Environmental Protection Act. Due to its toxicity and relatively high solubility (water solubility: 1700 mg/L), it represents a significant threat to human health as it can cause hematological effects, which may ultimately lead to aplastic anemia and acute myelogenous leukemia (ATSDR 2004; Dou et al. 2008a).

Various in situ and ex situ physico-chemical treatment methods are available for the treatment of these contaminants. Due to economic and environmental concern, the bioremediation approach is an emerging and more applicable technique for the treatment of any type of hazardous waste including volatile organic carbons (VOCs). Bioremediation of benzene under aerobic and anaerobic condition has proved a better technology for decontamination of benzene from environments. Aerobic bioremediation is faster than the anaerobic process, however, in sub-surface soil and groundwater environments, oxygen may be limiting due to faster consumption of available molecular oxygen by the aerobic microorganisms than replenishment. Therefore, anaerobic degradation of aromatic hydrocarbons may be the determining mechanisms and depends on the activity of bacteria capable of metabolizing hydrocarbons under anaerobic conditions at contaminated sites (Lovely 1997; Hu et al. 2007). Anaerobic metabolism is a vital process with respect to petroleum hydrocarbon biodegradation and bioremediation most significantly, through developments and applications of molecular techniques. The processes of hydrocarbon catabolism, genetic and biochemical bases of biodegradation and many novel catalytic mechanisms have been characterized. Molecular approaches contribute to a more detailed characterization of bacterial membrane structure (Van Hamme et al. 2003; Leutwein and Heider 1999; Hu et al. 2007). Less attention has been given to the same enriched bacteria under different electron acceptor conditions and differences in their degradation ability. Therefore, to support and develop anaerobic bioremediation clean-up technologies, more information is essential regarding the rates and effectiveness of organic substrates for in situ remediation of contaminated sites (Dou et al. 2008b). Isolation and identification of indigenous microorganisms from the contaminated sites and use of molecular techniques to assess biodegradation for the improvement of bioremediation technology of benzene degradation is an important aspect. The role of enzymes in metabolism, functional gene expression and recent trends in enhancing the rate of degradation by enriched and immobilized microbial culture is very useful for the better understanding of bioremediation technology.

1.1 Physico-chemical Properties of Benzene

Benzene is recovered from refinery streams during catalytic reformation and other petroleum processes. The benzene ring is represented by double bonds between alternate carbon atoms. It is a clear, colorless, highly flammable liquid at room temperature (Table 1). Benzene is an excellent solvent because of its lipophilic nature; used in paints, thinners, inks, adhesives etc. According to the USEPA PEL (permissible exposure limit), the time weighted average (TWA) in drinking water is 5 ppb and in air the PEL for 8-h TWA is 1 ppm and for 15 min is 5 ppm, It has a pleasant, aromatic odor detectable at concentrations of 1.5–4.7 parts per million (ppm). The workplace permissible exposure level PEL is 1 ppm (OSHA 1987; USEPA 2006).

1.1.1 Sources of Benzene in the Environment

Benzene is introduced into the environment in two ways (1) natural sources and (2) anthropogenic sources. Benzene emissions may result whenever carbon-rich materials undergo incomplete combustion. It is a principal product from the combustion of PVC (polyvinyl chloride). Natural sources of emission into the atmosphere are volcanoes and fires but they play a negligible role. However, human activities are the main source of benzene released in the environment (ATSDR 2007a; Piero et al. 2011). Industrial and domestic sources contributed 9.8 thousand tons of benzene emission to the atmosphere in 2012 (ONS 2014). Industrial source-emissions include different combustion sources such as exhaust emissions,

Table 1 Physico-chemical characteristic of benzene

Empirical formula	C_6H_6
Structure	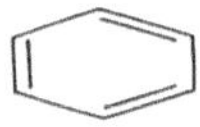 Or 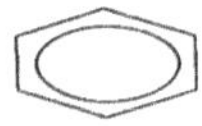
Physical state and appearance	Liquid
Odor	Aromatic. Gasoline-like
Molecular weight	78.11 g/mole
Color	Clear colourless
Boiling point	80.1 °C (176.2 F)
Melting point	5.5 °C (41.9 F)
Critical temperature	288.9 °C (552 F)
Specific gravity	0.8787 at 15 °C
Vapour pressure	10 kPa at 20 °C
Solubility	Miscible in alcohol, chloroform, carbon disulfide oils, carbon tetrachloride, glacial acetic acid, diethyl ether, acetone, very slightly soluble in cold water
Partition coefficients	Log K_{ow}: 2.13 (Hansch and Leo 1985), Log K_{oc}: 1.8 to 1.9 (HSDB 1997, USEPA (US Environmental Protection Agency) 2006)
Henry's law constant	5.5×10^{-3} atm m^3/mol (Mackay and Leinonen 1975) at 25 °C

Source: Material safety datasheet (MSD)

evaporation losses from motor vehicles, wood combustion, stationary fossil fuel combustion, handling, distribution and storage of petrol, accidental spillage and leakage from underground storage tank (WHO 2000). Indoor sources include cigarette smoke; smoke from wood burning fires and some household products which contain petroleum-based chemicals like glues, paints, furniture wax, and lubricants. Domestic sources currently contribute the greatest proportion of benzene emissions, principally through the combustion of fuels for cooking and heating and the operation of garden appliances such as lawn mowers and patio heaters (ONS 2006; Chilcott 2007).

1.1.2 Toxicity of Benzene and Its Metabolism

Toxicity of a chemical substance depends on exposure time and concentration. Toxico-kinetics (movement of toxicant into the body) of benzene includes (1) absorption (includes inhalation, dermal contact), (2) distribution, (3) metabolism and (4) elimination. Chilcott (2007) described various routes of exposure of the human body. The primary route of entry is via inhalation, dermal absorption and ingestion (Fig. 1). Occupational exposure to benzene suggests that absorption occurs mainly by inhalation and dermal routes in many workplaces (Tables 2, 3).

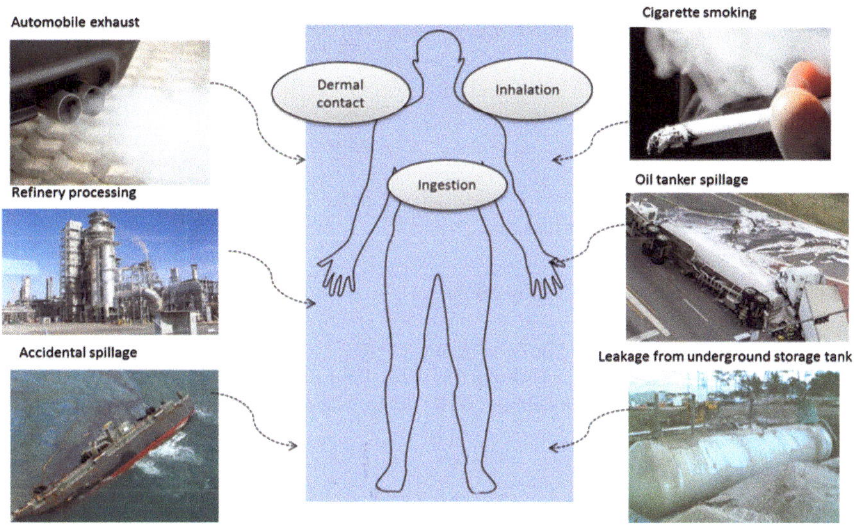

Fig. 1 Benzene exposure to human body by various routes

Table 2 Reported concentrations of benzene in air and water

	Concentration
Air	
Remote rural area	0.2–16 μg/m^3
Industrial centre with high traffic density	Up to 349 μg/m^3
Water	
Surface water	<0.1–0.2 μg/L
Contaminated surface water	Up to 100 μg/L
Ground water	<0.1–1.8 μg/L
Contaminated ground water	Up to 330 μg/L
Drinking water	<0.1–5 μg/L

Daily intake of benzene

Air breathing	90–1300 μg/day
Cigarette smoking	1800 μg/day
Food	Up to 250 μg/day
Drinking water	Up to 10 μg/day

People employed in the petrochemical industries, steel plants, printing, rubber industry, shoe makers, chemical laboratories, and gas station sites may be at high risk of benzene exposure. In contrast, several human studies have indicated that the respiratory absorption of benzene is approximately 50 %. Sabourin et al. (1987) studied benzene retention in rats and mice. During a 6-h exposure, retention of benzene was 33 % in rats and 50 % in mice. Benzene absorption through the human skin was also studied by applying 0.02–0.05 % of the benzene with the remainder volatilizing rapidly. An hourly absorption of 0.4 mg/cm^2 has been reported when the forearm was bathed in liquid benzene. It has been estimated that an adult working in ambient air containing benzene vapors at a concentration of 32 mg/m^3 (10 ppm) would absorb 7.5 μl/h via inhalation and 1.5 μl/h via whole body (2 m^2) dermal exposure. Hence, benzene exposure and absorption through

Table 3 Water and air quality guideline for benzene

	CASRN	WHO DWG	NAAQS
Water	71–43–2 μg/L	10 μg/L	–
Air	71–43–2 μg/L		–
Industrial, residential, rural and other areas	–	–	5 $\mu g/m^3$ annually
Ecological sensitive areas (notified by central government)	–	–	5 $\mu g/m^3$ annually

Source: ATSDR (2000, ATSDR 2007a, b, c), IPCS (1985, 1993, 1996, 1997), NTP (2005), WHO (2008), NHMRC (2004). Chemical Abstract Services Registry Number (CASRN), World Health Organization Drinking Water Guidelines (WHO 2008), National Ambient Air Quality Standards (NAAQS 2009)

inhalation is greater than dermal contact. Distribution of hazardous substances in the human body includes inhalation exposure and distribution of benzene. As in animals, benzene is distributed to several organs, with lipid-rich tissues containing the highest levels. It has been reported that levels of 0.38 mg% in blood (mg % = mg per 100 mL of blood or mg per 100 g of tissue), 1.38 mg% in the brain, and 0.26 mg% in the liver of a worker who died from exposure to very high air concentrations of the chemical (Cox 1991; Medeiros et al. 1997). Exposure to very high concentrations in air (10,000,000 ppb and above) can cause death (ATSDR 2007b). Lower levels (700,000–3,000,000 ppb) can cause drowsiness, dizziness, rapid heart rate, headaches, tremors, confusion, and unconsciousness. Long-term exposure to benzene can cause cancer of blood forming organs (Leukemia). A study revealed that workers exposed to 210–650 ppm for 1–15 years could result in preleukemia or acute leukemia (Rosenstock et al. 2004; ATSDR 2004).

Once benzene is absorbed in the body, it is initially metabolized in the liver and later in the bone marrow. Quantities of the metabolite is greater in blood than bone marrow, because it is a target organ for several chemicals, including benzene. It is currently hypothesized that the carcinogenic effects of benzene are predominantly mediated via metabolites and the radiomimetic nature of benzene and its ability to induce different sites of neoplasm indicates the formation of oxygen radicals is a major cause of benzene toxicity, while examples of metabolites formed by oxidation are phenol, quinol, catechol, transmuconic acid/2-phenyl mercapturic acid (Yardley-Jones et al. 1991; ATSDR 1992; DH 1998; Chilcott 2007). Mixed-function oxidases (MFOs) are the toxic metabolites and main cause of benzene toxicity in the liver and bone marrow. MFO-inducing drugs (e.g., phenobarbital and ethanol) and certain chemicals (e.g., chlordane and parathion) may increase the rate at which toxic metabolites of benzene are formed (Cliona et al. 2011). Benzene metabolism is catalyzed by specific cytochrome P450 enzymes; which are responsible for oxygenation of benzene and have a propensity to generate oxygen radicals (ATSDR 1992, 1997; Cliona et al. 2011). The first step in benzene metabolism is the formation of the epoxide, benzene oxide, catalyzed by cytochrome P450 2E1 (CYP2E1). After formation of benzene oxide, the metabolism of benzene branches into several alternative metabolic pathways (Lovern et al. 1997; EPA 2002). Further benzene oxide may react with glutathione (GSH) to form phenyl-mercapturic acid; undergo enzymatic conversion by epoxide hydrolase to benzene dihydrodiol with subsequent formation of catechol; or undergo an iron-catalyzed, ring-opening reaction to form trans-muconaldehyde (MUC) with subsequent metabolism to trans-muconic acid (MA). Phenol is further oxidized by CYP2E1 catalysis to hydroquinone. Further oxidation of hydroquinone to *p*-benzoquinone is catalyzed by myeloperoxidase (MPO) (Smith et al. 1989; EPA 2002). Recently, a study was conducted on the role of CYP2E1 expression in bone marrow and its intra- and interspecies variability in rats, rabbits, and humans, because it is a target organ for several chemicals, including benzene. The data demonstrated a presence of CYP2E1 in the bone marrow of all species investigated, thus supporting the hypothesis of CYP2E1-dependent local metabolism of benzene possibly contributing to myelotoxicity and hematotoxicity (Bernauer et al. 2000; EPA 2002).

1.2 Environmental Fate of Benzene

Benzene is released into the environment through industrial effluent discharge by various ways. In soils, the benzene is mainly subjected to volatilization near the surface and some part may leach to groundwater (Canadian Council of Ministers of The Environment 1999; Ritter et al. 2002). The major pathway that controls the fate and behavior of benzene in the environment is presented in Fig. 2.

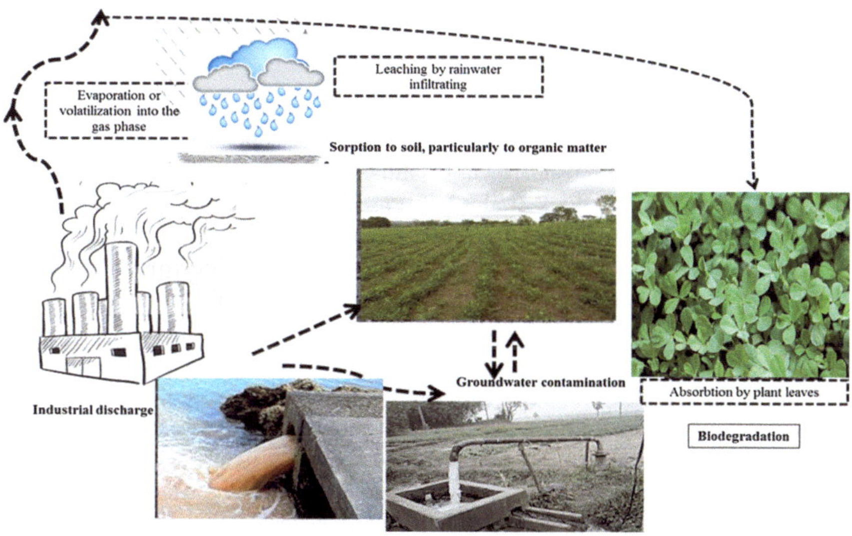

Fig. 2 Environmental fate of benzene

1.2.1 Soil

In natural environments, benzene is exposed to soils through industrial discharges, land disposal of benzene-containing wastes, and gasoline leaks from underground storage tanks. Benzene enters to soil surfaces, partitions to the atmosphere through volatilization, to surface water through runoff, and to groundwater as a result of leaching. The soil organic carbon sorption coefficient (K_{oc}) for benzene has been measured to be in a range of 60–83 (Karickhoff et al. 1979; Kenaga 1980), indicating that benzene is highly mobile in soil and readily leaches into groundwater. Other parameters that influence leaching potential include the soil type (e.g., sand versus clay), amount of rainfall, depth of the groundwater, and extent of degradation. In a study on adsorptive characteristics of benzene to groundwater aquifer solids, benzene showed a tendency to adsorb to aquifer solids; and it was concluded that soil adsorption increased with the increase in organic matter content (Uchrin and Mangels 1986). An investigation of the mechanisms governing the rates of adsorption and desorption of benzene by dry soil grains revealed that several hours are required to achieve equilibrium and that adsorption is much faster than desorption (Lin et al. 1994). The fate of benzene in planted and unplanted soils was investigated using high concentration of 620 μg/kg soil to determine separately root uptake and foliar uptake of radiolabeled benzene. ^{14}C-labeled VOC_s and $^{14}CO_2$ resulting from mineralization were measured at 12 h intervals. Less than 2 % of the recovered ^{14}C was associated with the plant shoots and between 2 to 8 % in the root fraction (root tissue with rhizosphere soil). The average total recovery of added radiolabel in all experiments was greater than 90 %. Comparisons of planted and unplanted soils indicated that alfalfa did not enhance the degradation of benzene (Ferro et al. 1996).

1.2.2 Water

Benzene is found in both polluted and unpolluted waters and rainwater. A substantial portion of any benzene in rainwater that is deposited to soil or water will be returned to the atmosphere via volatilization. Predicting the fate of benzene in aquatic environments, Wick et al. (2000) used a dynamic model, CHEMSEE to study benzene degradation rates in lakes with an initial concentration about 20 μM or 1.5 ppm into its anoxic hypolimnion via discharge of saline groundwater and found that benzene was degraded in three metalimnion samples at rates between 1 and 2.5 day^{-1} in summertime. An epilimnion sample yielded a similar result, but no degradation was observed in another epilimnion sample. Losses were ≤ 0.04 day^{-1} in a sulfate-rich hypolimnion sample. Since benzene loss could be inhibited by filtration or with a mixture of poisons and antibiotics, it was apparently being biodegraded. The whole-lake mass balance of benzene found that approximately 80 % of the benzene entering the lake was degraded during the water's residence in the lake. Vertical distributions of benzene in the water column indicated that the major sink of benzene was located in the metalimnion.

1.2.3 Biological Fate

There is evidence that aromatic hydrocarbons can be metabolised in plants, although the fate of the metabolites is not known. A vast literature is present on the effects of ethylene, because of its role as a plant hormone. The main responses are morphological, which may be reversible, and on the development of flowers and fruit. Effects on seed production may be positive or negative, depending on the exposure concentration (Cape 2003). For example, the uptake of benzene by cress and barley from soil established bioconcentration factors (BCF) for barley after 125 days and for cress after 79 days of 4.6 and 1.9, respectively (Scheunert et al. 1985; Topp et al. 1989). In a research study, toxicity of mixture of BTEX (benzene, toluene, ethylene, xylene) on terrestrial plants (*Sorghum biocolor* and *Cucumis staivus*) was demonstrated with shoots being more susceptible to increased concentrations of BTEX than roots. Above 90 mg/kg soil (as dry weight of BTEX) is toxic to *Sorghum biocolor* and *Cucumis staivus.* The study was limited to acute toxicity of BTEX to plants while lower levels of BTEX exposure may have chronic adverse effect on the plants (An 2004).

1.3 Treatment Technologies

Presently conventional treatment methods are used by industries for clean-up of hazardous aromatic compounds in order to avoid public health hazard. Several remediation technologies are known for treatment of aromatic hydrocarbons and organic pollutants such as physical and chemical methods have been developed. Some in situ and ex situ remediation technologies are listed in (Tables 4, 5). However, these treatment methods can lead to secondary effluent problems and other toxic gaseous emission. The ecologically acceptable disposal of organic waste remains a major challenge to the petrochemical industry. Bioremediation is a natural phenomenon to treat hazardous waste without creating any nuisance to the environment hence it is an accepted alternative, an economical, energy-efficient and environmentally sound approach to other remediation techniques.

1.4 Anaerobic Biodegradation: An Effective Biological Treatment

Although various physical–chemical treatment technologies discussed above for the remediation of the contaminant are used extensively, there are several drawbacks are associated with them. For example ex situ treatment requires pumping of groundwater, leading to increased costs and engineering for equipment, permission and material handling. To avoid these, a holistic approach is required viz. the use of biological treatment technology (aerobic and anaerobic bioremediation). Aerobic

Table 4 Remediation methods for water contaminated by oil

Category	In situ technologies	Ex situ technologies	References
Physical	Electro remediation	Electro remediation	Holliger et al. (1997)
	Capping	Electrodialysis	Daifullah and Girgis (2003)
	Barrier	Air stripping	Nyer (1998)
	Hydraulic containment	Surfactant-modified	Nyer (1998)
	Air sparging	Zeolite Carbon adsorption Filtration (membrane)	Khan et al. (2004), Arvin et al. (2005), Tiburtius et al. (2005), Ranck et al. (2005)
Chemical	Injection of chemical oxidation compounds (O_3, O_2, H_2O_2, Cl_2)	Photocatalysis remediation (nanoparticle TiO_2/UV) coagulation, flocculation and precipitation, application of chemical oxidation compounds (O_3, O_2, H_2O_2, Cl_2)	Wallace and Kadlec (2005)
Biological	Phytoremediation, in situ bioremediation	Ex situ bioremediation (aerobic and anaerobic bioreactors)	Farhadian et al. (2006), Farhadian (2008)

Table 5 Remediation methods for soil contaminated by oil

In situ technologies	Ex situ technologies	References
High vacuum and low vacuum extraction	Incineration	Farhadian et al. (2006), Farhadian (2008)
Volatilization	Solidification and Stabilization	Wallace and Kadlec (2005), Farhadian (2008)
Leaching or soil washing	Soil washing	
Isolation and containment		

in situ biodegradation has been studied extensively and applied in biological approaches to date, but it poses some significant challenges such as; seasonal variation in microbial activity due to direct exposure to changes in environmental factors that cannot be controlled and problematic application of treatment additives. Hence remediation using aerobic techniques becomes difficult owing to the highly reduced nature of the sediment (Pardieck et al. 1992; Coates and Anderson 2000). Presence of recalcitrant organic compounds in aquatic and terrestrial environments results in reduction of dissolved oxygen and microbial activity at the subsurface water whereas insoluble hydrocarbon fractions remain as such in the deep sea water or marshy sediment (Dubinsky et al. 2013). In this view, stimulation of anaerobic processes in situ or ex situ can be an effective treatment alternative for priority contaminants such as benzene in heavily contaminated sites. Generally, a sequence of redox zones develop near aerobic aquifers contaminated with organic pollutants such as landfill leachate. Often near the source of the organic pollutants, methanogenic conditions are observed (Weelink et al. 2010). Therefore, benzene and other mono-aromatic compounds are often present as contaminants in anaerobic zones of the environment (Lovely 1997; Weelink 2008). More recently, the number of anaerobic pure bacterial cultures shown to metabolize benzene with different electron acceptors has increased (Table 5). This diverse set of bacteria includes members of the α, β, γ and δ subclasses of the proteobacteria and forms an excellent framework from which to elucidate the underlying biochemical and molecular mechanisms driving anaerobic hydrocarbon metabolism. Moreover, BTEX degradation in underground water in anaerobic conditions is also actively investigated (Chakraborty and Coates 2005; Jahn et al. 2005; Botton and Parsons 2007). Namely *Dechloromonas* strains RCB and JJ and *Azoarcus* DN11 and AN9 strains are reported to be pure cultures capable of benzene mineralization with nitrate as an electron acceptor. *Dechloromonas* strain RCB is also capable of mineralizing benzene with nitrate, chlorate, perchlorate and oxygen as terminal electron acceptors (Chakraborty et al. 2005). Benzoate is a principle intermediate of benzene degradation in the presence of different electron acceptors and enzymatic activity (Fig. 3). Comprehensive lists of anaerobic bacteria studied for the biodegradation of benzene has been given in Tables 6 and 7.

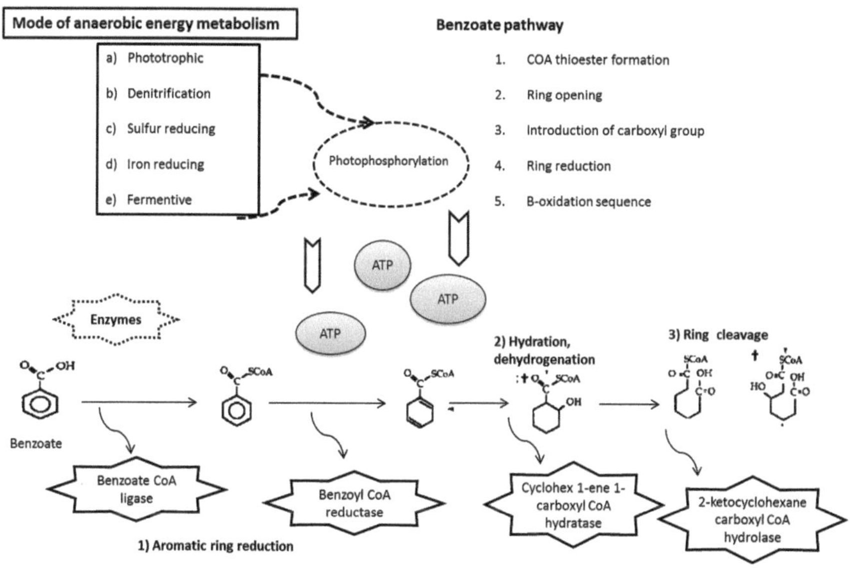

Fig. 3 Anaerobic benzoate degradation pathway conversion of the sequence of reaction includes are proposed by (Koch et al. 1993; Stupperich 1993; Hartel et al. 1993; Boll and Fuchs 1995; Harwood and Gibson 1997) based on the identification of cyclohex-1,5-diene-1-carboxyl CoA and 6-hydroxycyclohex-1-ene-1-carboxyl-CoA enzymatic activities that have been detected in benzoate grown *R. palustris* indicated hypothetical enzymatic reaction

Table 6 List of anaerobic bacteria for biodegradation of benzene

Bacterial strains	e-Acceptor	Incubation time	Intermediates	Concentration	Percentage removal	References
Azoarcus	Nitrate	14 day	Benzoate and phenol	200 μm	100 %	Yuki et al. (2006)
Geobacter fermantans	Fe(III)	85 days	ND	140 μM	60 %	Rooney-Varga et al. (1999)
V. parodoxus	Fe(III)	25 days	ND	140 μM	100 %	Rooney-Varga et al. (1999)
Geothrix	Fe(III)	85 days	ND	–	–	Rooney-Varga et al. (1999)
Clostridia (Pepto-coccaceae) and *Deltaproteo-bacteria*	Ferrihydrite	119 days	ND	900 μM	4.78 mM $^{13}CO_2$ and 20.5 mM Fe(II) produced	Kunapuli et al. (2007)
Therminocola	Sulfate	120 days	Phenol, 2-hydroxy benzoate	900 mM	5.5 $^{13}CO_2$ (mM) produced	Kunapuli et al. (2007), Weelink et al. (2010)
Pelotomoculum	Sulfate	100	Phenol, benzoate	150 μM	1.65 mM day^{-1}	Laban et al. (2009)
Methanogenic culture	Methanogenic	64 days	Phenol, cyclohexanone and propanoic acid	30 mM	60.3 % methane produced	Grbic-Galic and Vogel (1987)
Dechloromonas strain RCB	Nitrate	15 days	ND	35 μM	50–55 % oxidized	Chakraborty et al. (2005)
Bacillus cereus	Nitrate	25 days	Phenol, benzoate	150 mg/L	100 %	Dou et al. (2010)

(continued)

Table 6 (continued)

Bacterial strains	e-Acceptor	Incubation time	Intermediates	Concentration	Percentage removal	References
Alicycliphilus denitrificans	Chlorate	100 days	Phenol, catechol	150 mM	ND	Weelink et al. (2008)
Geobacter metallireducens	Fe(III) as the sole electron acceptor	10 days	Phenol, CO_2	2.78 × 109 Bq $mmol^{-1}$	ND	Zhang et al. (2014)
Ferroglobus placidus	Fe(III) as the sole electron acceptor	40 days	Benzoate, CO_2	200 μM	80 %, $^{14}CO_2$ recovered	Holmes et al. (2011)
Geobacter daltonii strain FRC-32	Fe(III), succinate, nitrate	70, 10 days respectively	Toluene, CO_2	300, 400 μM respectively	58.3, 75 % respectively	Kanak (2014)

ND not determine

Table 7 Benzene degradation by facultative anaerobic bacterial strains

Bacterial strains	e-Acceptor	Incubation time	Intermediates	Concentration	Percentage removal	References
Pseudomonas spp.	NO_3^-	35 h	Protocatechuate benzoate, p-hydroxy benzoate	5 mM	60–70 % CO_2 produced in the form of benzoate	Taylor et al. (1970)
Bacillus spp., *Pseudomonas* spp, *Micrococcus* spp.	–	6 days	ND	50 mg/L	66.20 %	Prakash et al. (2014)
Dechloromonas strain RCB	chlorate	15 days	ND	66 µM	104 %	Chakraborty et al. (2005)

ND not determine

1.4.1 Significance and Role of Anaerobic Fungi

Anaerobic fungi and their mode of growth involve hyphal extension and possession of powerful xylanolytic and cellulytic activities (Wood et al. 1986). Because of their fibrolytic activity, anaerobic fungi are now being considered for use in anaerobic digester and their hydrogen producing capability has been exploited in processes that combine biological hydrogen producing reactions with palladium (Pd) catalyzed hydrogenation (Mountfort and Kaspar 1986). There is also scope of future work in assessment of fibre digesting capability of fungi in relation to diet and possible environment implication (i.e. methane production) (Douglas and Colin 1994). Tabak and Cooke (1968) studied anaerobiosis of 13 fungal strains isolated from oxygen-limited environments, such as sewage sludge, polluted water, and organically enriched soil. Fungal denitrification and ammonia fermentation in response to hypoxic conditions are considered to be extensively distributed among many fungi (Takaya 2009). Wada (1976) reported that an unidentified spore-forming filamentous fungus colonized rice straws in submerged rice field soil where anaerobic conditions were sufficient for sulfate reduction. Recently, it has been suggested that *Fusarium oxysporum* likely acquires energy for growth by denitrification (Shoun and Tanimoto 1991) and by ammonia fermentation. Fungal denitrification occurs in an O_2 limited environment in which N_2O is generated as a final denitrified product because of the lack of N_2O reductase to generate N_2 from N_2O (Shoun and Tanimoto 1991). On the other hand, fungal ammonia fermentation occurs in an anaerobic environment in which NO_3^- is reduced to NH_4^+ and ethanol is oxidized to acetate (Zhou et al. 2002). These findings show the possibility that filamentous fungi capable of growth in the absence of oxygen are likely distributed in anaerobic environments in greater numbers than expected (Tonouchi 2010). Milton et al. (1994) reported that particles of various woodland soils were placed onto the surface of nutrient-free silica gel and incubated at 25 °C anaerobically in an atmosphere of CO_2 or $CO_2 + H_2$. Fine fungal mycelium grew from the soil particles onto the nutrient-free silica gel incubated in both atmospheres. Although all particles did not yield mycelium, growth of *Fagus sylvatica* was observed in all soils, most frequently those from below beech. The anaerobic fungus grew both aerobically or anaerobically on Czapek Dox agar and was identified as *Fusarium solani*. These observations show that woodland soils contain at least one fungus which can grow both on nutrient-free silica gel and anaerobically in an atmosphere of $CO_2 + H_2$ or CO_2 alone, using only nutrients from the soil or atmosphere. It seems, therefore, that at least some fungi are capable of growing anaerobically in soils (Milton et al. 1994). One of the examples of facultative anaerobe is the yeast *Saccharomyces cerevisiae* (Carlile et al. 2001). A novel filamentous fungus strain designated RB-1 was isolated from Japanese rice field soil through an anaerobic roll tube technique. Tonouchi (2010) studied a mitosporic fungus that grows in both aerobic and strict anaerobic conditions using various mono-, di-, tri-, and polysaccharides with acetate and ethanol productions. The amount of acetate produced was higher than that of ethanol in both aerobic and anaerobic cultures. The characteristic punctuate conidia of RB-1 closely resembled strains of the genus

Thermomyces, a thermophilic or mesophilic anamorphic ascomycete (Tonouchi 2010). Further studies should be carried out along these lines to establish unequivocally the role of anaerobic fungi in lignin degradation. To date no study has been carried out on cultivation of anaerobic fungi for the production of bioactive compounds and research into the capacity of anaerobic fungi to produce antibiotics and pharmaceuticals is wide open.

1.4.2 Advantages of Anaerobic Mechanism of Benzene Degradation

Anaerobic microorganisms found within the subsurface can transform a wide variety of organic and inorganic contaminants. Mono-aromatic compounds associated with petroleum hydrocarbon mixtures are some of the most prevalent contaminants found in aquifers. The relative mobility of these hydrocarbon components in heavily contaminated aquifers (e.g. BTEX) frequently produces plumes of contamination within groundwater (Seagren and Becker 2002b). Contaminant plumes often contain regions of highly reduced sediment dominated by anaerobic processes and these can be difficult to remediate using aerobic techniques because of the abundance of reduced products of anaerobic metabolism (e.g. ferrous iron, sulfide). A wide variety of inorganic contaminants (e.g. metals and metalloids) that can be anaerobically transformed to less toxic and/or immobilized forms suggests that the in situ stimulation of anaerobic processes has the potential for widespread application (White et al. 1995). In sediments in which anaerobic processes are established, stimulating the anaerobic community has the advantage of enhancing an active and acclimated microbial population. Various research studies showed that use of different electron acceptors like nitrate, sulphate, iron during anaerobic processes for removing a priority contaminants from groundwater in contaminated aquifers offers potential for anaerobic degradation. This is important because many contaminated aquifers contain large volumes that are dominated by anaerobic processes (Anderson and Lovely 1997; Coates and Anderson 2000).

1.5 Anaerobic Benzene Mineralization Pathways

The possible initial steps are hydroxylation, carboxylation and methylation, and subsequent transformation to the central aromatic intermediate benzoyl-coA which is further degraded to CO_2 (Weelink et al. 2010).

1.5.1 Benzene Hydroxylation

Phenol, cyclohexanone, and propionate were detected as putative intermediates, suggesting initial hydroxylation of benzene to phenol and cyclohexane. Experiments with ^{18}O-labeled water suggested that the oxygen incorporated into the

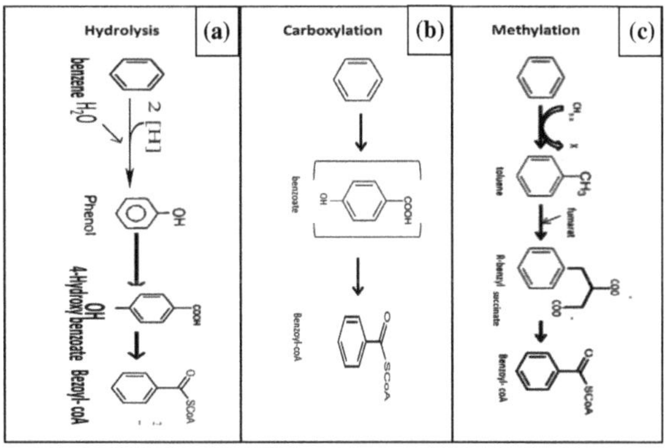

Fig. 4 Pathways of benzene mineralization

aromatic ring was derived from water. However, the proportion of ^{18}O-labeled phenol that was produced was less than one third of the amount expected if all of the phenol had originated from anaerobic benzene oxidation with water supplying the oxygen (Vogel and Grbic-Galic 1986). Weiner and Lovely (1998) found phenol, propionate, and acetate as intermediates in a benzene-degrading methanogenic enrichment. The enzyme responsible for hydrating benzene (lack of a functional group) is still unknown but a recently purified molybdenum-containing, iron-sulfur protein that adds a hydroxyl group to ethylbenzene may provide a model for the hydroxylation of benzene (Johnson et al. 2001). Evidence for benzene-degradation via phenol formation and subsequent conversion to benzoate was found in a benzene degrading methanogenic enrichment culture (Ulrich et al. 2005) and in a benzene-degrading iron reducing enrichment (Botton and Parsons 2007). Recently, the anaerobic benzene degradation pathway was studied in *Dechloromonas aromatic* RCB with nitrate as the terminal electron acceptor. Initial hydroxylation to phenol and subsequent carboxylation to p-hydroxybenzoate and the loss of hydroxyl group to form benzoate (benzoyl co-A) were reported for this organism. The pathway of benzene hydrolysis is shown in (Fig. 4a) (Chakraborty and Coates 2005).

1.5.2 Benzene Carboxylation

Carboxylation of benzene to form benzoate (or benzoyl-CoA) has been reported, however, in a sulfate-reducing enrichment culture; deuterated benzoate was detected when deuterated benzene was added to the enrichment but phenol was not detected (Caldwell and Suflita 2000; Phelps et al. 2001b). Kunapuli et al. (2008) demonstrated that in a benzene-degrading iron-reducing enrichment culture the carboxyl group of benzoate was derived from the bicarbonate buffer, indicative of direct carboxylation of benzene (Fig. 4b).

1.5.3 Benzene Methylation

Benzene is susceptible to substitution by strong electrophiles as in the Friedel Crafts alkylation. The electrophile could be the H_3C^+ unit derived from S-adenosyl-methionine. Benzene methylation is represented in (Fig. 4c). Biologically mediated methylation of benzene to toluene is energetically favourable using S-adenosyl-methionine as the methyl donor (Coates et al. 2002; Weelink 2008). The possibility of fumarate addition, as is seen in toluene degradation, is unlikely due to the absence of a methyl group. Hence, the carboxyl donor has yet to be identified (Phelps et al. 2001a). In a sulfate-reducing enrichment culture, the methyl group might possibly be generated from bicarbonate via a reverse co-dehydrogenase pathway (Safinowski and Meckenstock 2005). The first evidence of benzene transformation was observed in a nitrate-reducing and methanogenic enrichment culture [ring-^{13}C] in which toluene was detected by GC/MS (Ulrich et al. 2005; Weelink 2008).

1.6 Enhanced Benzene Degradation Using Enrichment Cultures and Co-cultures

Enriched microbial cultitvation relies on creating a condition in which the survival and growth rate of the bacterial culture with whatever trait is desired/favored and used to isolate the microorganisms which can survive in these conditions. Simultaneously, it is applicable to detoxify many xenobiotics. The substrate range of enzymes catalyzing a certain reaction can be expanded through the use of enrichment cultures (Doble et al. 2007). Weelink et al. (2010) reported benzene degradation at a rate of 1.65 mM benzene per day, which is similar to reported aerobic benzene degradation rates but 20–1650 times higher than reported for anaerobic benzene degradation by a stable anoxic enrichment culture *Alicycliphilus denitrificans* K601 in the presence of chlorate-reduction conditions. Similarly, Jahn et al. (2005) reported the anaerobic degradation of o-xylene and ethylbenzene in sediment-free iron-reducing enrichment cultures. Recently a nitrate-reducing enriched bacterial consortium included two strain of *Bacillus cereus, Citrobacter freundii* A11-1, *Brevibacterium linens* S402003 and was found to be effective in benzene degradation (Gupta et al. 2015). *Pseudomonas fluorescence* was isolated by using enrichment techniques and BTEX metabolism was also studied under hypoxic condition with nitrate-reducing conditions (Mark et al. 1993). Only a few stable enrichment cultures capable of benzene mineralization under anaerobic conditions are available. These include sulfate-reducing, nitrate-reducing and iron-reducing enrichment, methanogenic, perchlorate reduction, humic substances and other e-acceptors. Anaerobic benzene-degrading enrichment cultures performing methanogenesis were obtained from non-contaminated lotus field soil in which three dominant bacterial groups were found. Methane is believed to be produced from benzene by the sequential degradation of benzene by fermenting bacteria, hydrogen-producing acetogens, and methanogens. *Syntrophus gentianae* was the most prominent bacterium that assimilated labeled benzene early in the process of benzene degradation (Sakai et al. 2009). The different type of electron reducing conditions are discussed in detail below.

1.6.1 Sulfate Reduction

Sulfate addition to methanogenic zones of a petroleum-contaminated aquifer stimulated the growth and activity of sulfate-reducing microorganism and enhanced benzene removal with a lag period of 100 days (Lovely 2000). Anderson and Lovely (2000) reported benzene removal of up to 53 % under sulfate-reducing anaerobic conditions. Decrease in sulfate concentration along the ground water flow path, proved sulfate reduction as a predominant terminal electron acceptor process. Caldwell and Suflita (2000) proposed that benzoate is formed as a metabolite of benzene metabolism under anaerobic condition by using sulphidogenic cultures. The origin of the benzoate was confirmed by the use of stable-isotope studies in which deuterated benzoate was detected in cultures amended with deuterated benzene. Musat and Widdel (2008) reported delta-propeobacteria (sulfate-reducing bacteria) in synthetic seawater medium (with 28 mM sulfate) with marine sediment as a source of microorganisms and benzene (0.7 %, v/v) supplied in hepta-methylnonane as an inert hydrophobic carrier. After 6 months, incubations had formed approximately 15 mM sulfide in comparison with approximately 5 mM formed in benzene-free controls and it was proven to completely oxidize benzene under sulfate-reducing condition. Previous research studies suggested that benzene degradation by enriched sulfate-reducing bacteria is mediated by dehydrogenation and then hydroxylation to form phenol or methylation to form toluene. Kleinsteuber et al. (2008) proposed that the dominant organisms colonizing lava granules were related to *Magnetobacterium* sp., followed by a phylotype affiliated with the genera *Cryptanaerobacter/ Pelotomaculum* and several *delta-proteobacteria* syntrophically fermented benzene to acetate, as 0.24 mM benzene was consumed within 1100 h (46 days) coupled with sulfide production. A functional model for syntrophic benzene degradation under sulfate-reducing conditions based on the possible functions of the community members and thermodynamic calculations is also given (Schink 1997; Jackson and Mcinerney 2002; Kleinsteuber et al. 2008).

1. $C_6H_6 + 3H_2O + 3.75\ SO_4^{-2} \rightarrow 6HCO_3 + 1.875\ HS^- + 1:875H_2S + 0.375H +$ $\Delta G° - 185\ \text{kJ mol}^{-1}$
2. $\text{Benzene} + HCO_3 \rightarrow \text{Benzoate} + H_2O\Delta G° + 1.7\ \text{kJ mol}^{-1}$
3. $\text{Benzene} + 6H_2O \rightarrow 3\text{Acetate} + 3H^+ + 3H_2\Delta G° + 72.3\ \text{kJ mol}^{-1}$
4. $3H_2 + 0.75\ SO_4^{-2} + 1.125H^+ \rightarrow 3H_2O + 0.375HS + 0:375H_2S\Delta G°$ $116.3\ \text{kJ mol}^{-1}$

5. Benzene + 0.75 SO_4^{-2} + $3H_2O \rightarrow$ 3Acetate + 0.375HS + 0.375 H_2S + 1.875 H^+ $\Delta G°$ 44 kJ mol^{-1}

1.6.2 Nitrate Reduction

Several studies have suggested that the addition of nitrate might stimulate anaerobic benzene degradation in petroleum-contaminated aquifers and benzene degradation linked to nitrate reduction has been found recently with enrichment cultures developed from soil and groundwater microcosms (Ulrich and Edwards 2003; Weelink et al. 2010) and in pure cultures (Coates et al. 2001; Kasai et al. 2006; Weelink et al. 2010). A study, involving two microbial strains (facultatively anaerobes and members of the *Dechloromonas*) in the presence of nitrate as an e^- acceptor, showed that the oxidation of 163 ± 19 mM benzene by the strain RCB resulted in the reduction of 843 ± 64 mM nitrate. This represents greater than 86 % of the theoretical ratio for benzene oxidation coupled to nitrate reduction according to:

$$C_6H_6 + 6NO_3^{-2} + 6H^+ \rightarrow 6CO_2 + 3N_2 + 6H_2O$$

The study concluded that 92 % of the initial ^{14}C label, showing that benzene is completely oxidized to CO_2 (Coates et al. 2001). BTEX (benzene, toluene, ethylbenzene, o-, p-, m-xylene)-adapted consortia under nitrate-reducing conditions, degraded benzene at moderate rates within 70 days and nitrite accumulated due to nitrate reduction without inhibitory effects. The initial concentration of BTEX mixtures in soil was 30 mg/kg, viz. the concentration of each substrate was approximately 5 mg/kg, When the initial concentrations were increased to 90 and 150 mg/kg, BTEX was completely degraded after a period of 40 and 50 days, respectively. However, benzene and p-xylene showed a relatively slower degradation rate. While the degradation rates of benzene, o-xylene, and p-xylene were slow, benzene, o-xylene, and p-xylene were degraded with a lag phase of 20 days compared to toluene and ethylbenzene. For o-xylene, approximately 78 % was degraded after 50 days, whereas for benzene and p-xylene, only 31 and 48 % was degraded throughout the experimental period. It was concluded that mixed enriched bacterial culture developed from indigenous bacteria of soil could enhance the BTEX biodegradation ability (Dou et al. 2008a).

1.6.3 Perchlorate Reduction

Perchlorate and chlorate are electron acceptors whose reduction results in the formation of molecular oxygen. The produced oxygen can be used for activation of anaerobic persistent pollutants, like benzene. In a study conducted by Tan et al. (2006), chlorate was tested as a potential electron acceptor to stimulate benzene degradation in an anoxic polluted soil column. A chlorate-amended benzene-polluted soil column was operated over a period of 500 days. Benzene was immediately degraded in the column after start up, and benzene removal recovered completely after omission of chlorate due to a too high influent chlorate concentration (22 mM). Mass balance calculations showed at per mole of benzene, five mole of chlorate were reduced. At the end of the experiment, higher loading rates were applied to measure the maximal benzene degradation rate in this system; a breakthrough of benzene degradation was not observed. The average benzene degradation rate over this period was 31–78 mol L^{-1} h^{-1}. Denaturing gradient gel electrophoresis of part of the 16S rRNA gene, cloning and sequencing showed that the culture had a stable composition after the seventh transfer. Five bacterial clones were further analyzed. DGGE analysis of cultures grown with different electron donors and acceptors indicated that the bacterium related to *Alicycliphilus denitrificans* K601 is able to degrade benzene coupled to chlorate reduction (Weelink et al. 2007). It was discovered that *A. denitrificans* type strain K601, a cyclohexanol-degrading, nitrate-reducing beta-proteobacterium (99.7 % similarity based on 16S rRNA genes), strain BC could grow on some other aromatic compounds in the absence of oxygen with chlorate as the electron acceptor. This strain could degrade about 100 μM benzene per day and tolerates benzene concentrations as high as 1 mM. *Dechloromonas aromatic* strain RCB was tested at concentrations only as high as 100 μM, and its benzene degradation rate with chlorate as the electron acceptor was less than 10 μM per day (Weelink et al. 2008).

1.6.4 Iron Reduction

No pure cultures of dissimilatory iron-reducing bacteria are able to degrade benzene could be isolated up to now (Chakraborty and Coates 2004; Coates et al. 2002). Further research revealed that sediments, in which benzene degradation was detected, were enriched in bacteria belonging to the Geobacteraceae family (Anderson et al. 1998; Rooney-Varga et al. 1999; Botton and Parsons 2007). Benzene removal under iron-reducing conditions was also reported in sediment inoculations and estuary sediments contaminated with petroleum (Kazumi et al. 1997; Botton and Parsons 2007). Several benzene-mineralizing iron-reducing cultures have since been successfully established using amorphous iron hydroxide [Fe $(OH)_3$] as an electron acceptor (Villatoro-Monzon et al. 2003; Jahn et al. 2005; Kunapuli et al. 2007; Vogt et al. 2011). A study indicated that the ability of Fe(III)-reducing microorganisms to degrade benzene at sampling sites cannot be attribute to groundwater components that make Fe(III) more available for reduction or other potential factors (Anderson and Lovely 1999). Recently, benzene degradation was also verified directly in anoxic iron-reducing zones of plumes by an in situ microcosm approaches (Cozzarelli et al. 2010; Vogt et al. 2011). Jahn et al. (2005) conducted a research study on Iron(III) reduction by enrichment cultures grown with benzene as sole carbon and energy source and amorphous iron(III) hydroxide as the electron acceptor where enrichment cultures contained AQDS

(9,10-anthraquinone-2,6-disulfonic acid; 2 mM) as an additional agent to enhance microbial iron reduction. An initial concentration 150 μM of benzene was degraded up to ~80 % with simultaneous decrease in Fe(III) concentration from 4 to <2 mM. Enriched cultures showed recovery of electrons from benzene oxidation through ferric iron reduction of 82 and 74 %, respectively.

1.6.5 Methanogenic

Methane is believed to be produced from benzene by the sequential degradation of benzene by fermenting bacteria, hydrogen-producing acetogens, and methanogens. *Bacteroidetes*, *Euryarchaeota*, *Firmicutes*, and *Thermotogae* are related to members of dechlorinating microbial communities (Chang et al. 2005; Sakai et al. 2009; Cupples 2011). Weiner and Lovely (1998) presented the first report of rapid conversion of benzene to methane and carbon dioxide in methanogenic aquifer sediments. The fact that benzene was degraded without a lag phase suggests that the microbial population was adapted for benzene degradation and that benzene was being degraded in situ. [^{14}C] benzene (58.2 mCi/mmol) had been mineralized in sediments up to 53 %. The added [^{14}C] benzene was steadily converted to $^{14}CH_4$ and $^{14}CO_2$ over time. Parallel studies in which [2-^{14}C]acetate was added to the sediments further confirmed that methane production was the predominant terminal electron-accepting process (TEAP), as $^{14}CH_4$ accounted for 83 % of the radiolabel recovered as $^{14}CH_4$ and $^{14}CO_2$. The study also suggested that methanogenesis was the dominant TEAP, as the groundwater contained high concentrations of dissolved methane (675 mM). $^{14}CH_4$ and $^{14}CO_2$ were produced in equal proportions from [^{14}C]benzene. Benzene was converted solely to methane and carbon dioxide according to the following equation

$$4C_6H_6 + 27H_2O \rightarrow 15CH_4 + 9HCO_3^- + 9H^+.$$

It showed that total radiolabel recovered in methane and carbon dioxide, 62.5 % would be recovered as methane and recovery of $^{14}CH_4$ from [^{14}C] benzene was 80 %. Sakai et al. (2009) studied anaerobic benzene-degrading enrichment cultures performing methanogenesis, *Syntrophus gentianae* Hasda-A was the most prominent bacterium involved in the initial steps of benzene degradation under methanogenic conditions. Methanogenic flow through aquifer columns were used to investigate the potential of bioaugmentation to enhance anaerobic benzene–toluene–ethylbenzene–xylene (BTEX) degradation in groundwater contaminated with ethanol-blended gasoline. Two different methanogenic consortia (enriched with benzene or toluene and o-xylene) were used as inoculi. Significant benzene biodegradation (from 49 ± 14 to 88 ± 7 %) was observed only in a column bioaugmented with the benzene-enriched methanogenic consortium, and this removal efficiency was sustained for 1 year with no significant decrease in permeability due to bioaugmentation (Da Silva and Alvarez 2004). Studies of Chang et al. (2005); Sakai et al. (2009) and Cupples (2011) could be used to design targeted isolation strategies for anaerobic benzene-degrading strains under methanogenic conditions.

1.6.6 Manganese (IV) Reduction

Manganese (IV) was shown to be used as electron acceptor for anaerobic benzene degradation in sediment microcosms and columns. The benzene degradation rates were significantly higher for manganese (IV) compared with ferric iron, indicating that this process is relevant in the environment (Villatoro-Monzon et al. 2003; Vogt et al. 2011). The benzene mineralization was also studied by using sediment microorganisms using a graphite anode as an electron acceptor (Zhang et al. 2010; Vogt et al. 2011). Anaerobic benzene biodegradation was performed in batch experiments using Rhine River sediments as inoculum and amorphous Mn(IV) or Fe(III) as independent final electron acceptors. Benzene (4.5 μmol) was degraded in 80 and 710 days, respectively. Highest benzene degradation rates of 0.07 μmol/day, was obtained under Mn(IV) reducing conditions, with soluble Mn (II) and CO_2 recoveries of 71.5 and 93 % regarding to the stoichiometric values, respectively. Likewise, benzene biodegradation was performed in a continuous column coupled to the reduction of Mn(IV). Efficiency of benzene biodegradation was up to 97 % under steady state operation in a sediment column operated continuously for more than 160 days. The carbon dioxide and Mn(II) recoveries were 88 and 77 %, respectively, of the theoretical ratio according to the stoichiometry for benzene biodegradation (Villatoro-Monzón et al. 2008).

1.7 Role of Enzymes During Anaerobic Degradation of Benzene

Ability of microorganism to degrade aromatic compounds is dictated by their metabolic enzymes (Alexander 1999). Benzene ring reduction requires a large input of energy and this metabolic capability has so far, been reported only in bacteria. The abstraction of a hydrogen atom from benzene would need an activation energy of more than 460 kJ mol^{-1} which is roughly 100 kJ mol^{-1} more compared with the formation of a benzyl radical from alkylated benzene derivatives (Widdel and Rabus 2001; Musat and Widdel 2008; Vogt et al. 2011). These activation reactions of aromatic compounds degradation are generally divided into two-phase processes; in the first phase, the various aromatic substrates are converted to intermediates, which are eventually transformed to the central metabolite (e.g. benzoyl-CoA). The second phase involves de-aromatization of benzoyl-CoA to cyclohex-1,5-diene-1-carboxyl-CoA using the common key enzyme benzoylCoA reductase, followed by the further oxidation to acetyl-CoA and CO_2, in a sequence of reactions analogous to β-oxidation (Heider and Fuchs 1997). The gene encoding the ring-cleavage enzyme, 2-keto-cyclohexanecarboxyl-CoA hydrolase, was identified by assaying the enzymatic activity of the protein expressed in *Escherichia coli* (Schühle et al. 2003). An increase in information on the structure and function of catabolic enzymes and pathways offers further possibilities for their optimization. A rational site-directed mutagenesis approach to improve enzyme function is possible, if a detailed characterization of a given enzyme and at best the crystal structure is available. Alternatively, sequence alignments can help in identifying residues critical for enzyme activity or substrate specificity and a rational design can lead to improved biocatalysts (Pieper and Reineke 2000). Genes encoding benzoyl-CoA reductase, a novel enzyme able to overcome the resonance stability of the aromatic ring were identified by directed mutagenesis (Egland et al. 1997; Schühle et al. 2003). Anaerobic benzene oxidation coupled to the reduction of Fe(III) was studied in *Ferroglobus placidus* and a putative carboxylase protein potentially involved in the initial activation of benzene was discovered. The gene coding for ORF 138 from the benzene-degrading, Fe(III)-reducing enrichment culture was located within a cluster of genes containing three other potential carboxylase genes and a benzoate-CoA ligase gene (Schühle et al. 2003; Laban et al. 2010; Holmes et al. 2011). The enzyme was synthesized and induced when cells were grown anaerobically in the light with benzoate as the sole carbon source. The kinetic properties of the enzyme match the kinetics of substrate uptake by whole cells and confirm a role for benzoate-CoA ligase in maintaining entry of benzoate into cells as well as in catalyzing the first step in the anaerobic degradation of benzoate because benzoate is a principle intermediate of sequential degradation of benzene (Geissler et al. 1988).

1.8 Immobilized Biomass-Emerging Field

Immobilized microbial techniques (IMT) can shield microorganisms from harsh environmental conditions, mutually allowing higher biomass density, providing a greater prospect for retrieval and reuse (Zhang et al. 2007; Tsekova et al. 2010; El-Naas et al. 2014). Another advantage of this technique is increase in biodegradation rates through higher cell loading (Kumar et al. 2012; El-Naas et al. 2014). The immobilization of cells can be achieved by different methods: covalent coupling, cross-linking, physical entrapment, and the natural process of adhesion (El-Naas et al. 2014). Recently, one alternative option that turned out to be important is the production of natural fiber-polymer materials, with principle applications in the production of construction materials (Robledo-Ort et al. 2011; El-Naas et al. 2014). De Nardi et al. (2002) studied a bench-scale, horizontal-flow, anaerobic immobilized biomass (HAIB) reactor filled with polyurethane foam; a matrix containing immobilized anaerobic biomass was initially fed with synthetic substrate containing protein, carbohydrates and lipids. It was assayed to verify its degradation performance in a reactor fed with BTEX. Thereafter, BTEX degradation was evaluated in the presence of the co-solvents ethanol and lineal alkylbenzene sulphonate (LAS), in two sequential experiments. An active enriched microbial consortium was observed in the reactor, containing BTEX-degraders, and also acetogenic, acetotrophic and hydrogenotrophic microorganisms. Similarly in another research study, a denitrifying consortium was used as inoculum to form a biofilm in a horizontal-flow anaerobic immobilized biomass reactor for the biodegradation study. The reactor was fed with hydrocarbons, separately and also with BTEX solution. The hydrocarbons were dissolved in a solution containing ethanol. Hydrocarbons removal efficiencies were 99 % at an initial concentration of benzene of 26.5 mg/L, toluene 30.8 mg/L, m-xylene 32.1 mg/L, ethylbenzene 33.3 mg/L and BTEX 26.5 mg/L (Estebar et al. 2012; Ribeiro et al. 2013). Microbial diversity assessed by a small portion of 16S rDNA suggested the dominance of species related to the phylotypes *Pseudomonas, Paracoccus* and *Bacteroides*. This system showed to be an alternative for the treatment of wastewater contaminated with nitrate, ethanol and hydrocarbons (Gusmão et al. 2006; Estebar et al. 2012). Shim and Yang (1999) reported that a fibrous-bed bioreactor containing the co-culture of *Pseudomonas putida* and *P. fluorescens* was able to degrade benzene (B), toluene (T), ethylbenzene (E), and o-xylene (X) in synthetic waste streams. In general, the BTEX biodegradation rate increased with increasing substrate concentration and then decreased after reaching a maximum, showing substrate-inhibition kinetics. However, for immobilized cells, the degradation rate was much higher than that of free cells. Compared to free cells, immobilized cells in the bioreactor tolerated higher concentrations of benzene and toluene, and gave maximum degradation rates for benzene, ethylbenzene, and o-xylene. Complete and simultaneous degradation of the BTEX mixture was achieved in the bioreactor under hypoxic conditions (Shim and Yang 1999).

1.9 Effect of Physical and Chemical Factors on Microbial Degradation of Benzene

1.9.1 Effect of Chemical Solubility and Its Tolerance to Microorganisms

Organic compounds differ widely in their solubility from polar compounds (methanol, phenol) to non-polar compounds for example poly- and mono-aromatic hydrocarbons. The availability of compound to organisms reveals its biodegradability. Compounds with greater aqueous solubility are more available to degradative enzymes. Benzene solubility in water is 1791 mg/L and it is readily biodegradable. The existence of contaminated areas with highly toxic organic compounds, such as benzene has stimulated the search for solvent-tolerant microorganisms that have the catabolic potential to remove at high rates these toxic compounds (Sardessai and Bhosle 2004). Solvent toxicity is directly related to accumulation of solvent molecules inside the cell membrane (Torres et al. 2011). Solubility is one of the important factors promoting anaerobic benzene degradation specifically in the presence of iron-reducing conditions. For example, when chelators for Fe(III) were added to aquifer sediments, benzene could be degraded rapidly under Fe(III)-reducing conditions. Soluble Fe(III) is more readily available for microbial reduction than insoluble Fe(III) oxides (Anderson and Lovely 1999). Likewise, solvent utilization at high rates or solvent biotransformation to a less toxic product was observed in some tolerant bacteria (Vangnai et al. 2002). Dou et al. (2010) isolated *Bacillus cereus* from a gasoline-contaminated site and evaluated biodegradation of benzene. 150 mg/L had been removed in 30 days. The metabolites of nitrate-dependent benzene oxidation were identified as phenol and benzoate. While the metabolic pathways driving the activation of hydrocarbons into central metabolic pathways are well understood, the molecular and biochemical basis of microbial behavior and physiological responses to hydrocarbons and the impact of these responses on bioremediation have yet to be explored.

1.9.2 Surfactants

Surfactants have been reported to solubilize non-polar organic compounds added to aqueous solution in samples of soil and water. Due to solubilization, the accessibility of the substrates to microorganisms can be increased. The influence of the non-ionic surfactant Triton X-100 and another pollutant on BTEX was tested; Triton X-100 up to 400 mg/L did not enhance the solubility of the relatively soluble BTEX compounds significantly (Lee et al. 2000). Several research studies, on the other hand, showed that surfactants enhance the rate of degradation of BTEX or other organic contaminant.

1.9.3 Effect of Co-substrate Interaction

In a study by Laban et al. (2009), biodegradation of benzene with co-substrate phenol and toluene in the presence of sulphate as an e-acceptor, established a decrease in sulphate concentration (from 2.5 to 1 mM), in case of phenol (as co-substrate) at 500 μM concentration of substrate (benzene). In case of toluene, sulphate concentrations varied from (3–2.5 mM) and benzene degradation was found to be 100 % at 500 μM, governed by a dominant *Pelotomaculum* (Gram-positive bacterium) in the anaerobic degradation of benzene. Co-metabolic degradation experiments resulted in retarded rates of benzene degradation in the presence of phenol whereas toluene had no effect on benzene metabolism. Phenol, 2-hydroxybenzoate, 4-hydroxybenzoate, and benzoate were identified as putative metabolites in the enrichment culture (Laban et al. 2009). Few studies have been conducted in an effort to understand substrate interactions of mono-aromatics in a mixture. A mixed consortium and strain *Pseudomonas* sp. PPO1 were found to degrade benzene and toluene individually, p-xylene remained unaffected, while in a mixture of benzene, toluene, and p-xylene, respectively, benzene and toluene were degraded according to competitive inhibition kinetics, whereas p-xylene was partially removed by co-metabolism (Oh et al. 1994). Burback and Perry (1993) reported that *Mycobacterium vaccae* could individually catabolize benzene or toluene, and also a mixture of the compounds. Ethylbenzene was the most potent inhibitor of benzene, toluene and xylene degradation for *R. rhodochrous* cultured in all tested substrate mixtures (Deeb and Cohet 1999).

1.9.4 pH

pH is an important environmental factor which may increase or decrease microbial activity and metabolism. Delfino and Miles (1985) reported that under acidic conditions (pH 5.3, 20 °C) benzene was completely degraded by microbes in 16 days in groundwater taken from a shallow well. Likewise Jahn et al. (2005) studied benzene degradation with the help of enrichment cultures grown in a carbonate-buffered mineral medium with a pH between 7.2 and 7.4, resulting in enrichment of different iron-reducing cultures, which were able to oxidize all BTEX compounds, respectively. Purified benzene-oxidizing enrichment culture and mixed enrichment cultures showed recoveries of electrons from benzene oxidized by ferric iron reduction of 82 and 74 %, respectively.

1.9.5 Nitrogen Concentration

Laboratory studies on microbial degradation of benzene with mixed cultures of microorganisms in gasoline-contaminated groundwater revealed that nitrogen concentration is major controlling factor in the anaerobic biodegradation of benzene. Karlson and Frankenberger (1989) reported that nitrogen enhanced the biodegradation rate of benzene compared to innoculum-enriched water alone. More than 95 % of the benzene in groundwater was removed through microbial action within 73.5 h.

1.9.6 Nutrient

Nutrients including carbon, nitrogen, hydrogen, phosphorus, oxygen and other trace metals are commonly require for the growth and perpetuation of microbes. BTEX removal efficiency increased with nutrient feeding rates (NFR) (Lu et al. 2002).

1.9.7 Temperature

In general, rate of biological reaction at mesophilic condition (25–27 °C) provide a suitable environment for the metabolism of microbes while increased temperature induces enhanced volatilization rate of mono-aromatic hydrocarbons.

1.9.8 Hydraulic Retention Time

Retention time is a key component in controlling microbes for the degradation of hazardous compounds under anaerobic condition. Flexible handling of influent flow rates is required for effective control. Influent quality variation affects the efficiency of bioreactors, selection of flow direction could influence mixing and degree of contaminant removal (Farhadian 2008).

1.9.9 Effect of Metal on Anaerobic Microbial Growth

The toxicity of a heavy metal to anaerobic bacteria depends upon a number of factors, including chemical form, bioavailability, and the presence of bacterial species that are resistant to or can detoxify the metal (Hughes and Poole 1991; Kuo and Genthner 1996). Biocatalytic activity of hydrocarbon-degrading enzymes can be inhibited by the presence of other environmental pollutants such as heavy metals (Shen et al. 2006; Mathew et al. 2010; Fellie et al. 2012). The effects of metal ions on the biodegradation of benzene and toluene were investigated. Among 12 tested metal ions, Cu^{2+}, Ni^{2+}, Co^{2+} and As^{+} inhibited the degradation of benzene and toluene by *Alcaligenes xylosoxidans* Y234 severely. Cu^{2+} was found to inhibit catechol 1,2-dioxygenase in the degradation of benzene and toluene. Co^{2+} and Ni^{2+} were proposed to inhibit benzoate 1,2-dioxygenase as well, while Ag^{+} inhibited benzaldehyde dehydrogenase in the degradation of toluene. The inhibitory effect caused by these metal ions could be overcome by microbial adaptation and addition of specific aromatic compounds to the broth (Yeom and Yoo 1997). Similarly, a metal acclimated consortium culture (CC) consisting of *Agrobacterium* sp., *Chryseomonas* sp., *Flavobacterium* sp., *Pseudomonas* sp., *Serratia* sp., *Xanthomonas* sp., *Arthobacter* sp., *Bacillus* sp. and *Micrococcus* sp. biodegraded benzene (35.43 % in 24 h and 61.66 % in 48 h) (Fellie et al. 2012).

1.10 Molecular Tools for Assessing Bioremediation

There are some drawbacks linked to enzymatic bioremediation like; high production costs, low yields, and enzymatic inhibition are some of the problems that must be overcome. Therefore, molecular tools are being widely explored to provide competitive enzymatic bioremediation products. Advanced molecular tools can open unlimited windows of opportunity, as it is possible to detect genes related to degrading enzymes from cultivable or non-cultivable organisms (Zeyaullah et al. 2009; Alcalde et al. 2006; Peixoto et al. 2011; Gupta et al. 2014). These methods rely on the characterization of cellular constituents such as nucleic acids, proteins, fatty acids and other taxa-specific compounds (Rossello-Mora and Amann 2001; Malik et al. 2008). Such molecules can be extracted directly from environmental samples without the need for culturing and analysis of the molecular composition can be used to elucidate the composition of the microbial community (Amann et al. 1995; Greene and Voordouw 2003; Malik et al. 2008). Kim and Jeon (2009) and Peixoto et al. (2011) suggested that genetic manipulation would be useful to allow or improve petroleum degradation in extreme environments, such as cold or hypersaline sites. The use of free extremozymes would be advantageous in these environments, since it avoids some of the limitations of bioremediation using whole cells in extreme conditions, such as microbial competition. Some of the molecular tools along with their functions, which are useful in assessing benzene-degrading microbial communities, are highlighted in Tables 8 and 9.

Table 8 List of molecular tools for assessing bioremediation

Molecular tools	Functions	Target cellular component	References
Phospholipid-derived fatty acids (PLFA)	Quantify total biomass and assess the entire microbial population	Phospholipids	Dijkman and Kromkamp (2006)
Reverse transcription polymerase chain reaction (RT PCR)	Rapidly detect and quantify specific microbial population and processes	Nucleic acid	Joyce (2002)
Stable-isotope probing (SIP)	Definitively prove bioremediation of a specific compound	Nucleic acid	Radajewski et al. (2000)
Amplified Ribosomal DNA Restriction Analysis (ARDRA)	TO measure bacterial community structure	Nucleic acid	Malik et al. (2008)
Ribosomal Intergenic Spacer Analysis (RISA)	used to define microbial diversity and community composition in freshwater environments	rRNA genes in the rRNA operon	Fisher and Triplett (1999)
Bio-Sep Bead	To determine concentration of environmental contaminant in groundwater	–	Key (2013)
Reverse sample genome probing (RSGP)	It identifies and characterises bacteria	Nucleic acid	Voordouw et al. (1991), Malik et al. (2008)

(continued)

Table 8 (continued)

Molecular tools	Functions	Target cellular component	References
Microarray technologies (a) Functional gene arrays (FGA) (b) Community genome arrays (CGA) (c) Phylogenetic oligonucleotide arrays	Identify organism and define its ecological role (a) Measure genes (b) Encoding key enzymes in a metabolic process (c) Species-to-strain level differentiation simultaneously identifying many microorganisms from a sample	Nucleic acid	Zhou and Thompson (2002), Gentry et al. (2006)

Table 9 Role of some microbial strain identified by molecular tools

Strains	Molecular tools	Function	References
Azoarcus	Stable isotopes probing (SIP) with denaturing gradient gel electrophoresis (DGGE)	Benzene degradation under NO_3^- reducing condition	Kasai et al. (2006)
Acidovorax avenae isolate C1 and *A. denitri- ficans* type strain K601	Bead-beating and phenol-chloroform based DNA extraction	Genes encoding monooxygenase and dioxygenase	Weelink et al. (2008)
Geothrix fermentans Variovorax paradoxus	PLFA	Fe(III)-reducing enrichment culture	Rooney-Varga et al. (1999)
Geobacter metallireducens	Microarray analysis	Metabolizes many aromatic compound	Butler et al. (2007)
F. placidus	Transcriptomic analysis	Benzene metabolism to benzoate	Holmes et al. (2011)

With the rapid exploration of knowledge and understanding in this field, metabolomics can be an excellent tool for the study of the unique chemical fingerprints of specific cellular processes and for determining the phenotype caused by genetic manipulation used to characterize the interactions of organisms with their environment (Daviss 2005). Analytes in a metabolomic sample comprise highly complex mixtures. This complex mixture can be simplified prior to detection by separating some analytes from others using GC MS. Gieg and Suflita (2005) investigated a variety of signature metabolites associated with anaerobic hydrocarbon biodegradation in groundwater from monitoring wells at a former refinery site by using field metabolomics. It was used to assess the in situ anaerobic attenuation of hydrocarbons in a contaminated aquifer underlying a former refinery. Benzene, ethylbenzene,

2-methyl-naphthalene, 1,2,4- and 1,3,5-trimethyl-benzene were targeted as contaminants and low copy numbers of *Desulfobacterium* were found, a genus previously implicated in anaerobic benzene biodegradation (Parisi et al. 2009). Assessment of anaerobic benzene degradation potential using 16S rRNA gene-targeted real-time PCR and comparison with a 16S rRNA gene clone library of the benzene-enriched methanogenic consortium revealed the presence of an operational taxonomic unit, clone OR-M2, which was affiliated with *Desulfobacterium anilini, Desulfosarcina variabilis* and *Desulfobacula toluolica* (Ulrich and Edwards 2003). Mono-oxygenase genes were detected in bacterial strains obtained from hydrocarbon-polluted aquifers for degradation of BTEX (Hendricks et al. 2006; Malik et al. 2008). In attempts to understand transformation mechanisms in mixed cultures, molecular techniques have been developed to target the active organisms. A key method for this is called stable isotope probing (SIP) and has been applied only recently to study anaerobic BTEX degradation. *Geobacter metallireducens* is the first anaerobic benzene-degrading organism that can be genetically manipulated. This opens the possibility of definitively elucidating the pathway for anaerobic benzene degradation in this organism, identifying molecular signals that can be useful for monitoring rates of anaerobic benzene degradation in the subsurface, and developing new strategies for stimulating anaerobic benzene degradation informed by a better understanding of this metabolism (Zhang et al. 2012). Community analysis using terminal restriction fragment length polymorphism, 16S rRNA gene sequencing and FISH revealed 95 % dominance of one phylotype that is affiliated to the Gram-positive bacterial genus *Pelotomaculum* showing that sulfate-reducing bacteria are involved in anaerobic benzene degradation (Laban et al. 2009). In recent studies, nucleic acid-based techniques like amplified ribosomal DNA restriction analysis (ARDRA) has been combined with other molecular techniques such as T-RFLP and DGGE to characterize microbial communities from contaminated sources (Watts et al. 2001; Haack et al. 2004). Gich et al. (2000) used ARDRA to examine microbial differences in activated sludge from treatment plants fed on domestic or industrial wastewater. It was observed that the bacterial communities in activated sludge were different from industrial and domestic wastewater treatment plants (Malik et al. 2008). Phylotypes performing distinct functions related to benzene degradation in complex microbial biofilms from an aerated treatment pond containing coconut textile have been identified, benzene was degraded at rates of 11 μM per day and per gram coconut textile under oxic conditions. Carbon isotope fractionation with isotopic enrichment factors of 0.6–1 and no significant hydrogen isotope fractionation indicated a dihydroxylation reaction for the initial ring attack. The incubation with $[^{13}C_6]$-benzene led to $^{13}CO_2$ formation accompanied by ^{13}C-labeling of RNA and proteins of the active biomass (Jechalke et al. 2013). Key et al. (2013) examined the role of vadose zones in an aquifer; microorganisms in hydrocarbon attenuation. Reverse transcriptase-qPCR (RT-qPCR) analysis indicated that microbial activity in the vadose zone generally increased with depth. To address this problem, the molecular method of SIP was applied to identify anaerobic toluene and benzene degraders from a number of environments across the globe. For example, $^{13}C_6$-benzene provided direct evidence of benzene biodegradation in six of the eight wells studied. The highest levels of ^{13}C were detected in dissolved inorganic carbon (DIC) extracted from the two monitoring wells closest to the river. RT-qPCR and SIP results were used to refine the site's conceptual model by identifying areas where that genetic potential was actively being expressed and locations where biodegradation was lagging. The remediation of sites contaminated with gasoline has been limited by a lack of information on the microorganisms able to transform these chemicals under anaerobic conditions. A greater variety of organisms were identified as anaerobic benzene degraders, several studies linked anaerobic benzene degradation to the *Deltaproteobacteria* (e.g. *Desulfobacteraceae, Desulfobulbacea*) and the *Clostridia* (e.g. *Peptococcaceae*) (Cupples 2011).

2 Summary and Conclusion

Growing awareness of the harmful effects of environmental pollution has led to a marked increase in research into various strategies that might be useful to clean up contaminated environments. The field of anaerobic bioremediation is an attractive remediation technique and much of the literature to date is based on field and laboratory observations. Field trials account for the effects of multiple substrates, diverse terminal electron acceptors (TEAs), and mixed indigenous microbiota subject to shifts in species composition. Isolating strictly anaerobic benzene-degrading bacteria is still important for a better understanding of anaerobic benzene degradation. Biotechnology always places emphasis on ethical responsibilities before using such novel strategies for bioremediation. Exploiting microbial diversity for enhanced bioremediation involves the elucidation of efficient or novel catabolic pathways for detoxification and complete mineralization. Sequencing the genome of an organism may allow revealing the upper and lower degradation pathways; also more controlled physiological experiments would be possible for elucidating the anaerobic benzene activation mechanism. In addition, before releasing microorganism to contaminated sites, its genetic modifications should be understood. Furthermore, intrinsic capabilities and activities of the site of interest have to be characterized. Genetic engineering techniques offer the possibility to equip organisms to survive and be active when introduced into certain environments, with the desired catabolic potential.

Along with cultivable microbes, uncultivable microbial diversity can also be exploited to identify desirable degradative genes and construction of metagenomics libraries to transfer them subsequently to cultivable bacteria for enhanced bioremediation. As more is learned about the genes encoding anaerobic hydrocarbon degradation, the use of functional gene probes may accelerate the assignation of genes and the screening of diverse environments and enrichment cultures for marker genes. Evaluating gene regulation and enzyme specificity will provide information for modeling of complex biodegradation processes. However such future endeavors are based on the assumption that detected novel degraders can be cultivated for study. It may be feasible to use diagnostic techniques to determine which terminal electron acceptors (TEAs) would be most beneficial at a specific site given its particular contaminants and indigenous microbiota, thus taking laboratory observations to application in the field.

References

Alcalde M, Ferrer M, Plou FJ, Ballesteros A (2006) Environmental biocatalysis: from remediation with enzymes to novel green processes. Trends Biotechnol 24(6):281–287

Alexander M (1999) Biodegradation and bioremediation (2nd edition). Academic Press, San Diego, pp 335–365

Amann RI, Ludwig W, Schleifer KH (1995) Phylogenetic identification and in situ detection of individual microbial cells without cultivation. Microbiol Rev 59:143–169

An YJ (2004) toxicity of mixture of BTEX (benzene, toluene, ethylene, xylene) on the terrestrial plant (sorghum biocolor and *cucumis staivus*). Bull Environ Contam Toxicol 72:1006–1011

Anderson RT, Lovely D (1999) Naphthalene and benzene degradation under Fe (III) reducing condition in petroleum contaminated aquifer. Bioremediat J 3(2):121–135

Anderson RT, Lovely DR (2000) Anaerobic bioremediation of benzene under sulphate reducing condition in a petroleum contaminated aquifer. Environ Sci Technol 34:2261–2266

Anderson RT, Rooney-Varga JN, Gaw CV, Lovley DR (1998) Anaerobic benzene oxidation in the Fe(III) reduction zone of petroleum-contaminated aquifers. Environ Sci Technol 32:1222–1229

Arvin E, Engelsen P, Sebber U (2005) Biodegradation of gasoline compounds (BTEX) in a water works sand filter. Water Supply 4(5–6):29–33

ATSDR (1992) Case Studies in Environmental Medicine, benzene toxicity, ATSDR Publication No.: ATSDR-HE-CS-2001-0003

ATSDR (1997) Toxicological profile for benzene (update). US Department of Health and Human Services, Atlanta

ATSDR (2000) Toxicological profile for toluene. US Department of Health and Human Services, Agency for Toxic Substances and Disease Registry, Atlanta

ATSDR (Agency for Toxic Substances and Disease Registry) (2004) Interaction profile for benzene, toluene, ethylbenzene, and xylenes (BTEX). U.S. Department of Health and Human Services, Public Health Service, Atlanta

ATSDR (2007a) Toxicological profile for benzene. Agency for Toxic Substances and Disease Registry, US Department of Health and Human Services, Public Health Service, Atlanta

ATSDR (2007b) Draft toxicological profile for ethylbenzene. US Department of Health and Human Services, Agency for Toxic Substances and Disease Registry, Atlanta

ATSDR (2007c) Toxicological profile for xylene. US Department of Health and Human Services, Agency for Toxic Substances and Disease Registry, Atlanta

Bennett K (1999) In-situ treatment of soil contaminated by benzene (a BTEX compound) restoration and reclamation review. University of Minnesota, Department of Horticultural Science. Retrieved from the University of Minnesota Digital Conservancy, http://hdl.handle.net/11299/59450

Bernauer U, Vieth B, Ellrich R (2000) CYP2E1 expression in bone marrow and its intra- and interspecies variability: approaches for a more reliable extrapolation from one species to another in the risk assessment of chemicals. Arch Toxicol 73(12):618–624

Boll M, Fuchs G (1995) Benzoyl-coenzyme A reductase (de-aromatizing), a key enzyme of anaerobic aromatic metabolism. Eur J Biochem 234:921–933

Botton S, Parsons JR (2007) Degradation of BTX by dissimilatory iron-reducing cultures. Biodegradation 18:371–381

Burback B, Perry I (1993) Biodegradation and biotranslbrmation of groundwater pollutant mixtures by *Mycobacterium Vacce*. Appl Environ Microbiol 59:1025–1029

Butler JE, Qiang He, Nevin KP, Zhili He, Zhou J, Lovley DR (2007) Genomic and microarray analysis of aromatics degradation in *Geobacter metallireducens* and comparison to a *Geobacter* isolate from a contaminated field site. BMC Genom 8:180

Caldwell ME, Suflita JM (2000) Detection of phenol and benzoate as intermediates of anaerobic benzene biodegradation under different terminal electron-accepting conditions. Environ Sci Technol 34:1216–1220

Canadian Council of Ministers of The Environment (1999) Canadian soil quality guidelines for the protection of environmental and human health: benzene. Canadian Environmental Quality Guidelines, Winnipeg

Cape JN (2003) Effects of airborne volatile organic compounds on plants. Environ Pollut 122 (1):145–157

Carlile MJ, Watkinson SC, Gooday GW (2001) The fungi, 2nd edn. Academic Press, London, pp 85–105. ISBN 0-12-738446-4

cdn.intechopen.com

Chakraborty R, Coates JD (2004) Anaerobic degradation of monoaromatic hydrocarbons. Appl Microbiol Biotechol 64:437–446

Chakraborty R, Coates JD (2005) Hydroxylation and carboxylation-two crucial steps of anaerobic benzene degradation by *Dechloromonas strain RCB*. Appl Environ Microbiol 71:5427–5432

Chakraborty R, O'Connor SM, Chan E, Coates JD (2005) Anaerobic degradation of benzene, toluene, ethylbenzene, and xylene compounds by *Dechloromonas Strain RCB*. Appl Environ Microbiol 71(12):8649–8655

Chang W, Um Y, Pulliam HTR (2005) Molecular characterization of anaerobic microbial communities from benzene-degrading sediments under methanogenic conditions. Biotechnol Prog 21(6):1789–1794

Chilcott RP (2007) Benzene toxicological overview. Health Protection Agency, CHAPD HQ, HPA, Version 2

Cliona MM, Luoping Z, Martyn TS (2011) Current understanding of the mechanism of benzene-induced leukemia in humans: implications for risk assessment. Carcinogenesis 33 (2):240–252

Coates JD, Anderson RT (2000) Emerging techniques for anaerobic bioremediation of contaminated environments. Trends Biotechnol 18(10):408–412

Coates JD, Chakraborty R, Lack JG, O'Connor SM, Cole KA, Bender KS, Achenbach LA (2001) Anaerobic benzene oxidation coupled to nitrate reduction in pure culture by two strains of *Dechloromonas*. Nature 411:1039–1043

Coates JD, Chakraborty R, Mcinerney MJ (2002) Anaerobic benzene biodegradation—a new era. Res Microbiol 153:621–628

Cole GM (1994) Assessment and remediation of petroleum contaminated sites. CRC Press, Boca Raton

Cox LA (1991) Biological basis of chemical carcinogenesis: insights from benzene. Risk Anal 11(3):453–464

Cozzarelli IM, Bekins BA, Eganhouse RP, Warren E, Essaid HI (2010) In situ measurements of volatile aromatic hydrocarbon biodegradation rates in groundwater. J Contam Hydrol 111:48–64

Cupples AM (2011) The use of nucleic acid based stable isotope probing to identify the microorganisms responsible for anaerobic benzene and toluene biodegradation. J Microbiol Methods 85(2):83–91

Da Silva MLB, Alvarez PJJ (2004) Enhanced anaerobic biodegradation of benzene–toluene–ethylbenzene–xylene–ethanol mixtures in bioaugmented aquifer columns. Appl Environ Microbiol 70(8):4720–4726

Daifullah AAM, Girgis BS (2003) Impact of surface characteristics of activated carbon on adsorption of BTEX. Colloids Surf A Physicochem Eng Asp 214(1):181–193

Daviss B (2005) Growing pains for metabolomics. Scientist 19(8):25–28

De Nardi IR, Varesche MB, Zaiat M, Foresti E (2002) Anaerobic degradation of BTEX in a packed-bed reactor. Water Sci Technol 45(10):175–180

Deeb RA, Cohet IA (1999) Temperature effects and substrate interactions during the aerobic biotransformation of BTEX mixtures by toluene-enriched consortia and *Rhodochrous*. Biotechnol Bioeng 62:526–536
Delfino JJ, Miles CJ (1985) Aerobic and anaerobic degradation of organic contaminants in Florida ground water. In: Proceedings-Soil and Crop Science Society of Florida, vol 44, pp 9–14
denr.sd.gov
DH (1998) Annual report of the committees on toxicity mutagenicity carcinogenicity of chemicals in food, consumer products and the environment. In: Eighth joint annual report
Dijkman N, Kromkamp J (2006) Phospholipid-derived fatty acids as chemotaxonomic markers for phytoplankton: application for inferring phytoplankton composition. Mar Ecol Prog Ser 324: P113–P125
Doble M, Rollins K, Kumar A (2007) Green chemistry and engineering, 1st edn. Academic Press, London
Dou J, Liu X, Hu Z (2008a) Substrate interactions during anaerobic biodegradation of BTEX by the mixed cultures under nitrate reducing conditions. J Hazard Mater 158:264–272
Dou J, Liu X, Hu Z, Deng D (2008b) Anaerobic BTEX biodegradation linked to nitrate and sulfate reduction. J Hazard Mater 151:720–729
Dou J, Ding A, Liu X, Du Y, Deng D, Wang J (2010) Anaerobic benzene biodegradation by a pure bacterial culture of *Bacillus cereus* under nitrate reducing conditions. J Environ Sci 22(5):709–715
Douglas OM, Colin GO (1994) Anaerobic fungi, biology, ecology and function, mycology series, vol 12, 1st edn. CRC Press, ISBN 0-8247-8948-2
Dubinsky EA et al (2013) Succession of hydrocarbon-degrading bacteria in the aftermath of the deep-water horizon oil spill in the gulf of Mexico. Environ Sci Technol 47(19):10860–10867
Egland PG, Pelletier DA, Dispensa M, Gibson J, Harwood CS (1997) A cluster of bacterial genes for anaerobic benzene ring biodegradation. Proc Natl Acad Sci USA 94(12):6484–6489
El-Naas MH, Acio JA, El Telib AE (2014) Aerobic biodegradation of BTEX: progresses and prospects. J Environ Chem Eng 2:1104–1122
EPA (2002) Toxicological review of benzene (noncancer effects), in support of summary information on the integrated risk information system (IRIS), CAS No. 71-43-2,635/R-02/001F, U.S. Environmental Protection Agency, Washington, DC
Estebar M, Mehdi MA, Poursafa P, Ghasemian M, Jaafarzadeh HHN, Fatehizadeh A (2012) Biodegradation of benzene–toluene–xylene in petrochemical industries wastewater through anaerobic sequencing biofilm batch reactor in bench scale. Int J Env Health Eng 2012(1):22
Farhadian M (2008) Development of an effective bioremediation technology for volatile monoaromatics removal from contaminated water. Thesis, University Blaise Pascal, Clermont-Ferrand
Farhadian M, Larroche C, Borghei M, Troquet J, Vachelard C (2006) Bioremediation of BTEX-contaminated groundwater through bioreactors. 4ème Colloque Franco-Roumain de Chimie Appliquée, Université Blaise Pascal, Clermont-Ferrand, France, p. 438
Fellie EA, Sannasi P, Wong KK, Salmijah S, Kader J (2012) Tolerance and biodegradation of benzene, toluene, ethylbenzene and xylenes (BTEX) by a metal acclimatized bacterial consortium culture. Res J Biotechnol 7(3):52–58
Ferro A, Kennedy J, Doucette W, Nelson S, Jauregui G, McFarland B, Bugbee B (1996) Fate of benzene in soils planted with alfalfa: uptake, volatilization, and degradation. In: Phytoremediation f Soil and Water Contaminants Symposium, American Chemical Society, Washington, Dc 20036 (USA), pp 223–237
Fisher MM, Triplett EW (1999) Automated approach for ribosomal intergenic spacer analysis of microbial diversity and its application to freshwater bacterial communities. Appl Environ Microbial 65:4630–4636
Geissler JF, Harwood CS, Gibson J (1988) Purification and properties of benzoate-coenzyme a ligase, a *Rhodopseudomonas palustris* enzyme involved in the anaerobic degradation of benzoate. J Bacteriol 170:1709–1714

Gentry TJ, Wickham GS, Schadt CW, He Z, Zhou J (2006) Microarray applications in microbial ecology research. Microbiol Ecol 52:159–175

Gich FB, Amer E, Figueras JB, Abella CA, Balaguer MD, Poch M (2000) Assessment of microbial community structure changes by amplified ribosomal DNA restriction analysis (ARDRA). Int Microbiol 3:103–106

Gieg LM, Suflita JM (2005) Metabolic indicators of anaerobic hydrocarbon biodegradation in petroleum-laden environments, In: Ollivier B, Magot M (eds) Petroleum Microbiology. ASM Press, Washington, DC. doi: 10.1128/9781555817589.ch17

Grbic-Galic D, Vogel TM (1987) Transformation of toluene and benzene by mixed methanogenic cultures. Appl Environ Microbiol 53:254–260

Greene EA, Voordouw G (2003) Analysis of environmental microbial communities by reverse sample genome probing. J Microbiol Methods 53:211–219

Gupta S, Pathak B, Fulekar MH (2014) Molecular approaches for biodegradation of polycyclic aromatic hydrocarbon compounds. Rev Environ Sci Biotechnol. doi:10.1007/s11157-014-9353-3

Gupta S, Pathak B, Fulekar MH (2015) Biodegradation of benzene under anaerobic condition using enriched microbial culture. IJSRSET 1(4):207–215

Gusmão VR, Martins TH, Chinalia FA, Sakamoto IK, Henriquethiemann O, Varesche MBA (2006) BTEX and ethanol removal in horizontal-flow anaerobic immobilized biomass reactor, under denitrifying condition. Process Biochem 41(6):1391–1400, 1359–5113

Haack SK, Fogarty LR, West TG, Alm EW, McGuire JT, Long DT (2004) Spatial and temporal changes in microbial community structure associated with recharge-influenced chemical gradients in a contaminated aquifer. Environ Microbiol 6:438–448

Hansch C, Leo AJ (1985) Medchem project, Issue No. 26. Pomona College, Claremont, CA

Hartel U, Eckel E, Koch J, Fuchs G, Linder D, Buckel W (1993) Purification of glutaryl-CoA dehydrogenase from *Pseudomonas* sp., an enzyme involved in the anaerobic degradation of benzoate. Arch Microbiol 159:174–181

Harwood CS, Gibson J (1997) Shedding light on anaerobic benzene ring degradation: a process unique to prokaryotes? J Bacteriol 179:301–309

Heider J, Fuchs G (1997) Anaerobic metabolism of aromatic compounds. Eur J Biochem 243:577–596

Hendricks B, Dejonghe W, Faber F, Boenne W, Bastiaens L, Verstraete W (2006) PCR-DGGE method to assess the diversity of BTEX mono-oxygenase genes at contaminated sites. FEMS Microbiol Ecol 55:262–273

Holliger C, Gaspard S, Glod G, Heijman C, Schumacher W, Schwarzenbach RP, Vazquez F (1997) Contaminated environments in the subsurface and bioremediation: organic contaminants. FEMS Microbiol Rev 20(3–4):517–523

Holmes DE, Risso C, Smith JA, Lovley DR (2011) Anaerobic oxidation of benzene by the hyperthermophilic archaeon *ferroglobus placidus*. Appl Environ Microbiol 77(17):5926–5933

HSDB (Hazardous Substances Data Bank) (1997) National library of medicine, National toxicology program (via toxnet), Bethesda, MD

Hu Z, Dou J, Liu X, Zheng X, Deng D (2007) Anaerobic biodegradation of benzene series compounds by mixed cultures based on optional electronic acceptors. J Environ Sci 19:1049–1054

Hughes MN, Poole RK (1991) Metal speciation and microbial growth the hard (and soft) facts. J Gen Microbiol 137:725–734

IPCS (1985) Environmental health criteria 52: toluene. International Programme on Chemical Safety, World Health Organization, Geneva

IPCS (1993) Environmental health criteria 150: benzene. International Programme on Chemical Safety, World Health Organization, Geneva

IPCS (1996) Environmental health criteria 186: ethylbenzene. International Programme on Chemical Safety, World Health Organization, Geneva

IPCS (1997) Environmental health criteria 190: xylenes. International Programme on Chemical Safety, World Health Organization, Geneva

Jackson BE, McInerney MJ (2002) Anaerobic microbial metabolism can proceed close to thermodynamic limits. Nature 415:454–456

Jahn MK, Haderlein SB, Meckenstock RU (2005) Anaerobic degradation of benzene, toluene, ethylbenzene, and o-xylene in sediment-free iron-reducing enrichment cultures. Appl Environ Microbiol 71(6):3355–3358

Jechalke S, Franchini AG, Bastida F, Bombach P, Rosell M, Seifert J, von Bergen M, Vogt C, Richnow HH (2013) Analysis of structure, function, and activity of a benzene-degrading microbial community. FEMS Microbiol Ecol 85:14–26

Johnson HA, Pelletier DA, Spormann AM (2001) Isolation and characterization of anaerobic ethylbenzene dehydrogenase, a novel Mo–Fe–S enzyme. J Bacteriol 183:4536–4542

Joyce C (2002) Quantitative RT-PCR. A review of current methodologies. Methods Mol Biol 193:83–92

Kanak A (2014) Benzene and beyond: mechanisms of novel anaerobic aromatic degradation pathways in *Geobacter Daltonii*. Dissertation, Georgia State University

Karickhoff SW, Brown DS, Scott TA (1979) Sorption of hydrophobic pollutants on natural sediments. Water Res 13:241–248

Karlson U, Frankenberger WT (1989) Microbial degradation of benzene and toluene in groundwater. Bull Environ Contam Toxicol 43:505–510

Kasai Y, Takahata Y, Manefield M, Watanabe K (2006) Stable isotope probing and isolation of anaerobic benzene degrading bacteria from gasoline contaminated groundwater. Appl Environ Microbiol 72:3586–3592

Kazumi J, Caldwell ME, Suflita JM, Lovley DR, Young LY (1997) Anaerobic degradation of benzene in diverse anoxic environments. Environ Sci Technol 31:813–818

Kenaga EE (1980) Predicted bioconcentration factors and soil sorption coefficients of pesticides and other chemicals. Ecotoxicol Environ Safety 4:26–38

Key KC (2013) Applications of Bio-Sep(RTM) beads and molecular biological tools: assessment and stimulation of the biodegradation of petroleum hydrocarbons and fuel oxygenates. Ph.D Thesis, The University of Tulsa, p 3558746

Key KC, Sublette KL, Johannes TW, Ogles D, Baldwin B, Biernacki A (2013) Assessing BTEX biodegradation potential at a refinery using molecular biological tools. Groundwater Monitor Remediat 34(1 winter 2014):35–48

Khan FI, Husain T, Hejazi R (2004) An overview and analysis of site remediation technologies. J Environ Manag 71(2):95–122

Kim JM, Jeon CO (2009) Isolation and characterization of a new benzene, toluene, and ethylbenzene degrading bacterium *acinetobacter sp. b113*. Curr Microbiol 58(1):70–75

Kleinsteuber S, Schleinitz KM, Breitfeld J, Harms H, Richnow HH, Vogt C (2008) Molecular characterization of bacterial communities mineralizing benzene under sulfate-reducing conditions. FEMS Microbiol Ecol 66:143–157

Koch J, Eisenreich W, Bacher A, Fuchs G (1993) Products of enzymatic reduction of benzoyl-CoA, a key reaction in anaerobic aromatic metabolism. Eur J Biochem 211:649–662

Kumar SS, Kumar MS, Siddavattam D, Karegoudar TB (2012) Generation of continuous packed bed reactor with PVA–alginate blend immobilized *ochrobactrum sp.* dgvk1 cells for effective removal of *n, n*-dimethylformamide from industrial effluents. J Hazard Mater 199–200:58–63, 22079508

Kunapuli U, Lueders T, Meckenstock RU (2007) The use of stable isotope probing to identify key iron-reducing microorganisms involved in anaerobic benzene degradation. ISME J 1 (7):643–653

Kunapuli U, Griebler C, Beller HR, Meckenstock RU (2008) Identification of intermediates formed during anaerobic benzene degradation by ana iron reducing enrichment culture. Environ Microbiol 10:1703–1712

Kuo C-W, Genthner BRS (1996) Effect of added heavy metal ions on biotransformation and biodegradation of 2-chlorophenol and 3-chlorobenzoate in anaerobic bacterial consortia. Appl Environ Microbiol 62(7):2317–2323

Laban NA, Selesi D, Jobelius C, Meckenstock RU (2009) Anaerobic benzene degradation by gram positive sulfate reducing bacteria. FEMS Microbial Ecol 68:300–311

Laban A, Selesi D, Rattei T, Tischler P, Meckenstock RU (2010) Identification of enzymes involved in anaerobic benzene degradation by a strictly anaerobic iron-reducing enrichment culture. Environ Microbiol 12:2783–2796

Lee JF, Liao PM, Kuo CC, Yang HT, Chiou CT (2000) Influence of a nonionic surfactant (Triton X-100) on contaminant distribution between water and several soil solids. J Colloid Interface Sci 229(2):445–452

Leutwein C, Heider J (1999) Anaerobic toluene-catabolic pathway in denitrifying *Thauera aromatica*: activation and b-oxidation of the first intermediate, (R)-(M)-benzylsuccinate. Microbiology 145:3265–3271

Lin TF, Little JC, Nazaroff WW (1994) Transport and sorption of volatile organic compounds and water vapor within dry soil grains. Environ Sci Technol 28:322–330

Lovely DR (1997) Potential for anaerobic bioremediation of BTEX in petroleum-contaminated aquifer. J Ind Microbiol Biotechnol 18:75–81

Lovely DR (2000) Anaerobic benzene degradation. Biodegradation 11:107–116

Lovern MR, Turner MJ, Meyer M (1997) Identification of benzene oxide as a product of benzene metabolism by mouse, rat and human liver microsomes. Carcinogenesis 18:1695–1700

Lovley DR (1991) Dissimilatory Fe(III) and Mn(IV) reduction. Microbiol Rev 55:259–287

Lu C, Lin M-R, Chu C (2002) Effects of pH, moisture, and flow pattern on trickle-bed air biofilter performance for BTEX removal. Adv Environ Res 6:99–106

Mackay D, Leinonen PJ (1975) Rate of evaporation of low-solubility contaminants from water bodies to atmosphere. Environ Sci Technol 9:1178–1180

Malik S, Beer M, Megharaj M, Naidu R (2008) The use of molecular techniques to characterize the microbial communities in contaminated soil and water. Environ Int 34:265–276

Mark DM, Jerome JK, Ronald HO (1993) Metabolic diversity of aromatic hydrocarbon-degrading bacteria from petroleum-contaminated aquifer. Biodegradation 4:249–259

Mathew A, Thanga VSG, Reshma JK (2010) Simultaneous phenol degradation and chromium (VI) reduction by bacterial isolates. Res J Biotechnol 5(1):46–49

Medeiros AM, Bird MG, Witz G (1997) Potential biomarkers of benzene exposure. J Toxicol Environ Health 51:519–539

Milton W, Tasneem AA, Killham K (1994) Anaerobic growth of fungal mycelium from soil particles onto nutrient-free silica gel. Mycol Res 98(7):761–762

mmbr.asm.org

Mountfort DO, Kaspar HF (1986) Palladium mediated hydrogenation of unsaturated hydrocarbons with hydrogen gas release during anaerobic cellulase degradation. Appl Environ Microbial 52:744–750

Musat F, Widdel F (2008) Anaerobic degradation of benzene by a marine sulfate reducing enrichment culture and cell hybridization of dominant phylotype. Environ Microbial 10(1): 10–19

National Ambient Air Quality Standards (NAAQS) (2009) Central Pollution Control Board Notification. http://www.appcb.ap.nic.in/env-standards/cat_in_unitop/aaq_standards_09

NHMRC (2004) Australian drinking water guidelines. National Health and Medical Research Council and Natural Resource Management Ministerial Council, Canberra, Australia

NTP (2005) Report on carcinogens, Eleventh edn. US Department of Health and Human Services, Public Health Service, National Toxicology Program, Atlanta

Nyer EK (1998) Groundwater and soil remediation: practical methods and strategies. Sleeping Bear Press, Ann Arbor, pp 171–205

Office for National Statistics (ONS) (2014) Environmental accounts. Springer, Heidelberg

Oh YS, Shareefdeen Z, Baltzis BC, Bartha R (1994) Interactions between benzene, toluene and p-xylene (BTX) during their biodegradation. Biotechnol Bioeng 44:533–538

OSHA (Occupational Safety and Health Administration) (1987) Benzene. U.S. Department of Labor, Occupational Safety and Health Administration. Code of Federal Regulations. 29 CFR 1910.1028

Pardieck D et al (1992) Hydrogen peroxide use to increase oxidant capacity for in situ bioremediation of contaminated soils and aquifers: a review. J Contam Hydrol 9:221–242

Parisi VA, Brubaker GR, Zenker MJ, Prince RC, Gieg LM, Da Silva MLB, Alvarez PJ, Suflita JM (2009) Field metabolomics and laboratory assessments of anaerobic intrinsic bioremediation of hydrocarbons at a petroleum-contaminated site. Microbial Biotechnol 2(2):202–212

Peixoto RS, Vermelho AB, Rosado AS (2011) Petroleum-degrading enzymes: bioremediation and new prospects. Enz Res 475193:7

Phelps CD, Zhang X, Young LY (2001a) Use of stable isotopes to identify benzoate as a metabolite of benzene degradation in a sulphidogenic consortium. Environ Microbiol 3 (9):600–603

Phelps CD, Zhang X, Young LY (2001b) Use of stable isotopes to identify benzoate as a metabolite of benzene degradation in a sulphidogenic consortium. Environ Microbiol 3:600–603

Pieper DH, Reineke W (2000) Engineering bacteria for bioremediation. Curr Opin Biotechnol 11:262–270

Piero L, MN D'Errico, Fustinoni S, Drago I, Barbieri A, Sabatini L, Carrieri M, Apostoli P, Soleo L (2011) Assessment of environmental exposure to benzene: traditional and new biomarkers of internal dose. In: Popovic D (ed) Air quality-models and applications. ISBN: 978-953-307-307-1

Radajewski S, Ineson P, Parekh NR, Murrell C (2000) Stable-isotope probing as a tool in microbial ecology. Nature 403(6770):646–649

Ranck JM, Bowman RS, Weeber JL, Katz LE, Sullivan EJ (2005) BTEX removal from produced water using surfactant-modified zeolite. J Environ Eng 131(3):434–442

Ribeiro R, De Nardi IR, Fernandes BS, Foresti E, Zaiat M (2013) BTEX removal in a horizontal-flow anaerobic immobilized biomass reactor under denitrifying conditions. Biodegradation 2:269–278

Ritter L, Solomon K, Sibley P, Hall K, Keen P, Mattu G (2002) Sources, pathways and relative risks of contaminants in surface water and groundwater: a perspective prepared for the walkerton inquiry. J Toxicol Environ Health Part A 65:1–142

Robledo-Ort JR, Ram Irez-Arreola DE, Erez-Fonseca AAP, Omez CG, Gonz Alez-Reynoso O, Ramos-Quirarte J (2011) Benzene, toluene, and o-xylene degradation by free and immobilized *p. putida* f1 of postconsumer agave-fiber polymer foamed composites. Int Biodeteriorat Biodegrad 65:539–546

Rooney-Varga JN, Anderson RT, Fraga JL, Ringelberg D, Lovley DR (1999) Microbial communities associated with anaerobic benzene degradation in a petroleum-contaminated aquifer. Appl Environ Microbiol 65(7):3056–3063

Rosenstock L, Cullen MR, Brodkin CA, Redlich CA (2004) Textbook of clinical occupational and environmental medicine, 2nd edn. Elsevier, Philadelphia

Rossello-Mora R, Amann R (2001) The species concept for prokaryotes. FEMS Microbiol Rev 25:39–67

Sabourin PJ, Chen BT, Lucier G et al (1987) Effect of dose on the absorption and excretion of [^{14}C]benzene administered orally or by inhalation in rats and mice. Toxicol Appl Pharmacol 87:325–336

Safinowski M, Meckenstock RU (2005) Methylation is the initial reaction in anaerobic naphthalene degradation by a sulfate-reducing enrichment culture. Environ Microbiol 8:347–352

Sakai N, Kurisu F, Yagi O, Nakajima F, Yamamoto K (2009) Identification of putative benzene degrading bacteria in methanogenic enrichment cultures. J Biosci Bioeng 108(6):501–507

Sardessai Y, Bhosle S (2004) Industrial potential of organic solvent tolerant bacteria. Biotechnol Prog 20:655–660

Scheunert I, Topp E, Schmitzer J (1985) Formation and fate of bound residues on [^{14}C] benzene and [^{14}C] chlorobenzenes in soil and plants. Ecotox Environ Saf 9:159–170

Schink B (1997) Energetics of syntrophic cooperation in methanogenic degradation. Microbiol Mol Biol Rev 61:262–280

Schühle K, Gescher J, Feil U, Paul M, Jahn M, Schägger H, Fuchs G, Georg Fuchs Mikrobiologie, Institut FüR Biologie Ii, UniversitäT Freiburg, Freiburg, Zentrum (2003) Benzoate-coenzyme a ligase from thauera aromatica: an enzyme acting in anaerobic and aerobic pathways. J Bacteriol 185(16):4920–4929

Seagren EA, Becker JG (2002) Review of natural attenuation of BTEX and MTBE in Groundwater. Pract Period Hazard Toxic Radioact Waste Manag 6(3):156–172

Senthil Kumar M et al (2013) Quantification of benzene in groundwater sources and risk analysis in a popular South Indian Pilgrimage City—a GIS based approach. Arabian J Chem. doi:10.1016/j.arabjc.2013.09.022

Shen G, Lu Y, Hong J (2006) Combined effect of heavy metals and polycyclic aromatic hydrocarbons on urease activity in soil. Ecotoxicol Environ Saf 65:474–480

Shim H, Yang ST (1999) Biodegradation of benzene, toluene, ethylbenzene, and o-xylene by a coculture of *Pseudomonas putida* and *Pseudomonas fluorescens* immobilized in a fibrous-bed bioreactor. J Biotechnol 2267(2–3):99–112

Shoun H, Tanimoto T (1991) Denitrification by the fungus *fusarium oxysporum* and involvement of cytochrome p-450 in the respiratory nitrite reduction. J Biol Chem 266(17):11078–11082

Smith MT, Yager JW, Steinmetz K (1989) Peroxidase-dependent metabolism of benzene's phenolic metabolites and its potential role in benzene toxicity and carcinogenicity. Environ Health Perspect 82:23–29

Stupperich E (1993) Recent advances in elucidation of biological corrinoid functions. FEMS Microbiol Rev 12:349–366

Tabak HH, Cooke WB (1968) Growth and metabolism of fungi in an atmosphere of nitrogen. Mycologia 60(1):115–140

Takaya N (2009) Response to hypoxia, reduction of electron acceptors, and subsequent survival by filamentous fungi. Biosci Biotechnol Biochem 73(1):1–8

Tan NC, Van Doesburg W, Langenhoff AA, Stams AJ (2006) Benzene degradation coupled with chlorate reduction in a soil column study. Biodegradation 17:11–17

Taylor BF, Campbell WL, Chinoy I (1970) Anaerobic degradation of benzene nucleus by facultatively anaerobic microorganisms. J Bacteriol 102:430–437

Tiburtius ER, Peralta-Zamora P, Emmel A (2005) Treatment of gasoline-contaminated waters by advanced oxidation processes. J Hazard Mater 126(1–3):86–90

Tonouchi A (2010) Isolation and characterization of a novel facultative anaerobic filamentous fungus from Japanese rice field soil. Int J Microbiol. Article ID 571383, p 9, 571383

Topp E, Scheunert I, Korte F (1989) Kinetics of the uptake of ^{14}C-labeled chlorinated benzene from soil by plants. Ecotox Environ Saf 17:157–166

Torres S, Pandey A, Castro GR (2011) Organic solvent adaptation of Gram positive bacteria: applications and biotechnological potentials. Biotechnol Adv 29(4):442–452

Tsekova K, Todorova D, Ganeva S (2010) Removal of heavy metals from industrial wastewater by free and immobilized cells of *Aspergillus niger*. Int Biodeterior Biodegrad 64:447–451

Uchrin CG, Mangels G (1986) Chloroform sorption to new jersey coastal plain ground water aquifer solids. Environ Toxicol Chem 5(40):339–343

Ulrich AC, Edwards EA (2003) Physiological and molecular characterization of anaerobic benzene-degrading mixed cultures. Environ Microbiol 5(2):92–102

Ulrich AC, Beller HR, Edwards EA (2005) Metabolites detected during biodegradation of 13c6-benzene in nitrate-reducing and methanogenic enrichment cultures. Environ Sci Technol 39:6681–6691

USEPA (2005) National Emissions Inventory. http://www.epa.gov/ttn/ chief/net/ 2005inventory

USEPA (US Environmental Protection Agency) (2006) Edition of the drinking water standards and health advisories. EPA 822-r-06-013, Washington DC

Van Hamme JD, Singh A, Ward OP (2003) Recent advances in petroleum microbiology. Microbiol Mol Biol Rev 67(4):503–549

Vangnai AS, La Sayavedra-Soto, Arp DJ (2002) Roles for the two 1-butanol dehydrogenases of pseudomonas butanovora in butane and 1-butanol metabolism. J Bacteriol 184:4343–4350

Villatoro-Monzon WR, Mesta-Howard AM, Razo- Flores E (2003) Anaerobic biodegradation of BTEX using Mn(IV) and Fe(III) as alternative electron acceptors. Water Sci Technol 48:125–131

Villatoro-Monzón WR, Morales-Ibarria MG, Velázquez EK, Ramírez-Saad H, Razo-Flores E (2008) Benzene biodegradation under anaerobic conditions coupled with metal oxides reduction. Water Air Soil Pollut 192:165–172

Vogel TM, Grbic-Galic D (1986) Incorporation of oxygen from water into toluene and benzene during anaerobic fermentative transformation. Appl Environ Microbiol 52:200–202

Vogt C, Kleinsteuber S, Richnow HH (2011) Anaerobic benzene degradation by bacteria. Microbial Biotechnol 4(6):710–724

Wada H (1976) Decomposition processes of plant debrises in submerged soil—part 1. J Sci Soil Manure Japan 47(10):458–462

Wallace S, Kadlec R (2005) BTEX degradation in a cold-climate wetland system. Water Sci Technol 51(9):165–171

Watts JE, Wu Q, Schreier SB, May HD, Sowers KR (2001) Comparative analysis of polychlorinated biphenyl-dechlorinating communities in enrichment cultures using three different molecular screening techniques. Environ Microbiol 3:710–719

Weelink SAB (2008) Degradation of benzene and other aromatic hydrocarbons by anaerobic bacteria. Ph.D thesis, ISBN 978-90-8585-239-1

Weelink SAB, Tan NCG, Ten Broeke H, Van Doesburg W, Langenhoff AAM, Gerritse J, Stams AJM (2007) Physiological and phylogenetic characterization of a stable benzene-degrading, chlorate-reducing microbial community. FEMS Microbiol Ecol 60 (2):312–321 (Epub 26)

Weelink SAB, Tan NCG, Ten Broeke H, Van Den Kieboom C, Van Doesburg W, Langenhoff AAM, Gerritse J, Junca H, Stams AJM (2008) Isolation and characterization of alicycliphilus denitrificans strain bc, which grows on benzene with chlorate as the electron acceptor. Appl Environ Microbiol 74:216672–216681

Weelink SAB, Van Eekert MHA, Stams AJM (2010) Degradation of BTEX by anaerobic bacteria: physiology and application. Rev Environ Sci Biotechnol 9:359–385

Weiner JM, Lovley DR (1998) Rapid benzene degradation in methanogenic sediments from a petroleum-contaminated aquifer. Appl Environ Microbiol 64:1937–1939

White C, Wilkinson SC, Gadd GM (1995) The role of microorganisms in biosorption of toxic metals and radionuclides. Int Biodeteriorat Biodegrad 17–40

WHO (2000) Benzene, Chapter 5.2, air quality guidelines, second edition. WHO Regional Office for Europe, Copenhagen, Denmark

WHO (2008) Guidelines for drinking water quality. Third Edition incorporating the first and second addenda. World Health Organization, Geneva

Wick LY, McNeill K, Rojo M, Medilanski E, Gschwend PM (2000) Fate of benzene in a stratified lake receiving contaminated groundwater discharges from a superfund site. Environ Sci Technol 34(20):4354–4362

Widdel F, Rabus R (2001) Anaerobic biodegradation of saturated and aromatic hydrocarbons. Curr Opin Biotechnol 12:259–276

Wood TM, Wilson CA, Mccrare SI, Joblin KN (1986) A Highly active extracellular cellulase from the anaerobic rumen fungus *neocallimastix frontalis*. FEMS Microbial Lett 34:37–40

wrhsrc.oregonstate.edu

www.atsdr.cdc.gov

www.inchem.org

www.ncbi.nlm.nih.gov

www.bren.ucsb.edu

www.hpa.org.uk

www.environment.gov.ab.ca

www.ons.gov.uk

Yardley-Jones A, Anderson D, Parke DV (1991) The toxicity of benzene and its metabolism and molecular pathology in human risk assessment. Br J Ind Med 48(7):437–444

Yeom SH, Yoo YJ (1997) Overcoming the inhibition effects of metal ions the degradation of benzene and toluene by *Alcaligenes Xylosoxidans Y234*. Korean J Chem Eng 14(3):204–208

Yuki K, Takahata Y, Manefield M, Watanabe K (2006) RNA-based stable isotope probing and isolation of anaerobic benzene-degrading bacteria from gasoline-contaminated groundwater. Appl Environ Microbiol 3586–3592

Zeyaullah M, Kamli MR, Islam B (2009) Metagenomics—an advanced approach for non-cultivable micro-organisms. Biotechnol Mol Biol Rev 4(3):49–54

Zhang L-S, Wu W-Z, Wang J-L (2007) Immobilization of activated sludge us- ing improved polyvinyl alcohol (PVA) gel. J Environ Sci (China) 19:1293–1297

Zhang T, Gannon SM, Nevin KP, Franks AE, Lovley DR (2010) Stimulating the anaerobic degradation of aromatic hydrocarbons in contaminated sediments by providing an electrode as the electron acceptor. Environ Microbiol 12:1011–1020

Zhang T, Bain TS, Nevin KP, Barlett MA, Lovley DR (2012) Anaerobic benzene oxidation by *Geobacter* species. Appl Environ Microbiol 78(23):8304–8310

Zhang T, Tremblay P-L, Chaurasia AK, Smith JA, Bain TS, Lovley DR (2014) Identification of genes specifically required for the anaerobic metabolism of benzene in *geobacter metallireducens*. Front Microbiol 5:245

Zhou J, Thompson DK (2002) Challenges in applying microarrays to environmental studies. Curr Opin Biotechnol 13:204–207

Zhou Z, Takaya N, Nakamura A, Yamaguchi M, Takeo K, Shoun H (2002) Ammonia fermentation, a novel anoxic metabolism of nitrate by fungi. J Biol Chem 277(3):1892–1896

Catalytic Hydro-Cracking of Bio-oil to Bio-fuel

Amir Ahmad Forghani, David M. Lewis and Phillip Pendleton

Abstract Over the last hundred years, fossil fuels consumption has increased dramatically leading to a significant increase in greenhouse gas emissions, the depletion of natural reserves of fossil fuels and increase fuel production costs. Consequently, renewable and sustainable fuel sources such as bio-oil are receiving increased attention. In bio-oils, such as microalgae oil, triglycerides and fatty acids are sustainable resources with high energy densities that can be converted into liquid hydrocarbon fuels, efficiently. One of the efficient ways for bio-oil conversion to applicable fuels is catalytic hydro-cracking. This chapter presents research on the catalytic conversion of oleic acid (main component in all types of bio-oil) in bio-oil to liquid hydrocarbon fuels employing two catalysts. These catalysts include Ni-ZSM-5 and Ni-Zeolite β, which were prepared by impregnating cheap catalyst supports (ZSM-5 and Zeolite β) with $Ni(NO_3)_2 \cdot 6H_2O$ calcined at a temperature of 500 °C. The catalysts were characterized using the Brunauer–Emmet–Teller Nitrogen Adsorption technique, scanning electron microscopy (SEM) and SEM–EDX (energy-dispersive X-ray spectroscopy) to analyse nickel impregnation and measure surface areas and pore size distribution. Conversion rates of oleic acid and product yields of liquid hydrocarbon fuels using each catalyst sample were determined via hydro-cracking reactions run at a temperature range of 300–450 °C and under a 30 bar pressure.

Keywords Ni-Zeolite-based catalyst · Catalytic upgrading · Hydro-cracking · Biofuel production · Characterisation

A.A. Forghani (✉) · D.M. Lewis
Micro-Algae Engineering Research Group, School of Chemical Engineering,
The University of Adelaide, Adelaide, SA 5005, Australia
e-mail: amir.forghani@adelaide.edu.au

P. Pendleton
Centre for Molecular and Materials Sciences, Sansom Institute,
University of South Australia, Adelaide, SA 5000, Australia

K. Heimann et al. (eds.), *Biodegradation and Bioconversion of Hydrocarbons*,
Environmental Footprints and Eco-design of Products and Processes,
DOI 10.1007/978-981-10-0201-4_6

1 Introduction

Nowadays, dramatic increases in the atmospheric concentration of greenhouse gases and carbon dioxide, lead to worldwide concern about global warming (Metz 2005; Figueroa et al. 2008). The most abundant greenhouse gases in the Earth's atmosphere are: vapour of water (H_2O), carbon dioxide (CO_2), methane, oxides of nitrogen and ozone (O_3) and chlorofluorocarbons (CFCs) (Luque et al. 2008). However, CO_2 has the major contribution to global warming, due to its high concentration in the atmosphere (Yamasaki 2003; Caspeta et al. 2013). During the last decades, significant global attempts have been allocated to developing new technologies that enables preventing CO_2 emission. Among these technologies, alternative energy production ones through bio-fuel has gained more interest. This is mainly due to the synergy between the available power generation systems and the fuels produced from biosources. Currently, biofuels contribute about 2.7 % of the global energy consumption especially in the transport sector (Chum et al. 2011) and this mirrored in the evidence of high attention to bio-fuel technologies. Various resources exist for biofuel production which are categorised to edible and non-edible feedstocks (Luque et al. 2008; Brennan and Owende 2010). Non-edible feedstock such as Jatropha oil, Palm oil and algal oil is one of the major feed sources for economical biofuel production (Milne et al. 1990; Mortensen et al. 2011). Different types of bio-oil are produced in different places in the world. For instance, Malaysia and Indonesia are the key producers of more than 90 % of palm oil in the world (Sumathi et al. 2008). India is the most active country in Jatropha oil production (more than 400,000 ha plantation) and currently, several projects are active on bio-diesel production from Jatropha oil (Study 2008). Moreover, based on the Algae 2020 report, the governments of USA, Australia and New-Zealand allocated significant budgets for algal oil production (Demirbas and Fatih Demirbas 2011). But most of the bio-oil production plants for biofuel production are at research and development stage and have not been commercialised yet. For instance, Muradel, a company in South Australia, is one of the biggest algal oil production companies in Australia. This company can produce 30,000 L of green crude-oil per year (Vorrath 2014). The bio-oil from different feedstocks such as palm oil and algal oil, can be used directly in diesel engines, but the direct engine combustion causes carbon deposits and coking on the injectors and high viscosity of bio-oil interferes with engine operation (Bezergianni and Kalogianni 2009; Kumar et al. 2010). The afore-mentioned problems prove that bio-oil requires upgrading. Recently, bio-oil upgrading to applicable fuels has been highlighted due to low heating values, high viscosity and poor stability of crude bio-oil (Sharif Hossain and Salleh 2008; Sharma et al. 2012; Pragya et al. 2013). The overall aim of this chapter is focused on bio-oil upgrading towards biofuels with cost-effective catalysts. In the following sections, the different upgrading methods are investigated in details.

1.1 Bio-oil Upgrading Processes and Applied Catalysts

Bio-oil upgrading to applicable fuels is mandated due to low heating values, high viscosity and poor stability of crude bio-oil (Czernik and Bridgwater 2004; Elliott 2007; Wang et al. 2012). Bio-oil upgrading began with vegetable oil conversion over acidic and basic catalysts in the 1920s (Idem et al. 1997). Al_2O_3 and $AlCl_3$ were the early cracking acidic catalysts which were operated in batch reactors between temperatures of 450–500 °C (Leung et al. 1995; Idem et al. 1997). Esterification, hydro-treating and hydro-cracking are the three major existing processes for bio-oil conversion to fuels (Xiu and Shahbazi 2012). Esterification is the catalytic reaction between bio-oil and short chain alcohols as solvent such as methanol and ethanol and it is very popular in bio-diesel production but requires large amounts of solvent (Shi et al. 2012; Xiu and Shahbazi 2012). Hydro-treating is a simple hydrogenation process which has been applied in refineries to convert aromatics to naphthenes at temperatures above 500 °C and atmospheric pressure (760 torr), and has recently been applied for bio-crude conversion over sulfided Co and Mo/Al_2O_3, Ni and Mo/Al_2O in order to remove oxygen from triglyceride (Nava et al. 2009; Yang et al. 2009). Yields from hydro-treating processes are low due to sulfided catalyst applications as this produces significant amounts of char and coke in the process which can result in reactor clogging and catalyst deactivation (Xiu and Shahbazi 2012). Hydro-cracking is a combination of different cracking reactions with hydrogenation at high temperatures (above 350 °C) and high pressures (above 7 bar) (James and Jorge 2007; Mortensen et al. 2011) over dual-function catalysts. Dual-function catalysts have a zeolite or silica alumina base which provides the cracking function and metal oxides such as nickel, molybdenum which are catalysing the reactions towards light hydrocarbons (Milne et al. 1990; Nava et al. 2009; Mortensen et al. 2011; Tiwari et al. 2011; Verma et al. 2011; Shi et al. 2012; Xiu and Shahbazi 2012).

One of the recent upgrading methods is catalytic cracking of bio-oil to liquid hydrocarbon fuels (Benson et al. 2008; Bezergianni et al. 2009b; Kumar et al. 2010; Anand and Sinha 2012; Sharma et al. 2012; Pragya et al. 2013). The main advantage of using catalytic bio-oil cracking is that it is a single step process consuming less energy than other upgrading processes (Sharif Hossain and Salleh 2008; Verma et al. 2011; Shi et al. 2012). Catalytic upgrading of bio-oil consists of several complex reactions such as cracking, de-carbonylation, de-carboxylation, hydro-cracking, hydro-deoxygenation and hydrogenation. These complex reactions have been reported to take place on zeoilitic supported catalysts (Adjaye and Bakhshi 1995a, b; Wildschut et al. 2009). Hence, bio-oil catalytic upgrading over zeolite catalysts became an alternative method for bio-oil and triglyceride conversion to usable fuels (Twaiq et al. 1999). Zeolitic catalysts are emerging as effective materials for efficient bio-oil cracking due to their high heating stability and strong acidity (Verma et al. 2011; Shi et al. 2012; Saxena and Viswanadham 2014). In 1995 and 1999, three types of zeolites such as HZSM-5, HBEA and USY zeolites were applied in a fixed bed reactor for the cracking of palm oil (Katikaneni et al.

1995; Twaiq et al. 1999). These three types of zeolite catalysts were operated in the temperature range of 350–450 °C yielding 99, 82 and 53 % conversions with gasoline selectivity of 28, 22 and 7 %, respectively (Katikaneni et al. 1995; Twaiq et al. 1999). Also HZSM-5 led to produce aromatic hydrocarbons especially benzene and toluene, USY favoured the formation of diesel range of hydrocarbons while BEA formed fewer diesel (Katikaneni et al. 1995; Idem et al. 1997; Twaiq et al. 1999). The main disadvantage of HBEA and USY catalysts in catalytic cracking of bio-oil is coke formation. The rate of coke formation for these two catalysts are higher than for HZSM-5, an issue correlating with pore diameter. Increased pore diameters intensifies coke formation inside the catalyst (Katikaneni et al. 1995; Zhao et al. 2013), but ZSM-5 and HZSM-5 catalysts have a great potential to produce gasoline- and jet fuel-range of hydrocarbons with less coke formation [14, 26, 27]. Application of ZSM-5, HZSM-5 and MCM-41 as catalyst for fatty acids and bio-oil conversion have been investigated (Twaiq et al. 2004; Ooi et al. 2005), demonstrating gasoline selectivity increases to the range of 38–47 % (Twaiq et al. 2004; Ooi et al. 2005). The most important section in bio-oil upgrading to biofuels is oxygen extraction from bio-oil. Recent investigations has shown that oxygen removal can be performed with a number of parallel reactions such as hydro-deoxygenation which produces water, de-carboxylation which produces CO_2 and de-carbonylation which generates carbon monoxide (CO) (Morgan et al. 2012). Based on this approach, several researchers developed a single step process with different ranges of zeolitic catalysts to convert bio-oil to paraffinic fuels. The single step process of bio-oil upgrading has been termed catalytic hydro-treating. Since, this process is performed in only one step, it is energy efficient and it can be performed over metal based catalysts.

In Table 1, the list of various catalysts which were applied in single step hydro-processing of bio-oil from different feedstocks is provided. In this table, the bio-oil feedstocks were categorised as vegetable oil, Jatropha oil and Algal oil. All of the listed catalysts were analysed in continues flow reactors.

In the next part, the application of various metal-based catalysts and their advantages and disadvantages for bio-oil hydro-treating are discussed in detail.

1.2 Application of Supported Metal Sulfide Catalysts

Metal-based catalysts for bio-oil and triglyceride conversion can be categorised as two main types: metal-sulfide catalysts and sulfur-free metal catalysts. The common sulfided catalysts for bio-oil conversion are conventional Ni–Mo and Co–Mo sulphide catalysts over zeolites. They are used in petroleum refineries to produce straight-chain alkanes ranging from C_{12} to C_{18} at temperatures of 350–450 °C in the presence of hydrogen with pressures of 40–150 bar (Donnis et al. 2009; Choudhary and Phillips 2011). The sulfided catalysts have been analysed widely in hydro-treating of bio-oil to produce diesel and middle range of hydrocarbons. For example, Bezergianni et al. analysed catalytic hydrocracking of waste cooking oil

Table 1 Hydro-processing of different bio-oil feedstocks over a range of catalysts (Verma et al. 2011; Zhao et al. 2013)

Feedstock	Catalyst	T (°C)	t (h)	Conversion (%)
Vegetable oil	1 % Pt/SiO_2	300	12	2.4
	1 % $Pt/SiO_2–Al_2O_3$	300	12	5.6
	1 % Pt/HZSM-5	300	12	68
	1 % Pt/HZSM-5	300	12	43
	1 % Pt/USY	300	12	90
	1 % Pt/USY	300	12	75
	1 % Pt/BEA	300	12	25
	1 % Pt/HY	300	12	30
	1 % Pt/H-MOR	300	12	2.3
	1 % Pt/FER	300	12	52
	1 % Pt/L	300	12	1
	5 % Pd/C	350	10	23
	1 % Pt/C	350	10	30
	20 % Ni/C	350	10	92
	RANEY® Ni	340	9	85
	27 % Ni/Al_2O_3	340	9	90
	5 % Pt/Al_2O_3	325	5	46
Jatropha oil	1 % Pt/HZSM-5	270	12	100
	1 % Pt/USY	270	12	31
	1 % Pt/HZSM-5	270	12	22
	4 % NiO, 18 % MoO_3/Al_2O_3	370	–	99
Algal oil	10 % Ni/HBEA	260	8	100
	10 % Ni/ZrO_2	270	8	100
	4 % NiO, 18 % MoO_3/Al_2O_3	370	–	99
	5 % NiO, 18 % MoO_3/H-ZSM-5	410	–	98

with sulfided Ni–Mo/γ-Al_2O_3 to produce white diesel. Hydrocracking increases the H/C ratio, and removes hetero-atoms S, N, O, and metals from the bio-oil feedstock (Bezergianni et al. 2010a, b). In their studies, temperature increase from 330 to 398 °C led to an increase in the fraction of iso-paraffins from 5 to 38 %, and the lighter alkanes (C_8–C_{14}) increased also to 22 % at 398 °C. A main disadvantage of this liquid fuel is the poor flow properties at lower temperatures (Bezergianni et al. 2010a, b). Gusmao et al. investigated hydrocracking of soy bean and babassu oils in a batch reactor at a temperature range of 350–400 °C and at H_2 partial pressures of 10–200 bar. They applied sulfided Ni–Mo/γ-Al_2O_3 and reduced Ni/SiO_2 catalysts, and found that only double bonds were hydrogenated in soy bean oil at a reaction temperatures <200 °C. The decomposition of triglycerides accompanied by hydrogenation of the decomposition products was detected above a temperature of 290 °C. Compared to reduced Ni catalyst, sulfided Ni–Mo catalysts required higher reaction temperatures (230–280 °C) to hydrogenate double bonds. On the other hand, transformation of the carboxylate groups was achieved at around temperature of 300 °C on both catalysts (Gusmão et al. 1989).

Kubička et al. researched hydro-processing of rapeseed oil for the production of components in the diesel fuel range (Smejkal et al. 2009; Kubička and Kaluža 2010; Šimácek et al. 2010). Rapeseed oil was hydro-processed at 260–340 °C under a hydrogen pressure of 70 bar in a continuous trickle bed reactor. Three Ni-Mo/Al_2O_3 sulfide hydro-treating catalysts were evaluated. Reaction products included water, H_2-rich gas, and an organic liquid product. The main components of the organic liquid product were C_{17} and C_{18} n-alkanes. At low reaction temperature, organic liquid components contained also free fatty acids and unconverted triglycerides. At reaction temperatures higher than 310 °C, organic liquid components contained only green hydrocarbons similar to the diesel range of hydrocarbons (Smejkal et al. 2009; Kubička and Kaluža 2010; Šimácek et al. 2010). Furthermore, Ni/Mo-based catalysts were applied in hydro-processing of Jatropha oil, Palm oil, Canola oil and sunflower oil (Topsøe et al. 1996; Huber et al. 2007; Gutiérrez et al. 2011). Corma et al. analysed the hydro-processing of pure sunflower oil over sulfided Ni–Mo/Al_2O_3 at 350 °C and achieved complete oil conversion with 71 % yield of C_{15}–C_{18} alkanes (Huber et al. 2007). Also Jatropha, Palm and Canola oil were tested over Ni/Mo-based catalysts in a high pressure fixed bed flow reactor at a temperature of 350 °C and hydrogen pressure of 40 bar (Topsøe et al. 1996; Gutiérrez et al. 2011). Among the Ni/Mo-based catalysts, Ni–Mo/H-Y and Ni–Mo/H-ZSM-5 formed a high yield of gasoline-range hydrocarbons due to the strong acid sites of zeolites [41]. Although sulfided metal support catalysts can improve the yields of fuel productivity and bio-oil conversion, but they contaminate products via sulfur leaching, and deactivate the catalysts due to S removal from the surface by a reverse Mars van Krevelen mechanism [40].

1.3 Application of Supported Sulfured-Free Metal Catalysts

Supported noble metal catalysts such as Pd/C are frequently analysed for de-carboxylation and de-carbonylation of fatty acids in different bio-oil feedstocks (Maier et al. 1982; Kubicková et al. 2005; Snåre et al. 2006; Immer et al. 2010; Immer and Lamb 2010; Ping et al. 2010, 2011; Arend et al. 2011; Fu et al. 2011; Ford et al. 2012). For instance, 5 wt% Pd/C can yield 100 % C_{17} and C_{18} from stearic acid at a temperature of 300 °C (Snåre et al. 2006). Noble metal catalysts showed high activities and selectivity for fatty acid conversion in a comprehensive analysis by Murzin et al. (Mäki-Arvela et al. 2006; Snåre et al. 2007, 2008; Lestari et al. 2008; Lestari et al. 2009a, b; Simakova et al. 2009). However, activities and selectivities for the targeted alkanes for triglycerides conversion were considerably lower (Simakova et al. 2008; Morgan et al. 2010). The high price of noble metals limits their industrial applications as catalysts.

Furthermore, Peng et al. (2012) used base metal catalysts such as Ni supported on zeolites for hydro-deoxygenation of fatty acids. Ni/HZSM-5 (10 wt%) catalysed complete conversion of stearic acid to dodecane at a temperature of 260 °C for 6 h in the presence of 40 bar hydrogen pressure. Zhang et al. (2013) prepared Ni-based catalysts over γ-Al_2O_3 and HZSM-5 (Si/Al = 38) for hydro-treatment of bio-oil. Their catalysts were analysed in a temperature range of 160–240 °C achieving a conversion of 91.8 %. Yakovlev et al. (2009) investigated nickel-based catalysts due to their cost-effective property and found that applying nickel in the catalytic upgrading system of bio-oil could conduct the upgrading reactions at lower temperatures and cost and produce biofuel with low sulphur content.

In conclusion, the upgrading of bio-oil is expected to be commercialised, so the application of noble metal catalysts over zeolite could significantly raise biofuel production costs. On the other hand, sulfided metal-based zeolitic catalysts led to product contamination with sulfur and deactivation of the catalyst by S removal. Consequently, application of non-sulfur metal-based catalyst over zeolite is the most cost-effective option for bio-oil upgrading which is investigated in this research. In this chapter two types of inexpensive non-sulfided Ni-based bi-functional catalysts are introduced characterised and then analysed in a bio-oil upgrading process.

Sulfur present in catalyst structures caused bio-oil contamination and catalyst deactivation (Bui et al. 2011; Xiu and Shahbazi 2012). As mentioned before, several researchers have published hydro-cracking of bio-oil based on non-sulfided catalysts with zeolite or Al_2O_3 as a base and metal-oxide function such Pd and Pt (Díaz et al. 2007; Fisk et al. 2009; Zhao et al. 2009; Zhang et al. 2013; Forghani et al. 2014; Forghani and Lewis 2015). Among different metal oxides, application of nickel on zeolitic structures is cost-effective and efficient in hydro-processing of oxygenated compounds such as oleic acid in bio-oil. Therefore, in this presentation two types of inexpensive sulfur-free, Ni-based bi-functional catalysts are introduced, characterised and then analysed in a lab scale hydro-cracking unit. Temperature, residence time, and catalyst structure are investigated as effective parameters on reaction conversion and production yield.

2 Materials and Methods

2.1 Catalyst Preparation

Ni/ZSM5 and Ni/β-Zeolite were prepared using a conventional impregnation method. ZSM-5 and β-Zeolite were purchased from ACS Materials Company and Zeolyst International, respectively (with SiO_2/Al_2O_3 molar ratio of 38 for both of them). The impregnation of $Ni(NO_3)_2 \cdot 6H_2O$ (Sigma Aldrich) solution on ZSM-5 and β-Zeolite supports was performed, by mixing the supports with the nickel impregnation solution under stirring for 2 h, followed by drying in an oven. Materials were then calcined in a furnace with starting temperature increases of 10 °C/min to 500 °C and kept at this temperature for 5 h.

2.2 *Characterisation of Catalyst*

2.2.1 Nitrogen Adsorption Isotherm Measurements

Nitrogen gas adsorption experiments were carried out at 77 K using a Belsorp-Max automated manometric gas adsorption apparatus. Samples were degassed prior to the experiments at 300 °C and a background vacuum of 0.1 MPa for 4 h. Ultra-high purity (99.999 %) helium and nitrogen from BOC Gases Australia were used for dead-space measurements and adsorption experiments, respectively.

2.2.2 Scanning Electron Microscopy (SEM) Assisted with Energy Dispersive X-Ray Analysis (EDAX)

SEM–EDAX analysis was carried out to investigate the morphology of the catalyst surface and also to analyse the atomic composition on the examined surface. SEM analyses were carried out on a Philips XL-20-FEG microscope with an accelerating voltage of 5 kV, equipped with an energy dispersive X-ray spectrometer (EDXS) analyser (Adelaide Microscopy, model 6587). Specimens were prepared by gold sputtering of catalyst samples deposited as powders on aluminium pin flat stubs.

2.3 *Hydrocracking Reaction*

To analyse the prepared catalyst structures in hydro-cracking reactions, a commercial micro-scale trickle bed reactor (Autoclave Engineers' BTRS-Jr) was applied. For each run of the reactor, one gram (1 g) of catalyst was weighted (based on the reactor design) and then loaded into the stainless steel tubular reactor with an inside diameter of 1.2 cm and a catalytic bed of 20 cm. The reactor was situated in a furnace to run at high temperatures. Hydrogen pressure was controlled by a back pressure regulator. An HPLC pump was used to pump the liquid into the reactor and maintained the liquid flow through the catalyst bed. Oleic acid was selected as a feed because it is one of the main components of plant-based oil especially micro-algae oil. The oleic acid percentage in algae oil varies from 39–60 % (Demirbas and Fatih Demirbas 2011; Verma et al. 2011). Oleic acid was injected together with a large volume of hydrogen into the reactor and, after passing through the catalyst bed, the products, unreacted reactant and excess hydrogen passed through a separator and the liquid samples were obtained from the bottom of the separator. The schematic of the hydro-cracking lab-scale process is shown in Fig. 1. Furthermore, the hydrocracking reactions were performed at different reaction bed temperatures and the operating conditions are mentioned in Table 2.

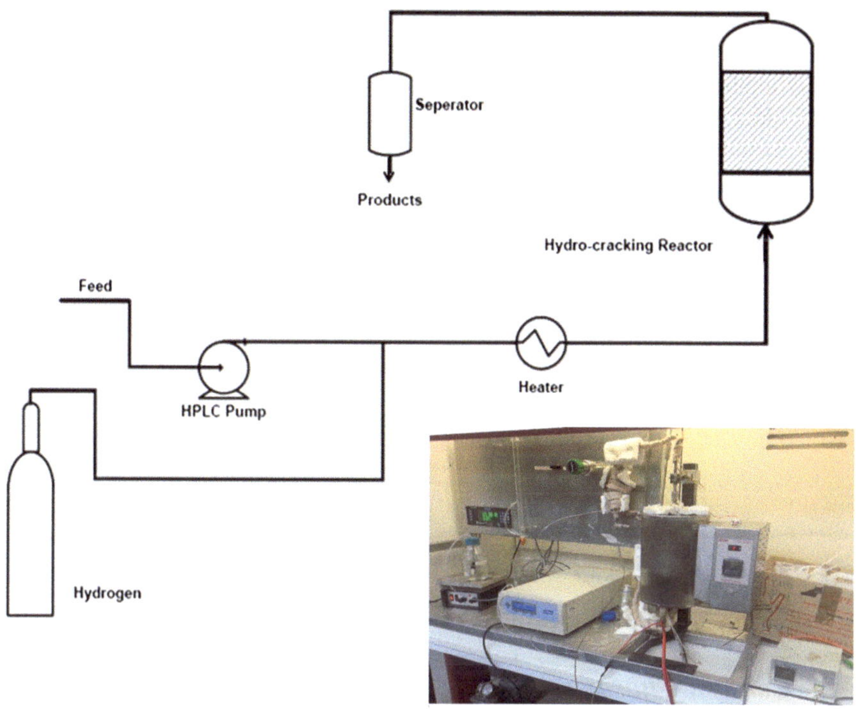

Fig. 1 Hydro-cracking lab-scale set up to analyse performance of the prepared catalyst

Table 2 Catalyst loading and operating conditions

Catalyst mass	1 g
Catalyst shape	Powder
Bed length	10 cm
Reactor diameter	1.1 cm
Reaction temperature range	300–450 °C
Reaction pressure	30 bar
Oleic acid flow	0.1 mL/min
Hydrogen/feed	1000 Nl/l

2.4 *Product Analysis*

The product samples were collected at the end of the separator and they were analysed by gas chromatography (GC) [Shimadzu GC2010 with a flame ionization detector (FID)] and a Varian 3800-GC column, 30 m, 0.25 mm, 0.1 µm, was applied for product analysis. Helium was selected as carrier gas with high purity. Nonane (C_9), Decane (C_{10}) and Dodecane (C_{12}) were purchased from Sigma Aldrich as standards to analyse and measure the outlet products of the hydro-cracking reactions. C_9, C_{10} and C_{12} are the main components of jet-fuel and by using them as standards, a calibration curve was used to measure the yields of C_9, C_{10} and C_{12} for each catalyst. The GC analysis was performed for all samples at an FID temperature of 200 °C and the oven temperature program was increased from 50 to 150 °C at the rate of 10 °C/min and then from 150 to 200 °C at the rate of 15 °C/min. The performance of each type of catalyst was evaluated based on their efficiency for conversion of oleic acid and yields of the jet fuel range of hydrocarbons. Conversion and yields of C_9, C_{10} and C_{12} were calculated based on calibration curves of the feed (oleic acid) and products which were defined with the following equations:

$$Conversion\,(\%) = \frac{Mass\ of\ oleic\ acid\ in\ the\ feed\,(g) - Mass\ of\ oleic\ acid\ in\ the\ product\,(g)}{Mass\ of\ oleic\ acid\ in\ the\ feed\,(g)} \tag{1}$$

$$Yield\ of\ hydrocarbon\,(wt\%) = \frac{Mass\ of\ hydrocarbon\ in\ the\ product\,(g)}{Mass\ of\ feed\,(g)} \tag{2}$$

3 Results and Discussions

3.1 Catalyst Characterisation

Figure 2 shows scanning electron microscopy images of Ni-Zeolite β (a) and Ni-ZSM-5 (b) catalysts. Particles are agglomerated in some areas, shown as grey rectangular structures. Black areas correspond to inter-particle spacing and bright white spots correspond to impregnated Ni on catalyst surface. Images clearly show presence of Ni agents on the surface of both catalyst samples. To make sure bright white spots are impregnated nickel, energy dispersive X-ray spectroscopy (EDXS) tests were randomly selected bright white spots. The average spectrum is shown in Fig. 2c. Nickel has K and L emission energy rates of 7.477 and 0.851 keV, respectively. Figure 2c shows two pronounced peaks for Ni at corresponding Ni energy rates for Ni-Zeolite β. Figure 2d also shows a single peak at K energy rate of Ni. Analyses of EDXS results together with electron microscopy images confirmed the presence of nickel on the catalyst surface.

The nitrogen adsorption isotherms, pore size distribution, and textural properties of the Ni-ZSM5 and Ni-Zeolite β are provided in Fig. 3 and Table 3. Both catalysts show similar types of adsorption isotherms including a relatively large contribution of micropores, significant contribution of mesopores and a type H4 hysteresis loop in the desorption branch. The hysteresis loop is more pronounced for Ni-ZSM-5 due to large contribution of mesopores than for Ni-Zeolite β, where it can hardly be seen in the adsorption isotherm.

Calculated pore size distribution (PSD) for both adsorbents show a narrow sharp peak in the micropore range (around 5 Å) and small but wider peaks in the larger

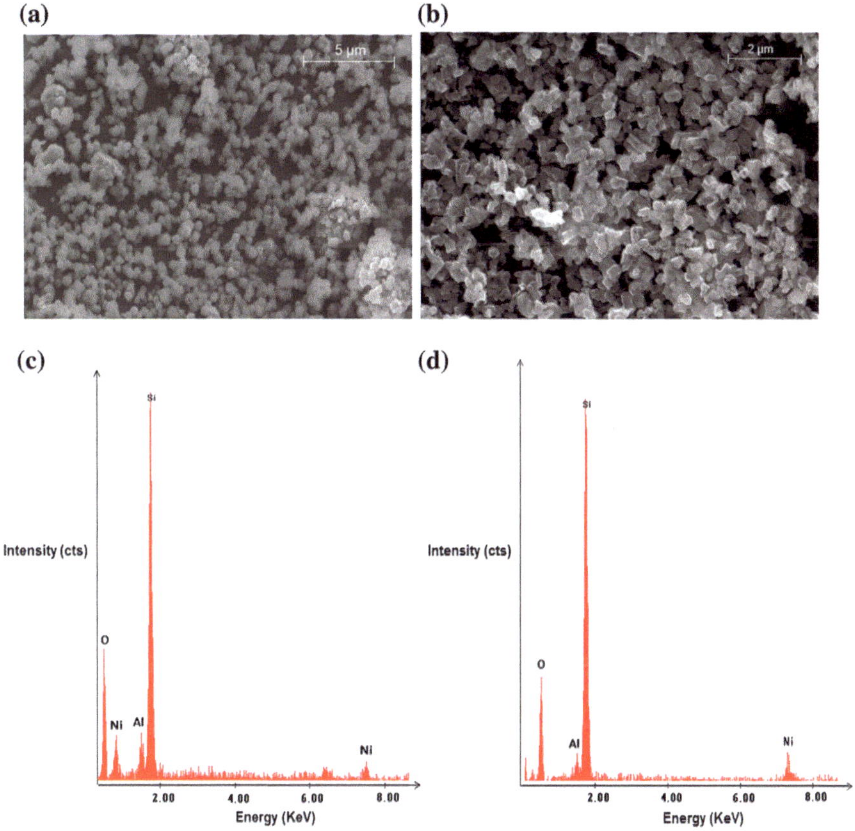

Fig. 2 Scanning electron microscopy (SEM) images of Ni-Zeolite β (**a**) and Ni-ZSM-5 (**b**) and Energy-dispersive X-ray spectroscopy for Ni-Zeolite β (**c**) and Ni-ZSM-5 (**d**)

micropore and mesopore ranges (around 14 and 28 Å for Ni-ZSM-5 and a wide peak extended from 13 to 25 Å for Ni-Zeolite β). Both samples had relatively the same pore volume (≈0.28 cm^3/g). Ni-Zeolite β was more microporous and had a larger contribution of micropores; 86 % of total porosity was due to microporosity (see Table 3 for comparison between two catalysts). Due to the large microporosity in Ni-Zeolite β, this sample had a larger surface area and if the surface area is accessible for reactants and nickel is impregnated uniformly all over the external surface and porosity, this catalyst is more desirable for hydro-cracking reactions.

Based on the porosity of the catalysts, two different reaction mechanisms are expected; if the reactant (oleic acid) and the products (C_9–C_{12}) can diffuse within and across the entire porosity of samples including micropores and mesopores, the entire surface area would be available for the reaction, reaction yields would be larger for the catalyst with the higher surface area (Ni-Zeolite β). If the size of the reactant and products are larger than micropores (≈5Å), micropore surface area

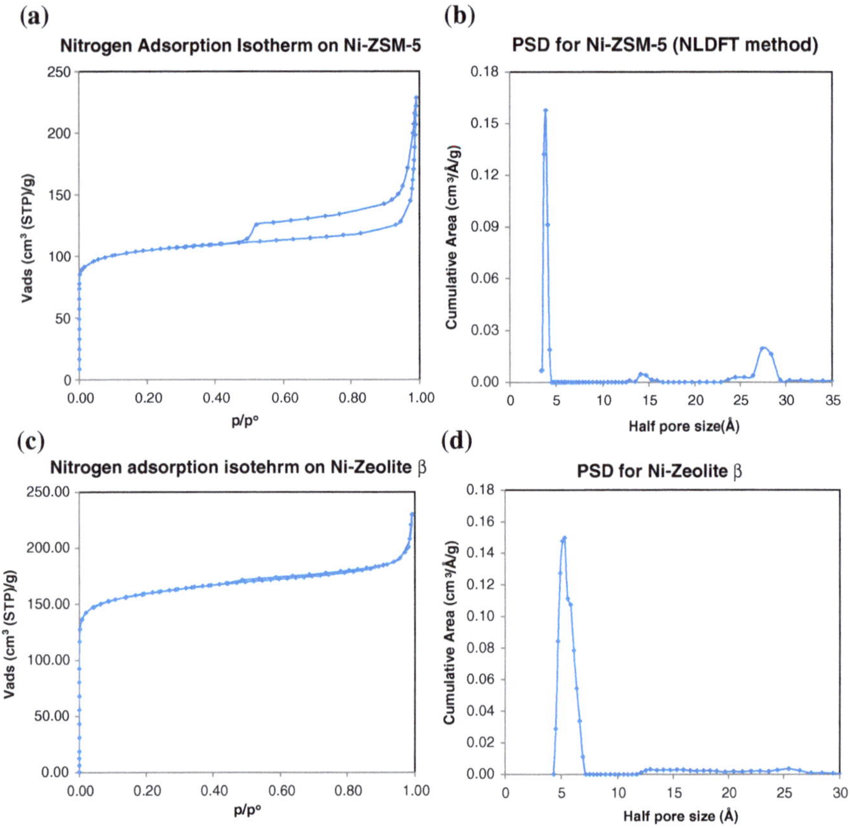

Fig. 3 Nitrogen adsorption isotherm and pore size distribution (PSD) results of Ni-ZSM-5 (**a**) and Ni-Zeolite β (**b**) catalysts

Table 3 Textural properties of Ni-ZSM-5 and Ni-Zeolite β

Catalyst	BET equivalent area (m^2/g)[a]	Micro-pore volume (cm^3/g)[b]	Total pore volume (cm^3/g)[c]
Ni-ZSM-5	396.743	0.157	0.278
Ni-Zeolite β	591.053	0.247	0.287

[a]Surface area is calculated based on BET method (Brunauer et al. 1938)
[b]Micropore volume is calculated based on α_S method (Sing 1968)
[c]Total pore volume is calculated based on nitrogen amount adsorbed at near saturation pressure converted to liquid volume

would not be accessible for reaction and reaction yields for the sample with higher mesoporosity would be larger (Ni-ZSM-5). Since the reactant (Oleic acid) and products (C_9–C_{12}) are all linear shaped hydrocarbons, their kinetic diameter is in the order of the kinetic diameter for linear hydrocarbons (normally less than 5 Å). For instance, kinetic diameter for n-Nonane (C_9) is reported 4.3 Å in the literature (Grillet et al. 1993). Hence, we expect the reactant and products to be able to diffuse throughout the porosity and across the total surface area for both samples accessible for reaction. Results in the next section shows higher reaction yields for Ni-Zeolite β which is consistent with our expectation.

3.2 *Hydro-Cracking of Oleic Acid*

The hydro-cracking of oleic acid over Ni-ZSM-5 and Ni-Zeolite β are presented in Figs. 4 and 5. Figure 5 represents the oleic acid conversion versus time over Ni-ZSM-5 and Ni-Zeolite β. The results of Fig. 4 were obtained at a temperature of 425 °C and other operating conditions were maintained as per Table 2. Conversion of each type of catalyst was obtained based on the GC calibration curve which was achieved from the standard samples of oleic acid. Oleic acid conversion, as a key parameter in the hydro-cracking reactions, remained stable after 30 min and the variation of conversion was not significant over time. This conversion rates were obtained after all system specifications such as temperature and pressure reached steady state conditions. The conversion rate of oleic acid was ~94 and ~87 % for Ni-Zeolite β and Ni-ZSM-5, respectively at a temperature of 425 °C. Figure 4 proves that steady state for the hydro-cracking reactions was reached after 30 min in the reactor and the variation in oleic acid conversion was insignificant. The retention time based on feed flowrate and reactor volume was ~9.5 min.

Figure 5a shows the oleic acid conversion over Ni-ZSM-5 and Ni-Zeolite β at different reaction bed temperatures. Temperature, as a key parameter of hydro-cracking reactions, has a great influence on catalyst effectiveness and activity. The effectiveness of hydro-cracking reactions is measured with oleic acid

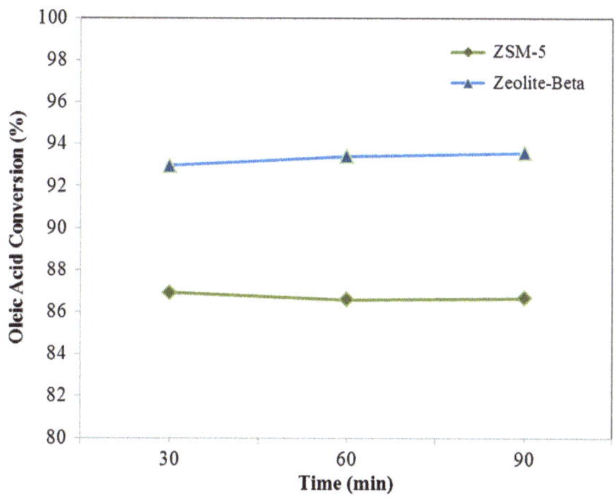

Fig. 4 Oleic acid conversion as a function of time over Ni-ZSM-5 and Ni-Zeolite β

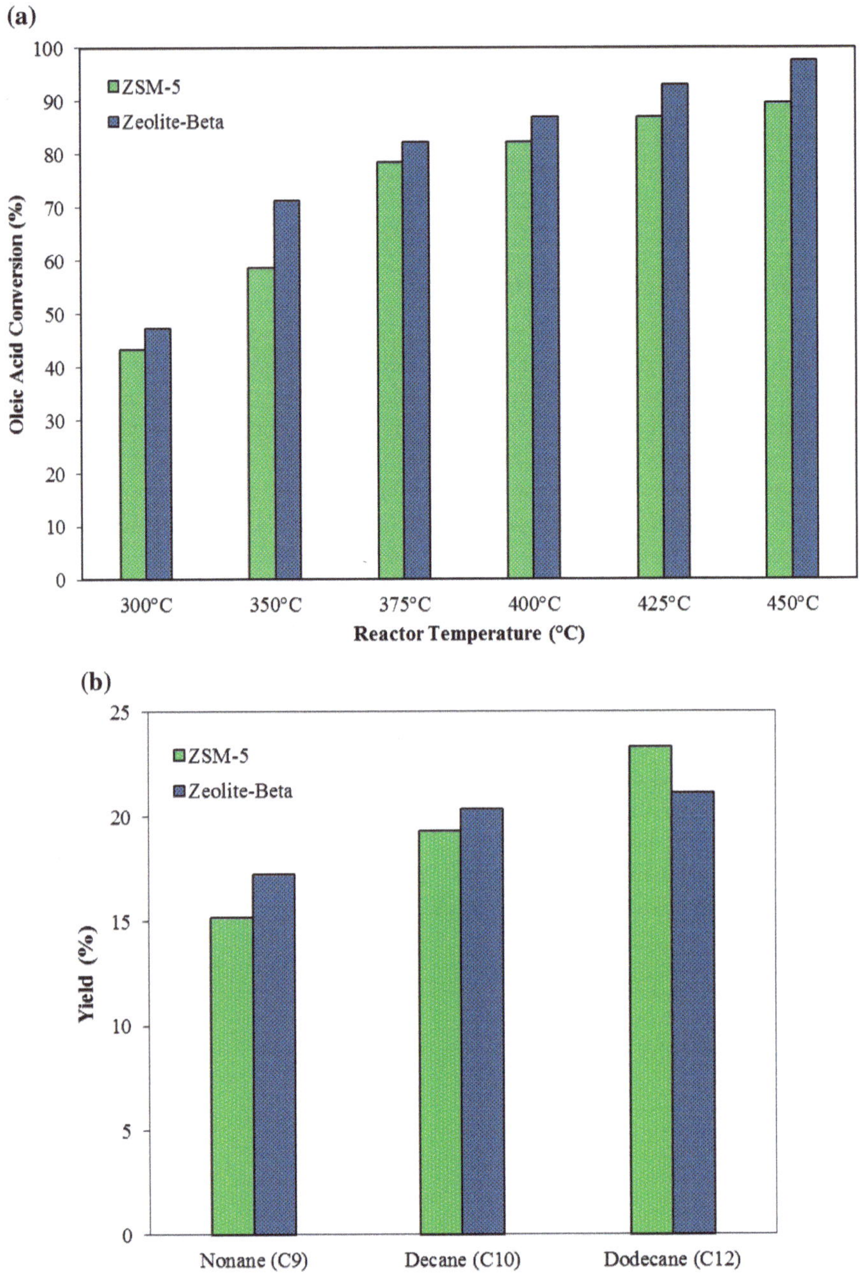

Fig. 5 Dependence of oleic acid conversion on reaction bed temperature for Ni-ZSM-5 and Ni-Zeolite β (**a**) and production yields of Nonane (C_9), Decane (C_{10}) and Dodecane (C_{12}) over Ni-ZSM-5 and Ni-Zeolite β at a temperature 400 °C

conversion which is defined in Eq. 1. The hydro-cracking reactions of oleic acid were dependent to reaction temperature, it is clearly observed that with temperature increment, oleic acid conversion increased over both catalyst samples but Ni-Zeolite β has greater conversion rates in comparison with Ni-ZSM-5 due to its larger total surface area. Maximal oleic acid conversion occurred at temperatures of 300 to 350 °C which were 35.4 and 50.5 % for Ni-ZSM-5 and Ni-Zeolite β, respectively. In addition, previous studies showed that hydro-cracking reactions especially deoxygenation of oxygenated compounds such as oleic acid are more active in higher temperatures (Huber et al. 2007; Bezergianni and Kalogianni 2009; Bezergianni et al. 2009a).

Yields of production over both catalyst types are presented in Fig. 5b. To measure production yields, hydro-cracking reactions were performed at 400 °C and a pressure of 30 bar with a Hydrogen/Feed of 1000 Nl/l. The hydro-cracking temperature was chosen and maintained at 400 °C based on the literature—described as an optimum temperature for hydro-cracking (Tiwari et al. 2011; Verma et al. 2011). Yields of three targeted components (C_9, C_{10} and C_{12}) were calculated based on Eq. 2. It is clearly seen that Nonane (C_9) and Decane (C_{10}) production yields over Ni-Zeolite β were greater than Ni-ZSM-5 but for Dodecane, Ni-ZSM-5 showed better performance. The reason for higher production yield for Nonane and Decane in Ni-Zeolite β was higher surface area and micropore concentration in Ni-Zeolite β but higher yield production of C_{12} over Ni-ZSM-5 is related to high mesopore concentration of Ni-ZSM-5 which can be appropriate for production of larger molecules such as Dodecane and improved selectivity of deoxygenation reaction towards C_{12} in comparison with Ni-Zeolite β. The difference of Dodecane production yield was 10.5 %.

Hence, better performance of Ni-Zeolite β was evident by the higher oleic acid conversion rates and higher production yields of C_9 and C_{10} but Ni-ZSM-5 showed better performance for C_{12} yields of production. Nonetheless, Ni-Zeolite β has a great potential to be applicable in large-scale hydro-cracking reactions for the conversion of bio-oil to C_9, C_{10} and C_{12} hydrocarbons, the main components of jet fuel.

4 Summary

Hydro-cracking of oleic acid with two samples of nickel impregnated catalysts on ZSM-5 and Zeolite β was performed to produce bio-fuel. Nickel impregnation of ZSM-5 and Zeolite β was characterised by BET Nitrogen Adsorption techniques, SEM and SEM-EDX and surface areas and pore size distribution was measured. The SEM and SEM-EDX analysis confirmed that nickel was impregnated properly in the two catalyst samples. The BET Nitrogen Adsorption results and pore size distribution analysis confirmed that Ni-Zeolite β had a larger surface area in comparison with the Ni-ZSM-5, but Ni-ZSM-5 had a higher mesopore concentration in its structure. The large surface area of Ni-Zeolite β played as a significant parameter in conversion of oleic acid. Ni-Zeolite β had higher production yields of Nonane and Decane, but Ni-ZSM-5 achieved higher Dodecane yields due to larger pore sizes compared to the Ni-Zeolite β. Therefore, Ni-Zeolite β and Ni-ZSM-5 are cheap to utilise as catalysts of hydro-cracking reactions with appropriate efficiency in bio-oil conversion towards C_9, C_{10} and C_{12} hydrocarbons, which the main components of jet fuel.

References

Adjaye JD, Bakhshi NN (1995a) Catalytic conversion of a biomass-derived oil to fuels and chemicals I: model compound studies and reaction pathways. Biomass Bioenergy 8(3):131–149

Adjaye JD, Bakhshi NN (1995b) Production of hydrocarbons by catalytic upgrading of a fast pyrolysis bio-oil. Part I: conversion over various catalysts. Fuel Process Technol 45(3):161–183

Anand M, Sinha AK (2012) Temperature-dependent reaction pathways for the anomalous hydrocracking of triglycerides in the presence of sulfided Co–Mo-catalyst. Bioresour Technol 126:148–155

Arend M et al (2011) Catalytic deoxygenation of oleic acid in continuous gas flow for the production of diesel-like hydrocarbons. Appl Catal A 399(1–2):198–204

Benson TJ et al (2008) Heterogeneous cracking of an unsaturated fatty acid and reaction intermediates on H+ ZSM-5 catalyst. CLEAN–Soil Air Water 36(8):652–656

Bezergianni S et al (2010a) Hydrotreating of waste cooking oil for biodiesel production. Part I: effect of temperature on product yields and heteroatom removal. Bioresour Technol 101 (17):6651–6656

Bezergianni S et al (2010b) Hydrotreating of waste cooking oil for biodiesel production. Part II: effect of temperature on hydrocarbon composition. Bioresour Technol 101(19):7658–7660

Bezergianni S, Kalogianni A (2009) Hydrocracking of used cooking oil for biofuels production. Bioresour Technol 100(17):3927–3932

Bezergianni S et al (2009a) Hydrocracking of vacuum gas oil-vegetable oil mixtures for biofuels production. Bioresour Technol 100(12):3036–3042

Bezergianni S et al (2009b) Catalytic hydrocracking of fresh and used cooking oil. Ind Eng Chem Res 48(18):8402–8406

Brennan L, Owende P (2010) Biofuels from microalgae—a review of technologies for production, processing, and extractions of biofuels and co-products. Renew Sustain Energy Rev 14 (2):557–577

Brunauer S et al (1938) Adsorption of gases in multimolecular layers. J Am Chem Soc 60(2):309–319

Bui VN et al (2011) Hydrodeoxygenation of guaiacol with CoMo catalysts. Part I: promoting effect of cobalt on HDO selectivity and activity. Appl Catal B 101(3–4):239–245

Caspeta L et al (2013) The role of biofuels in the future energy supply. Energy Environ Sci 6 (4):1077–1082

Choudhary TV, Phillips CB (2011) Renewable fuels via catalytic hydrodeoxygenation. Appl Catal A 397(1–2):1–12

Chum H, et al (2011) Bioenergy. IPCC special report on renewable energy sources and climate change mitigation. In: O Edenhofer, R Pichs-Madruga, Y Sokona et al (eds) Cambridge University Press, Cambridge

Czernik S, Bridgwater AV (2004) Overview of applications of biomass fast pyrolysis oil. Energy Fuels 18(2):590–598

Demirbas A, Fatih Demirbas M (2011) Importance of algae oil as a source of biodiesel. Energy Convers Manag 52(1):163–170
Díaz E et al (2007) Hydrogenation of phenol in aqueous phase with palladium on activated carbon catalysts. Chem Eng J 131(1–3):65–71
Donnis B et al (2009) Hydroprocessing of bio-oils and oxygenates to hydrocarbons. understanding the reaction routes. Top Catal 52(3):229–240
Elliott DC (2007) Historical developments in hydroprocessing bio-oils. Energy Fuels 21(3):1792–1815
Figueroa JD et al (2008) Advances in CO2 capture technology—the U.S. Department of Energy's Carbon Sequestration Program. Int J Greenh Gas Control 2(1):9–20
Fisk CA et al (2009) Bio-oil upgrading over platinum catalysts using in situ generated hydrogen. Appl Catal A 358(2):150–156
Ford J et al (2012) Palladium catalysts for fatty acid deoxygenation: influence of the support and fatty acid chain length on decarboxylation kinetics. Top Catal 55(3–4):175–184
Forghani AA et al (2014) Mathematical modelling of a hydrocracking reactor for triglyceride conversion to biofuel: model establishment and validation. Int J Energy Res 38(12):1624–1634
Forghani AA, Lewis DM (2015) Hydro-conversion of oleic acid in bio-oil to liquid hydrocarbons: an experimental and modeling investigation. J Chem Technol Biotechnol doi:10.1002/jctb.4618
Fu J et al (2011) Hydrothermal decarboxylation and hydrogenation of fatty acids over Pt/C. ChemSusChem 4(4):481–486
Grillet Y et al (1993) Evaluation of the n-nonane preadsorption method with a well characterized model adsorbent: Silicalite-l. Pure Appl Chem 65(10):2157–2167
Gusmão J et al (1989) Utilization of vegetable oils as an alternative source for diesel-type fuel: hydrocracking on reduced Ni/SiO_2 and sulphided $Ni–Mo/\gamma$-Al_2O_3. Catal Today 5(4):533–544
Gutiérrez OY et al (2011) Influence of potassium on the synthesis of methanethiol from carbonyl sulfide on sulfided Mo/Al_2O_3 catalyst. ChemCatChem 3(9):1480–1490
Huber GW et al (2007) Processing biomass in conventional oil refineries: Production of high quality diesel by hydrotreating vegetable oils in heavy vacuum oil mixtures. Appl Catal A 329:120–129
Idem RO et al (1997) Catalytic conversion of canola oil to fuels and chemicals: roles of catalyst acidity, basicity and shape selectivity on product distribution. Fuel Process Technol 51(1–2):101–125
Immer JG et al (2010) Catalytic reaction pathways in liquid-phase deoxygenation of C18 free fatty acids. Appl Catal A 375(1):134–139
Immer JG, Lamb HH (2010) Fed-batch catalytic deoxygenation of free fatty acids. Energy Fuels 24(10):5291–5299
James GS, Jorge A (2007) Hydroprocessing chemistry. Hydroprocessing of heavy oils and residua. CRC Press, Boca Raton, pp 35–50
Katikaneni SPR et al (1995) Catalytic conversion of canola oil to fuels and chemicals over various cracking catalysts. Can J Chem Eng 73(4):484–497
Kubička D, Kaluža L (2010) Deoxygenation of vegetable oils over sulfided Ni, Mo and NiMo catalysts. Appl Catal A 372(2):199–208
Kubičková I et al (2005) Hydrocarbons for diesel fuel via decarboxylation of vegetable oils. Catal Today 106(1–4):197–200
Kumar R et al (2010) Hydroprocessing of jatropha oil and its mixtures with gas oil. Green Chem 12(12):2232–2239
Lestari S et al (2009a) Catalytic Deoxygenation of stearic acid in a continuous reactor over a mesoporous carbon-supported Pd catalyst. Energy Fuels 23(8):3842–3845
Lestari S et al (2009b) Catalytic deoxygenation of stearic acid and palmitic acid in semibatch mode. Catal Lett 130(1–2):48–51
Lestari S et al (2008) Synthesis of biodiesel via deoxygenation of stearic acid over supported Pd/C catalyst. Catal Lett 122(3–4):247–251

Leung A et al (1995) Pathway for the catalytic conversion of carboxylic acids to hydrocarbons over activated alumina. Energy Fuels 9(5):913–920
Luque R et al (2008) Biofuels: a technological perspective. Energy Environ Sci 1(5):542–564
Maier WF et al (1982) Hydrogenolysis, IV. Gas phase decarboxylation of carboxylic acids. Chem Ber 115(2):808–812
Mäki-Arvela P et al (2006) Catalytic deoxygenation of fatty acids and their derivatives. Energy Fuels 21(1):30–41
Metz B (2005) Carbon dioxide capture and storage: special report of the intergovernmental panel on climate change. Cambridge University Press, Cambridge
Milne TA et al (1990) Catalytic conversion of microalgae and vegetable oils to premium gasoline, with shape-selective Zeolites. Biomass 21(3):219–232
Morgan T et al (2010) Conversion of triglycerides to hydrocarbons over supported metal catalysts. Top Catal 53(11–12):820–829
Morgan T et al (2012) Catalytic deoxygenation of triglycerides to hydrocarbons over supported nickel catalysts. Chem Eng J 189–190:346–355
Mortensen PM et al (2011) A review of catalytic upgrading of bio-oil to engine fuels. Appl Catal A 407(1–2):1–19
Nava R et al (2009) Upgrading of bio-liquids on different mesoporous silica-supported CoMo catalysts. Appl Catal B 92(1–2):154–167
Ooi Y-S et al (2005) Catalytic conversion of fatty acids mixture to liquid fuel and chemicals over composite microporous/mesoporous catalysts. Energy Fuels 19(3):736–743
Peng B et al (2012) Inside back cover: towards quantitative conversion of microalgae oil to diesel-range alkanes with bifunctional catalysts. Angew Chem Int Ed 51(9):2253
Ping EW et al (2011) On the nature of the deactivation of supported palladium nanoparticle catalysts in the decarboxylation of fatty acids. Appl Catal A 396(1–2):85–90
Ping EW et al (2010) Highly dispersed palladium nanoparticles on ultra-porous silica mesocellular foam for the catalytic decarboxylation of stearic acid. Microporous Mesoporous Mater 132(1–2):174–180
Pragya N et al (2013) A review on harvesting, oil extraction and biofuels production technologies from microalgae. Renew Sustain Energy Rev 24:159–171
Saxena SK, Viswanadham N (2014) Selective production of green gasoline by catalytic conversion of Jatropha oil. Fuel Process Technol 119:158–165
Sharif Hossain A, Salleh A (2008) Biodiesel fuel production from algae as renewable energy. Am J Biochem Biotechnol 4(3):250–254
Sharma RK et al (2012) Jatropha-oil conversion to liquid hydrocarbon fuels using mesoporous titanosilicate supported sulfide catalysts. Catal Today 198(1):314–320
Shi F et al (2012) Recent developments in the production of liquid fuels via catalytic conversion of microalgae: experiments and simulations. RSC Adv 2(26):9727–9747
Šimáček P et al (2010) Fuel properties of hydroprocessed rapeseed oil. Fuel 89(3):611–615
Simakova I et al (2009) Deoxygenation of palmitic and stearic acid over supported Pd catalysts: effect of metal dispersion. Appl Catal A 355(1–2):100–108
Simakova IL et al (2008) Hydrogenation of vegetable oils over pd on nanocomposite carbon catalysts. Ind Eng Chem Res 47(19):7219–7225
Sing KSW (1968) Empirical method for analysis of adsorption isotherms. Chem Ind 44:1520–1521
Smejkal Q et al (2009) Thermodynamic balance in reaction system of total vegetable oil hydrogenation. Chem Eng J 146(1):155–160
Snåre M et al (2008) Catalytic deoxygenation of unsaturated renewable feedstocks for production of diesel fuel hydrocarbons. Fuel 87(6):933–945
Snåre M et al (2006) Heterogeneous catalytic deoxygenation of stearic acid for production of biodiesel. Ind Eng Chem Res 45(16):5708–5715
Snåre M et al (2007) Production of diesel fuel from renewable feeds: kinetics of ethyl stearate decarboxylation. Chem Eng J 134(1–3):29–34
Study GGJ (2008) Global market study on Jatropha—final report. GEXSI Global Jatropha Study

Sumathi S et al (2008) Utilization of oil palm as a source of renewable energy in Malaysia. Renew Sustain Energy Rev 12(9):2404–2421
Tiwari R et al (2011) Hydrotreating and hydrocracking catalysts for processing of waste soya-oil and refinery-oil mixtures. Catal Commun 12(6):559–562
Topsøe H, et al (1996) Hydrotreating catalysis. Catalysis. In: J Anderson, M Boudart (ed) Springer, Berlin 11:1–269
Twaiq FA et al (1999) Catalytic conversion of palm oil to hydrocarbons: performance of various Zeolite catalysts. Ind Eng Chem Res 38(9):3230–3237
Twaiq FAA et al (2004) Performance of composite catalysts in palm oil cracking for the production of liquid fuels and chemicals. Fuel Process Technol 85(11):1283–1300
Verma D et al (2011) Aviation fuel production from lipids by a single-step route using hierarchical mesoporous Zeolites. Energy Environ Sci 4(5):1667–1671
Vorrath S (2014) Algae oil test plant launched in South Australia. Renew Economy, 3 Nov 2014
Wang S et al (2012) Catalytic conversion of carboxylic acids in bio-oil for liquid hydrocarbons production. Biomass Bioenergy 45:138–143
Wildschut J et al (2009) Hydrotreatment of fast pyrolysis oil using heterogeneous noble-metal catalysts. Ind Eng Chem Res 48(23):10324–10334
Xiu S, Shahbazi A (2012) Bio-oil production and upgrading research: a review. Renew Sustain Energy Rev 16(7):4406–4414
Yakovlev VA et al (2009) Development of new catalytic systems for upgraded bio-fuels production from bio-crude-oil and biodiesel. Catal Today 144(3–4):362–366
Yamasaki A (2003) An overview of CO_2 mitigation options for global warming-emphasizing CO_2 sequestration options. J Chem Eng Jpn 36(4):361–375
Yang Y et al (2009) Hydrodeoxygenation of bio-crude in supercritical hexane with sulfided CoMo and CoMoP catalysts supported on MgO: a model compound study using phenol. Appl Catal A 360(2):242–249
Zhang X et al (2013) Hydrotreatment of bio-oil over Ni-based catalyst. Bioresour Technol 127:306–311
Zhao C et al (2013) Catalytic deoxygenation of microalgae oil to green hydrocarbons. Green Chem 15(7):1720–1739
Zhao C et al (2009) Highly selective catalytic conversion of phenolic bio-oil to alkanes. Angew Chem Int Ed 48(22):3987–3990
(2011) Algae 2020, vol 2

HC-0B-05: Two-Phase Partitioning Bioreactors for Treatment of Volatile Hydrocarbons

Guillermo Quijano, José A. Miguel-Romera, Luis M. Bonilla-Morte and Ivonne Figueroa-González

Abstract Nowadays air pollutants are becoming more widespread as the pace of industrial activity accelerates. Emission inventories reveal that atmospheric pollutant emissions have continuously increased since the beginning of the twentieth century, with volatile organic compounds (VOCs) representing about 7 % of these emissions. Despite this relatively low emission share, VOC emissions represent a major environmental and human health problem since most VOCs can be toxic depending on the concentration and exposure time and they also contribute to substantial damage to natural ecosystems. The negative effects of VOCs on human health and natural ecosystems have therefore led to stricter environmental regulations worldwide. VOCs emitted from industrial facilities are usually volatile hydrocarbons (VHs), which are extensively used in manufacturing processes. Among the available technologies for VHs treatment, biological processes in many cases constitute the most cost-effective technology for treating low pollutant concentrations and their implementation at industrial scale is growing exponentially. Unfortunately, several VHs can produce toxic effects to the microbial communities, leading to inhibition issues as in the case of aromatic compounds. Furthermore, some VHs used as monomers in the plastics industry or as industrial solvents exhibit a very low aqueous solubility, leading to mass transfer issues and poor removal performance. Two-phase partitioning bioreactors (TPPBs) emerged as innovative multiphase systems capable of overcoming the key limitations of traditional biological technologies such as the low mass transfer rates of hydrophobic

---

G. Quijano (✉) · I. Figueroa-González
Department of Chemical Engineering and Environmental Technology,
Agrarian Engineering School, University of Valladolid,
Soria Campus, 42004 Soria, Spain
e-mail: gquijano@iq.uva.es

J.A. Miguel-Romera
Department of Agroforestry Sciences, Agrarian Engineering School,
University of Valladolid, Soria Campus, 42004 Soria, Spain

L.M. Bonilla-Morte
Department of Agricultural and Forestry Engineering, Agrarian Engineering School,
University of Valladolid, Soria Campus, 42004 Soria, Spain

K. Heimann et al. (eds.), *Biodegradation and Bioconversion of Hydrocarbons*,
Environmental Footprints and Eco-design of Products and Processes,
DOI 10.1007/978-981-10-0201-4_7

VOCs and microbial inhibition at high pollutant loading rates. This work presents an updated state-of-the-art on the advances of TPPB technology for the treatment of VHs. The fundamentals of TPPB design, operation, microbiology and mass transfer are reviewed. Niches for future research, opportunities for TPPB optimization and challenges towards full-scale applications are identified and discussed.

Keywords Atmospheric pollution · Biological air treatment · Two-phase partitioning bioreactors · Volatile hydrocarbons · Volatile organic compounds

1 Introduction

Atmospheric pollution has traditionally occupied a secondary position in environmental concern since it apparently represents a lower risk for human health and natural ecosystems when compared to water and soil pollution (Stuetz et al. 2001; Lebrero et al. 2011). Nevertheless, this perception is rapidly changing, as recent investigations consistently confirm a key role of atmospheric pollution on problems of global relevance such as climate change and a variety of human diseases (Kampa and Castanas 2008). Volatile organic compounds (VOCs) constitute a common group of air pollutants, with some being classified as known or probable human carcinogens and some having neurologic, developmental, or reproductive effects at high levels (Wu et al. 2011). Many VOCs emitted from industrial facilities are volatile hydrocarbons (VHs) extensively used in manufacturing processes as raw commodities, solvents or fuels. VOCs are defined by the European EC-Directive 1999/13/EC as organic compounds with initial boiling points ≤ 250 °C measured at a standard pressure of 101.3 kPa (EUPO 2007). Thus, in this chapter the term VH will refer to organic compounds constituted exclusively by hydrogen and carbon that fit the VOC definition.

Pollutant release inventories demonstrate that air pollution has experienced a continuous increase since the beginning of the twentieth century, the industrial sector alone being responsible for ~ 13 % of these emissions with a VOC share ranging from ~ 7 to ~ 15 % (EEA 2014; Delhoménie and Heitz 2005). Canada, Europe-28 and the USA yearly release about 41, 45 and 150 Mt of pollutants to the atmosphere, respectively, including particulate matter, SO_X, NO_X, CO_X and VOCs (EEA 2014; Delhoménie and Heitz 2005). Approximately 53 % of the total VOC emissions are released from stationary industrial sources (non-combustion processes) (Kennes and Veiga 2013). In this regard, many VHs find their applications in industrial processes as solvent thinners, degreasers, cleaners, lubricants, and fuels (Khan and Ghoshal 2000). One of the main concerns about air pollution by VHs (and in general by VOCs) arise from their key role on the formation of ground-level ozone, which is a significant component of photochemical smog with adverse effects on human respiratory function (Lippmann 1989; Kennes and Veiga 2013). Degradation reactions of VHs in the troposphere lead to the formation of intermediate alkyl peroxy ($R\dot{O}_2$) and HO_2 radicals, which react with NO yielding NO_2. Then, NO_2 in the presence of O_2 photolizes to form O_3 and NO, establishing an autocatalytic cycle (Atkinson 2000). Moreover, besides playing an important role on ground-level ozone formation, VHs have also been linked to specific health issues ranging from respiratory tract irritation and headaches to cancer (Table 1).

The harmful consequences derived from polluted air emissions have resulted in increasing public concern and the enforcement of stricter environmental legislations worldwide (Estrada et al. 2015). Thus, several treatment technologies are nowadays available for the treatment of VHs. These technologies can be divided into two main groups, namely physical–chemical technologies and biotechnologies. The selection of the adequate technology will depend on the characteristics of the VH emission such as the air flow rate, pollutant concentration, pollutant solubility in water and biodegradability (Khan and Ghoshal 2000; Delhoménie and Heitz 2005; Ravindra et al. 2008).

2 Air Treatment Systems

2.1 Key Parameters in Air Pollution Control Systems

In order to quantify and characterize polluted air streams and to evaluate the performance of air treatment systems under a standardized framework, several fundamental parameters must be defined. The gas residence time (GRT) is a critical parameter related to the time that the air spends in the treatment system (Revah and Morgan-Sagastume 2005), being defined as follows:

$$GRT = \frac{V}{Q} \tag{1}$$

where V and Q are the treatment working volume and the air flow rate, respectively. It is important to stress that in the case of packed bed systems, V is the empty bed volume and GRT becomes the *empty bed residence time*. Since the GRT is related to the volume of the treatment system, it strongly impacts on the size of the system, and therefore, on the economy of the process. The GRT may take values from a few seconds to several minutes, depending on the characteristics of the pollutant emission (Estrada et al. 2015). The amount of pollutant entering the treatment system is usually quantified in terms of pollutant loading rate (L), defined as:

$$L = C_{G,in} \times \frac{Q}{V} \tag{2}$$

where $C_{G,in}$ is the pollutant (or O_2) concentration at the inlet of the treatment system. Moreover, the performance of the treatment systems is commonly

Table 1 Common volatile hydrocarbons emitted by industrial processes and their effects on human health and environment (UNEP 2006; USEPA 2015)

VH	Chemical formula	Industrial application/occurrence	Effects on human health and the environment
Methane	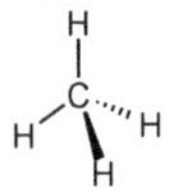	Mainly emitted from coal mining, organic waste treatment activities such as landfilling, composting, wastewater treatment and livestock farming. CH_4 is also the main constituent of natural gas	Greenhouse gas with a global warming potential 23 times higher than that of CO_2 (excluding additional harmful effects of water vapor production from CH_4 breakdown). Represents ~14 % of the total greenhouse gases emissions worldwide
Ethylene	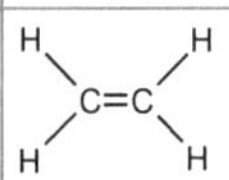	Raw material used in the manufacture of polymers such as polyethylene (PE), polyethylene terephthalate (PET), polyvinyl chloride (PVC) and polystyrene (PS). These polymers are used in the packaging, transportation, electrical/electronic, textile and construction industries	Ethylene is metabolized to ethylene oxide, a compound with carcinogenic and mutagenic effects. Vegetation has proven highly susceptible to this gas, inhibiting the growth of some plant species relevant for agriculture
n-Pentane		Colorless volatile hydrocarbon obtained during the distillation of crude oil. Pentane is used as a blowing (foaming) agent during the manufacture of polystyrene and polyurethane foams. Pentane is also widely used as a solvent in aerosol cosmetics, aerosol paints and car care products	Inhalation of n-pentane might produce lung irritation. Inhalation of high concentrations of pentane may affect the nervous system or cause irritation of the

(continued)

Table 1 (continued)

VH	Chemical formula	Industrial application/occurrence	Effects on human health and the environment
			nose and throat. There are no human or animal studies indicating pentane has a potential to be a human carcinogen
n-Hexane		Colorless volatile hydrocarbon obtained during the distillation of crude oil. Most n-hexane used in industry is mixed with other organic solvents. The major use for solvents containing n-hexane is to extract vegetable oils from crops. It is also used as a cleaning agent in the printing, textile, furniture, and shoemaking industries. Certain kinds of glues used in the roofing, shoe and leather industries also contain n-hexane. Gasoline contains about 1–3 % n-hexane	Inhalation of n-hexane may cause respiratory tract irritation. Exposure produces central nervous system depression, dizziness or suffocation. Prolonged or repeated exposure may cause adverse reproductive effects, nervous system abnormalities with muscle weakness and damage as well as motor incoordination. Chronic exposure produces peripheral neuropathy. Not classifiable as human carcinogen
Isooctane	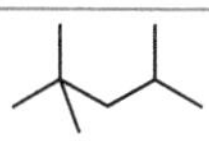	Important component of gasoline used to increase the knock resistance of the fuel. Isooctane defines the 100 point on the octane rating scale. Isooctane is produced on a massive scale in the petroleum industry, usually as a mixture with related hydrocarbons	Inhalation of isooctane may produce cardiac dysrhythmias, depression of the central nervous system as well as

(continued)

Table 1 (continued)

VH	Chemical formula	Industrial application/occurrence	Effects on human health and the environment
			irritation of nose, throat and lungs. Effects on the central nervous system include fatigue, headache and light-headedness
Benzene		Chemical building block for the production of many important industrial compounds such as styrene, phenol, cyclohexane, aniline, alkylbenzenes, and chlorobenzenes. These intermediates are used to produce pharmaceuticals, specialty chemicals, plastics, glazing materials, resins, dyes and pesticides	Benzene is toxic to both humans and experimental animals via all routes of administration. Repeated exposures to benzene in humans lead to depression of white and red blood cells, being classified as a human carcinogen
Toluene	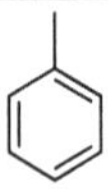	Added to gasoline to improve octane ratings and it is widely used as a solvent in paints, coatings, synthetic fragrances, adhesives, inks, and cleaning agents. Toluene is also used in the production of nylon, polyurethanes, pharmaceuticals, dyes, cosmetic nail products, and the synthesis of organic chemicals	The central nervous system is the primary target organ for toluene toxicity in humans. Symptoms after inhalation include fatigue, sleepiness, headaches, and nausea. Cardiac arrhythmia has also been reported in humans acutely exposed to toluene. Limited or no evidence of the carcinogenic potential of toluene has been reported.

(continued)

Table 1 (continued)

VH	Chemical formula	Industrial application/occurrence	Effects on human health and the environment
			Available epidemiological studies have failed to demonstrate increased risk of cancer due to inhalation exposure to toluene
Xylene	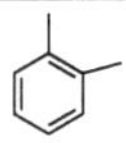 o-xylene 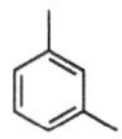 m-xylene 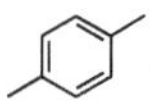 p-xylene	Commercial xylene contains about 40–65 % m-xylene and up to 20 % each of o-xylene and p-xylene. p-Xylene is the principal precursor to terephthalic acid and dimethyl terephthalate, both monomers used in the production of polyethylene terephthalate (PET) plastic bottles and polyester clothing. The mixture of o-, m- and p- isomers is often referred to as xylenes or xylol and their areas of application include the printing, rubber, and leather industries	Symptoms after inhalation of xylene vapors include depression of the central nervous system, headache, dizziness, nausea and vomiting. Exposure to xylene can irritate the lungs, causing chest pain and shortness of breath. No conclusive evidence on the carcinogenicity of xylene in humans has been reported
Ethylbenzene	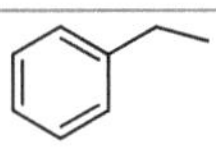	Mainly used as an intermediate for the manufacture of styrene monomer, one of the most important large-volume commodity chemicals, or in the co-production of styrene monomer with propylene oxide. In less extent, ethylbenzene is used as a solvent for paints, lacquers and varnishes	Breathing of ethylbenzene vapours over prolonged periods can cause depression of the central nervous system resulting in dizziness, headache, drowsiness, nausea and loss of coordination.

(continued)

Table 1 (continued)

VH	Chemical formula	Industrial application/occurrence	Effects on human health and the environment
			Ethylbenzene is currently classified as possible human carcinogen
Styrene	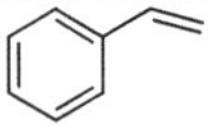	Widely used in the manufacture of resins, plastics, synthetic rubber and polymers. The most popular use for styrene monomer is the production of solid polystyrene. The styrene monomer is also used in the development of expandable polystyrene (EPS), styrene butadiene latex (SBL), acrylonitrile-butadiene-styrene/terpolymer (ABS), unsaturated polyester resins (UPR), and styrene-butadiene rubber (SBR)	Styrene vapor is irritating to the eyes and respiratory tract of humans. Workers with chronic exposure might present low erythrocyte counts and altered liver enzyme profiles. The evidence for carcinogenicity of styrene is limited, being currently classified as possible human carcinogen
α-Pinene	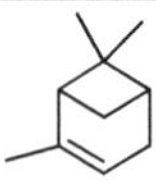	Volatile compound typically found in waste gases from wood-related industries. α-Pinene is used in the manufacture of Camphor, insecticides, solvents, perfumes and synthetic pine oil	Symptoms after inhalation of α-pinene include nose and throat irritation causing coughing and wheezing. Prolonged exposure to α-pinene can cause headache, nausea and vomiting. High exposure may affect the nervous system causing loos of coordination, dizziness and confusion

evaluated in terms of elimination capacity (EC) and removal efficiency (RE), defined as follows:

$$EC = \left(C_{G,in} - C_{G,out}\right) \times \frac{Q}{V} \tag{3}$$

$$RE = \frac{C_{G,in} - C_{G,out}}{C_{G,in}} \times 100\,\% \tag{4}$$

where $C_{G,out}$ represents the pollutant concentration at the outlet of the treatment system. Pollutant concentrations in air streams are usually reported in terms of mass or volumetric concentrations: g m^{-3} or ppm_v, respectively. Conversion of volumetric to mass concentrations can be done using the ideal gas law, which under normal conditions can be expressed as (Cox and Deshusses 2001):

$$[\text{g m}^{-3}] = \frac{[\text{ppm}_v] \times \text{ molecular weight of the pollutant (g mol}^{-1})}{22776} \tag{5}$$

The parameters and concentrations defined in Eqs. 1–5 constitute a basic common terminology, which is very useful for comparing the performance of air treatment technologies.

2.2 Conventional Physical–Chemical Treatment Technologies

In general terms, physical–chemical technologies for the treatment of VHs can be classified as *destructive* and *recovery* techniques (Fig. 1). Thermal and catalytic oxidation processes are the most common destructive techniques, which are recommended for high air flow rates (e.g. 8×10^3–1×10^5 m^3 h^{-1}) and a wide range of VH concentrations in gas phase (e.g. 3–100 g m^{-3}) (Revah and Morgan-Sagastume 2005; Delhoménie and Heitz 2005). In thermal oxidation processes, VHs are combusted at temperatures of 650–800 °C, which allows the oxidation of virtually all VOCs (RE = 95–99 %) at GRTs of 0.5–2.0 s (Khan and Ghoshal 2000). Catalytic oxidation processes operate at temperatures ranging from 250 to 500 °C, which are significantly lower than that used in thermal oxidation systems. This temperature reduction can be done by using platinum- or palladium-based catalysts that reduce the combustion energy requirements (Revah and Morgan-Sagastume 2005). The inlet air stream is heated and passed through a support structure coated with catalyst. Catalytic oxidation systems can treat similar air flow rates and VH concentrations than that treated by thermal oxidation, supporting VOC removal efficiencies of 90–95 % at GRTs of 0.5–2.0 s (Khan and Ghoshal 2000).

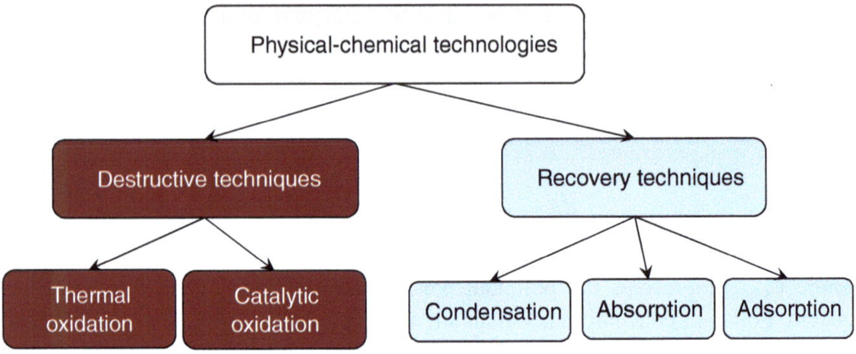

Fig. 1 Conventional physical–chemical technologies for the treatment of VHs

On the other hand, the most common recovery techniques include condensation, absorption and adsorption processes. VHs condensation can be performed by lowering the gas stream temperature at constant pressure or increasing the gas stream pressure at constant temperature (Revah and Morgan-Sagastume 2005). This technique is applied to the treatment of VHs with high boiling points (>38 °C) present at gas concentrations greater than 30–50 g m^{-3} (Schnelle and Brown 2001; Delhoménie and Heitz 2005). Removal efficiencies range from 50 to 90 %, the upper end of efficiencies are achievable for VH concentrations of 50 g m^{-3} and higher (Schnelle and Brown 2001).

Absorption processes remove VHs from gas streams by contacting the contaminated air with a liquid solvent (usually water with adjusted pH). As VHs are transferred from the gas to the liquid phase, the air stream is scrubbed. Thus, absorption towers are designed to provide enough gas/liquid interfacial areas for the removal of soluble VHs. These systems are recommended for a wide range of air flow rates (e.g. 100–100,000 m^3 h^{-1}) and for VH concentrations ranging from 10 to 50 g m^{-3} (Delhoménie and Heitz 2005). An absorber can achieve removal efficiencies of 90–98 % and, depending on the type of pollutants, the GRT can take values from few seconds to minutes (Khan and Ghoshal 2000; Delhoménie and Heitz 2005). This technique is not recommendable for VHs with low water solubility.

Adsorption processes are based on the contact of pollutant molecules with the surface of a solid adsorbent, being retained via weak intermolecular forces (van der Waals interactions). Carbon is the most common adsorbent in use today for VOC/VH treatment. However, silica gel, alumina and zeolite can also be used (Khan and Ghoshal 2000; Revah and Morgan-Sagastume 2005). Adsorbents are activated by heat under controlled conditions to remove the volatile non-carbon constituents and to increase the gas/solid surface area. Likewise, the adsorbent can be regenerated by applying heat or by decreasing the pressure in the packed bed in order to volatilize the adsorbed compounds. Continuous air treatment by adsorption can be done by means of the so-called "regenerative adsorption" processes, which operate two or more activated carbon beds in parallel. In general terms, non-regenerative adsorption processes are recommended for VH concentrations lower than 1 g m^{-3} and air flow rates of 10–100 m^3 h^{-1}, while regenerative adsorption processes are suitable for VH concentrations ranging 1–10 g m^{-3} and air flow rates of 100–100,000 m^3 h^{-1} (Delhoménie and Heitz 2005). Removal efficiencies of 90–99 % can be achieved with adsorption treatment systems (Schnelle and Brown 2001).

2.3 Conventional Biological Treatment Technologies

Biotechnologies for the treatment of VHs are based on the ability of microbial communities to oxidize these pollutants, yielding CO_2, H_2O and biomass. Microorganisms are able to metabolize a wide variety of organic compounds, which can be directly used as single carbon and energy sources by bacteria or fungi as in the case of VHs (Cox and Deshusses 2001; Kennes et al. 2013). Activated sludge from wastewater treatment plants is one of the most common sources of inocula for biotechnologies as it contains a large variety of microorganisms with the potential ability to degrade a wide variety of pollutants (Wagner et al. 2002; Khammar et al. 2005; Malhautier et al. 2005). From an industrial point of view, the use of activated sludge as inoculum is advantageous due to the ease of implementation, high availability and low cost (Pérez et al. 2014). However, peat, soil, compost, leachates from landfills and pure bacterial or fungal strains have also been used for inoculation purposes (Khammar et al. 2005; Pérez et al. 2014; López et al. 2014).

Biotechnology is considered as the best option for the treatment of relatively high air flow rates (e.g. 100–500,000 m^3 h^{-1}) and pollutant concentrations below 5 g m^{-3} (Cox and Deshusses 2001; Delhoménie and Heitz 2005). Since biotechnologies can be operated at atmospheric pressure and temperatures between 10 and 40 °C, they are usually less expensive, simpler to operate, and exhibit lower carbon footprints compared with physical–chemical technologies (Kennes and Thalasso 1998; Noyola et al. 2006; Estrada et al. 2011; Alfonsín et al. 2015). In general terms, the investment costs of biotechnologies can be up to 70, 22, 10 and 13 % lower than that for adsorption, catalytic oxidation, absorption and condensation technologies (Delhoménie and Heitz 2005). The higher cost-effectiveness of biotechnologies relative to physical–chemical treatments is even sharper when considering the operating costs, which can be up to 250, 700, 1000 and 980 % lower for biotechnologies than that for adsorption, catalytic oxidation, absorption and condensation technologies, respectively (Delhoménie and Heitz 2005). Therefore, it is not surprising that the implementation of biotechnologies for air pollution control at industrial scale is growing exponentially (Popov et al. 2004; Revah and Morgan-Sagastume 2005; Easter et al. 2005; Rene et al. 2013).

Pioneering studies on biological air treatment during the years 1960–1970 showed that the biological reaction was the main mechanism for the elimination of air pollutants, the first commercial applications being established in The

Netherlands, West Germany and the United States (Revah and Morgan-Sagastume 2005). Initially, simple open biological filters (commonly referred as *biofilters*) based on a compost packed bed and inoculated with wastewater-activated sludge were used for the removal of odors and organic compounds from waste gases (Ottengraf and Van Den Oever 1983). Biofiltration research progressively intensified in the following years motivated by the enactment of stringent regulations on odorous emissions from wastewater treatment plants, improving the operation and design of the pioneering biofilters (better irrigation systems and new packing materials) (Rene et al. 2013). Nowadays, besides biofilters, there are several bioreactor configurations that can be regarded as *conventional* biological technologies with a proven air treatment performance at industrial scale (Fig. 2). The fundamentals and application spectra of these conventional biological technologies are summarized in the following lines.

2.3.1 Biofilters

In biofilters, air emissions are usually humidified before passing through a packed bed supporting the active biofilm responsible for pollutant biodegradation (Fig. 2a). These systems are designed with enough GRT to allow pollutant mass transfer from

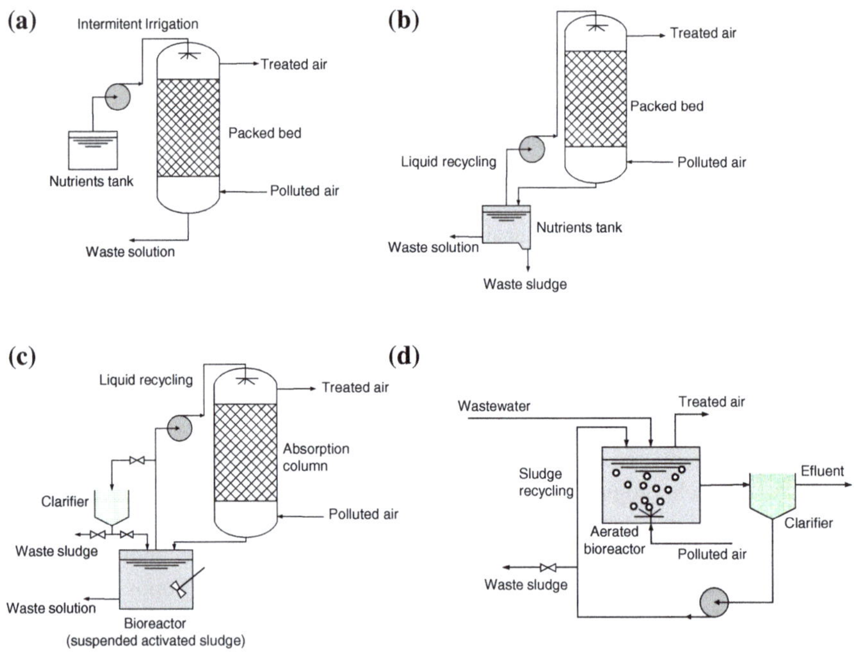

Fig. 2 Schematic representation of conventional biotechnologies for air pollution control: **a** biofilter, **b** biotrickling filter, **c** bioscrubber and **d** activated sludge diffusion system

the gas phase to the biofilm formed mainly by microorganisms, exo-enzymes, exo-polysacharides and water (Revah and Morgan-Sagastume 2005). For moderately hydrophilic pollutants, which are compounds with dimensionless Henry's law constant (DHLC) of 0.1–0.99, biofilters support efficient removal since the biofilm is not totally covered by a water layer (Kraakman et al. 2011). Biofilters also exhibit large gas-biofilm interfacial areas, a rapid start-up and low operating costs (≈2 € $m^{-3} \times h$) (Estrada et al. 2012). On the contrary, biofilters require both a high GRT for efficient pollutant removal (20 s–2 min) and a limited height (<1.5 m) in order to maintain low pressure drops. Typical pressure drops in biofilters range from 1000 to 1500 Pa m^{-1} (Estrada et al. 2011). Other limitations are a rapid bed compaction when using organic packed beds, biomass overgrowth, a poor control over environmental conditions such as pH, T and humidity, and the potential accumulation of inhibitory metabolites. Biofilters exhibit typical VOC removal efficiencies of 90–100 % and scale-dependent investment costs of 5–28 € m^{-3} h^{-1} (Estrada et al. 2011).

2.3.2 Biotrickling Filters

In biotrickling filters, the polluted air emission passes through a biofilm immobilized on a packed bed continuously irrigated with a nutrient solution (Fig. 2b). The gas pollutants initially absorbed in the trickling aqueous solution are subsequently degraded in the fixed biofilm using the nutrients contained in the solution (Revah and Morgan-Sagastume 2005). Full scale biotrickling filters can support almost a complete pollutant removal at GRTs as low as 1.6–2.2 s (Gabriel and Deshusses 2003; Witherspoon et al. 2008), which makes them a good alternative in scenarios with limitation in land availability. This biotechnology supports an efficient control of pH, temperature, moisture content and nutrient supply. However, the presence of a continuous water layer over the biofilm often entails a limited abatement performance for hydrophobic gas pollutants as a result of their poor gas-liquid mass transport (as in the case of some VHs such as hexane, pentane, methane, etc.). The investment costs of this biotechnology range from 10 to 41 € $m^{-3} \times h$, while presenting very low investment costs (1.2 € $m^{-3} \times h$) due to its low pressure drops (50–500 Pa m^{-1}) and low liquid recycling rates (5–10 L $m^{-3}{}_{reactor}$ min^{-1}). The largest environmental impact is caused by water consumption (an empirical rule of 2.5 L per gram of pollutant removed is applied), although the use of treated wastewater has been proposed as a potential solution to increase biotrickling filter sustainability (Estrada et al. 2011; Kraakman and Cesca 2012).

2.3.3 Bioscrubbers

Bioscrubbers are 2-stage systems composed of an absorption tower where the air pollutants are absorbed into an aqueous solution that is further transferred to an interconnected suspended growth bioreactor where the biological oxidation of the pollutant occurs aerobically (Fig. 2c) (Kennes and Thalasso 1998). Most hydrophilic air pollutants (holding DHLC < 0.099) are efficiently removed (RE $\approx$ 99 %) in bioscrubbers (Hansen and Rindel 2000). The main advantages of this biotechnology are its proficient control of pH, temperature, biofilm moisture content and accumulation of inhibitory metabolites, its low pressure drop and its low operating costs (2–8 € $m^{-3} \times h$). Some bioscrubbers are constructed with an inert packed bed to increase the contact between the waste air stream and the absorbent aqueous solution. Apart from their low abatement performance for hydrophobic VOCs, bioscrubbers exhibit the highest investment costs among conventional biotechnologies (23–92 € $m^{-3} \times h$), which might explain why this biotechnology has been gradually displaced by biotrickling filters (more compact and economic).

2.3.4 Activated Sludge Diffusion Systems

Activated sludge diffusion systems are mainly focused on odor control in wastewater treatment plants (Burgess et al. 2001). These systems are based on biological oxidation in liquid media (Fig. 2d). Polluted air emissions are collected at source and transferred to an activated sludge tank provided with gas diffusers. Pollutant removal efficiencies achieved by activated sludge diffusion systems are usually greater than those obtained with conventional chemical scrubbers or biofilters (Ostojic 1992; Bowker 2000). These systems are particularly advantageous in wastewater treatment plants already using diffused aerated systems due to the low footprint of this technology and to the negligible operation cost, since they can be included in the general costs involved in wastewater treatment. Unfortunately, this technology, as the rest of conventional biological technologies, exhibits poor removal efficiencies of hydrophobic air pollutants (compounds with DHLC > 0.99, as in the case of several VHs). Most authors agree on the fact that mass transfer limitations increase with pollutant hydrophobicity as a result of the lower driving force available for mass transfer from the gas to the liquid phase (Muñoz et al. 2012).

3 Two-Phase Partitioning Bioreactors for Removal of VHs

Biological technologies for air pollution control perform poorly when treating VHs (or their metabolites) exhibiting a high toxicity towards the microorganisms due to cell inhibition issues (Muñoz et al. 2012). Likewise, the low mass transfer of poorly water-soluble VHs from the gas to the microorganisms limits the performance of biotechnologies (Quijano et al. 2009a; Yeom et al. 2010). As shown in Fig. 3, a general classification of VHs as a function of their DHLC values can be done in order to anticipate cell inhibition or mass transfer limitations issues. Although this classification is arbitrary, it is useful to identify those VHs whose biological removal is intrinsically limited by toxicity issues or by mass transfer limitations as

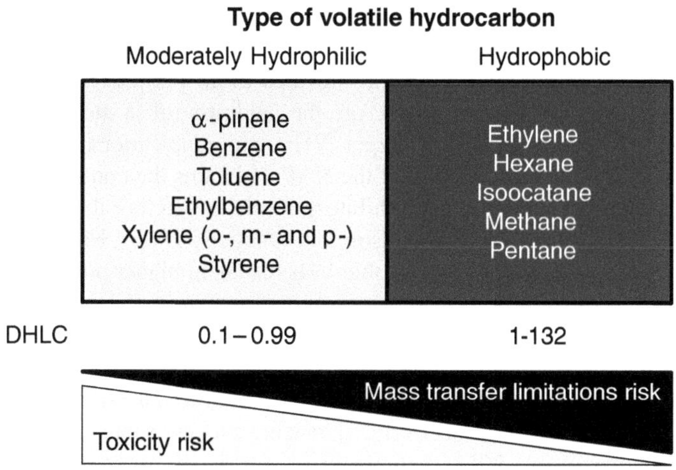

Fig. 3 Classification of VHs in terms of their dimensionless Henry's law constant, mass transfer limitation and toxicity risks in biological treatment systems

confirmed by empirical observations (Quijano et al. 2009a). Air pollutants exhibiting DHLC values below 0.099 are regarded as hydrophilic compounds and include air pollutants such as acetone, methanol, methyl ethyl ketone, among others (Muñoz et al. 2012). Although VHs of industrial or environmental interest are not classified as hydrophilic compounds, toxicity risks are higher for those VHs with DHLCs between 0.1 and 0.99 (moderately hydrophilic ones), while mass transfer limitation risks are higher for those VHs with DHLCs $\geq$ 1 (hydrophobic ones) (Table 2).

Table 2 DHLC values for volatile hydrocarbons and oxygen (values based on concentration ratios at 25 °C) (Sander 1999)

Compound	DHLC
Styrene	0.117
o-Xylene	0.160
Benzene	0.227
m-Xylene	0.270
Toluene	0.273
p-Xylene	0.290
Ethylbenzene	0.340
α-Pinene	0.835
Ethylene	9
Methane	30
Oxygen	32
Pentane	51
Hexane	70
Isooctane	132

Two-phase partitioning bioreactors (TPPBs) emerged in the early 1990s as innovative multiphase systems capable of overcoming the above mentioned limitations of conventional biotechnologies devoted to air pollution control (Fam and Daugulis 2013). TPPBs are based on the addition of a non-aqueous phase (NAP) with a high affinity for target VHs into conventional biotechnologies (Muñoz et al. 2012). The presence of the NAP maintains the concentration of toxic VHs in the aqueous phase at sub-inhibitory levels, protecting the cells from inhibition issues (Daugulis 2001; Davidson and Daugulis 2003). Moreover, the high affinity of the NAP towards hydrophobic VHs results in higher pollutant absorption and driving forces for mass transfer, improving therefore pollutant removal (Clarke et al. 2006; Hernández et al. 2011b; Dumont and Andrès 2012). NAP addition can also avoid microbial inhibition issues due to pollutant loading surges and starvation periods by temporarily decreasing the VH concentration in the aqueous phase or by acting as a VH reservoir, respectively, enhancing process robustness (Bailón et al. 2009; Kraakman et al. 2011; Lebrero et al. 2014). Furthermore, many reports showed that most NAPs used in TPPBs also exhibit a high affinity towards O_2 and, consequently, the O_2 mass transfer rate is also improved (Dumont et al. 2006; Kraakman et al. 2011). Thus, an increase in the VH mass transfer in TPPBs is concomitant with an increase in the O_2 transfer rate, enhancing opportunities for complete pollutant mineralization (Muñoz et al. 2012). Consequently, VH biodegradation performance and process robustness in TPPBs are usually superior to that recorded in conventional biological systems.

The VH removal performance of a TPPB fundamentally depends on the selection of the NAP, the process microbiology and the bioreactor configuration. Intensive research on TPPBs carried out in the last decade has helped to provide a better understanding of these fundamental aspects, which are addressed below.

3.1 NAP Selection

The selection of the NAP for implementing an air treatment process in a TPPB is of paramount importance since the performance of the TPPB depends to a large extent on the adequate selection of the NAP (Muñoz et al. 2007, 2008). Table 3 summarizes the characteristics that a NAP must fulfill for the implementation of air treatment processes. Traditionally, liquid organic solvents such as alkanes and perfluorocarbons have been used as NAPs due to the high diffusivity of the target substrates into these NAPs (Muñoz et al. 2008; Fam and Daugulis 2013). However, the use of hexadecane, tetradecane, undecane and diethyl sebacate as NAPs is hindered by their biodegradability (MacLeod and Daugulis 2005; Arriaga et al. 2006). Similarly, the high volatility of perfluorocarbons together with their ozone depleting potential discourages their use as a NAP in air treatment applications (Muñoz et al. 2012). Ionic liquids also have been studied as potential NAPs for air treatment processes. However, their use is still limited by their high cost, toxicity towards microbial cells and their high solubility in water (up to 2×10^4 g m^{-3} for

Table 3 Characteristics to be fulfilled by a NAP for TPPB implementation in air pollution control processes

Characteristic	Rationale	References
Biocompatible	Avoiding decreased metabolic activity	Bruce and Daugulis (1991)
Non-biodegradable	Avoiding preferential NAP uptake over VH biodegradation and NAP losses	
High affinity for the target VH	High VH absorption in the TPPB	
Non-hazardous	Avoiding toxic effects for operators and the environment	
Available in bulk quantities	Key characteristic for full scale applications	
Low cost	Facilitating the implementation of a cost-effective process	
Density different from that of water	Facilitating phase separation	Cesario et al. (1992)
Immiscible in water	Avoiding NAP losses and facilitating phase separation	
Low vapor pressure	Avoiding NAP losses and release to the environment	
Odorless	Avoiding unpleasant working conditions or even toxic effects to operators	Van Groenestijn and Lake (1999)
High VH/O_2 diffusivity	Improving Gas/NAP and NAP/Water mass transfer characteristics	Muñoz et al. (2008)

some imidazolium-based ionic liquids) (Quijano et al. 2010b). It has been reported that imidazolium ionic liquids might inhibit toluene biodegradation at levels as low as 5 % (v/v) (Quijano et al. 2011). In contrast, some phosphonium ionic liquids were biocompatible with *Pseudomonas putida*, *Achromobacter xylosoxidans*, or *Sphingomonas aromaticivorans* (Baumann et al. 2005), while ammonium ionic liquids improved the conversion of ketones by *Escherichia coli* TOP10 pQR239 (Melgarejo-Torres et al. 2015). This indicates that the potential of specific ionic liquids as NAPs for air pollution control in TPPBs deserve further investigation. Another aspect worth taking into account when using ionic liquids as NAPs is their thermal decomposition in the presence of water, which has been reported to occur at temperatures above 50 °C, leading to the formation of toxic compounds (e.g. hydrogen fluoride) (Jacquemin et al. 2006). Thus, the use of ionic liquids in biological air treatment systems must be carefully considered.

Most authors agree on the fact that silicone oil is the only liquid NAP that fulfills all the expected characteristics listed in Table 3 (Arriaga et al. 2006; Quijano et al. 2010c; Darracq et al. 2010b; Dumont and Andrès 2012; Dumont et al. 2014). Silicone oil at percentages of 5–20 % (v/v) has been the most commonly used NAP in air treatment applications over the last 10 years. Operational problems relating to foaming and adhesion to reactor internals have been observed when using silicone oil as a NAP (Daugulis et al. 2011). However, these operational issues have been reported only in stirred tanks, a bioreactor configuration without full-scale application potential in off-gas treatment processes due to its high energy demand (Hernández et al. 2011b; Muñoz et al. 2012). Recently, solid inert polymers based on copolymers of polyurethane, polyester, vinyl acetate, polyamide, ethyleneoctene and ethylene (e.g. commercial trademarks Kraton, Elvax, Desmopan, Hytrel, Zytrel, Engage) as well as silicone rubber and recycled rubber tires have been tested as NAPs (Fam and Daugulis 2013). The main advantages of solid NAPs are their low cost and facilitated separation and recycling during pollutant treatment (Morrish and Daugulis 2008; Daugulis et al. 2011).

The evaluation of the NAP affinity towards target VHs and O_2 is quantified in terms of the dimensionless Gas/NAP partition coefficient ($H_{G/NAP}$), which is defined as:

$$H_{G/NAP} = \frac{C_G}{C_{NAP}} \tag{6}$$

where C_G and C_{NAP} are the VH (or O_2) concentration in the gas phase and the NAP, respectively. Thus, NAPs with high affinity for target VHs and O_2 exhibit $H_{G/NAP}$ values $\ll$ DHLC (Muñoz et al. 2008; Quijano et al. 2010c; Darracq et al. 2010b). The Gas/NAP partition coefficient is one of the most important criteria for NAP selection (Daugulis 2001; Muñoz et al. 2008). As shown in Table 4, several solid NAPs exhibit $H_{G/NAP}$ values comparable with those reported for liquid NAPs (comparable affinities towards VHs and O_2).

Experimental studies on VH abatement in TPPBs implemented with solid and liquid NAPs have demonstrated the superior performance of liquid NAPs. For instance, the hexane removal in the presence of Kraton was similar to that recorded in a control reactor without NAP, while the TPPB implemented with silicone oil supported a twofold higher hexane removal compared with the control without NAP (Hernández et al. 2010). Likewise, solid polymers failed in improving the removal of a mixture of methyl ethyl ketone, toluene and hexane under steady conditions, slight improvements on the removal of these pollutants being observed only under transient conditions (Hernández et al. 2011a). Therefore, up to date there is not enough experimental evidence to encourage the use of solid NAPs in air pollution control processes. Research on VH removal in low energy-demanding bioreactor configurations (e.g. biofilters, biotrickling filters, etc.) under realistic operational conditions is still required.

The low performance of solid NAPs in air treatment processes is likely due to the low diffusivity of air pollutants into the solid polymers. Diffusivity of VHs, oxygen and other organic compounds into solid NAPs can be up to 500 times lower than that in liquid NAPs (Muñoz et al. 2008; Montes et al. 2011). At this point, it is important to highlight that $H_{G/NAP}$ is obtained with pollutant concentrations in gas and NAP phases at equilibrium and, therefore, the time scales of the mass transfer processes are not taken into account in this parameter. Although it has not received much attention in the literature, the VH/O_2 diffusivity in the NAP is an important criterion for selecting the optimum NAP among candidates with similar $H_{G/NAP}$

Table 4 NAPs reported in the literature and their affinity towards volatile hydrocarbons and oxygen

NAP	NAP characteristics	Target compounds	$H_{G/NAP}$ (C_G C_{NAP}^{-1})	References
Silicone oil	Liquid polymer of dimethylsiloxane (kinematic viscosities from 5–200 cSt)	Oxygen	3.6	Quijano et al. (2010c)
		Methane	2.0	Rocha-Rios et al. (2011)
		Hexane	0.0058	Hernández et al. (2010)
		Toluene	0.00064	Darracq et al. (2010a)
		α-pinene	0.00018	Muñoz et al. (2008)
		Styrene	0.00044	Dumont et al. (2014)
Hexadecane	Linear liquid alkane	Oxygen	6.6	Dumont et al. (2006)
		Hexane	0.0040	Arriaga et al. (2006)
		Toluene	0.0010	Vuong et al. (2009)
		α-pinene	0.000071	Muñoz et al. (2008)
		Benzene	0.0021	Yeom and Daugulis (2001)
FC40	Liquid perfluorocarbon	Oxygen	3.0	Quijano et al. (2010c)
		Hexane	0.027	Hernández et al. (2010)
		Toluene	0.012	Cesário et al. (1997)
		α-pinene	0.0014	Muñoz et al. (2008)
Diethyl sebacate	Liquid ester	Hexane	0.015	Arriaga et al. (2006)
		α-pinene	0.000074	Muñoz et al. (2008)

(continued)

Table 4 (continued)

NAP	NAP characteristics	Target compounds	$H_{G/NAP}$ (C_G C_{NAP}^{-1})	References
Heptamethylnonane	Liquid and highly branched alkane	Oxygen	3.4	Quijano et al. (2010c)
		Methane	2.1	Rocha-Rios et al. (2011)
		Hexane	0.0027	Hernández et al. (2010)
		α-pinene	0.000056	Muñoz et al. (2008)
Heptamethyltrisiloxane	Liquid polyalkyleneoxide with three siloxane units	Hexane	0.0029	Hernández et al. (2010)
$[C_4mim][PF_6]$	Liquid imidazolium ionic liquid	Toluene	0.00096	Quijano et al. (2011)
$[C_4mim][NTf_2]$	Liquid imidazolium ionic liquid	Toluene	0.00061	Quijano et al. (2011)
Rubber tires	Solid copolymer of polyisoprene, styrene-butadiene copolymer and polybutadiene	α-pinene	0.0005	Montes et al. (2011)
Kraton D4150K	Solid styrene/butadiene triblock copolymer	α-pinene	0.0003	Montes et al. (2011)
Kraton G1657	Solid styrene/butadiene triblock copolymer	Methane	4.0	Rocha-Rios et al. (2011)
		Oxygen	3.0	Quijano et al. (2010c)
		Hexane	0.0057	Hernández et al. (2010)
Pebax 2533	Solid polyether block amide	α-pinene	0.0009	Montes et al. (2011)
Hytrel G3548L	Solid copolymer of poly (butylene terephthalate) and polyether	α-pinene	0.0013	Montes et al. (2011)
Silicone rubber	Solid polymer of dimethylsiloxane	α-pinene	0.0007	Montes et al. (2011)
Desmopan 9370A	Solid polyurethane of poly (oxytetramethylene) glycol and methyldiisocyanate	Methane	5.4	Rocha-Rios et al. (2011)
		Oxygen	5.1	Quijano et al. (2010c)
		Hexane	0.0168	Hernández et al. (2010)
Elvax 360	Solid poly (ethyleneco-vinyl acetate)	Oxygen	3.5	Quijano et al. (2010c)

(continued)

Table 4 (continued)

NAP	NAP characteristics	Target compounds	$H_{G/NAP}$ (C_G C_{NAP}^{-1})	References
		Hexane	0.0135	Hernández et al. (2010)
Elvax 880	Solid poly (ethyleneco-vinyl acetate)	Methane	9.7	Rocha-Rios et al. (2011)

values. NAPs exhibiting higher VH/O_2 diffusivities are expected to support higher Gas/NAP and NAP/Water mass transfer coefficients and therefore enhanced TPPB performance (Muñoz et al. 2008).

3.2 Process Microbiology and Mass Transfer Considerations

Mixed and pure microbial cultures have been used in TPPBs devoted to air treatment. However, activated sludge from wastewater treatment plants is by far the most common inoculum source in TPPBs focused on environmental applications (Van Groenestijn and Lake 1999; Guieysse et al. 2001; Darracq et al. 2012; Pérez et al. 2014). Although activated sludge constitutes a cheap and widely available inoculum source, other sources can be considered when targeting the removal of specific VHs. For instance, mixed methane-degrading communities (methanotrophs) can be obtained from landfill covers and leachates (López et al. 2014). The preference on using mixed cultures over pure strains for TPPB inoculation comes from the fact that mixed cultures confer microbial diversity, high resilience and robustness to biological air treatment systems (Khammar et al. 2005; Cabrol and Malhautier 2011). Furthermore, when bioreactors devoted to air treatment are inoculated with pure strains it is common to find that other microbial populations become dominant after long-term operation (Bailón et al. 2009; Pérez et al. 2014). TPPBs as a technology for air treatment are open (non-sterile) systems that are intended to cope with fluctuating VOC loadings and pollutant compositions. Therefore, gradual changes in the microbial community structure are expected (Muñoz et al. 2012). In this regard, Bailon et al. (2009) studied the changes in the microbial populations in a TPPB degrading dichloromethane. The authors found that although the originally inoculated *Hyphomicrobium* strains were present, they were not dominant after long-term operation and new bacterial populations were detected. However, no deterioration of the gas pollutant removal due to the change in the bacterial population was observed. This work clearly showed that microbial diversity cannot be tightly controlled in open TPPBs over long periods of time. However, such variations in the microbial population structure do not necessarily imply deterioration of the biodegradation performance.

Conventional TPPB operation has generally considered that metabolic activity occurs exclusively in the aqueous phase (Daugulis 2001) (Fig. 4). Therefore, the central mechanisms underlying enhanced removal of VHs in TPPBs have been traditionally explained in terms of the establishment of an additional Gas/NAP/Water mass transfer pathway and the prevalence of sub-inhibitory pollutant concentrations in the aqueous phase mediated by the high affinity of the NAP for the target pollutants (Fig. 4b) (Daugulis 2001; Dumont et al. 2006; Quijano et al. 2009b, 2010e; Kraakman et al. 2011). However, recent studies revealed that other complex uptake mechanisms might take place within the NAP or in the NAP/water interface, highlighting the relevance of microbiology on the performance and optimization of TPPBs (Fig. 4c). The key role of microbial characteristics on the performance of TPPBs was noted early from biodegradation experiments of chlorobenzenes and ethyl butyrate using silicone oil as a NAP. In these studies, a significant amount of active biomass grew adhered to the NAP/water interface and a direct pollutant uptake from the NAP/water interface was suggested (Ascon-Cabrera and Lebeault 1993, 1995). Later, it was reported that biodegradation of poly aromatic hydrocarbons by a *Mycobacterium* strain occurred exclusively at the NAP/water interface (MacLeod and Daugulis 2005). Further studies on substrate uptake using silicone oil as a NAP showed that cell adhesion to the NAP/water interface was not induced by a higher availability of the carbon source in the NAP, cell affinity for the NAP/water interface being observed even when the carbon source was not soluble in the NAP (glucose) (Quijano et al. 2010d). Cell affinity for liquid organic phases was reported to be an inducible and species-dependent characteristic (Hernández and Muñoz 2011). It is important to stress that water and nutrients can diffuse into liquid NAPs such as silicone oil, and therefore, biomass growth can be supported within these organic phases (Harz and Knoche 2001; Yoshino and Schmidt 2014).

The benefits from operating TPPBs with hydrophobic microorganisms growing within liquid NAPs on the removal of hydrophobic gas pollutants was reported for the first time by Hernández et al. (2012). They showed that hydrophobic microorganisms growing immersed in silicone oil were able to take up both oxygen

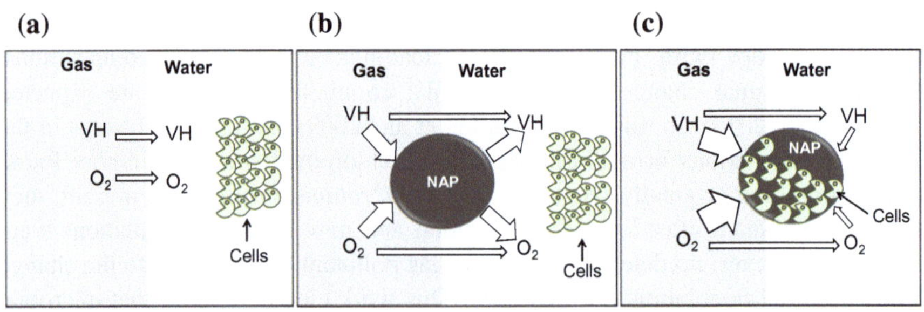

Fig. 4 VH and O_2 mass transfer pathways and microbial uptake mechanisms in **a** conventional biological air treatment system, **b** TPPB considering uptake exclusively in the aqueous phase, and **c** TPPB considering uptake exclusively in the NAP

and hexane from the NAP, supporting a hexane removal performance eightfold higher relative to the same TPPB operated with microorganisms dispersed in the aqueous phase. Further studies confirmed that a stable long-term removal of hexane can be achieved by using hydrophobic microorganisms confined within silicone oil (Muñoz et al. 2013). Despite changes in the microbial populations were not ruled out in this study, macroscopic behavior of the microorganisms (e.g. cell hydrophobicity) as well as hexane removal performance remained stable. A recent mathematical modeling work on TPPBs devoted to air pollution control showed that prediction performance of the model was significantly improved when pollutant and oxygen uptake from the NAP was considered, while neglecting this uptake mechanism (e.g. considering VH/O_2 uptake exclusively in aqueous phase) led to underestimations of the VH removal performance (Dorado et al. 2015).The experimental assessment of the main uptake mechanism in a TPPB is a challenging task since VH/O_2 uptake may occur simultaneously in the aqueous phase and in the NAP, depending on the microbiology of the system, the bioreactor configuration and the operating conditions. However, theoretical considerations can be done in order to select and promote the best uptake mechanism according to the specific characteristics of the target VHs to be treated. Hence, at steady state: (1) if VOC/O_2 uptake occurs in the NAP, the Gas/Water mass transfer rate ($F_{G/W}$) equals the Water/NAP mass transfer rate ($F_{W/NAP}$) and (2) if VOC/O_2 uptake occurs in the aqueous phase, the Gas/NAP mass transfer rate ($F_{G/NAP}$) equals the NAP/Water mass transfer rate ($F_{NAP/W}$). $F_{G/W}$ and $F_{G/NAP}$ can be written as follows (Muñoz et al. 2007):

$$F_{G/W} = k_L^{G/W} a \left(\frac{C_G}{DHLC} - C_W \right) \tag{7}$$

$$F_{G/NAP} = k_L^{G/NAP} a \left(\frac{C_G}{H_{G/NAP}} - C_{NAP} \right) \tag{8}$$

where $k_L^{G/W}a$ and $k_L^{G/NAP}a$ are the volumetric mass transfer coefficients between the gas and aqueous phases and between the gas and the NAP, respectively. Considering the concentration gradients for mass transfer in Eqs. 7 and 8, taking hexane and silicone oil as examples of VH and NAP, respectively (DHLC = 70; $H_{G/NAP} = 0.0058$), and assuming a hexane concentration in the gas phase of $1\ g\ m^{-3}$, the maximum Gas/Water and Gas/NAP concentration gradients are:

$$\left(\frac{C_G}{DHLC} - C_W \right) = \left(\frac{1 \frac{g}{m^3_{Gas}}}{70} - 0 \right) = 0.014 \frac{g}{m^3_{Water}}$$

$$\left(\frac{C_G}{H_{G/NAP}} - C_{NAP} \right) = \left(\frac{1 \frac{g}{m^3_{Gas}}}{0.0058} - 0 \right) = 172 \frac{g}{m^3_{NAP}}$$

The maximum Gas/Water concentration gradient ($C_W = 0$) can be achieved when VH/O_2 uptake occurs in the aqueous phase, while the maximum Gas/NAP concentration gradient ($C_{NAP} = 0$) can be achieved when VH/O_2 uptake occurs in the NAP. Therefore, since the maximum Gas/NAP concentration gradient is much higher than the maximum Gas/Water concentration gradient, the removal performance will be optimized with microbial uptake in the NAP when the pollutant mass transfer is the limiting step (e.g. when dealing with hydrophobic VHs). On the contrary, the removal performance of toxic and water-soluble gas pollutants will be optimized with microbial uptake in the aqueous phase since the Gas/NAP transfer rate (and consequently the NAP/Water transfer rate) will not operate at its maximum capacity ($C_{NAP} > 0$). Moreover, pollutant loading surges will be buffered by the NAP avoiding toxicity issues for the cells residing in the aqueous phase.

Biomass growth in the NAP can be promoted by establishing an aqueous dilution rate high enough to wash out progressively those microorganisms unable to attach to the NAP/Water interface or colonize the NAP. For instance, silicone oil colonization by hexane-degrading microorganisms was achieved with aqueous dilution rates ranging 0.025–0.2 day^{-1} (Hernández et al. 2012; Muñoz et al. 2013). However, different dilution rates could be required for other VHs. It is also important to take into account that the apparent kinetic parameters of the microorganisms growing immersed in the liquid organic phases might significantly differ from those obtained for the same microorganisms growing dispersed in the aqueous phase only. For instance, silicone oil-mediated changes the kinetic parameters of *Methylosinus sporium* during methane biodegradation, with variations of 60, 120, and 150 % in the oxygen uptake rate, half-saturation constant and the maximum specific growth rate compared with the parameters determined in a control system without NAP, respectively (Ordaz et al. 2014). Thus, such significant changes in the kinetic parameters mediated by the NAP must be considered for the design, operation and modeling of TPPBs devoted to air pollution control.

3.3 Reactor Configurations

TPPBs for air pollution control have been implemented in conventional setups for air pollution control such as biotrickling filters and biofilters (Table 5). However, other less conventional bioreactor configurations such as stirred tanks, airlifts and bubble columns have been also investigated (Fig. 5). Air treatment processes based on TPPBs were initially operated as two-stage systems coupling an abiotic VOC absorption step with VOC biodegradation in separated devices (Cesario et al. 1992; Yeom and Daugulis 2001). However, two-stage systems were rapidly replaced by one-step processes where absorption and biodegradation were performed in the same reactor (Muñoz et al. 2012). Difficulties in separating the active biomass from the NAP after the biodegradation stage likely prompted this change in the concept of TPPBs. Several studies indeed confirmed that biomass separation from the NAP is extremely difficult once hydrophobic microorganisms have colonized the organic

Table 5 Bioreactor configurations used for TPPB implementation

Reactor configuration	NAP	Target gas pollutant	References
Stirred tank	FC40	Toluene	Cesário et al. (1997)
Stirred tank	Silicone oil	Hexane	Muñoz et al. (2006)
Stirred tank	Silicone oil	Hexane	Arriaga et al. (2006)
Stirred tank	Hexadecane	Benzene	Nielsen et al. (2006)
Stirred tank	Hexadecane	Toluene	Boudreau and Daugulis (2006)
Stirred tank	FC40, heptamethylnonane	α-pinene	Muñoz et al. (2008)
Stirred tank	Silicone oil	Dichloromethane	Bailón et al. (2009)
Stirred tank	Silicone oil	Methane	Rocha-Rios et al. (2009)
Stirred tank	Silicone oil	Hexane	Hernández et al. (2010)
Stirred tank	Silicone oil	Hexane	Hernández et al. (2011b)
Stirred tank	Silicone oil	Methane	Ordaz et al. (2014)
Biotrickling filter	Silicone oil	Hexane	Van Groenestijn and Lake (1999)
Biotrickling filter	Silicone oil	Styrene	Djeribi et al. (2005)
Biotrickling filter	Silicone oil	dichloromethane	Bailón et al. (2009)
Biotrickling filter	Silicone oil	Methane	Rocha-Rios et al. (2009)
Biotrickling filter	Silicone oil	α-pinene	Montes et al. (2010)
Biotrickling filter	Silicone oil	Hexane	Hernández et al. (2011b)
Airlift	Silicone oil and rubber beads	Benzene, toluene, ethylbenzene, xylene	Littlejohns and Daugulis (2009)
Airlift	Silicone oil, heptamethylnonane, Elvax 880, Desmopan9370A, Kraton G1657	Methane	Rocha-Rios et al. (2011)

(continued)

Table 5 (continued)

Reactor configuration	NAP	Target gas pollutant	References
Airlift	Silicone oil	Hexane	Hernández et al. (2011b)
Biofilter	Silicone oil	Hexane	Fazaelipoor and Shojaosadati (2002)
Biofilter	Silicone oil	Hexane	Lebrero et al. (2014)
Biofilter	Silicone oil	Hexane	Arriaga et al. (2006)
Bubble column	Silicone oil	Styrene	Dumont et al. (2014)
Bubble column	Silicone oil	Hexane	Hashemi et al. (2012)

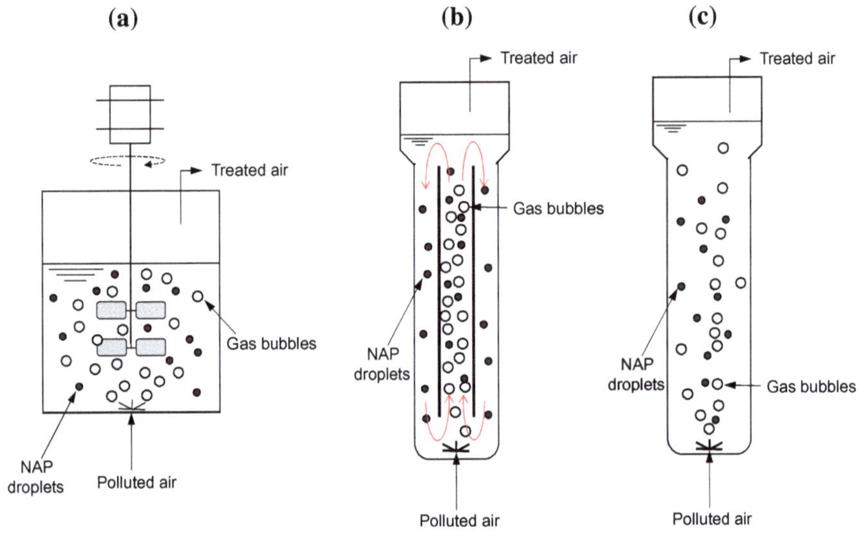

Fig. 5 Two-phase partitioning bioreactors operated as: **a** stirred tank, **b** airlift and **c** bubble column

phase (Ascon-Cabrera and Lebeault 1995; Hernández et al. 2012; Muñoz et al. 2013). On the other hand, NAP pumping represents additional energy requirements and results in more complex process control.

Stirred tank is the most common TPPB configuration used in lab-scale investigations. However, it is also the most energy-demanding reactor configuration

(Littlejohns and Daugulis 2009; Hernández et al. 2011b). Therefore, although the stirred tank is a good model for fundamental studies, this reactor configuration has no real potential for air treatment applications in TPPBs at full-scale (Lebrero et al. 2014). TPPBs operated as biotrickling filters have been successfully used for the removal of moderately hydrophilic and hydrophobic gas pollutants (Van Groenestijn and Lake 1999; Bailón et al. 2009). A recent study reported that biotrickling filters might support comparable pollutant mass transfer performance as stirred tanks (Hernández et al. 2011b). These authors showed that the amount of VOC transferred per volumetric power required was superior for the biotrickling filter configuration compared with the stirred tank using silicone oil as a NAP in both configurations. Likewise, TPPBs operated as biofilters exhibited an enhanced hexane abatement and also a more stable performance compared with the control biofilters without NAP (Arriaga et al. 2006; Lebrero et al. 2014). However, the packed bed of biofilters are particularly prone to clogging by excess of NAP, leading to decreases in the removal performance. Therefore, the amount of NAP added in biofilters must be carefully selected (Lebrero et al. 2014). As a rule of thumb, no more than 10 % of NAP relative to the working biofilter volume should be added.

Removal of benzene, toluene, ethylbenzene, xylene and methane has been performed in TPPBs operated as airlift reactors (Littlejohns and Daugulis 2009; Rocha-Rios et al. 2011). Although these authors concluded that the pollutant removal performance was superior in their stirred tank counterparts, the amount of VOC removed per volumetric power required was superior for the airlift configuration. TPPBs have also been implemented in bubble columns as a low energy-intensive system to treat hexane (Hashemi et al. 2012). These authors reached RE values of up to 76 % using silicone oil as the NAP, which constitutes a high biodegradation performance compared with other reactor configurations and further studies should be done to confirm such promising results. At this point it is important to stress that the development of TPPB configurations exhibiting a pollutant biodegradation-to-operating cost ratio comparable with conventional biofiltration systems such biotrickling filters and biofilters is an important research area to be pursued.

The volumetric power of the different air treatment technologies is a tool that can be used to compare quantitatively their energy requirements. Several studies report the volumetric power input (P_G/V) values for TPPBs implemented in several bioreactor configurations. However, an additional correction factor based on the liquid column and the gas holdup (ε) must be made in those bioreactors where a liquid phase (or liquid phases if a NAP is added) occupies most the working volume as in the case of stirred tanks, airlifts and bubble columns.

The reported P_G/V values for TPPBs implemented in stirred tanks range from 0.2 to 15 kW $m_{reactor}^{-3}$ (Littlejohns and Daugulis 2009; Quijano et al. 2010a). However, these P_G/V values only consider the energy required for the mechanical agitation and therefore, this value will be significantly increased if the power required for overcoming the liquid column height is considered. Thus, it must be taken into account that 1 m of water column equals 9.8 kPa of pressure drop (PD). Then, the PD value must be further corrected with the specific ε value as follows:

$$PD \times (1 - \varepsilon) \quad (9)$$

Likewise, the reported P_G/V values for TPPBs implemented in airlifts range of 0.02–0.06 kW $m^{-3}_{reactor}$ (Littlejohns and Daugulis 2009; Hernández et al. 2011b). However, these values must be corrected by considering the PD of the liquid column according to Eq. 9. Once the power consumption required to overcome the liquid columns is considered, fair comparisons on the energy requirements among the several bioreactor configurations can be done. Therefore, the development of innovative bioreactor configurations holding P_G/V values attractive for full-scale implementation (e.g. $P_G/V < 1$ kW $m^{-3}_{reactor}$) deserves further investigation (Muñoz et al. 2012).

4 Conclusions and Future Challenges

TPPBs constitute an innovative technology for the treatment of volatile organic compounds, being particularly suitable for the removal of volatile hydrocarbons. TPPBs support a proven superior performance and stability relative to their conventional biological counterparts. The full potential of TPPBs can be achieved by selecting adequate NAPs, process microbiology and reactor configuration, which must be based on the characteristics of off-air emission and the target VHs to be removed. Recent findings on microbial growth within liquid NAPs opens the possibility to further enhance the removal potential of hydrophobic VHs in TPPBs. However, research is still necessary to assess the performance of TPPBs operated with NAPs loaded with active biomass for the removal of mixtures of VHs containing both moderately hydrophilic and hydrophobic gas pollutants. Finally, the development of innovative bioreactor configurations and operating strategies focused on reducing the energy consumption is one of the most challenging aspects to be addressed in order to boost the application of TPPBs at full scale. Once TPPBs achieve energy requirements comparable to those of well-established biological systems such as biofilters or biotrickling filters, they will move from an innovative to a mature and widely applied technology for air treatment.

References

Alfonsín C, Lebrero R, Estrada JM et al (2015) Selection of odour removal technologies in wastewater treatment plants: a guideline based on Life Cycle Assessment. J Environ Manage 149:77–84. doi:10.1016/j.jenvman.2014.10.011

Arriaga S, Muñoz R, Hernández S et al (2006) Gaseous hexane biodegradation by *Fusarium solani* in two liquid phase packed-bed and stirred-tank bioreactors. Environ Sci Technol 40:2390–2395. doi:10.1021/es051512m

Ascon-Cabrera M, Lebeault J-M (1993) Selection of xenobiotic-degrading microorganisms in a biphasic aqueous-organic system. Appl Envir Microbiol 59:1717–1724

Ascon-Cabrera MA, Lebeault J-M (1995) Cell hydrophobicity influencing the activity/stability of xenobiotic-degrading microorganisms in a continuous biphasic aqueous-organic system. J Ferment Bioeng 80:270–275. doi:10.1016/0922-338X(95)90828-N

Atkinson R (2000) Atmospheric chemistry of VOCs and NO_x. Atmos Environ 34:2063–2101. doi:10.1016/S1352-2310(99)00460-4

Bailón L, Nikolausz M, Kästner M et al (2009) Removal of dichloromethane from waste gases in one- and two-liquid-phase stirred tank bioreactors and biotrickling filters. Water Res 43:11–20. doi:10.1016/j.watres.2008.09.031

Baumann MD, Daugulis AJ, Jessop PG (2005) Phosphonium ionic liquids for degradation of phenol in a two-phase partitioning bioreactor. Appl Microbiol Biotechnol 67:131–137. doi:10.1007/s00253-004-1768-2

Boudreau NG, Daugulis AJ (2006) Transient performance of two-phase partitioning bioreactors treating a toluene contaminated gas stream. Biotechnol Bioeng 94:448–457. doi:10.1002/bit.20876

Bowker RP (2000) Biological odour control by diffusion into activated sludge basins. Water Sci Technol 41:127–132

Bruce LJ, Daugulis AJ (1991) Solvent selection strategies for extractive biocatalysis. Biotechnol Prog 7:116–124. doi:10.1021/bp00008a006

Burgess JE, Parsons SA, Stuetz RM (2001) Developments in odour control and waste gas treatment biotechnology: a review. Biotechnol Adv 19:35–63. doi:10.1016/S0734-9750(00)00058-6

Cabrol L, Malhautier L (2011) Integrating microbial ecology in bioprocess understanding: the case of gas biofiltration. Appl Microbiol Biotechnol 90:837–849. doi:10.1007/s00253-011-3191-9

Cesario MT, Beeftink HH, Tramper J (1992) Biological treatment of waste gases containing poorly-water soluble compounds. In: Dragt AJ, van Ham J (eds) Biotech. Air Pollut. Abat. Odour Control Policies. Elsevier Science Publishers, Amsterdam, pp 135–140

Cesário MT, Beverloo WA, Tramper J, Beeftink HH (1997) Enhancement of gas-liquid mass transfer rate of apolar pollutants in the biological waste gas treatment by a dispersed organic solvent. Enzyme Microb Technol 21:578–588. doi:10.1016/S0141-0229(97)00069-0

Clarke KG, Williams PC, Smit MS, Harrison STL (2006) Enhancement and repression of the volumetric oxygen transfer coefficient through hydrocarbon addition and its influence on oxygen transfer rate in stirred tank bioreactors. Biochem Eng J 28:237–242. doi:10.1016/j.bej.2005.11.007

Cox HHJ, Deshusses MA (2001) Biotrickling Filters. In: Kennes C, Veiga MC (eds) Bioreact. Waste Gas Treat. Kluwer, Dordrecht, pp 99–131

Darracq G, Couvert A, Couriol C et al (2010a) Integrated process for hydrophobic VOC treatment-solvent choice. Can J Chem Eng. doi:10.1002/cjce.20325

Darracq G, Couvert A, Couriol C et al (2010b) Silicone oil: an effective absorbent for the removal of hydrophobic volatile organic compounds. J Chem Technol Biotechnol 85:309–313. doi:10.1002/jctb.2331

Darracq G, Couvert A, Couriol C et al (2012) Activated sludge acclimation for hydrophobic VOC removal in a two-phase partitioning reactor. Water Air Soil Pollut 223:3117–3124. doi:10.1007/s11270-012-1094-8

Daugulis AJ (2001) Two-phase partitioning bioreactors: a new technology platform for destroying xenobiotics. Trends Biotechnol 19:457–462. doi:10.1016/S0167-7799(01)01789-9

Daugulis AJ, Tomei MC, Guieysse B (2011) Overcoming substrate inhibition during biological treatment of monoaromatics: recent advances in bioprocess design. Appl Microbiol Biotechnol 90:1589–1608. doi:10.1007/s00253-011-3229-z

Davidson CT, Daugulis AJ (2003) Addressing biofilter limitations: a two-phase partitioning bioreactor process for the treatment of benzene and toluene contaminated gas streams. Biodegradation 14:415–421. doi:10.1023/A:1027363526518

Delhoménie M-C, Heitz M (2005) Biofiltration of air: a review. Crit Rev Biotechnol 25:53–72. doi:10.1080/07388550590935814

Djeribi R, Dezenclos T, Pauss A, Lebeault J-M (2005) Removal of styrene from waste gas using a biological trickling filter. Eng Life Sci 5:450–457. doi:10.1002/elsc.200520092

Dorado AD, Dumont E, Muñoz R, Quijano G (2015) A novel mathematical approach for the understanding and optimization of two-phase partitioning bioreactors devoted to air pollution control. Chem Eng J 263:239–248. doi:10.1016/j.cej.2014.11.014

Dumont E, Andrès Y (2012) Styrene absorption in water/silicone oil mixtures. Chem Eng J 200–202:81–90. doi:10.1016/j.cej.2012.06.028

Dumont E, Andrès Y, Le Cloirec P (2006) Effect of organic solvents on oxygen mass transfer in multiphase systems: application to bioreactors in environmental protection. Biochem Eng J 30:245–252. doi:10.1016/j.bej.2006.05.003

Dumont E, Andrès Y, Le Cloirec P (2014) Mass transfer coefficients of styrene into water/silicone oil mixtures: new interpretation using the "equivalent absorption capacity" concept. Chem Eng J 237:236–241. doi:10.1016/j.cej.2013.10.021

Easter C, Quigley C, Burrowes P et al (2005) Odor and air emissions control using biotechnology for both collection and wastewater treatment systems. Chem Eng J 113:93–104. doi:10.1016/j.cej.2005.04.007

EEA, European Environmental Agency (2014) European Union emission inventory report 1990–2012 under the UNECE Convention on Long-range Transboundary Air Pollution (LRTAP). http://www.eea.europa.eu/publications/lrtap-2014

Estrada JM, Kraakman NJRB, Lebrero R, Muñoz R (2012) A sensitivity analysis of process design parameters, commodity prices and robustness on the economics of odour abatement technologies. Biotechnol Adv 30:1354–1363. doi:10.1016/j.biotechadv.2012.02.010

Estrada JM, Kraakman NJRB, Muñoz R, Lebrero R (2011) A comparative analysis of odour treatment technologies in wastewater treatment plants. Environ Sci Technol 45:1100–1106. doi:10.1021/es103478j

Estrada JM, Lebrero R, Quijano G et al (2015) Odour abatement technologies in WWTP: energetic and economic efficiency. In: Tsagarakis KP, Stamatelatou K (eds) Sewage Treatment Plants: economic evaluation of innovative technologies for energy efficiency. IWA Publishing, London, pp 163–187

EUPO (2007) Directive 2004/42/CE of the European Parliament and the Council. In: Eur. Union Publ. Off. http://eur-lex.europa.eu/LexUriServ/LexUriServ.do?uri=OJ:L:2004:143:0087:0096:EN:PDF

Fam H, Daugulis AJ (2013) Two-phase partitioning bioreactors. In: Kennes C, Veiga MC (eds) Air pollution prevention and control: bioreactors and bioenergy. Wiley, Chichester, UK, pp 185–205

Fazaelipoor MH, Shojaosadati SA (2002) The effect of silicone oil on biofiltration of hydrophobic compounds. Environ Prog 21:221–224. doi:10.1002/ep.670210410

Gabriel D, Deshusses MA (2003) Performance of a full-scale biotrickling filter treating H_2S at a gas contact time of 1.6 to 2.2 seconds. Environ Prog 22:111–118. doi:10.1002/ep.670220213

Guieysse B, Cirne MDTG, Mattiasson B (2001) Microbial degradation of phenanthrene and pyrene in a two-liquid phase-partitioning bioreactor. Appl Microbiol Biotechnol 56:796–802. doi:10.1007/s002530100706

Hansen N, Rindel K (2000) Bioscrubbing, an effective and economic solution to odour control at wastewater treatment plants. Water Sci Technol 41:155–164

Harz M, Knoche M (2001) Droplet sizing using silicone oils. Crop Prot 20:489–498. doi:10.1016/S0261-2194(01)00014-X

Hashemi SF, Goharrizi AS, Fazaelipoor MH (2012) Two liquid-phase bubble column bioreactors for the removal of volatile organic compounds from air streams. Asia-Pacific J Chem Eng 7:442–447. doi:10.1002/apj.591

Hernández M, Muñoz R (2011) Long-term influence of the presence of a non-aqueous phase on the cell surface hydrophobicity of Pseudomonas in two-phase partitioning bioreactors. Appl Microbiol Biotechnol 89:1573–1581. doi:10.1007/s00253-010-2975-7

Hernández M, Muñoz R, Daugulis AJ (2011a) Biodegradation of VOC mixtures of different hydrophobicities in two-phase partitioning bioreactors containing tailored polymer mixtures. J Chem Technol Biotechnol 86:138–144. doi:10.1002/jctb.2496

Hernández M, Quijano G, Muñoz R, Bordel S (2011b) Modeling of VOC mass transfer in two-liquid phase stirred tank, biotrickling filter and airlift reactors. Chem Eng J 172:961–969. doi:10.1016/j.cej.2011.07.008

Hernández M, Quijano G, Muñoz R (2012) Key role of microbial characteristics on the performance of VOC biodegradation in two-liquid phase bioreactors. Environ Sci Technol 46:4059–4066. doi:10.1021/es204144c

Hernández M, Quijano G, Thalasso F et al (2010) A comparative study of solid and liquid non-aqueous phases for the biodegradation of hexane in two-phase partitioning bioreactors. Biotechnol Bioeng 106:731–740. doi:10.1002/bit.22748

Jacquemin J, Husson P, Padua AAH, Majer V (2006) Density and viscosity of several pure and water-saturated ionic liquids. Green Chem 8:172–180. doi:10.1039/B513231B

Kampa M, Castanas E (2008) Human health effects of air pollution. Environ Pollut 151:362–367. doi:10.1016/j.envpol.2007.06.012

Kennes C, Abubackar HN, Veiga MC (2013) Biodegradation and bioconversion of volatile pollutants. In: Kennes C, Veiga MC (eds) Air pollution prevention and control: bioreactors and bioenergy. Wiley, Chichester, pp 19–30

Kennes C, Thalasso F (1998) Review: waste gas biotreatment technology. J Chem Technol Biotechnol 72:303–319. doi:10.1002/(SICI)1097-4660(199808)72:4<303:AID-JCTB903>3.0.CO;2-Y

Kennes C, Veiga MC (2013) Introduction to air pollution. In: Kennes C, Veiga MC (eds) Air pollution prevention and control: bioreactors and bioenergy. Wiley, Chichester, UK, pp 3–18

Khammar N, Malhautier L, Degrange V et al (2005) Link between spatial structure of microbial communities and degradation of a complex mixture of volatile organic compounds in peat biofilters. J Appl Microbiol 98:476–490. doi:10.1111/j.1365-2672.2004.02474.x

Khan FI, Ghoshal AK (2000) Removal of volatile organic compounds from polluted air. J Loss Prev Process Ind 13:527–545. doi:10.1016/S0950-4230(00)00007-3

Kraakman NJR, Cesca J (2012) Evaluating odour control technology using reliability and sustainability criteria. Water, November: 65–70

Kraakman NJR, Rocha-Rios J, van Loosdrecht MCM (2011) Review of mass transfer aspects for biological gas treatment. Appl Microbiol Biotechnol 91:873–886. doi:10.1007/s00253-011-3365-5

Lebrero R, Bouchy L, Stuetz R, Muñoz R (2011) Odor assessment and management in wastewater treatment plants: a review. Crit Rev Environ Sci Technol 41:915–950. doi:10.1080/10643380903300000

Lebrero R, Hernández M, Quijano G, Muñoz R (2014) Hexane biodegradation in two-liquid phase biofilters operated with hydrophobic biomass: effect of the organic phase-packing media ratio and the irrigation rate. Chem Eng J 237:162–168. doi:10.1016/j.cej.2013.10.016

Lippmann M (1989) Health effects of ozone a critical review. JAPCA 39:672–695. doi:10.1080/08940630.1989.10466554

Littlejohns JV, Daugulis AJ (2009) A two-phase partitioning airlift bioreactor for the treatment of BTEX contaminated gases. Biotechnol Bioeng 103:1077–1086. doi:10.1002/bit.22343

López JC, Quijano G, Pérez R, Muñoz R (2014) Assessing the influence of CH4 concentration during culture enrichment on the biodegradation kinetics and population structure. J Environ Manage 146:116–123. doi:10.1016/j.jenvman.2014.06.026

MacLeod CT, Daugulis AJ (2005) Interfacial effects in a two-phase partitioning bioreactor: degradation of polycyclic aromatic hydrocarbons (PAHs) by a hydrophobic Mycobacterium. Process Biochem 40:1799–1805. doi:10.1016/j.procbio.2004.06.042

Malhautier L, Khammar N, Bayle S, Fanlo J-L (2005) Biofiltration of volatile organic compounds. Appl Microbiol Biotechnol 68:16–22. doi:10.1007/s00253-005-1960-z

Melgarejo-Torres R, Castillo-Araiza CO, Dutta A et al (2015) Mathematical model of a three phase partitioning bioreactor for conversion of ketones using whole cells. Chem Eng J 260:765–775. doi:10.1016/j.cej.2014.08.097

Montes M, Daugulis AJ, Veiga MC, Kennes C (2011) Characterization of absorbent polymers for the removal of volatile hydrophobic pollutants from air. J Chem Technol Biotechnol 86:47–53. doi:10.1002/jctb.2517

Montes M, Veiga MC, Kennes C (2010) Two-liquid-phase mesophilic and thermophilic biotrickling filters for the biodegradation of alpha-pinene. Bioresour Technol 101:9493–9499. doi:10.1016/j.biortech.2010.07.101

Morrish JLE, Daugulis AJ (2008) Improved reactor performance and operability in the biotransformation of carveol to carvone using a solid-liquid two-phase partitioning bioreactor. Biotechnol Bioeng 101:946–956. doi:10.1002/bit.21957

Muñoz R, Arriaga S, Hernández S et al (2006) Enhanced hexane biodegradation in a two phase partitioning bioreactor: overcoming pollutant transport limitations. Process Biochem 41:1614–1619. doi:10.1016/j.procbio.2006.03.007

Muñoz R, Chambaud M, Bordel S, Villaverde S (2008) A systematic selection of the non-aqueous phase in a bacterial two liquid phase bioreactor treating alpha-pinene. Appl Microbiol Biotechnol 79:33–41. doi:10.1007/s00253-008-1400-y

Muñoz R, Daugulis AJ, Hernández M, Quijano G (2012) Recent advances in two-phase partitioning bioreactors for the treatment of volatile organic compounds. Biotechnol Adv 30:1707–1720. doi:10.1016/j.biotechadv.2012.08.009

Muñoz R, Gan EIHH, Hernández M, Quijano G (2013) Hexane biodegradation in two-liquid phase bioreactors: High-performance operation based on the use of hydrophobic biomass. Biochem Eng J 70:9–16. doi:10.1016/j.bej.2012.09.009

Muñoz R, Villaverde S, Guieysse B, Revah S (2007) Two-phase partitioning bioreactors for treatment of volatile organic compounds. Biotechnol Adv 25:410–422. doi:10.1016/j.biotechadv.2007.03.005

Nielsen DR, Sask KN, McLellan PJ, Daugulis AJ (2006) Benzene vapor treatment using a two-phase partitioning bioscrubber: an improved steady-state protocol to enhance long-term operation. Bioprocess Biosyst Eng 29:229–240. doi:10.1007/s00449-006-0071-2

Noyola A, Morgan-Sagastume JM, López-Hernández JE (2006) Treatment of biogas produced in anaerobic reactors for domestic wastewater: odor control and energy/resource recovery. Rev Environ Sci Bio/Technol 5:93–114. doi:10.1007/s11157-005-2754-6

Ordaz A, López JC, Figueroa-González I et al (2014) Assessment of methane biodegradation kinetics in two-phase partitioning bioreactors by pulse respirometry. Water Res 67C:46–54. doi:10.1016/j.watres.2014.08.054

Ostojic NL (1992) Activated sludge treatment for odor control. Biocycle 33:74

Ottengraf SP, Van Den Oever AH (1983) Kinetics of organic compound removal from waste gases with a biological filter. Biotechnol Bioeng 25:3089–3102. doi:10.1002/bit.260251222

Pérez MC, Alvarez-Hornos FJ, Portune K, Gabaldón C (2014) Abatement of styrene waste gas emission by biofilter and biotrickling filter: comparison of packing materials and inoculation procedures. Appl Microbiol Biotechnol. doi:10.1007/s00253-014-5773-9

Popov VO, Bezborodov AM, Cavanagh M, Cross P (2004) Evaluation of industrial biotrickling filter at the flexographic printing facility. Environ Prog 23:39–44. doi:10.1002/ep.10005

Quijano G, Chávez-Avila R, Muñoz R et al (2010a) KLa measurement in two-phase partitioning bioreactors: new insights on potential errors at low power input. J Chem Technol Biotechnol 85:1407–1412. doi:10.1002/jctb.2460

Quijano G, Couvert A, Amrane A (2010b) Ionic liquids: applications and future trends in bioreactor technology. Bioresour Technol 101:8923–8930. doi:10.1016/j.biortech.2010.06.161

Quijano G, Couvert A, Amrane A et al (2011) Potential of ionic liquids for VOC absorption and biodegradation in multiphase systems. Chem Eng Sci 66:2707–2712. doi:10.1016/j.ces.2011.01.047

Quijano G, Hernandez M, Thalasso F et al (2009a) Two-phase partitioning bioreactors in environmental biotechnology. Appl Microbiol Biotechnol 84:829–846. doi:10.1007/s00253-009-2158-6

Quijano G, Revah S, Gutiérrez-Rojas M et al (2009b) Oxygen transfer in three-phase airlift and stirred tank reactors using silicone oil as transfer vector. Process Biochem 44:619–624. doi:10.1016/j.procbio.2009.01.015

Quijano G, Hernandez M, Villaverde S et al (2010c) A step-forward in the characterization and potential applications of solid and liquid oxygen transfer vectors. Appl Microbiol Biotechnol 85:543–551. doi:10.1007/s00253-009-2146-x

Quijano G, Ordaz A, Muñoz R, Thalasso F (2010d) New insights on O2 uptake mechanisms in two-phase partitioning bioreactors. Biotechnol Lett 32:223–228. doi:10.1007/s10529-009-0146-7

Quijano G, Rocha-Ríos J, Hernández M et al (2010e) Determining the effect of solid and liquid vectors on the gaseous interfacial area and oxygen transfer rates in two-phase partitioning bioreactors. J Hazard Mater 175:1085–1089. doi:10.1016/j.jhazmat.2009.10.020

Ravindra K, Sokhi R, Van Grieken R (2008) Atmospheric polycyclic aromatic hydrocarbons: source attribution, emission factors and regulation. Atmos Environ 42:2895–2921. doi:10.1016/j.atmosenv.2007.12.010

Rene ER, Veiga MC, Kennes C (2013) Biofilters. In: Kennes C, Veiga MC (eds) Air pollution prevention and control: bioreactors and bioenergy. Wiley, Chichester, pp 59–119

Revah S, Morgan-Sagastume JM (2005) Biotechnology for odor and air pollution control. In: Shareefdeen Z, Singh A (eds) Biotechnology for odor and air pollution control. Springer, Berlin, pp 29–63

Rocha-Rios J, Bordel S, Hernández S, Revah S (2009) Methane degradation in two-phase partition bioreactors. Chem Eng J 152:289–292. doi:10.1016/j.cej.2009.04.028

Rocha-Rios J, Quijano G, Thalasso F et al (2011) Methane biodegradation in a two-phase partition internal loop airlift reactor with gas recirculation. J Chem Technol Biotechnol 86:353–360. doi:10.1002/jctb.2523

Sander R (1999) Compilation of Henry's law constants for inorganic and organic species of potential importance in environmental chemistry. http://www.henrys-law.org/henry.pdf. Accessed 6 Nov 2014

Schnelle JKB, Brown CA (2001) Air pollution control technology handbook. Florida, Boca Raton

Stuetz RM, Gostelow P, Burgess JE (2001) Odour perception. In: Stuetz RM, Frechen FB (eds) Odours wastewater treat Meas. Model. Control. IWA Publishing, London, pp 1–13

UNEP (2006) Screening Information Datasets (SIDS) for high volume chemicals. In: United Nations Environ. Program.—Chem. Branch. http://www.chem.unep.ch/irptc/sids/OECDSIDS/. Accessed 20 May 2003

USEPA (2015) US Environmental Protection Agency - Air Toxics Web Site. http://www.epa.gov/airtoxics/. Accessed 1 Mar 2015

Van Groenestijn JW, Lake ME (1999) Elimination of alkanes from off-gases using biotrickling filters containing two liquid phases. Environ Prog 18:151–155. doi:10.1002/ep.670180310

Vuong M-D, Couvert A, Couriol C et al (2009) Determination of the Henry's constant and the mass transfer rate of VOCs in solvents. Chem Eng J 150:426–430. doi:10.1016/j.cej.2009.01.027

Wagner M, Loy A, Nogueira R et al (2002) Microbial community composition and function in wastewater treatment plants. Antonie Van Leeuwenhoek 81:665–680. doi:10.1023/A:1020586312170

Witherspoon J, Easter C, Voigt R et al (2008) An odor control master planning approach to public outreach programs. Proceeding of 3rd IWA international conference of Odour VOCs. Meas. Regul. Control Tech. 2006:

Wu XM, Apte MG, Maddalena R, Bennett DH (2011) Volatile organic compounds in small- and medium-sized commercial buildings in California. Environ Sci Technol 45:9075–9083. doi:10.1021/es202132u

Yeom SH, Daugulis AJ (2001) Development of a novel bioreactor system for treatment of gaseous benzene. Biotechnol Bioeng 72:156–165

Yeom SH, Daugulis AJ, Nielsen DR (2010) A strategic approach for the design and operation of two-phase partitioning bioscrubbers for the treatment of volatile organic compounds. Biotechnol Prog 26:1777–1786. doi:10.1002/btpr.481

Yoshino K, Schmidt WF (2014) Ion mobilities in poly-dimethylsiloxane silicone oils. IEEE international Conference on Liquids Dielectric Curran Associates, Inc., Bled, Slovenia, pp 123–126

Biosurfactant Producing Bacteria from Hydrocarbon Contaminted Environment

Sam Joy, Tanvi Butalia, Shashi Sharma and Pattanathu K.S.M. Rahman

Abstract As a result of global industrialization and increasing population there has been an alarming increase in the global demands for energy which is being fulfilled by exploiting various natural resources significantly hydrocarbons. As a result enormous amounts of hydrocarbons and hydrocarbon-based products have been released into the environment, threatening health and sustainability of the ecosystem. These different types of hydrocarbon-contaminated environments vary in their microbial composition and serve as an excellent reservoir of microbial flora, with a potential to degrade hydrocarbons and produce biosurfactants. In this chapter, an overview of biosurfactant-producing microorganisms from hydrocarbon-contaminated environments and their role in utilisation and degradation of hydrocarbon compounds is presented. Micro-organisms growing in hydrocarbon-rich environments undergo many adaptations, such as production of biosurfactants, which increases access to these hydrophobic substrates. Industrially, biosurfactants, which constitute as a group of surface-active amphiphilic compounds, are of great significance as they are biodegradable and nontoxic compared to synthetic chemical surfactants. Thus, biosurfactants have found wide applications and are used in bioremediation, oil exploration and enhanced recovery, health care, oil and food processing industries.

Keywords Biosurfactant · Hydrocarbon · Microbial diversity · Production · Applications of biosurfactant

---

S. Joy · T. Butalia · S. Sharma (✉)
Amity Institute of Biotechnology, Amity University, Sector-125, Noida, UP 201303, India
e-mail: ssharma@amity.edu

S. Joy
e-mail: sjoy@amity.edu

T. Butalia
e-mail: tanvi.butalia@gmail.com

P.K.S.M. Rahman
School of Science and Engineering, Teesside University, Middlesbrough, UK
e-mail: p.rahman@tees.ac.uk

K. Heimann et al. (eds.), *Biodegradation and Bioconversion of Hydrocarbons*,
Environmental Footprints and Eco-design of Products and Processes,
DOI 10.1007/978-981-10-0201-4_8

1 Introduction

In the last 10 years (2004–2014), the alarming increase in the demand of energy by 22.4 % [10,556.6 mega tonne oil equivalent (MTOE) in 2004 and 12,928.4 MTOE in 2014] has led to an increased exploration, production, and utilisation of hydrocarbons and hydrocarbon-based products (BP Statistical Review of World Energy June 2015). These compounds therefore continuously enter the environment; either as natural products originating from animals, plants and microorganisms such as steroids, terpenes and waxes or as pollutants by anthropogenic activities from the oil and petrochemical industries. The chemical composition and physical properties of the hydrocarbon contaminants may vary significantly depending on: the stability of the contaminant, site of contamination, anthropogenic activities, applications and the physico–chemical conditions of the environment present at the site of contamination (Christofi and Ivshina 2002; Valentin et al. 2013).

Microbial degradation is the major and ultimate natural mechanism to clean up hydrocarbon pollutants from the environment, as they colonize almost all ecological niches as a result of their metabolic versatility and adaptability to different carbon sources. Many of these hydrocarbon pollutants are characterized by low aqueous solubility and high solid–water distribution ratios, which limits their interactions with microbial cells which use molecules dissolved in the water phase (Johnsen et al. 2005; Cameotra and Makkar 2010). Substrate bioavailability is therefore a significant factor, affected by complex processes involving: physical parameters, chemical substrate characteristics, the environment or matrix (i.e. water, soil, sediment, organic matter), kinetic parameters (i.e. diffusion and flow rate, mass transfer, spatial separation between cells and substrate) and the intrinsic physiological properties of the cells (Johnsen et al. 2005).

The susceptibility of different hydrocarbons to microbial attack and their subsequent utilisation depends upon the nature and composition of the hydrocarbon present, and is variable from 6 to 82 % for soil fungi, 0.13 to 50 % for soil bacteria, and 0.003 to 100 % for marine bacteria (Das and Chandran 2011). It has been reported that mixed populations and adapted communities result in almost complete mineralisation of complex mixtures of hydrocarbons in soil, fresh and sea water compared to individual strains or communities with no history of hydrocarbon contamination.

Certain polycyclic aromatic hydrocarbons (PAHs) which are recalcitrant to microbial degradation require solubilisation by surfactants and emulsifiers having both hydrophilic and hydrophobic domains to interact with the complex hydrocarbon to further facilitate and enhance the degradation the hydrocarbon. The production of biosurfactants facilitates the dispersion of hydrophobic materials into aqueous phases leading to bioavailability of the hydrocarbon for the surfactant-producing microorganisms (Assadi et al. 2004; Adams et al. 2011).

Because of concern in relation to environmental issues, the use of microbiologically produced surfactants or biosurfactants has steadily increased, offering numerous advantages such as: low toxicity; high biodegradability; tolerance to broad temperature, pH, and salinity ranges; excellent surface and interfacial activity at low concentrations (Soberon-Chavez and Maier 2011).

Biosurfactants are a heterogeneous group of surface-active molecules produced by microorganisms either directly on microbial cell surfaces or by extracellular secretion. They have an amphipathic configuration. The hydrophilic moiety consists of mono, oligo or polysaccharides, peptides or proteins and the hydrophobic moiety usually contains saturated, unsaturated and hydroxylated fatty acids or fatty alcohols. This structural uniqueness allows them to spontaneously associate between liquid interfaces (oil/water and water/oil) to form micelles and to exhibit specific properties as detergency, emulsification, lubrication, foaming ability, solubilisation and phase dispersion (Banat et al. 2000). The physiological role of biosurfactants include: (1) increasing the surface area of hydrophobic growth substrates; by reducing the surface tension and interfacial tension between the cell wall and surrounding environment (2) desorbing hydrophobic compounds from surfaces by increasing their apparent solubility (3) induce membrane modifications like increasing the hydrophobicity by reducing the lipopolysaccharide content of the cell wall; enhancing the dispersion of hydrocarbons by encapsulation into micelles (Aparna et al. 2011).

2 Classification and Properties of Biosurfactants

Microbial surface-active compounds (SACs) are amphiphilic metabolites consisting of hydrophobic and hydrophilic moieties which allow them to partition at the interface between fluid phases with different degrees of polarity. These compounds comprise a wide variety of structures, and can be divided into two main groups: low molecular weight and high molecular weight SACs. Low molecular weight molecules reduce surface and interfacial tension as their primary activity. They include lipopeptides and glycolipids among others, and their molecular masses range from 500 to 1500 Da. On the other hand, high molecular weight polymers, so-called bioemulsifiers, are more effective in forming and stabilizing oil in water or water in oil emulsions, but do not necessarily reduce surface or interfacial tension.

Low molecular weight biosurfactants are usually categorized on the basis of their chemical composition and their microbial origin. The hydrophilic moiety consists of amino acids or peptides anions or cations; mono-, di-, or polysaccharides; and the hydrophobic moiety consists of unsaturated, saturated, or fatty acids. Accordingly, the major classes of biosurfactants include glycolipids, lipopeptides and lipoproteins, phospholipids, polymeric surfactants, and particulate surfactants (Hausmann and Syldatk 2014; Gudina et al. 2015a).

2.1 Glycolipids

Glycolipids are the most common types of biosurfactants. The carbohydrate part is composed of mono, di, tri and tetrasaccharides as glucose, mannose, galactose, glucuronic acid, rhamnose, and galactose-sulphate linked to the fatty acid component comprised of long chain aliphatic acids or hydroxyl-aliphatic acids via an ether or ester group (Jennings and Tanner 2000). The glycolipids can be categorized as follows:

2.1.1 Rhamnolipids

Bacteria of the genus *Pseudomonas* are known to produce glycolipid surfactant containing rhamnose and β-hydroxy decanoic acid. The –OH group of β-hydroxydecanoic acid is involved in glycosidic linkage with the reducing end of the rhamnose sugar and the –OH group of the second β-hydroxydecanoic acid is involved in ester formation. Rhamnolipids produced by *Pseudomonas aeruginosa* have been widely studied and reported as a mixture of homologous species RL1 ($RhC_{10}C_{10}$), RL2 (RhC_{10}), RL3 ($Rh_2C_{10}C_{10}$) and RL4 (Rh_2C_{10}) (Rahman et al. 2002b; Li et al. 2011; Gudina et al. 2015b). They possess a low average minimum surface tension (30–32 mN/m); high average emulsifying activity (10.4–15.5 U/ml filtrate), low critical micelle concentration (CMC) (5–65 mg/L) and a high affinity for hydrophobic organic molecules (Van-Dyke et al. 1993) (Fig. 1a, 1b).

2.1.2 Sophorolipids

These are the complex mixtures of 6–9 varied hydrophobic units consisting of a dimeric sugar (sophorose) and a hydroxyl fatty acid linked by a β-glycosidic bond. They are commonly produced by *Torulopsis* sp. and exist in two forms: the acid form which has free carboxylic acid as a tail and the lactone form which consists of a lactone ring (macrolactones) formed by the esterification between the carboxylic end of the fatty acid and 4″—hydroxyl group of the sophorose sugar (VanBogaert et al. 2014).

Generally lactonic sophorolipids have better surface tension lowering properties and can reduce the interfacial tension of n-hexadecane and water from 40 to 5 mN/m, whereas the acidic sophorolipids have better potential to form foam and possess superior solubility properties (Mulligan and Gibbs 2004).

2.1.3 Trehalose Lipids

Trehalose lipids, generally produced by Gram-positive bacteria are comprised of a disaccharide, trehalose, linked by an ester bond to α-branched *β*-hydroxy fatty acids at the C6 and C6′ of the carbohydrate structure in the case of the trehalose dimycolates and at C6 for the monomycolates (Lang and Philp 1998). Trehalose biosynthesis has been mainly explored in *Mycobacterium*, *Escherichia coli*, *Corynebacterium* and *Propionibacterium freudenreichii* (Ruhal et al. 2013) (Fig. 1c).

2.2 Lipoproteins and Lipopeptides

A great number of cyclic lipopetides, including decapeptide antibiotics (gramicidins) and lipopeptide antibiotics (polymyxins) are produced by micro-organisms. Microbial lipopeptides are generally cyclic peptides that are acylated with fatty acids. These contain a lipid linked to a polypeptide chain and are produced by *Bacillus* and *Pseudomonas* sp. *Bacillus* cyclic lipopeptides consist of three major groups known as the surfactin, iturin and fengycin. One of the most promising potential biosurfactants is surfactin synthesised by *Bacillus subtilis* ATCC 21332. It is made up of a cyclic heptapeptide (Glu-Leu-D-Leu-Val-Asp-D-Leu-Leu) linked to a fatty acid moiety of 13–15 carbons by lactone linkage and has the ability to reduce water surface tension from 72 to 27 mN/m at a critical micelle concentration (CMC) as low as 20 μM at a concentration as low as 0.005 % (Peypoux et al. 1999; Mnif and Ghribi 2015) (Fig. 1e).

2.3 Fatty Acids and Phospholipids

The genus *Thiobacillus*, *Aspergillus*, *Candida*, *Cornynebacterium*, *Micrococcus* and *Acinetobacter* have been reported as producers of phospholipids and fatty acids (Liu et al. 2015). The hydrophilic and lipophilic balance (HLB) is directly related to the length of the hydrocarbon chain in their structures. In *Acinetobacter* sp. strain HO1-N, phosphatidyl ethanolamine-rich vesicles are produced, which form optically clear micro-emulsions of alkanes in water. Phosphatidyl-ethanolamine produced by *Rhodococcus erythropolis* grown on alkane causes a lowering of interfacial tension between water and hexadecane to less than 1 mN/m at a CMC of 30 mg/L (Sanket and Yagnik 2013).

2.4 Polymeric Biosurfactants

Polymeric biosurfactants are high-weight molecular biopolymers consisting of polysaccharides, proteins, lipopolysaccharides, lipoproteins or mixtures of these. The best known polymeric biosurfactants are emulsan, liposan, mannoprotein and biodispersan produced by *Acinetobacter* sp., *Candida lipolytica*, *Schizonella malanogramma, Ustilago maydis, Pseudomonas* sp. and *Saccharomyces cerevisiae*. These polymeric biosurfactants exhibit certain physical properties, viz. high viscosity, tensile strength and resistance to shearing (Fig. 1d).

2.5 Particulate Biosurfactants

Extracellular membrane vesicles partition hydrocarbons to from a microemulsion, which plays a very important role in alkane uptake by microbial cells. Vesicles of *Acinetobacter* sp. H01-N have a diameter of 20–50 nm and a buoyant density of 1.158 g/cm^3, which consists of protein, phospholipids and lipopolysaccharide (Gharaei-Fathabad 2011).

3 Distribution of Biosurfactant Producing Microorganisms in the Hydrocarbon-Contaminated Environment

Biosurfactants seem to confer an essential evolutionary advantage allowing microorganisms to grow under specific conditions, as evidenced by their wide distribution across the eubacterial and archeal domains (Bodour et al. 2003; Lenchi et al. 2013). Biosurfactant-producing microorganisms are ubiquitous in nature, inhabiting both water (sea, freshwater, groundwater) and land (soil, sediment, sludge) as well as extreme environments (e.g. oil reservoirs, shallow low-temperature shale reservoirs) and thrive at a wide range of temperatures, pH values and salinity. Of the screened population, 25 % is contained within polluted soils as compared to 2–3 % in uncontaminated soils. Enrichment culture techniques specific for hydrocarbon-degrading bacteria leads to a much higher detection of biosurfactant producers, with estimates of up to 80 % (Rahman et al. 2002a) with 95 % being Gram-positive isolates from undisturbed and metal-contaminated soils and 83 % being Gram-negative isolates from hydrocarbon-contaminated and co-contaminated soils (Bodour et al. 2003). In soils and sediments; dominant genera are *Pseudomonas*, *Bacillus*, *Sphingomonas* and in marine ecosystems they include *Actinobacteria* and *Pseudoalteromonas*, *Halomonas*, *Alcanivorax* and *Acinetobacter*. The occurrence of these common micro-organisms in different environments can be explained on the basis that the diversity of a microbial community, which is a sensitive index of pollution, decreases in communities under stress, such as soils and sediments contaminated with hydrocarbon compounds (Bento et al. 2005). Table 1 highlights the isolation of biosurfactant-producing micro-organisms from different hydrocarbon-contaminated environments.

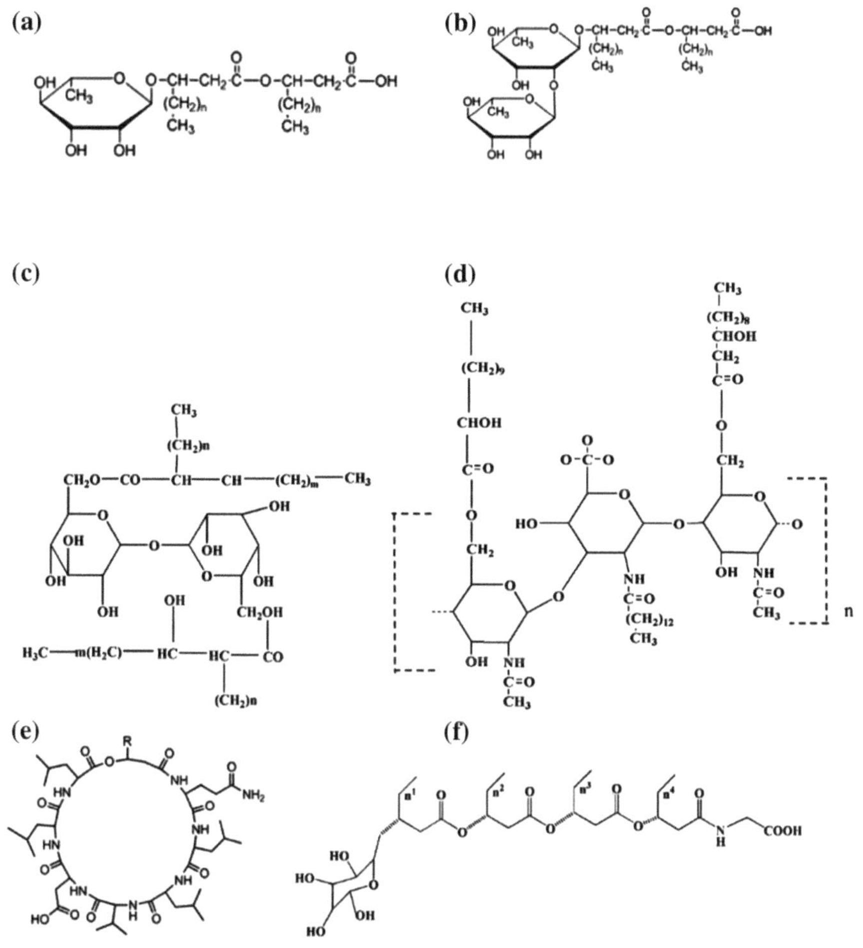

Fig. 1 Structures of biosurfactants and bioemulsifiers from the most representative hydrocarbon-degrading microorganisms. **a** L-rhamnosyl-3-hydroxydecanoyl-3-hydroxydecanoate (mono-rhamnolipid-Cl0–C10) from *Psendomonas aeruginosa*, where n = (6–18). **b** L-rhamnosyl-L-rhamnosyl-3-hydroxydecanoyl-3-hydroxydecanoate (di-rhamnolipid-Cl0–C10) from *Pseudomonas aeruginosa*, where n = (6–18). **c** Trehalosedicorynomycolate from *Rhodococcus* sp. where m + n = (27–31). **d** Emulsan from *Acinetobacter* RAG-l. **e** Surfactin from *Bacillus* sp., **f** Glucolipid from *Alcanivorax borkumensis* (Perfumo et al. 2010)

4 Metabolism of Biosurfactant Production

The structural diversity of biosurfactants produced by different genera implies molecular differences with regards to biosynthetic capacity. Studies on molecular genetics and biochemistry of the synthesis of several biosurfactants have revealed the operons, enzymes and metabolic pathways responsible for biosurfactant

Table 1 Isolation of biosurfactant-producing micro-organisms from different hydrocarbon environments

Micro-organisms	Sources of isolation	References
Pseudomonas aeruginosa HATH	Petroleum-contaminated garage site	Amani (2015)
Bacillus amyloliquefaciens An6	Soil sample from detergent industry	Ayed et al. (2015)
Bacillus thuringiensis pak2310	Petroleum-contaminated soil	Deepak and Jayapradha (2015)
Bacillus licheniformis MTCC 5514	Marine samples	Kavitha et al. (2014)
Pseudomonaas aeruginosa ATCC 33988 *Marinobacter hydrocarbonoclasticus*	Fuel storage tank and hydrocarbon-polluted marine environment	Striebich et al. (2014)
Acinetobacter calcoaceticus BS *Alcanivorax dieselolei* PG-12	Oil contaminated sites of the Persian Gulf, and the Caspian Sea	Hassanshahian et al. (2012)
*Gallaecimonas pentaromativorans*O15 *Brevibacterium epidermidis* N3 *Psychrobacter namhaensis* I36 *Pseudomonas fluorescens* T26 *Shewanella alga* N4 *Shewanella upenei* E14 *Vibrio furnissii* U15	Sediments and seawater samples located near to petrochemical plants and oil refinery	Hassanshahian (2014)
Alcanivorax sp. *Exiguobacterium* sp. *Halomonas* sp. *Rhodococcus* sp. *Bacillus* sp. *Acinetobacter* sp. *Pseudomonas* sp. *Streptomyces* sp.	Cold marine environments contaminated by petroleum hydrocarbon in the northern Atlantic, Canada	Cai et al. (2014)
Pseudomonas xanthomarina KMM 1447 *Pseudomonas stutzeri* ATCC 17588	Oil-polluted area in Gemsa Bay, the Red Sea, Egypt	El-Sheshtawy et al. (2014)
Bacillus methylotrophicus USTBa	Hydrocarbon-contaminated aqueous medium	Chandankere et al. (2014)
Pseudomonas sp. J4AJ	From roots of plant (*Scirpus triqueter*) growing on diesel-contaminated phytoremediation site	Zhang et al. (2014)
Pseudomonas sp. WJ6	Heavy oil-contaminated soil from oil well	Xia et al. (2014)
Serratia marcescens EU555434 *Aeromonas hydrophila* KF049214 *Bacillus cereus* KJ605415	Oil- and grease-contaminated industrial waste waters	Affandi et al. (2014)

(continued)

Table 1 (continued)

Micro-organisms	Sources of isolation	References
Cellulophaga sp, *Cobetia* sp. *Thalassospira* sp. *Pseudovibrio* sp. *Idiomarina* sp. *Cohaesibacter* sp.	Crude oil enrichments polychaetes	Rizzo et al. (2013)
Pseudomonas sp. *strain* GBS.5	Dye industry	Singh and Cameotra (2013)
Bacillus sp. I-15	Contaminated water, soil samples from refinery	Ismail et al. (2013)
Bacillus methylotrophicus	Petroleum reservoir samples	Chandankere et al. (2013)
Bacillus amyloliquefaciens BZ-6	Oily sludge, petroleum-contaminated soil	Liu et al. (2012)
Cronobacter sakazakii *Klebsiella strain* RJ-03	Oil-contaminated wastewaters	Jain et al. (2012; 2013)
Pseudomonas sp. BP10 *Rhodococcus* sp. NJ2	Crude oil-contaminated site	Kumari et al. (2012)
Staphylococcus sp. 1E	Crude oil-contaminated soil	Eddouaouda et al. (2012)
Bacillus subtilis LSFM-05	Hydrocarbon-contaminated site of a petroleum refinery	De Faria et al. (2011)
Stenotrophomonas acidaminiphila BB5 *Bacillus megaterium* BB6 *Bacillus cibi* *Pseudomonas aeruginosa* *Bacillus cereus* BS20	Petrochemical residues	Cerqueira et al. (2011)
Ochrobactrum sp. 1C *Brevibacterium* sp. 7G	Crude-oil-contaminated soil	Ferhat et al. (2011)
Escherichia fergusonii KLU01	Oil-contaminated soil	Sriram et al. (2011)
Paenibacillus alvei	Oil well sample	Najafi et al. (2011)
Pseudomonas aeruginosa S6	Oil-containing wastewater	Yin et al. (2009)
Pseudomonas aeruginosa strain NY3	Petroleum-contaminated soil samples	Nie et al. (2010)
Bacillus mycoides	Oil field	Najafi et al. (2010)
Pseudomonas aeruginosa sp.	Petroleum-contaminated soil samples	Bordoloi and Konwar (2009)
Gordonia amicalis LH3	Oil contaminated water samples	Hao et al. (2008)
Bacillus subtilis PT2 *Pseudomonas aeruginosa* SP4	Sludge oil and petroleum-contaminated soil	Pornsunthorntawee et al. (2008a)
Bacillus circulans	Marine biosurfactant producing strain	Das et al. (2008b)
Bacillus subtilis DM-04	Petroleum-oil-contaminated soil from North-East India	Das and Mukherjee (2007)

(continued)

Table 1 (continued)

Micro-organisms	Sources of isolation	References
Bacillus strain JF2	oil well injection brine site	Folmsbee et al. (2006)
Aeromonas sp.	Lagoon site contaminated with crude oil and diesel	Ilori et al. (2005)
Pseudoxanthomonas kaohsiungensis sp.	Waste water from oil-polluted site	Chang et al. (2005)
Bacillus thermoleovorans	Deep subterranean petroleum reservoirs	Kato et al. (2001)

secretion. Detailed research of biosurfactant production would facilitate the development of cost-effective methods for industrial-scale production, and to genetically engineer organisms for specific customised applications (Das et al. 2008a; Banat et al. 2014b). Among all biosurfactants reported to date, the earliest and best understood molecular biosynthetic regulation of all biosurfactants are those of rhamnolipids and surfactin.

4.1 Biosynthesis of Rhamnolipids

Two primary metabolic pathways are involved in the synthesis of the hydrophobic and hydrophilic moieties of biosurfactants. The pathways for the synthesis of these two groups of precursors are diverse and utilise specific sets of enzymes. According to Syldatk and Wagner (1987), it was initially hypothesised that the synthesis of different moieties of biosurfactant (especially of glycolipids) and their linkage follows one of the four possible pathways (Hommel and Ratledge 1993; Desai and Banat 1997).

- Both moieties are independently synthesised by two different de novo pathways
- Hydrophilic moiety synthesised by de novo pathway and hydrophobic induced by substrate
- Hydrophobic moiety synthesised by de novo pathway and hydrophilic moiety induced by substrate
- Synthesis of both biosurfactant moieties dependent upon substrates

The precursors for both the rhamnosyl and lipid moieties moieties are diverted from central metabolic pathways to the rhamnolipid biosynthesis pathways. The donor of the rhamnosyl moiety is activated deoxy-thymidine di-phospho-L-rhamnose (dTDP-L-rhamnose), produced from D-glucose and the lipid moiety precursor is activated 3-(3-hydroxyalkanoyloxy) alkanoate (HAA) diverted from de novo fatty acid synthesis (Zhang et al. 2012). The rhlG gene which is homologous to

NADPH-dependent keto-acyl reductase is responsible for obtaining the fatty acid precursors from the normal fatty acid biosynthetic pathway for synthesis of HAA. The rhamnolipids are synthesised from these two precursors by the action of three enzymes: RhlA, RhlB, and RhlC. A diversity of rhamnolipid congeners have been reported in *Pseudomonas aeruginosa*, varying in chain length and/or extent of saturation in the lipid moiety (Abdel-Mawgoud et al. 2010; Marchant and Banat 2012). Rhamnolipids biosynthesis occurs through three sequential reactions. The first reaction involves dimerization of two β-hydroxydecanoic acid chains catalyzed by RhlA. The dimer then undergoes two sequential rhamnosylation reactions with two different rhamnosyltransferases: rhamnosyltransferase 1 (Rt-1) and rhamnosyltransferase 2 (Rt-2). Rhamnosyltransferases RhlB and RhlC catalyse the transfer of dTDP-L-rhamnose to either HAA, or a previously generated mono-rhamnolipid, respectively (Deziel et al. 2003; Reis et al. 2011). The rhamnosyltransferase II (RhlC) adds another rhamnose moiety to the monorhamnolipid resulting in a di-rhamnolipid (Fig. 2a).

4.2 *Biosynthesis of Surfactins*

Surfactin biosynthesis is catalyzed non-ribosomally by a large multi-enzyme peptide synthetase complex called the surfactin synthetase. This surfactin synthetase consists of three protein subunits: SrfA, ComA (earlier known as SrfB) and SrfC.

The peptide synthetase required for amino acid moiety of surfactin is encoded by four ORFs in the srfA operon namely, SrfAA, SrfAB, SrfAC, SrfAD or SrfA-TE. SrfAA (SrfORF1) of the surfactin peptide synthetase contains the amino acid activating domains for the addition of Glu, Leu and D-Leu. SrfAB (SrfORF2) contains the amino acid activating domains for the addition of Val, Asp, and D-Leu; SrfAC contains the amino acid activating domain for Leu, and also encodes a thioesterase of a type I motif responsible for peptide termination. SrfAD has sequence homology with a thioesterase type II motif but is not found essential for surfactin biosynthesis, whereas the other three ORFs are absolutely essential for biosurfactant production.

The biosynthesis of surfactin is carried out by the thio-template mechanism, a step-by-step enzyme-catalysed reaction that involves attachment of the alpha-carboxyl of each amino acid to an SH group within the enzyme. The aminoacylation reaction is carried out by the adenylation domain followed by a thioesterification reaction according to the thiotemplate mechanism. The specific condensation domains will carry out aminoacylation at the N-terminus and additional condensations, leading to linear surfactin. The last steps are catalysed by the thioesterase (Te) domain to give the lipo-cyclic-heptapeptide (Shaligram and Singhal 2010) (Fig. 2b).

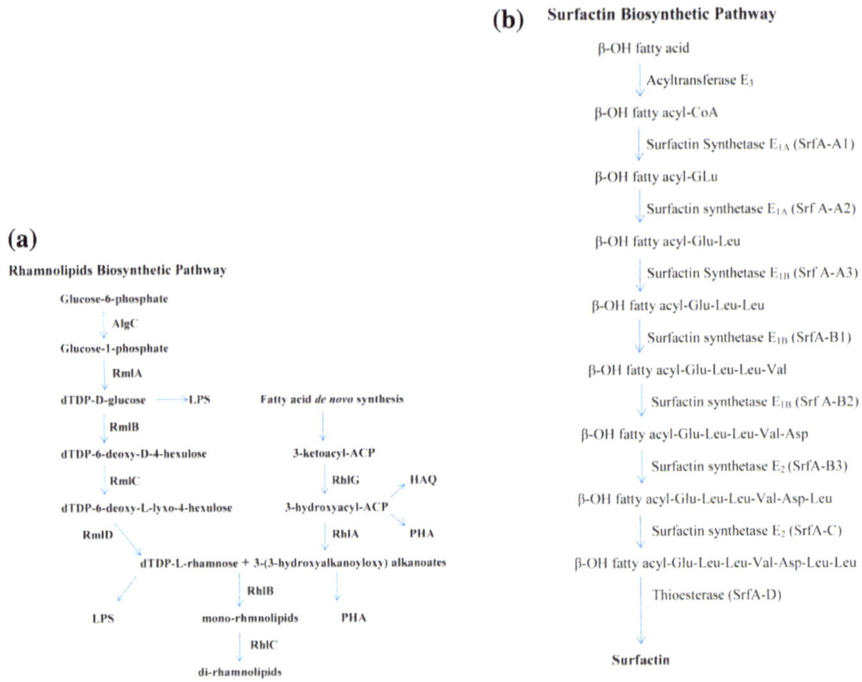

Fig. 2 **a** Rhamnolipid biosynthetic pathway (Zhang et al. 2012), **b** Surfactin biosynthetic pathway (Shaligram and Singhal 2010)

5 Quorum Sensing Associated Biosurfactant Production

In environments contaminated with hydrocarbons, cell densities are presumably low but the biosurfactant production is correspondingly high, which is presumably due to the creation of its own microenvironment by each cell within each micelle; thus stimulating biosurfactant production independent of increased cell density (Sullivan 1998). Bacteria behave as single cellular organism at low cell densities; however, once a threshold level in population density is reached, they shift their behavior to a 'multicellular' type (Atkinson and Williams 2009). At this stage, bacterial cells communicate via an intercellular signaling system termed 'quorum sensing' a mechanism by which bacterial cells regulate specific genes in response to critical concentrations of signal molecules called auto-inducers, produced as a function of cell density (Diggle et al. 2002). The detection of a minimal threshold stimulatory concentration of an auto-inducer leads the whole population to initiate a concerted action with an alteration in gene expression.

5.1 Quorum Sensing System Regulated Rhamnolipid Synthesis

Pseudomonas aeruginosa is one of the most extensively studied and reported strain for rhamnolipid production and quorum sensing systems (Reis et al. 2011). Two interrelated quorum sensing systems (QSS)—las and rhl operate in *P. aeruginosa* to regulate rhamnolipid synthesis and are present on two different regions of the chromosome. The first QSS regulating the production of the rhamnolipid in *P. aeruginosa* involves the *rhl* operon which contains four genes viz. rhlA, rhlB, RhlR and rhlI transcribed in 5′-rhlABRI-3′ direction. Genes rhlA and rhlB are located upstream while rhlR and rhlI are located downstream of the structural genes, and is transcribed in the same direction by a different promoter. Regulation of rhamnolipids is under the coordinated guidance of *rhlAB* genes that encodes a group of enzymes termed the rhamnosyltransferases 1 and rhamnosyltransferase 2. Genes rhlA, rhlB are positively regulated by rhlR, RhlA is presumably considered to be involved in the synthesis or transport of rhamnosyltransferase precursor substrates and in the stabilisation of the RhlB protein. The second rhamnosyltransferase is encoded by rhlC, and is reported to be coordinately regulated by rhlAB using a quorum sensing mechanism (Reis et al. 2011). Gene rhlI produces two signal molecules viz. *N*-butanoyl-L-homoserine (PAI-2) and hexanoyl-L-homoserine lactone. The transcriptional activator produced by rhlR binds to the auto-inducer PAI-2 and this active complex causes transcriptional activation of rhlA and rhlB that encodes rhmanosyltransferase I.

The second QSS contains two genes namely lasR and lasI. In this system, the auto-inducer, namely *N*-(3-oxododecanoyl) 1-homoserine lactone (3-oxo-C12-HSL), PAI-1 is encoded by lasI. RhlR regulatory protein requires the auto-inducers *N*-butryl-HSL and *N*-(3-oxohexanoyl)-HSL for its activity. Induction of the second QSS occurs by auto-induction from its own system and is dependent on cyclic AMP levels. The transcription of rhlR is positively regulated by the las system by hindering PAI-2 to bind to RhlR by PAI-1. The PAI-1 blocks PAI-2 from binding to RhlR until enough PAI-2 and/or RhlR is produced to overcome the blocking effect. Vfr induces lasR and the auto-inducer PAI-1 (encoded by lasI) binds to LasR and forms a LasR-PAI-1 complex (Dusane et al. 2010). This complex induces a negative regulator gene *rsaL*, *rhlR* and *pqsH*, required for PQS [2-heptyl-3-hydroxy-4-quinolone] production (Mashburn and Whiteley 2005). PQS directly or indirectly induces the *rhlI* gene and directs the production of PAI-2 (C4-HSL), the quorum sensing molecule that binds to and activates RhlR. The RhlR–PAI-2 complex interacts with the rhlA promoter (lux box) to begin transcription of the rhamnolipid producing rhlAB gene (Fig. 3a).

5.2 QSS Regulating Surfactin Synthesis

Bacillus subtilis is also known to regulate surfactin production by utilising a peptide pheromone, ComX based on a cell density-responsive mechanism and not on homoserine lactone (Menkhaus et al. 1993). The ComQ-XPA QS system of *B. subtilis* is a typical QS system of Gram-positive bacteria which controls the expression of nearly 200 genes (Comella and Grossman 2005). This QS system involves four proteins: the ComQ isoprenyl transferase, the ComX signal peptide, the ComP histidine kinase and the ComA response regulator. Quorum sensing controls the expression of srfA by ComX, which then interacts with ComP and ComA to activate the signal transduction system. In *Bacillus subtilis* the srfA gene is controlled by quorum sensing via a signal peptide (ComX). ComX accumulates in the medium when at high cell densities. ComX serves as a signal peptide after being modified by the gene product of comQ. ComX interacts at the cell membrane with a two-component signal transduction system, consisting of the membrane bound histidine kinase ComP and the response regulator ComA. Phosphorylation of the response regulator ComA by the histidine proteinkinase ComP results in its activation and binding to the promoter of srfA as a tetramer. This binding of ComA stimulates the transcription of the srf operon. The signaling peptide ComX is initially synthesized as a 55-residue pro-peptide which is processed and modified by the isoprenyl transferase ComQ (Bacon et al. 2002). The isoprenylated ComX is then secreted and upon reaching a critical concentration, it activates auto-phosphorylation of the membrane bound ComP, which then phosphorylates the transcriptional activator ComA (Ansaldi et al. 2002). Phosphorylated ComA directly modulates the expression of the srfA operon needed for non-ribosomal synthesis of the major lipopeptide surfactin including several other genes. Srf transcription is also activated by a pheromone CSF by inhibiting the ComA-phosphate phosphatase RapC. CSF is an extracellular peptide factor which has to be imported inside the cell accomplished by the oligopeptide permease Spo0K. ComR and SinR also influence srfA expression—ComR post-transcriptionally enhances srfA expression and SinR negatively controls srfA possibly by regulating comR (Shaligram and Singhal 2010) (Fig. 3b).

6 Degradation of Hydrocarbons

Biodegradation of hydrocarbons, both aliphatic and aromatic compounds, may occur under anaerobic or aerobic conditions (Peixoto et al. 2011). Depending on the chain length, enzyme systems are required to introduce oxygen in the substrate to initiate biodegradation. Cytochrome P450 alkane hydroxylases constitute a super family of ubiquitous heme-thiolate mono-oxygenases (introduce one oxygen atom) which play an important role in the aerobic microbial degradation of oil, chlorinated hydrocarbons, fuel additives, and many other compounds (Das and Chandran 2011). Under

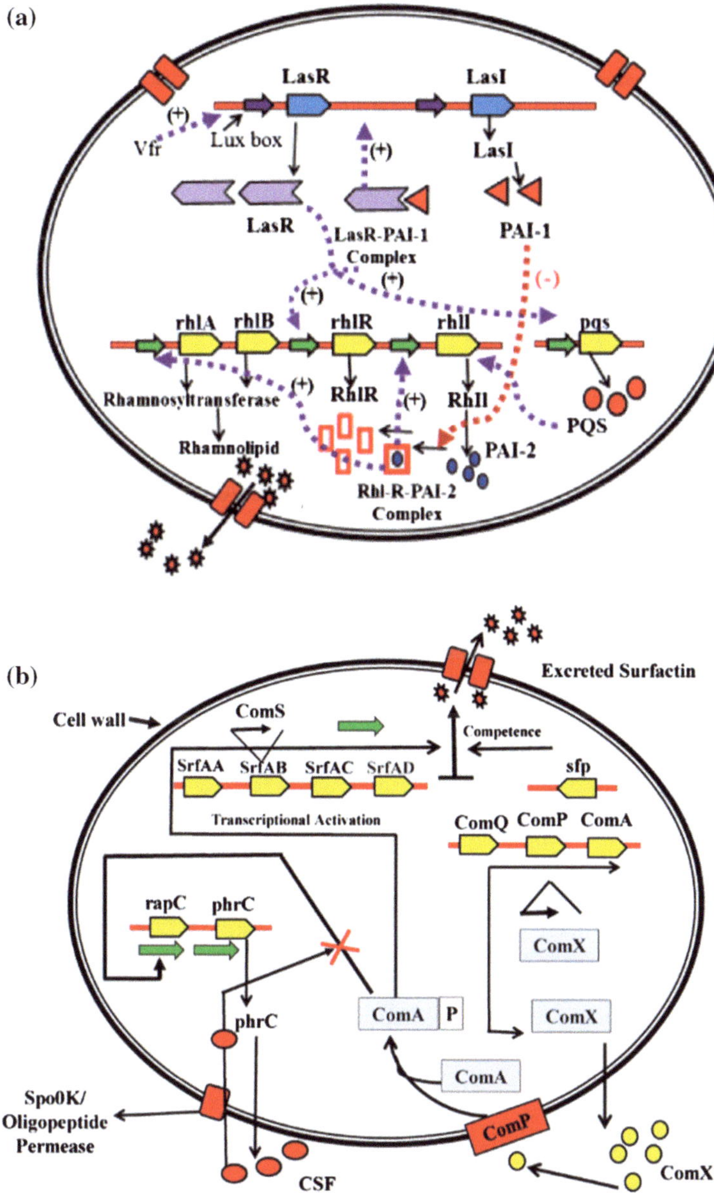

Fig. 3 **a** Quorum sensing regulation of rhamnolipid synthesis (Dusane et al. 2010). **b** Quorum sensing regulation of surfactin synthesis (Shaligram and Singhal 2010)

aerobic conditions the degradation of organic pollutants is rapid and almost complete. Peripheral degradation pathways convert organic pollutants step-by-step into intermediates of the central intermediary metabolism, as the tricarboxylic acid cycle (TCA). The final product of the oxidation of saturated aliphatic hydrocarbons; acetyl-CoA is catabolised in the TCA cycle together with the production of electrons in the electron transport chain. This chain is repeated, further oxidising the hydrocarbons fully to CO_2 (Sierra-Garcia et al. 2014). Biosynthesis of cell biomass occurs from the central precursor metabolites, as acetyl-CoA, succinate and pyruvate while the sugars required for various biosynthesis and growth are synthesised by gluconeogenesis. Aromatic hydrocarbons, such as benzene, toluene, xylene and naphthalene are also degraded in aerobic conditions to catechol or structurally related compounds in the initial steps. The catechol dioxygenase, classes of bacterial-iron containing enzymes, catalyse the addition of molecular oxygen atoms to 1, 2-dihydroxybenzene (catechol) and its derivatives, with subsequent cleavage of the aromatic ring (Cao et al. 2009). This is introduced into the TCA cycle and finally completely degraded to CO_2. In oxygen-limiting conditions as in mangroves, aquifers and sludge digesters, anaerobic bacteria catalyse the conversion of aromatic compounds into benzoyl-CoA which is reduced and de-aromatised in the presence of

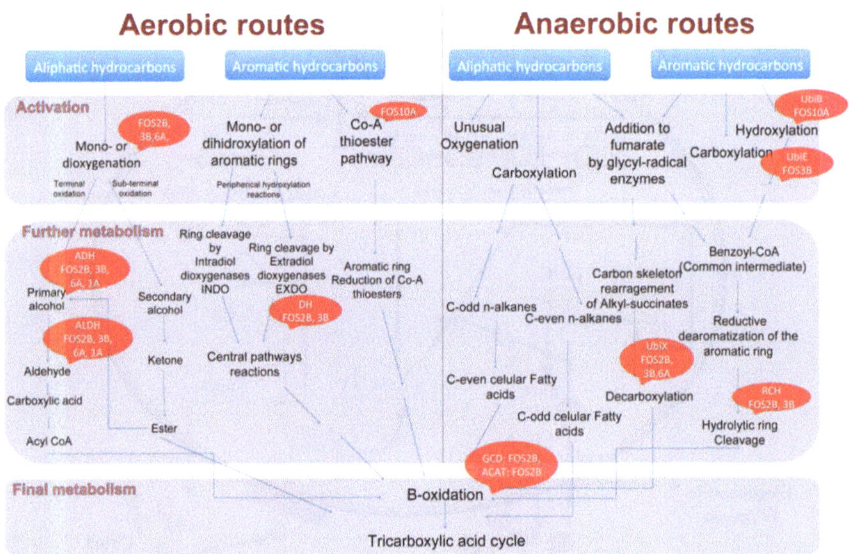

Fig. 4 Comprehensive map of biochemical pathways for aerobic and anaerobic bacterial degradation of hydrocarbon compounds. *Red bubbles* show which genes of these routes are present in one or more fosmids. Two consecutive *arrows* represent more than one reaction. (UbiB) Ubiquinone biosynthesis protein UbiB; (UbiE) Ubiquinone biosynthesis protein C-methyltransferase UbiE; (ADH) Alcohol dehydrogenase; (ALDH) Aldehyde dehydrogenase: (DH) Dienelactone hydrolase; (UbiX) 3-octoprenyl-4-hydroxybenzoate; (RCH) Ring cleavage hydrolase; (CCD) Glutaryl-CoA dehydrogenase; (ACAT) Acetyl-CoA acyltransferase (Sierra-Garcia et al. 2014)

benzoyl-CoA reductase (BCR) and terminal electron acceptors such as nitrate, sulphate or Fe (III), finally entering the TCA cycle after ring cleavage (Peixoto et al. 2011) (Fig. 4).

7 Microbial Growth Kinetics Related to Biosurfactant Production

The kinetics of biosurfactant production depends on the substrate used for the microbial growth. Biosurfactant production may occur simultaneously during the microbial active growth period or during the stage of limited growth under substrate scarcity (Desai and Banat 1997; Maciel et al. 2007). Accordingly, the kinetics of biosurfactant production has been categorised as:

- Growth-associated production
- Production under growth-limiting conditions
- Production by resting or immobilized cells
- Production with precursor supplementation

7.1 *Growth-Associated Production*

A parallel correlation of growth, substrate utilisation, and biosurfactant production is observed in case of growth-associated biosurfactant production. In growth-dependent biosurfactant production, the increase in cell density and the accumulation of metabolic products are strongly influenced by medium composition such as carbon and nitrogen sources (Mnif et al. 2011; Khopade et al. 2012; Hachaichi et al. 2014; Rufino et al. 2014). A mixed pattern of biosurfactant production i.e. growth-associated as well as non-growth have also been observed in some organisms. Goldman et al. (1982) and Gutnick et al. (1980) reported a mixed pattern of biosurfactant production in *Acinetobacter calcoaceticus* RAG-1 wherein emulsan or emulsan-like precursors accumulate as capsular material during the exponential growth phase and are subsequently released into the medium during the stationary phase.

7.2 *Production Under Growth-Limiting Conditions*

The production of biosurfactants under growth-limiting conditions is characterised by a sharp increase in biosurfactant levels as a result of limitation in one or more medium components in the late logarithmic and the stationary growth phase, in which specific enzymes are induced or depressed. A number of studies have reported enhanced production of biosurfactants under stress and nutrient-limiting conditions like nitrogen, phosphorus and iron limitations (Amezcua-Vega et al. 2004; Glick et al. 2010; Amani 2015). Hudak and Cassidy (2004) reported enhanced rhamnolipid accumulation of 1 g/L after 144 h with twenty times the CMC under nitrogen-limiting conditions. Chayabutra et al. (2001) reported fourfold to fivefold higher specific productivity of rhamnolipid from *Pseudomonas aeruginosa* under phosphorus limitation than the nitrogen limitation in non-growth associated production. Other than nitrogen, phosphorus, multivalent ions such as iron, magnesium, calcium, potassium, sodium and other trace elements also play an important role in growth limiting types of biosurfactant production (Desai and Banat 1997; Hemlata et al. 2015; Ismail et al. 2015). Iron forms a key microelement for biosurfactant production and is linked to activation of enzymes such as isocitrate lyase under iron-limiting conditions; this enzyme is vital in conversion of acetyl-CoA to a C4 unit during biosurfactant synthesis when grown on hydrophobic substrates (Maqsood and Jamal 2011). Schmidberger et al. (2014) reported an increase in relative expression of the genes rhlA, rhlC as well as the genes of the pqs quorum sensing regulon which are involved in rhamnolipid synthesis under iron-limiting conditions. Glick et al. (2010) demonstrated that under iron limitation, not only the production of rhamnolipids was enhanced but rather the time course of rhamnolipid expression shifted to the initial stages versus to the later stages in iron-depleted conditions.

7.3 Production by Resting or Immobilised Cells

Biosurfactant production by resting or immobilised cells requires no cell multiplication but the carbon source is continuously utilised for the synthesis of biosurfactants, thereby reducing the cost of product recovery. Since biosurfactant synthesis occurs as a secondary metabolite, the use of resting cells under suitable conditions of media compositions and physical parameters can result in the significant increase in the biosurfactant levels by maintaining stationary phase. Some examples of biosurfactant production by resting cells include: production of rhamnolipid by *Pseudomonas* sp. and *P. aeruginosa* CFTR-6; sophorolipid production by *Torulopsis bombicola* and *Candida apicola*; cellobiolipid production by *Ustilago maydis*; trehalose tetraester production by *Rhodococcus erythropolis* and mannosylerythritol lipid production by *Candida antarctica* (Desai and Banat 1997). Chtioui et al. (2010) reported the enhanced production of surfactin by immobilised cells of *Bacillus subtilis* ATCC 21332 in particles of polypropylene foamed with powder-activated carbon.

7.4 Production with Precursor Supplementation

The addition of biosurfactant precursors to the growth medium results in both qualitative and quantitative changes of the product. The precursors may be of any compound which can either induce or act as precursor molecules for the synthesis of either hydrophobic or hydrophilic moieties. Zhang et al. (2012) reported that the addition of octadecanoic fatty acid as co-substrate with glucose resulted in twofold increase in rhamnolipid yields. The increase in the production is suggested due to an activation of a "bypass route" through which an octanoyl-CoA intermediate of β-oxidation is incorporated into rhamnolipids without de novo fatty acid biosynthesis thereby enhancing the yield. Lotfabad et al. (2009) reported two-and-a-half-fold increase in the biosurfactant production (2.1 g/L) by incorporating (0.1 % w/v) isoleucine in MS medium from *P. aeruginosa* MR01 compared to the control without amino acid (0.8 g/L). It can be suggested that the observed increase may be as a result of direct uptake of amino acid as precursors for the surfactant biosynthesis. Similarly, supplementation of amino acids proved to be excellent nitrogen sources for the production of surfactin by *Bacillus subtilis* (Sandrin et al. 1990).

8 Parameters Influencing Biosurfactant Production

8.1 Nutritional Parameters

Biosurfactants are produced by a number of microorganisms, predominantly during their growth on water-immiscible substrates. The composition and characteristics of biosurfactants are influenced by the nature of carbon and nitrogen sources and the presence of iron, magnesium, manganese, phosphorus and sulphur ions in the media.

8.1.1 Carbon Source

The type of carbon sources which are generally reported to be used in the medium for the production of biosurfactants are soluble carbohydrates, lipids, vegetable oils, hydrocarbons oils, industrial effluents and lignocellulose. Gogotov and Khodakov (2008) reported that factors such as medium composition, the nature of the carbon source and oxygen supply significantly affect the production of biosurfactant. Singh et al. (2014a) suggested that substrate dependency has an influence on the structural and compositional diversity on biosurfactants produced; as in the presence of dextrose, sucrose and glycerol, lipopeptide biosurfactants were produced as a mixture of surfactin, iturin and fengycin; while only iturin was produced in presence of maltose, lactose and sorbitol. Wang et al. (2013) reported the dependence of biosurfactant production and cell surface hydrophobicity upon the chain lengths of

Table 2 Yield of microbial biosurfactants (BS) using different substrates

Micro-organism	Substrate	BS type	Properties	Yield	References
Hydrocarbons					
Acinetobacter sp. D3-2	Hydrocarbon substrates	Lipopeptide	26.3 mN/m	0.52 g/L	Bao et al. (2014)
Bacillus siamensis	Crude oil	Lipopeptide	36.1 mN/m	0.64 g/L	Varadavenkatesan and Murty (2013)
Pseudomonas aeruginosa C450R	Crude oil (2 %)	Rhamnolipid	35.1 mN/m	1.23 g/L	Mnif et al. (2011)
Sphingobacterium sp.	C_{11-13} (2 %)	ND	32.0 mN/m	0.19 g/L	Burgos-Diaz et al. (2011)
Rhodococcus erythropolis LMG 5359 *Rhodococcusruber* 5366	Naphthalene (1 %), Sodium nitrate (0.9 %) Diesel (1 %), sodium nitrate (0.7 %)	ND	ND	5.60 g/L 6.90 g/L	Haddadin et al. (2009)
Pseudomonas aeruginosa MTCC7812	Hexadecane (2 %)	Rhamnolipid	36.2 mN/m	4.06 g/L	Bordoloi and Konwar. (2008)
Pseudomonas aeruginosa EBN-8	Paraffin oil (1 %)	Rhamnolipid	29.0 mN/m	6.30 g/L	Raza et al. (2006)
Pseudomonas aeruginosa J4	Diesel	Rhamnolipid	31.0 mN/m, E_{24}(%) = 70	1.30 g/L	Wei et al. (2005)
Pseudomonas aeruginosa OBP1	n-hexadecane (2 %)	Rhamnolipid	31.1 mN/m; CMC = 45 mg/L	4.80 g/L	Bharali and Konwar (2011)
Vegetable oil					
Pseudomonas aeruginosa DSM50071	Waste fat from fish	Rhamnolipid	31.3 mN/m	0.42 g/L	Sana et al. (2015)

(continued)

Table 2 (continued)

Micro-organism	Substrate	BS type	Properties	Yield	References
Bacillus clausii BS02	Sunflower oil soap Stock (3 %)	Lipopeptide	30.0 mN/m; E_{24}(%) = 82 CMC = 45 mg/L	2.60 g/L	Hazra et al. (2014)
Aneurinibacillus thermoaerophilus MK01	Sunflower oil, sodium nitrate, yeast extract	Lipopeptide	43.0 mN/m; CMC = 1.2 mg/ml	6.75 g/L	Sharafi et al. (2014)
Cyberlindnera samutprakarnensis JP 52 (T)	Glucose (2 %w/v), Palm oil (2 %)	Sophorolipid	20.0 mN/m; CMC = 46 mg/L E_{24}(%) = 80	1.89 g/L	Poomtein et al. (2013)
Geobacillus stearothermophilus UCP 986	Palm Oil (7.5 %), CSL (4.5 %)	ND	27.0 mN/m; CMC = 40 mg/L	2.30 g/L	Jara et al. (2013)
Pseudomonas alcaligenes	Palm oil	Glycolipid	ND	2.30 g/L	Oliveira et al. (2009)
Pseudomonas fluorescens Migula 1895-DSMZ	Olive oil (2 %), Ammonium nitrate (10 %) and C/N 10	Rhamnolipid	38.0 mN/m, E_{24}(%) = 49	2.0 g/L	Abouseoud et al. (2008)
Pseudomonas aeruginosa SP4	Palm oil (2 %)	Rhamnolipid	29.0 mN/m, CMC = 200 mg/ml	5.20 g/L	Pornsunthorntawee et al. (2008b)
Pseudomonas aeruginosa EMS1	Acidified soyabean oil (2 %)	Rhamnolipid	ND	5.0 g/L	Cha et al. (2008)
Pseudomonas aeruginosa A41	Olive oil (2 %)	Glycolipid	ND	6.58 g/L	Thaniyavarn et al. (2006)
Pseudomonas aeruginosa DS10-129	Soyabean oil (0.6 %) + glucose (2 %)	Rhamnolipid	28.0 mN/m; E_{24}(%) = 70	4.31 g/L; 0.71 g/g Product yield	Rahman et al. (2002b)
Tsukamurella sp. DSM44370	Sunflower oil (1.1 %)	Glycolipid	30.0 mN/m	30.0 g/L; 0.27 g/g Product yield	Vollbrecht et al. (1999)

(continued)

Table 2 (continued)

Micro-organism	Substrate	BS type	Properties	Yield	References
Pseudomonas aeruginosa EBN-8 (mutant)	Soybean waste frying oil	Rhamnolipid	29.1 mN/m	9.30 g/L; 2.70 g/g/h	Raza et al. (2006)
Glycerol					
Pseudomonas aeruginosa BS161R	Bioglycerol (2.5 %)	Rhamnolipid	ND	3.31 g/L	Kumar et al. (2012)
Bacillus subtilis LAMI005	Glycerol	Lipopeptide	27.1 mN/m	0.44 g/L	Sousa et al. (2012)
Bacillus subtilis LSFM-05	Raw glycerol (5 %)	Lipopeptide	29.0 mN/m	0.93 g/L; 11.42 mg/L/h	De Faria et al. (2011)
Pseudomonas aeruginosa EM1	Glycerol (4 %)	Rhamnolipid	ND	4.9 g/L	Wu et al. (2008)
Soluble sugars and molasses					
B. subtilis 573	CSL (10 % v/v)	ND	30.7 mN/m	4.80 g/L	Gudina et al. (2015c)
Starmerella bombicola	Corn cob hydrolysate	Sophorolipid	ND	49.20 g/L; 12.3 g/L/day (vol. productivity)	Konishi et al. (2015)
Bacillus subtilis DSM 10 (T)	Glucose (25 %)	Surfactin	28.0 mN/m	0.278 (Yp/x g/g); 0.033 g/g (Yp/s)	Willenbacher et al. (2015)
Bacillus subtilis BTCC-TUSA-A	Maldex-15—substrate from corn starch	Surfactin	ND	36.10 g/L; 0.27 g/g Substrate yield; 2.58 g/L/h (VRP) Vol. productivity	Amin (2014)
Bacillus amylofaciens AR 2	Sucrose	Lipopeptide	30.0 mN/m; CMC = 80 mg/ml; $E_{24}(\%) = 66$	0.64 g/L; Productivity 10.6	Singh et al. (2014a)

(continued)

Table 2 (continued)

Micro-organism	Substrate	BS type	Properties	Yield	References
Pseudomonas. aeruginosa NITT6L	Sucrose	Glycolipid	39.8 mN/m	4.6 0 g/L	Vanavil et al. (2013)
Pseudomonas nitroreducens	Glucose	Rhamnolipid	37.0 mN/m; CMC = 28 mg/L	5.28 g/L	Onwosi and Odibo (2012)
Oleomonas Sagaranensis AT18	Molasses	Glycolipid	ND	5.30 g/L	Saimmai et al. (2012)
Pseudomonas aeruginosa GIM32	Molasses distillery waste	Rhamnolipid	ND	2.60 g/L	Li et al. (2011)
Bacillus subtilis BS5	Molasses	ND	ND	1.12 g/L	Abdel-Mawgoud et al. (2008)
Pseudomonas aeruginosa EM1	Glucose (4 %)	Rhamnolipid	ND	7.50 g/L	Wu et al. (2008)
Pseudomonas aeruginosa S2	Glucose (4 %)	Rhamnolipid	ND	2.37 g/L	Chen et al. (2007)
Dairy					
Lactobacillus helveticus	Deprotonized cheese whey	Glycolipid	39.5 mN/m; CMC = 2.5 mg/L	0.80 g/L	Sharma et al. (2014)
Pseudomonas aeruginosa ATCC145	Milk whey	ND	ND	10.50 g/L	Colak and Kahraman (2013)
Candida bombicola	Synthetic dairy waste	ND	ND	5.0 g/L	Daverey et al. (2009)
Pseudomonas aeruginosa BS2	Whey waste	Rhamnolipid	ND	0.92 g/L	Dubey and Juwarkar (2001)

(continued)

Table 2 (continued)

Micro-organism	Substrate	BS type	Properties	Yield	References
Renewable substrates					
Pleurotus djamor	Sunflower seed shell, grape wastes or potato peels	ND	28.8 mN/m	10.20 g/L	Velioglu and Ozturk Urek (2015)
Halobacteriaceae archaeon AS 65	Banana peel (35 % w/v)	ND	25.5 mN/m; CMC = 10 mg/L	5.30 g/L	Chooklin et al. (2014)
Providencia alcalifaciens	Palm Kernel cake	Glycolipid	33.8 mN/m	8.30 g/L	Jamal et al. (2014)
Ohrobactrum anthropic 2/3	Palm oil decanter cake (25 % v/v)	Glycolipid	25.0 mN/m	4.52 g/L	Noparat et al. (2014)
Candida lipolytica UCP 0988	Soyabean Oil refiner residue (6 %)	Lipopeptide	25.0 mN/m	8.0 g/L	Rufino et al. (2014)
Klebsiella sp. RJ03	Corn powder substrate	ND	ND	15.4 g/L	Jain et al. (2013)
Lactobacillus delbrueckii	Peanut oil cake (2.0 % w/v)	GLycolipid	ND	5.35 g/L	Thavasi et al. (2011)
Pseudomonas fluorescens	Cashew Apple juice	Rhamnolipid	ND	9.35 g/L	Parthasarathi and Sivakumaar (2009)
Bacillus subtilis LB5a	Cassava water	Lipopeptide	27.0 mN/m	2.40 g/L	Barros et al. (2008)
Lactobacillus pentosus	Distilled grape marc	ND	ND	0.004 g/L, 0.60 mg/g of sugar	Portilla-Rivera et al. (2008)

ND not defined

the n-alkanes used as carbon sources. Table 2 shows the effect of different carbon sources on biosurfactant yields obtained from different micro-organisms.

8.1.2 Nitrogen Source

Nitrogen is important in biosurfactant production because it is essential for microbial growth, as protein and enzyme synthesis depends on it. During stationary phase of cell growth, with the depletion of a nitrogen source, the specific activities of NAD- and NADP-dependent isocitrate dehydrogenases, which catalyse the oxidation of isocitrate to 2-oxoglutarate in the citric acid cycle decline. This decline in activities results in the accumulation of isocitrate and citrate in the mitochondria, which are then transported to the cytosol where citrate is converted to acetyl-coA by citrate synthase. Acetyl-coA being the precursor of fatty acid synthesis results in increased production of biosurfactants (Maneerat 2005).

Inorganic nitrogenous compounds reported to influence biosurfactant production include sodium nitrate (Gogotov and Khodakov 2008; Onwosi and Odibo 2012; Abbasi et al. 2012; Saikia et al. 2012), ammonium nitrate (Chen et al. 2007; Fonseca et al. 2007; Abouseoud et al. 2008) and ammonium sulphate (Mata-Sandoval et al. 2001). A maximum reduction in the surface tension was observed in medium supplemented with ammonium chloride (69.7–39.3 mN/m) at a concentration of 0.5 % (w/v) by *Bacillus pumilus* IJ-1 (Park and Kim 2015).

Different commonly used organic nitrogen sources are urea, peptone and yeast extract. Among the organic nitrogen sources, yeast extract proved to be the best (Manivasagan et al. 2014). However, Bharali and Konwar (2011) reported urea along with ammonium sulfate (2 g/L each) as the best nitrogen sources for rhamnolipid production by *Pseudomonas aeruginosa* strain OBP1.

8.1.3 C/N Ratio

One of the most important factors influencing biosurfactant production is the carbon to nitrogen ratio (C/N) ratio in the growth medium. Studies have reported that the carbon to nitrogen present in the medium not only induces biosurfactant production but also determines the composition and type of biosurfactant. Saikia et al. (2014) reported maximum rhamnolipid (0.80 g/L) production by *Pseudomonas aeruginosa* RS29, in a medium supplemented with glycerol and sodium nitrate at a C/N ratio of 12.5 with a 55 % reduction in the surface tension of the medium. Khopade et al. (2012) reported maximum biosurfactant production by the actinomyces *Nocardiopsis* sp. B4 for a medium containing olive oil and phenylalanine as carbon and nitrogen source respectively at a C/N ratio of 20:1. Ghribi and Ellouze-Chaabouni (2011) demonstrated enhanced biosurfactant production of up to 0.9 g/L by *Bacillus subtilis* SPBI at a C/N ratio of 7:1 using glucose and ammonium chloride. Earlier research by Fonseca et al. (2007) showed the interactive effect of C/N ratio and agitation rate using a complete factorial design for

Bacillus subtilis and achieved the highest reduction in surface tension at a C/N (sucrose: ammonium nitrate) ratio of 3 and an agitation rate of 250 rpm.

8.1.4 Multivalent Ions

Park and Kim (2015) found a maximum reduction in surface tension (60.8–33.6 mN/m) for the biosurfactant produced by *Bacillus pumilus* IJ-1 in the presence of potassium chloride. Pirog et al. (2013) reported enhanced concentrations of surface-active substances by 1.2–1.6 times by *Acinetobacter calcoaceticus* IMV B-7241 by incorporating 0.21 mmol/L of potassium chloride, 38 mmol/L of zinc sulfate and 0.16 mmol/L of copper sulfate in a medium containing glycerol, as compared to initial media which contained yeast autolysate and microelements. Roldan-Carrillo et al. (2011) demonstrated the optimal effect on the reduction of surface tension and biosurfactant production in the interaction between C/N and C/Mg with a yield of 4.1 g/L of biosurfactant in a 3 L bioreactor from *Serratia marcescens* using a Box-Behnken experimental design. Makkar and Cameotra (2002) reported that the effect of the metal cations, used together, is greater than when they are used individually. Biosurfactant production almost doubled by the addition of metal supplements. Glick et al. (2010) described how the translation of *rhl*AB and the production of rhamnolipids are inversely proportional to the concentration of Iron (Fe) available to the bacterial cells. This result was in accordance with those reported by Persson et al. (1990) where omission of Fe (III) from the medium increased the product yield of biosurfactant by 120 % independent of C/N ratio and the phosphorus content in the medium.

8.2 Environmental Parameters

Environmental parameters are considered to be vital in the yield and characteristics of the biosurfactant produced. Various physical parameters, i.e. pH, temperature, aeration and agitation are reported to be prerequisite influential factors and are needed to be optimised for higher biosurfactant yields.

8.2.1 pH

Optimum biosurfactant production is mostly reported in the pH range from 6.0 to 9.0 (pH 9.0, Park and Kim 2015); (7–8 pH, Saikia et al. 2012) and (6.0–8.0, Xia et al. 2012). Chen et al. (2012) observed the pH shift from initial neutral pH (pH 7.0) to alkaline during biosurfactant production from *Bacillus licheniformis* TKU004 resulting in reduced surface tension of 25.42 mN/m in 24 h. Najafi et al. (2010) reported biosurfactant production by *Bacillus mycoides* in a pH range of 5.0–9.0 with a maximum yield at pH 7.37 using a combination of central composite design (CCD) and response surface methodology (RSM). Makkar and Cameotra (1997) reported biosurfactant production from thermophilic *Bacillus subtilis* for a wide range of pH (4.5–10.5) at 45 °C.

8.2.2 Temperature

Most biosurfactant production studies reported a temperature range of 25–30 °C (Vanavil et al. 2013; Park and Kim 2015). However, few studies examined production for extreme temperatures (both thermal and non-thermal) conditions. Sharafi et al. (2014) determined 45 °C to be optimal for maximum lipopetide biosurfactant production by the thermophilic bacterium *Aneurinibacillus*. Hori et al. (2011) reported control over the product ratios of rhamnolipids and polyhydroxyalkanoates from *Pseudomonas aeruginosa* IFO3924 by changing the culture temperature. Joshi et al. (2008) reported a maximum production of biosurfactant by *Bacillus licheniformis* K51, *Bacillus subtilis* 20B, *Bacillus subtilis* R1 and *Bacillus strain* HS3 using molasses under thermophilic conditions at 45 °C. Rapp and Gabriel-Jurgens (2003) reported the production of biosurfactants from *Rhodococcus* sp. strain MS11 using a number of hydrocarbons as carbon sources over a range of from 5 to 30 °C, with optimal growth at 20 °C. Makkar and Cameotra (1997) confirmed 45 °C as optimal temperature for biosurfactant production for the thermophilic *Bacillus subtilis* using 2 % sucrose with a maximum reduction of surface tension of the medium from 68 to 28 mN/m. Gurjar et al. (1995) investigated the production of biosurfactant by *Bacillus stearothermophilus* VR-8 isolate in 4 % crude oil over a temperature range of 45–70 °C with a maximum recovery (0.6 g/L) at 50 °C. Very few reports are available for biosurfactant production at lower temperatures from psychrophilic organisms. Pruthi and Cameotra (1997) observed growth and production of biosurfactants from a psychrophilic strain of *Arthrobacter protophormiae* at 10 °C.

8.2.3 Aeration and Agitation

Aeration and agitation are important factors that influence the production of biosurfactants as both facilitate oxygen transfer from the gas phase to the aqueous phase. Pansiripat et al. (2010) described a slight reduction in surface tension during the first 4 h of aeration period followed by a drastic decrease from 66 to 35 mN/m, at an aeration time of 6 h in sequencing batch reactors. Ghribi and Ellouze-Chaabouni (2011) reported that aeration of the medium strongly affect growth-regulated biosurfactant synthesis by *B. subtilis*. They observed that in a fully automated 2.6 L fermenter, low or high aerations decreased biosurfactant yields but dissolved oxygen saturation of the medium at 30 %, produced biosurfactants of 4.92 g/L, indicating that oxygen transfer is one of the key parameters for the process productivity. Ray (2012) demonstrated an optimum agitation and aeration rate of 150 rpm and 1.0 vvm respectively for production of biosurfactant in a

3 L medium volume with a maximum oxygen transfer rate (KLa) to the aerobic cell of 129.76/h which proved significant oxygen transfer efficiency.

9 Feasibility of Biosurfactant Production from Industrial Wastes

Carbon sources such as glucose, glycerol, acetates and other organic acids, as well as pure alkanes and hydrocarbons are quite expensive and hence an approach to decrease production costs partially or completely relies on replacing pure chemicals with industrial/agricultural mixtures. Babu et al. (1996) proposed industrial wastes (distillery and whey) to be better substrates for biosurfactant production than synthetic media, as specific growth rates (μ_{max}) and specific product formation rates (V_{max}) were found to be comparatively higher.

9.1 Biosurfactant Production Using Byproducts of Vegetable Oil Industries

Oils and fats are a lipidic carbon source and are mostly composed of saturated or unsaturated fatty acids with varying carbon atoms. Numerous studies have reported the use of canola, corn, sunflower, safflower, olive, rapeseed, grape seed, palm, coconut, fish and soybean oil for biosurfactant production. These by-products generated by the oil processing industries are found to be rich in free fatty acids, mono-, di-, and triacylglycerides, phosphatides, pigments, sterols, tocopherols, glycerol, hydrocarbons, glycolipids, resinous and mucilaginous materials, which can value-added by using them as substrates for biosurfactant production (Dumont and Narine 2007).

9.2 Biosurfactant Production from Dairy Waste

Dairy industries produce large quantities of waste products that include curd whey, whey waste, cheese whey, lactic acid, lactose, butter sediment, unclarified oils etc. These products are rich in proteins, lipids, soluble vitamins, minerals and carbohydrates and therefore serve as a potent feedstock for biosurfactant production.

9.3 Biosurfactant from the Sugar Industry

Several studies have reported that sugars as carbon sources serve as excellent substrates for biosurfactant production, but the sugar composition varies among various substrates, sources, processes and industries involved. The food-processing industry, especially sugar processing plants, produces by-products and waste-streams containing a variety of sugars; molasses being the most conventional sugar substrates for biosurfactant production.

9.4 Biosurfactant Production from Glycerol and Glycerol-Containing Waste

A number of studies have reported the use of glycerol as a carbon source for biosurfactant production. Glycerol, a major component of lipids, is abundant in nature and is obtained as a by-product of the biodiesel industry by transesterification of *Jatropha curcas*—or rapeseed oil using methanol. Crude glycerol consists of 85–95 % glycerol. The yield (g/L) and the properties of the biosurfactants produced from different organisms using different substrates is presented in Table 2.

10 Commercial Applications of Biosurfactants

Microbial biosurfactants span across almost all major industrial sectors (Table 3).

10.1 Agriculture

Many rhizosphere and plant-associated microbes produce biosurfactants; these biomolecules play a vital role in motility, signaling, and biofilm formation, indicating that biosurfactants govern plant–microbe interaction. In agriculture, biosurfactants can be used for plant pathogen elimination, for increasing the bioavailability of nutrients for beneficial plant-associated microbes, and to improve the agricultural soil quality by soil remediation. Thus, these biomolecules can replace harsh synthetic surfactants presently used in the million dollar pesticide industries (Dhara et al. 2013).

10.2 Biomedical Applications

Biosurfactants have recently emerged as promising molecules in both clinical and hygienic sectors. Their ability to form pores and destabilise biological membrane permits their use as antimicrobial, hemolytic, antiviral, antitumor, and insecticide agents or as constituents of drug delivery systems (Bhadoriya et al. 2013; Banat et al. 2014a). Joshi-Navare and Prabhune (2013) reported the synergistic action of sophorolipids-cefaclor mixture which damages the cell membrane with pore formation and thus proves to be efficient against *Escherichia coli* with approx. 48 % more inhibition within 2 h of exposure as compared to cefaclor alone. Some types of microbial surfactants, such as lipopeptides and glycolipids, have been shown to selectively inhibit the proliferation of cancer cells causing their lysis through apoptosis pathways by inducing increased membrane permeability of biosurfactant-treated cancer cells and by incorporating its lipid moiety into the plasma membrane leading to formation of pores and membrane disruption. Moreover, the uses of biosurfactants as drug delivery vehicles offer commercially attractive and scientifically novel applications (Rodrigues 2015). Biosurfactants also find applications in biotechnology and cosmetics due to their multi-functional profile i.e. detergency, emulsifying, foaming and skin-hydrating properties (Varvaresou and Iakovou 2015).

10.3 Biosurfactants in the Food Industry

In the food industry, microbial biosurfactants control texture and shelf-life of starch-containing products, improve consistency and texture of fat-based products, agglomerate fat globules, stabilise aerated systems and modify rheological properties of wheat dough. In bakery and ice cream formulations, biosurfactants have been reported to improve consistency, delaying staling and solubilising flavor oils and as fat stabilisers and anti-spattering agents. L-Rhamnose has considerable potential as a precursor for flavoring and it is already used industrially as a precursor of high-quality flavour components like furaneol (strawberry furanone) (Campos et al. 2013). Muthusamy et al. (2008) stated its use as anti-adhesive agents to prevent food spoilage and disease transmission. The pre-conditioning of stainless steel surfaces with biosurfactant from *Pseudomonas fluorescens* has also been reported to inhibit the adhesion of *L. monocytogenes* LO28 strain.

10.4 Application of Biosurfactants in the Cosmetic Industry

In the cosmetic industry, due to its emulsification, foaming, water-binding capacity, spreading and wetting properties, biosurfactants have been proposed to replace

Table 3 Applications of microbial biosurfactants

Type of Biosurfactant	Microorganism involved	Application	References
Agriculture			
Orfamide	*Pseudomonas protegens* F6	Insecticidal against Myzus persicae	Jang et al. (2013)
Rhamnolipid	ND	control of zoosporic plant pathogens	Miao et al. (2015)
Mannosylerythritol lipids	ND	Potential Pesticides: inhibition of conidial germination in powdery mildew in wheat plants	Yoshida et al. (2015)
Rhamnolipid	*Pseudomonas aeruginosa* DS9	Antifungal agents against *F. sacchari* in pokkah boeng disease	Goswami et al. (2014)
Biomedical applications			
Antimicrobial agents			
Sophorolipids	ND	Antimicrobial properties and biofilm disruption	Diaz De Rienzo et al. (2015)
Lipopeptides	ND	Antifungal agents	Mnif and Ghribi (2015)
Sophorolipid	*Rhodotorula bogoriensis*	Antimicrobial property against *Propionibacterium acnes*	Solaiman et al. (2015)
Biosurfactants	ND	Nano sized vectors from micro emulsions, for drug delivery	Rodrigues (2015)
Lipopeptide	*Bacillus natto* TK-1	Anti-inflammatory activity	Zhang et al. (2015)
Lipopeptide	*Rhodococcus erythropolis* IMB Ac-5017	Anti- adhesive property	Pirog et al. (2014)
Lipopeptide Rhamnolipid	*Bacillus subtilis* *Pseudomonas aeruginosa*	Bio-film and anti-adhesive properties on medical implants	Prabhawathi et al. (2014)

(continued)

Table 3 (continued)

Type of Biosurfactant	Microorganism involved	Application	References
Rhamnolipid	*Pseudomonas aeruginosa*	Antioxidant properties	Singh et al. (2014b)
Exopolysaccharides	*Paenibacillus macerans*	Antimicrobial activity	Liang et al. (2014)
Di-rhamnolipid	ND	Stimulates adult neural stem cells, promotes nervous tissue regeneration	Stipcevic et al. (2013)
Sophorolipid	ND	Synergistic effect with antibiotics sophorolipids-cefaclor mixture proved to be efficient against *Escherichia coli*	Joshi-Navare and Prabhune. (2013)
Pseudofactin II	*Pseudomonas fluorescens BD5*	Anti-adheshive activity and disinfectant	Janek et al. (2012)
Anticancerous agents			
Sophorolipids	*Starmerella bombicola*	Cytotoxic effect on MDA-MB-321 breast cancer cell line	Ribeiro et al. (2015)
Glycoprotein (BioEG)	*Lactobacillus paracasei*	Anti-tumour potential on breast cancer cells T47D, MDA-MB-231	Duarte et al. (2014)
Lipopeptides	*Lysinibacillus fusiformis* S9 and *Bacillus tequilensis* CH	Cytotoxic effect and apoptosis in human embryonic kidney cancerous cell: HEK-293	Pradhan et al. (2014)
Rhamnolipid	*Pseudomonas aeruginosa* BN10	Inhibition and proliferation of BV-173 pre-B human leukemia cells (apoptosis)	Christova et al. (2013)
Lipopeptide	*Bacillus* sp. BS3	anticancer activity against mammary epithelial carcinoma cells, and antiviral properties against (white spot syndrome virus (WSSV)	Donio et al. (2013)
Pseudofactin II	*Pseudomonas fluorescens* BD5	Anti-melanoma agent	Janek et al. (2013)

(continued)

Table 3 (continued)

Type of Biosurfactant	Microorganism involved	Application	References
Food Industry			
Bioemulsifiers	*Candida utilis*	Emulsifiers	Campos et al. (2014)
Lipopeptide	*Citrobacter* sp. and *Enterobacter* sp.	Biopreservative	Mandal et al. (2013)
Bacteriocin-Biosurfactants	*Lactobacillus casei* MRTL3	Biopreservative or food ingredient	Sharma and Singh (2014)
Bioremediation			
Rhamnolipid	ND	As a chelating agent in the degradation of heavy metals and polycyclic aromatic hydrocarbons (PAHs)	Ammami et al. (2015)
Viscosin	Bacterial consortium	Stimulation and degradation of alkane (n-hexadecane)	Bak et al. (2015)
Rhamnolipid	Co-culture of (*Pseudomonas aeruginosa* ATCC 9027 and *Pseudomonas* sp. R)	Bio-augmentation Degradation of PAH (phenanthrene)	Chang et al. (2015)
Lipopeptide	*Bacillus subtilis*	Biodegradation of petroleum products	Montagnolli et al. (2015)
Mannosylerythritol lipids	ND	Petroleum degradation, bioconversion of chemical wastes	Yu et al. (2015)
Rhamnolipid	ND	Biodegradation of phenanthrene, pyrene	Yu et al. (2014)
Mixed biosurfactant systems	ND	Removal of heavy metals, radionuclides from soils	Mao et al. (2014)
Rhamnolipids	ND	Bioremediation of crude oil contaminated sand	Nikolopoulou et al. (2013)
Lipopeptide	*Bacillus subtilis* A21	Removal of petroleum hydrocarbons, heavy metals	Singh and Cameotra (2013)

(continued)

Table 3 (continued)

Type of Biosurfactant	Microorganism involved	Application	References
Rhamnolipid	*P. aeruginosa* ATCC 55925	Biodegradation of commercial C_{12}–C_{19} alkanes, Pristane (C_{19} branched alkanes)	Rocha et al. (2011)
Rhamnolipid	ND	Solubilization of phenanthrene and pyrene	Yu et al. (2011)
Rhamnolipid	*Pseudomonas aeruginosa* P-CG3	Desorption of phenanthrene, pyrene	Cheng et al. (2004)
Microbially enhanced oil recovery (MEOR)			
Lipopeptide	*Bacillus subtilis* 573	Enhanced oil recovery/bioremediation	Gudiña et al. (2015)
Lipopeptides	*Bacillus subtilis*	Petroleum recovery and spill mitigation	Liu et al. (2015)
Rhamnolipid	*Pseudomonas aeruginosa*	Enhanced oil recovery through anaerobic production of rhamnolipid	Zhao et al. (2015)
Lipopeptides	*Bacillus subtilis* B30	Ex situ microbial enhanced oil recovery processes	Al-Wahaibi et al. (2014)

ND not defined

chemically synthesised surfactants. These surfactants are used in insect repellents, antacids, bath products, acne pads, anti-dandruff products, contact lens solutions, baby products, mascara, lipsticks, toothpaste, dentine cleansers etc. (Gharaei-Fathabad 2011).

10.5 Application of Biosurfactant in Bioremediation

The capability of biosurfactants and biosurfactant-producing bacterial strains to enhance organic contaminants availability and biodegradation rates has been reported by many authors (Pacwa-Plociniczak et al. 2011).

10.6 Application of Biosurfactants in Microbial Enhanced Oil Recovery (MEOR)

Biosurfactants are novel group of molecules making powerful and versatile products with applications also as petroleum up-grading (Perfumo et al. 2010), enhanced oil recovery by contributing to the metabolism of viscous oil through microorganism that release lighter hydrocarbon fraction making oil more fluid (Pacwa-Plociniczak et al. 2011).

11 Conclusion

Biosurfactants have emerged as a popular choice for boosting the global biosurfactants market which was 344,068.40 tons in 2013 and is expected to reach 461,991.67 tons by 2020, growing at a CAGR of 4.3 % from 2014 to 2020 (Grand View Research 2014). The most important limitation for the commercial use of biosurfactants is the complexity and high cost of production, which has limited the development of their use on a large scale (Soberon-Chavez and Maier 2011). Prerequisites for making biosurfactant production more profitable and economically feasible include optimised growth/production conditions and novel and efficient multi-step downstream processing methods as well as the use of recombinant varieties of microorganisms or selected hyper-producing mutants, which can grow on a wide range of cheap renewable substrates. Thus, with a feasible and efficient production technology, the biosurfactant industry will not only provide innovative solutions to face the challenges of the competitive surfactant market but also will serve the purpose to contribute to a solution of the interconnected crisis of environmental pollution and greener production strategies.

References

Abbasi H, Hamedi MM, Lotfabad TB, Zahiri HS, Sharafi H, Masoomi F, Moosavi-Movahedi AA, Ortiz A, Amanlou M, Noghabi KA (2012) Biosurfactant producing bacterium, *Pseudomonas aeruginosa* MA01 isolated from spoiled apples: physicochemical and structural characteristics of isolated biosurfactant. J Biosci Bioeng 113(2):211–219

Abdel-Mawgoud AM, Aboulwafa MM, Hassouna NAH (2008) Optimization of surfactin production by *Bacillus subtilis* isolate BS5. Appl Biochem Biotechnol 150(3):305–325

Abdel-Mawgoud AM, Lepine F, Deziel E (2010) Rhamnolipids: diversity of structures, microbial origins and roles. Appl Microbiol Biotechnol 86(5):1323–1336

Abouseoud M, Maachi R, Amrane B, Nabi A (2008) Evaluation of different carbon and nitrogen sources in production of biosurfactant by *Pseudomonas fluorescens*. Desalination 223(1–3):143–151

Adams RH, Castillo-Acosta O, Escalante-Espinosa E, Zavala-Cruz J (2011) Natural attenuation and phytoremediation of petroleum hydrocarbon impacted soil in tropical wetland environments. In: Torres LG, Bandala ER (eds) Remediation of soils and aquifers. Nova Publishers, New York, pp 1–24

Affandi IE, Suratman NH, Abdullah S, Ahmad WA, Zakaria ZA (2014) Degradation of oil and grease from high-strength industrial effluents using locally isolated aerobic biosurfactant-producing bacteria. Int Biodeterior Biodegradation. doi:10.1016/j.ibiod.2014.04.009

Al-Wahaibi Y, Joshi S, Al-Bahry S, Elshafie A, Al-Bemani A, Shibulal B (2014) Biosurfactant production by *Bacillus subtilis* B30 and its application in enhancing oil recovery. Colloids Surf B Biointerfaces 114:324–333

Amani H (2015) Study of enhanced oil recovery by rhamnolipids in a homogeneous 2D micromodel. J Pet Sci Eng 128:212–219

Amezcua-Vega C, Ferrera-Cerrato R, Esparza-Garcia F, Rios-Leal E, Rodriguez-Vazquez R (2004) Effect of combined nutrients on biosurfactant produced by *Pseudomonas putida*. J Environ Sci Health A Tox Hazard Subst Environ Eng. 39(11–12):2983–2991

Amin GA (2014) Exponential fed-batch strategy for enhancing biosurfactant production by *Bacillus subtilis*. Water Sci Technol 70(2):234–240

Ammami MT, Portet-Koltalo B, Duclairoir-Poc W, LeDerf F (2015) Application of biosurfactants and periodic voltage gradient for enhanced electrokinetic remediation of metals and PAHs in dregraded marine sediments. Chemosphere 125:1–8

Ansaldi M, Marolt D, Stebe T, Mandic-Mulec I, Dubnau D (2002) Specific activation of the *Bacillus* quorum sensing systems by isoprenylated pheromone variants. Mol Microbiol 44:1561–1573

Aparna A, Srinikethan G, Smitha H (2011) Effect of Addition of Biosurfactant Produced by *Pseudomonas* sp. on biodegradation of crude oil. In: Second international conference on environmental science and technology, vol 6, IPCBEE Singapore, IACSIT Press

Assadi M, Rostamza M, Noohi AS, Levin M, Shahamati M (2004) Rhamnolipid production by *Pseudomonas aerogiosa* MM1011 from sugar beet molasses. Asian J Microbiol Biotech Environ Sci 6(2):203–207

Atkinson S, Williams P (2009) Quorum sensing and social networking in the microbial world. J R Soc Int 6(40):959–978

Ayed HB, Jemil N, Maalej H, Bayoudh A, Hmidet N, Nasri M (2015) Enhancement of solubilization and biodegradation of diesel oil by biosurfactant from *Bacillus amyloliquefaciens* An6. Int Biodeterior Biodegradation 99:8–14

Babu PS, Vaidya AN, Bal AS, Kapur R, Juwarkar A, Khanna P (1996) Kinetics of biosurfactant production by *Pseudomonas aeruginosa* strain BS2 from industrial wastes. Biotech Lett 18:263–268

Bacon SK, Palmer TM, Grossman AD (2002) Characterization of comQ and comX, two genes required for production of ComX pheromone in *Bacillus subtilis*. J Bacteriol 184:410–419

Bak F, Bonnichsen L, Jorgensen NO, Nicolaisen MH, Nybroe O (2015) The biosurfactant viscosin transiently stimulates n-hexadecane mineralization by a bacterial consortium. Appl Microbiol Biotechnol 99(3):1475–1483

Banat IM, De Rienzo MA, Quinn GA (2014a) Microbial biofilms: biosurfactants as antibiofilm agents. Appl Microbiol Biotechnol 98(24):9915–9929

Banat IM, Makkar SR, Cameotra SS (2000) Potential commercial application of microbial surfactants. Appl Microbiol Biotechnol 53(5):495–508

Banat IM, Satpute SK, Cameotra SS, Patil R, Nyayanit NV (2014b) Cost effective technologies and renewable substrates for biosurfactants' production. Front Microbiol 5:697. doi:10.3389/fmicb.2014.00697

Bao M, Pi Y, Wang L, Sun P, Li Y, Cao L (2014) Lipopeptide biosurfactant production bacteria *Acinetobacter* sp. D3-2 and its biodegradation of crude oil. Environ Sci Process Impacts 16 (4):897–903

Barros FF, Ponezi AN, Pastore GM (2008) Production of biosurfactant by *Bacillus subtilis* LB5a on a pilot scale using cassava wastewater as substrate. J Ind Microbiol Biotechnol 35(9):1071–1078

Bento MF, Camargo FA, Okeke BC, FrankenbergerJr WT (2005) Diversity of biosurfactant producing microorganisms isolated from soils contaminated with diesel oil. Microbiol Res 160 (3):249–255

Bhadoriya SS, Madoriya N, Shukla K, Parihar MS (2013) Biosurfactants: a new pharmaceutical additive for solubility enhancement and pharmaceutical development. Biochem Pharmacol. doi:10.4172/2167-0501.1000113

Bharali P, Konwar BK (2011) Production and physico-chemical characterization of a biosurfactant produced by *Pseudomonas aeruginosa* OBP1 isolated from petroleum sludge. Appl Biochem Biotechnol 164(8):1444–1460

Bodour AA, Drees KP, Maier RM (2003) Distribution of biosurfactant producing bacteria in undisturbed and contaminated arid southwestern soils. Appl Environ Microbiol 69(6):3280–3287

Bordoloi NK, Konwar BK (2008) Microbial surfactant-enhanced mineral oil recovery under laboratory conditions. Colloids Surf B Biointerfaces 63:73–82

Bordoloi NK, Konwar BK (2009) Bacterial biosurfactant in enhancing solubility and metabolism of petroleum hydrocarbons. J Hazard Mater 170(1):495–505

BP Statistical Review of World Energy June (2015) http://www.bp.com/en/global/corporate/about-bp/energy-economics/statistical-review-of-world-energy.html

Burgos-Diaz C, Pons R, Espuny MJ, Aranda FJ, Teruel JA, Manresa A, Ortiz A, Marques AM (2011) Isolation and partial characterization of a biosurfactant mixture produced by *Sphingobacterium* sp. isolated from soil. J Colloid Interface Sci 361(1):195–204

Cai Q, Zhang B, Chen B, Zhu Z, Lin W, Cao T (2014) Screening of biosurfactant producers from petroleum hydrocarbon contaminated sources in cold marine environments. Mar Pollut Bull 86 (1–2):402–410

Cameotra SS, Makkar RS (2010) Biosurfactant-enhanced bioremediation of hydrophobic pollutants. Pure Appl Chem 82(1):97–116

Campos JM, Stamford TL, Sarubbo LA, DeLuna JM, Rufino RD, Banat IM (2013) Microbial biosurfactants as additives for food industries. Biotechnol Prog 29(5):1097–1108

Campos JM, Stamford TL, Sarubbo LA (2014) Production of a bioemulsifier with potential application in the food industry. Appl Biochem Biotechnol 172(6):3234–3252

Cao B, Nagarajan K, Loh KC (2009) Biodegradation of aromatic compounds: current status and opportunities for biomolecular approaches. Appl Microbiol Biotechnol 85(2):207–228

Cerqueira VS, Hollenbach EB, Maboni F, Camargo FA, Peralba Mdo C, Bento FM (2011) Bio-prospection and selection of bacteria isolated from environments contaminated with petrochemical residues for application in bioremediation. World J Microbiol Biotechnol 28 (3):1203–1222

Cha M, Lee N, Kim M, Kim M, Lee S (2008) Heterologous production of *Pseudomonas aeruginosa* EMS1 biosurfactant in *Pseudomonas putida*. Bioresour Technol 99(7):2192–2199

Chandankere R, Yao J, Cai M, Masakorala K, Jain AK, Choi MMF (2014) Properties and characterization of biosurfactant in crude oil biodegradation by bacterium *Bacillus methylotrophicus* USTBa. Fuel 122:140–148
Chandankere R, Yao J, Choi MMF, Masakorala K, Chan Y (2013) An efficient biosurfactant-producing and crude-oil emulsifying bacterium *Bacillus methylotrophicus* USTBa isolated from petroleum reservoir. Biochem Eng J 74:46–53
Chang JS, Cha DK, Radosevich M, Jin Y (2015) Effects of biosurfactant-producing bacteria on biodegradation and transport of phenanthrene in subsurface soil. J Environ Sci Health A Tox Hazard Subst Environ Eng 50(6):611–616
Chang JS, Chou CL, Lin GH, Sheu SY, Chen WM (2005) *Pseudoxanthomonas kaohsiungensis*, sp. nov, a novel bacterium isolated from oil-polluted site produces extracellular surface activity. Syst Appl Microbiol 28(2):137–144
Chayabutra C, Wu J, Ju LK (2001) Rhamnolipid production by *Pseudomonas aeruginosa* under denitrification: effects of limiting nutrients and carbon substrates. Biotechnol Bioeng 72(1):25–33
Chen SY, Lu WB, Wei YH, Chen WM, Chang JS (2007) Improved production of biosurfactant with newly isolated *Pseudomonas aeruginosa* S2. Biotechnol Prog 23:661–666
Chen YC, Chiang TJ, Liang TW, Wang IL, Wang SL (2012) Reclamation of squid pen by *Bacillus licheniformis* TKU004 for the production of thermally stable and antimicrobial biosurfactant. Biocatal Agric Biotechnol 1(1):62–69
Cheng KY, Zhao ZY, Wong JW (2004) Solubilization and desorption of PAHs in soil-aqueous system by biosurfactants produced from *Pseudomonas aeruginosa* P-CG3 under thermophilic condition. Environ Technol 25(10):1159–1165
Chooklin CS, Maneerat S, Saimmai A (2014) Utilization of banana peel as a novel substrate for biosurfactant production by *Halobacteriaceae archaeon* AS65. Appl Biochem Biotechnol 173(2):624–645
Christofi N, Ivshina IB (2002) Microbial surfactants and their use in field studies of soil remediation. J Appl Microbiol 93(6):915–929
Christova N, Tuleva B, Kril A, Georgieva M, Konstantinov S, Terziyski I, Nikolova B, Stoineva I (2013) Chemical structure and in vitro antitumor activity of rhamnolipids from *Pseudomonas aeruginosa* BN10. Appl Biochem Biotechnol 170(3):676–689
Chtioui O, Dimitrov K, Gancel F, Nikov I (2010) Biosurfactants production by immobilized cells of *Bacillus subtilis* ATCC 21332 and their recovery by pertraction. Process Biochem 45 (11):1795–1799
Colak AK, Kahraman H (2013) The use of raw cheese whey and olive oil mill wastewater for rhamnolipid production by recombinant *Pseudomonas aeruginosa*. Environ Exp Biol 11:125–130
Comella N, Grossman AD (2005) Conservation of genes and processes controlled by the quorum response in bacteria: characterization of genes controlled by the quorum-sensing transcription factor ComA in *Bacillus subtilis*. Mol Microbiol 57(4):1159–1174
Das K, Mukherjee AK (2007) Crude petroleum oil biodegradation efficiency of *Bacillus subtilis* and *Pseudomonas aeruginosa* strains isolated from a petroleum-oil contaminated soil from North East India. Bioresour Technol 98(7):1339–1345
Das N, Chandran P (2011) Microbial degradation of petroleum hydrocarbon contaminants: an overview. Biotechnol Res Int. doi:10.4061/2011/941810
Das P, Mukherjee S, Sen R (2008a) Genetic regulations of the biosynthesis of microbial surfactants: an overview. Biotechnol Gen Eng Rev 25:165–186
Das P, Mukherjee S, Sen R (2008b) Antimicrobial biosurfactants from marine *Bacillus circulans*: extracellular synthesis and purification. Lett Appl Microbiol 48(3):281–288
Daverey A, Pakshirajan K, Sangeetha P (2009) Sophorolipids production by *Candida bombicola* using synthetic dairy wastewater. Int J Civ Environ Eng 1:4
De Faria AF, Teodoro-Martinez DS, Barbosa GND, Vaz BG, Silva IS, Garcia JS, Totola MR, Eberlin MN, Grossman M, Alves OL, Durrant LR (2011) Production and structural

characterization of surfactin (C 14/Leu7) produced by *Bacillus subtilis* isolate LSFM-05 grown on raw glycerol from the biodiesel industry. Process Biochem 46(10):1951–1957

Deepak R, Jayapradha R (2015) Lipopeptide biosurfactant from *Bacillus thuringiensis* pak 2310 A potential antagonist against *Fusarium oxysporum*. J Med Mycol 25(1):15–24

Desai JD, Banat IM (1997) Microbial production of surfactants and their commercial potential. Microbiol Mol Biol Rev 61:47–64

Deziel E, Lepine F, Milot S, Villemur R (2003) *rhlA* is required for the production of a novel biosurfactant promoting swarming motility in *Pseudomonasaeruginosa*:3-(3- hydroxyalkanoyloxy) alkanoic acids (HAAs), the precursors of rhamnolipids. Microbiol 149:2005–2013

Swaranjit DP, Cameotra SS (2013) Biosurfactants in agriculture. Appl Microbiol Biotechnol 97:1005–1016

Diaz De Rienzo MA, Banat IM, Dolman B, Winterburn J, Martin PJ (2015) Sophorolipid biosurfactants: Possible uses as antibacterial and antibiofilm agent. N Biotechnol. doi:10.1016/j.nbt.2015.02.009

Diggle SP, Winzer K, Lazdunski A, Williams P, Camara M (2002) Advancing the quorum in *Pseudomonas aeruginosa*: MvaT and the regulation of *N*-acylhomeserine lactone production and virulence gene expression. J Bacteriol 184(10):2576–2586

Donio MB, Ronica SF, Viji VT, Velmurugan S, Jenifer JA, Michaelbabu M, Citarasu T (2013) Isolation and characterization of halophilic *Bacillus* sp. BS3 able to produce pharmacologically important biosurfactants. Asian Pac J Trop Med 6(11):876–883

Duarte C, Gudina EJ, Lima CF, Rodrigues LR (2014) Effects of biosurfactants on the viability and proliferation of human breast cancer cells. AMB Express 4:40. doi:10.1186/s13568-014-0040-0

Dubey K, Juwarkar A (2001) Distillery and curd whey wastes as viable alternative sources for biosurfactant production. World J Microbiol Biotechnol 17(1):61–69

Dumont MJ, Narine SS (2007) Soapstock and deodorizer distillates from North American vegetable oils: review on their characterization, extraction and utilization. Food Res Int 40 (8):957–974

Dusane DH, Zinzarde SS, Venugopolan VP, Mclean RJC, Weber MM, Rahman PKSM (2010) Quorum sensing: implication on rhamnolipid biosurfactant production. Biotechnol Gen Eng Rev 27:159–184

Eddouaouda K, Mnif S, Badis A, Younes SB, Cherif S, Ferhat S, Mhiri N, Chamkha M, Sayadi S (2012) Characterization of a novel biosurfactant produced by *Staphylococcus* sp. 1E with potential application on hydrocarbon bioremediation. J Basic Microbiol 52(4):408–418

El-Sheshtawy HS, Khalil NM, Ahmed W, Abdallah RI (2014) Monitoring of oil pollution at Gemsa Bay and bioremediation capacity of bacterial isolates with biosurfactants and nanoparticles. Mar Pollut Bull 87(1–2):191–200

Ferhat S, Mnif S, Badisa A, Eddouaoudaa K, Alouaouic R, Boucherita A, Mhirib N, Moulai-Mostefac N, Sayadi S (2011) Screening and preliminary characterization of biosurfactants produced by *Ochrobactrum* sp. 1C and *Brevibacterium* sp. 7G isolated from hydrocarbon-contaminated soils. Int Biodeterior Biodegradation 65:1182–1188

Folmsbee M, Duncan K, Han SO, Nagle D, Jennings E, McInerney M (2006) Re-identification of the halotolerant, biosurfactant-producing *Bacillus licheniformis* strain JF-2 as *Bacillus mojavensis* strain JF-2. Syst Appl Microbiol 29(8):645–649

Fonseca RR, Silva AJ, DeFranca FP, Cardoso VL, Servulo EF (2007) Optimizing carbon/nitrogen ratio for biosurfactant production by a *Bacillus subtilis* strain. Appl Biochem Biotechnol 137–140(1–12):471–486

Gharaei-Fathabad E (2011) Biosurfactants in pharmaceutical industry: a mini review. Am J Drug Discov Dev 1(1):58–69

Ghribi D, Ellouze-Chaabouni S (2011) Enhancement of *Bacillus subtilis* lipopeptide biosurfactants production through optimization of medium composition and adequate control of aeration. Biotechnol Res Int. doi:10.4061/2011/653654

Glick R, Gilmour C, Tremblay J, Satanower S, Avidan O, Deziel E, Greenberg EP, Poole K, Banin E (2010) Increase in rhamnolipid synthesis under iron limiting conditions influences

surface motility and biofilm formation in *Pseudomonas aeruginosa*. J Bacteriol 192(12):2973–2980
Gogotov IN, Khodakov RS (2008) Surfactant production by the *Rhodococcus erythropolis* sH-5 bacterium grown on various carbon sources. Prikl Biokhim Mikrobiol 44:207–212
Goldman S, Shabtai Y, Rubinovitz C, Rosenberg E, Gutnick DL (1982) Emulsan in *Acinetobacter calcoaceticus* RAG-1: distribution of cell-free and cell-associated cross-reacting Material. Appl Environ Microbiol 44(1):165–170
Goswami D, Handique PJ, Deka S (2014) Rhamnolipid biosurfactant against *Fusarium* sacchari the causal organism of pokkah boeng disease of sugarcane. J Basic Microbiol 54(6):548–557
Grand View Research (2014) Biosurfactants market analysis by product (Rhamnolipids, Sophorolipids, MES, APG, Sorbitan Esters, Sucrose Esters) and segment forecast To 2020 Published: April 2014|ISBN Code: 978-1-68038-012-5. http://www.grandviewresearch.com/industry-analysis/biosurfactants-industry
Gudina EJ, Fernandes EC, Rodrigues AI, Teixeira JA, Rodrigues LR (2015a) Biosurfactant production by *Bacillus subtilis* using corn steep liquor as culture medium. Front Microbiol 6:59. doi:10.3389/fmicb.2015.00059
Gudina EJ, Pereira JF, Costa R, Evtuguin DV, Coutinho JA, Teixeira JA, Rodrigues LR (2015b) Novel bioemulsifier produced by a Paenibacillus strain isolated from crude oil. Microb Cell Fact 14:14. doi:10.1186/s12934-015-0197-5
Gudina EJ, Rodrigues AI, Alves E, Domingues MR, Teixeira JA, Rodrigues LR (2015c) Bioconversion of agro-industrial by-products in rhamnolipids toward applications in enhanced oil recovery and bioremediation. Bioresour Technol 177:87–93
Gurjar M, Khire JM, Khan MI (1995) Bioemulsifier production by *Bacillus stearothermophilus* VR8 isolate. Lett Appl Microbiol 21:83–86
Gutnick DL, Bayer EA, Rubinovitz C, Pines O, Shabtai Y, Goldman S, Rosenberg E (1980) Emulsan production in *Acinetobacter* RAG-1. Adv Biotechnol 11:455–459
Hachaichi Z, Tazerouti A, Hacene H (2014) Growth kinetics study of a bacterial consortium producing biosurfactants, constructed with six strains isolated from an oily sludge. Adv Biosci Biotechnol 5:418–425
Haddadin MSY, AbouArqoub A, AbuReesh I, Haddadin J (2009) Kinetics of hydrocarbon extraction from oil shale using biosurfactant producing bacteria. Energy Convers Manag 50:983–990
Hao DH, Lin JQ, Lin XSJ, Su YJ, Qu YB (2008) Isolation, identification, and performance studies of a novel paraffin-degrading bacterium of *Gordonia amicalis* LH3. Biotechnol Biopro Eng 13 (1):61–68
Hassanshahian M (2014) Isolation and characterization of biosurfactant producing bacteria from Persian Gulf (Bushehr provenance). Mar Pollut Bull 86(1–2):361–366
Hassanshahian M, Emtiazi G, Cappello S (2012) Isolation and characterization of crude-oil-degrading bacteria from the Persian Gulf and the Caspian Sea. Mar Pollut Bull 64 (1):7–12
Hausmann R, Syldatk C (2014) Types and classification of microbial surfactants. In: Kosaric N, Sukan FV (eds) Biosurfactant production and utilization—processes, technologies, and economics, vol 159. CRC Press, London, pp 3–18
Hazra C, Kundu D, Chaudhari A (2014) Lipopeptide biosurfactant from *Bacillus clausii* BS02 using sunflower oil soapstock: evaluation of high throughput screening methods, production, purification, characterization and its insecticidal activity. Res Adv 5:2974–2982
Hemlata B, Selvin J, Tukaram K (2015) Optimization of iron chelating biosurfactant production by *Stenotrophomonas maltophilia* NBS-11. Biocatal Agric Biotechnol 4:135–143
Hommel RK, Ratledge C (1993) Biosynthetic mechanisms of low molecular weight surfactants and their precursor molecules. In: Kosaric N (ed) Biosurfactants: production, properties, applications. Marcel Dekker Inc, New York, pp 3–63
Hori K, Ichinohec R, Unnoc H, Marsudid S (2011) Simultaneous syntheses of polyhydroxyalkanoates and rhamnolipids by *Pseudomonas aeruginosa* IFO3924 at various temperatures and from various fatty acids. Biochem Eng J53:196–202

Hudak AJ, Cassidy DP (2004) Stimulating in-soil rhamnolipid production in a bioslurry reactor by limiting nitrogen. Biotechnol Bioeng 88(7):861–868

Ilori MO, Amobi CJ, Odocha AC (2005) Factors affecting biosurfactant production by oil degrading *Aeromonas* sp isolated from a tropical environment. Chemosphere 61(7):985–992

Ismail W, Al-rowaihi IS, Al-humam AA, Hamza RY, El AM, Bououdina M (2013) Characterization of a lipopeptide biosurfactant produced by a crude-oil-emulsifying *Bacillus* sp. I-15. Int Biodeterior Biodegradation 84:168–178

Ismail W, Shammary SA, El-Sayed WS, El-Sayed WS, Obuekwec C, El-Nayal AM, Raheema ASA, Al-Humam A (2015) Stimulation of rhamnolipid biosurfactants production in *Pseudomonas aeruginosa* AK6U by organosulfur compounds provided as sulfur sources. Biotechnol Reports 7:55–63

Jain RM, Mody K, Mishra A, Jha B (2012) Isolation and structural characterization of biosurfactant produced by an alkaliphilic bacterium *Cronobacter sakazakii* isolated from oil contaminated wastewater. Carbohydr Polym 87(3):2320–2326

Jain RM, Mody K, Joshi N, Mishra A, Jha B (2013) Production and structural characterization of biosurfactant produced by an alkaliphilic bacterium, *Klebsiella* sp. evaluation of different carbon sources. Colloids Surf B Biointerfaces 108:199–204

Jamal P, Mir S, Alam MZ, Wan Nawawi WM (2014) Isolation and selection of new biosurfactant producing bacteria from degraded palm kernel cake under liquid state fermentation. J Oleo Sci 63(8):795–804

Janek T, Krasowska A, Radwanska A, Lukaszewicz M (2013) Lipopeptide biosurfactant pseudofactin II induced apoptosis of melanoma A 375 cells by specific interaction with the plasma membrane. PLoS ONE 8(3):e57991. doi:10.1371/journal.pone.0057991

Janek T, Lukaszewicz M, Krasowska A (2012) Antiadhesive activity of the biosurfactant pseudofactin II secreted by the Arctic bacterium *Pseudomonas fluorescens* BD5. BMC Microbiol 12:24. doi:10.1186/1471-2180

Jang JY, Yang SY, Kim YC, Lee CW, Park MS, Kim JC, Kim IS (2013) Identification of orfamide A as an insecticidal metabolite produced by *Pseudomonas protegens* F6. J Agric Food Chem 61(28):6786–6791

Jara AM, Andrade RF, Campos-Takaki GM (2013) physicochemical characterization of tensio-active produced by *Geobacillus stearothermophilus* isolated from petroleum-contaminated soil. Colloids Surf B: Biointerfaces 101:315–318

Jennings EM, Tanner RS (2000) Biosurfactant-producing bacteria found in contaminated and uncontaminated soils. In: Proceedings of 2000 conference hazardous waste research, University of Oklahoma, pp 299–306

Johnsen AR, Wick LY, Harms H (2005) Principles of microbial PAH-degradation in soil. Environ Poll 133(1):71–84

Joshi S, Bharucha C, Jha S, Yadav S, Nerurkar A, Desai AJ (2008) Biosurfactant production using molasses and whey under thermophilic conditions. Bioresour Technol 99:195–1999

Joshi-Navare K, Prabhune A (2013) A biosurfactant-sophorolipid acts in synergy with antibiotics to enhance their efficiency. Biomed Res Int. doi:10.1155/2013/512495

Kato T, Haruki M, Imanaka T, Morikawa M, Kanaya S (2001) Isolation and characterization of long-chain-alkane degrading *Bacillus thermoleovorans* from deep subterranean petroleum reservoirs. J Biosci Bioeng 91(1):64–70

Kavitha V, Mandal AB, Gnanamani A (2014) Microbial biosurfactant mediated removal and or solubilization of crude oil contamination from soil and aqueous phase: an approach with *Bacillus licheniformis* MTCC 5514. Int Biodeterior Biodegradation 94:24–30

Khopade A, Biao R, Liu X, Mahadik K, Zhang L, Kokare C (2012) Production and stability studies of the biosurfactant isolated from marine *Nocardiopsis* sp. B4. Desalination 3:198–204

Konishi M, Yoshida Y, Horiuchi J (2015) Efficient production of sophorolipids by *Starmerella bombicola* using a corncob hydrolysate medium. J Biosci Bioeng 119(3):317–322

Kumar CG, Mamidyala SK, Sujitha P, Muluka H, Akkenapally S (2012) Evaluation of critical nutritional parameters and their significance in the production of rhamnolipid biosurfactants from *Pseudomonas aeruginosa* BS-161R. Biotechnol Prog 28(6):1507–1516

Kumari B, Singh SN, Singh DP (2012) Characterization of two biosurfactant producing strains in crude oil degradation. Process Biochem 47(12):2463–2471

Lang S, Philp JC (1998) Surface-active lipids in *rhodococci*. Antonie Van Leeuwenhoek 74(1–3):59–70

Lenchi N, Inceoglu O, Kebbouche-Gana S, Gana ML, Lliros M, Servais P, Garcia-Armisen T (2013) Diversity of microbial communities in production and injection waters of algerian oilfields revealed by 16S rRNA gene amplicon 454 pyrosequencing. PLoS ONE 8(6):e66588. doi:10.1371/journal.pone.0066588

Li AH, Xu MY, Sun W, Sun GP (2011) Rhamnolipid production by *Pseudomonas aeruginosa* GIM 32 using different substrates including molasses distillery wastewater. Appl Biochem Biotechnol 163(5):600–611

Liang TW, Wu CC, Cheng WT, Chen YC, Wang CL, Wang IL, Wang SL (2014) Exopolysaccharides and antimicrobial biosurfactants produced by *Paenibacillus macerans* TKU029. Appl Biochem Biotechnol 172(2):933–950

Liu JF, Mbadinga SM, Yang SZ, Gu JD, Mu BZ (2015) Chemical structure, property and potential applications of biosurfactants produced by *Bacillus subtilis* in petroleum recovery and spill mitigation. Int J Mol Sci 16(3):4814–4837

Liu W, Wang X, Wu L, Chen M, Tu C, Luo Y, Christie P (2012) Isolation, identification and characterization of *Bacillus amyloliquefaciens* BZ-6, a bacterial isolate for enhancing oil recovery from oily sludge. Chemosphere 87(10):1105–1110

Lotfabad TB, Shourian M, Roostaazad R, Najafabadi AR, Adelzadeh MR, Noghabi KA (2009) An efficient biosurfactant-producing bacterium *Pseudomonas aeruginosa* MR01, isolated from oil excavation areas in south of Iran. Colloids Surf B Biointerfaces 69(2):183–193

Maciel BM, Dias JCT, Santos ACF, Argolo Filho RC, Fontana R, Loguercio LL, Rezende RP (2007) Microbial surfactant activities from a petrochemical land farm in a humid tropical region of Brazil. Can J Microbiol 53(8):937–943

Makkar RS, Cameotra SS (1997) Biosurfactant production by a thermophilic *Bacillus subtilis* strain. J Ind Microbiol Biotechnol 18(1):37–42

Makkar RS, Cameotra SS (2002) Effects of various nutritional supplements on biosurfactant production by a strain of *Bacillus subtilis* at 45 C. J Surfactants Deterg 5(1):11–7

Mandal SM, Sharma S, Pinnaka AK, Kumari A, Korpole S (2013) Isolation and characterization of diverse antimicrobial lipopeptides produced by *Citrobacter* and *Enterobacter*. BMC Microbiol 13:152. doi:10.1186/1471-2180-13-152

Maneerat S (2005) Production of biosurfactant from renewable resources. Songklanakarin J Sci Technol 27(3):675–683

Manivasagan P, Sivasankar P, Venkatesan J, Sivakumar K, Kim SK (2014) Optimization, production and characterization of glycolipid biosurfactant from the marine actinobacterium, *Streptomyces* sp. MAB36. Bioprocess Biosyst Eng 5:783–797

Mao X, Jiang R, Xiao W, Yu J (2014) Use of surfactants for the remediation of contaminated soils: a review. J Hazard Mater 285:419–435

Maqsood MI, Jamal A (2011) Factors affecting the rhamnolipid biosurfactant production. Pak J Biotechnol 8(1):1–5

Marchant R, Banat IM (2012) Biosurfactants: a sustainable replacement for chemical surfactants? Biotechnol Lett 34(9):1597–1605

Mashburn LM, Whiteley M (2005) Membrane vesicles traffic signals and facilitates group activities in a prokaryote. Nature 437:422–425

Mata-Sandoval JC, Karns J, Torrents A (2001) Effect of nutritional and environmental conditions on the production and composition of rhamnolipids by *P. aeruginosa* UG2. Microbiol Res 155 (4):249–256

Menkhaus M, Ullrich C, Kluge B, Vater J, Vollenbroich D, Kamp RM (1993) Structural and functional organization of the surfactin synthetase multienzyme system. J Biol Chem 268:7678–7684

Miao S, Dashtbozorg SS, Callow NV, Ju LK (2015) Rhamnolipids as platform molecules for production of potential anti-zoospore agrochemicals. J Agric Food Chem 63:3367–3376
Mnif I, Ghribi D (2015) Review lipopeptides biosurfactants: mean classes and new insights for industrial, biomedical, and environmental applications. Biopolymers 104(3):129–147
Mnif S, Chamkha M, Labat M, Sayadi S (2011) Simultaneous hydrocarbon biodegradation and biosurfactant production by oilfield selected bacteria. J Appl Microbiol 111(3):525–536
Montagnolli RN, Lopes PR, Bidoia ED (2015) Assessing *Bacillus subtilis* biosurfactant effects on the biodegradation of petroleum products. Environ Monit Assess 187(1):4116
Mulligan CN, Gibbs BF (2004) Types, production and applications of biosurfactants. Proc Indian Natl Sci Acad B 70(1):31–55
Muthusamy K, Gopalakrishnan S, Ravi TK, Sivachidambaram P (2008) Biosurfactants: Properties, commercial production and application. Curr Sci 94:736–747
Najafi AR, Rahimpour MR, Jahanmiri AH, Roostaazad R, Arabian D (2011) Colloids and Surfaces B: Biointerfaces Interactive optimization of biosurfactant production by Paenibacillus alvei ARN63 isolated from an Iranian oil well. Colloids Surf B Biointerfaces 82(1):33–39
Najafi AR, Rahimpour MR, Jahanmiri AH, Roostaazad R, Arabian D, Ghobadi Z (2010) Enhancing biosurfactant production from an indigenous strain of *Bacillus mycoides* by optimizing the growth conditions using a response surface methodology. Chem Eng J 163 (3):188–194
Nie M, Yin X, Ren C, Wang Y, Xu F, Shen Q (2010) Novel rhamnolipid biosurfactants produced by a polycyclic aromatic hydrocarbon-degrading bacterium *Pseudomonas aeruginosa* strain NY3. Biotechnol Adv 28(5):635–643
Nikolopoulou M, Pasadakis N, Kalogerakis N (2013) Evaluation of autochthonous bioaugmentation and biostimulation during microcosm-simulated oil spills. Mar Pollut Bull 72(1):165–173
Noparat P, Maneerat S, Saimmai A (2014) Utilization of palm oil decanter cake as a novel substrate for biosurfactant production from a new and promising strain of *Ochrobactrum anthropi* 2/3. World J Microbiol Biotechnol 3:865–877
Oliveira FJS, Vazquez L, De Campos NP, De Franca FP (2009) Production of rhamnolipids by a *Pseudomonas alcaligenes* strain. Process Biochem 44:383–389
Onwosi CO, Odibo FJ (2012) Effects of carbon and nitrogen sources on rhamnolipid biosurfactant production by *Pseudomonas nitroreducens* isolated from soil. World J Microbiol Biotechnol 28:937–942
Pacwa-Plociniczak M, Plaza GA, Piotrowska-Seget Z, Cameotra SS (2011) Environmental applications of biosurfactants: recent advances. Int J Mol Sci 12:633–654
Pansiripat S, Pornsunthorntawee O, Rujiravanit R, Kitiyanan B, Somboonthanate P, Chavadej S (2010) Biosurfactant production by *Pseudomonas aeruginosa* SP4 using sequencing batch reactors: effect of oil-to-glucose ratio. Biochem Eng J 49:185–191
Park E, Kim J (2015) Characteristics of culture conditions for the production of biosurfactant by *Bacillus pumilus* IJ-1. J Appl Biol Chem 58:81–88
Parthasarathi R, Sivakumaar PK (2009) Effect of different carbon sources on the production of biosurfactant by pseudomonas fluorescens isolated from mangrove forests (Pichavaram) Tamil Nadu India. Glob J Environ Res 3(2):99–101
Peixoto RS, Vermelho AB, Rosado AS (2011) Petroleum-degrading enzymes: bioremediation and new prospects. Enzyme Res. doi:10.4061/2011/475193
Perfumo A, Rancich I, Banat IM (2010) Possibilities and challenges for biosurfactants use in petroleum industry. Adv Exp Med Biol 672:135–145
Persson A, Molin G, Andersson N, Sjoholm J (1990) Biosurfactant yields and nutrient consumption of *Pseudomonas fluorescens* 378 studied in a microcomputer controlled multifermentation system. Biotechnol Bioeng 36(3):252–255
Peypoux F, Bonmatin JM, Wallach J (1999) Recent trends in the biochemistry of surfactin. Appl Microbiol Biotechnol 51(5):553–563

Pirog TP, Konon AD, Beregovaya KA, Shulyakova MA (2014) Antiadhesive properties of the surfactants of *Acinetobacter calcoaceticus* IMB B-7241, *Rhodococcus erythropolis* IMB Ac-5017, and *Nocardia vaccinii* IMB B-7405. Microbiol 83(6):732–739

Pirog TP, Shevchuk TA, Oiu M, Parfeniuk SA, Iutinskaia GA (2013) Effect of growth factors and some microelements on biosurfactant synthesis of *Acinetobacter calcoaceticus* IMV B-7241. Mikrobiol Z 75(5):18–26

Poomtein J, Thaniyavarn J, Pinphanichakarn P, Jindamorakot M (2013) Production and characterization of a biosurfactant from *Cyberlindnera samutprakarnensis* JP52. Biosci Biotechnol Biochem 77(12):2362–2370

Pornsunthorntawee O, Arttaweeporn N, Paisanjit S (2008a) Isolation and comparison of biosurfactants produced by *Bacillus subtilis* PT2 and *Pseudomonas aeruginosa* SP4 for microbial surfactant-enhanced oil recovery. Biochem Eng J 42(2):172–179

Pornsunthorntawee O, Wongpanit P, Chavadej S, Abe M, Rujiravanit R (2008b) Structural and physicochemical characterization of crude biosurfactant produced by *Pseudomonas aeruginosa* SP4 isolated from petroleum-contaminated soil. Bioresour Technol 99(6):1589–1595

Portilla-Rivera O, Torrado A, Domilnguez JM, Moldes AB (2008) Stability and emulsifying capacity of biosurfactants obtained from lignocellulosic sources using *Lactobacillus pentosus*. J Agric Food Chem 56:8074–8080

Prabhawathi V, Thirunavukarasu K, Doble M (2014) A study on the long term effect of biofilm produced by biosurfactant producing microbe on medical implant. Mater Sci Eng C Mater Biol Appl 40:212–218. doi:10.1016/j.msec.2014.03.050

Pradhan AK, Pradhan N, Mohapatra P, Kundu CN, Panda PK, Mishra BK (2014) Cytotoxic effect of microbial biosurfactants against human embryonic kidney cancerous cell: HEK-293 and their possible role in apoptosis. Appl Biochem Biotechnol 174(5):1850–1858

Pruthi V, Cameotra SS (1997) Production and properties of a biosurfactant synthesized by *Arthrobacter protophormiae* an Antarctic strain. World J Microbiol Biotechnol 13:137–139

Rahman KSM, Rahman TJ, McClean S, Marchant R, Banat IM (2002a) Rhamnolipid biosurfactant production by strains of *Pseudomonas aeruginosa* using low-cost raw materials. Biotechnol Prog 6:1277–1281

Rahman KSM, Thahira-Rahman J, Lakshmanaperumalsamy P, Banat IM (2002b) Occurrence of crude oil degrading bacteria in gasoline and diesel station soils. J Basic Microbiol 42:286–293

Rapp P, Gabriel-Jurgens LH (2003) Degradation of alkanes and highly chlorinated benzenes, and production of biosurfactants, by a psychrophilic *Rhodococcus* sp. and genetic characterization of its chlorobenzene dioxygenase. Microbiol 149:2879–2890

Ray S (2012) Optimization of process conditions for biosurfactant production from mutant strain of *Bacillus* sp. (m28) in a 5l laboratory fermenter. J Microbiol Biotech Res 2(3):431–439

Raza ZA, Khan MS, Khalid ZM, Rehman A (2006) Production of Biosurfactant using different hydrocarbons by *Pseudomonas aeruginosa* EBN-8 mutant. Z Naturforsch 61(1–2):87–94

Reis RS, Pereira AG, Neves BC, Freira DMG (2011) Gene regulation of rhamnolipid production in *Pseudomonas*: a review. Biores Technol 102:6377–6384

Ribeiro IA, Faustino CM, Guerreiro PS, Frade RF, Bronze MR, Castro MF, Ribeiro MH (2015) Development of novel sophorolipids with improved cytotoxic activity toward MDA-MB-231 breast cancer cells. J Mol Recognit. doi:10.1002/jmr.2403

Rizzo C, Michaud L, Hormann B, Gerçe B, Syldatk C, Hausmann R, De Domenico E, Lo Giudice A (2013) Bacteria associated with sabellids (Polychaeta: Annelida) as a novel source of surface active compounds. Mar Pollut Bull 70(1–2):125–133

Rocha CA, Pedregosa AM, Laborda F (2011) Biosurfactant-mediated biodegradation of straight and methyl-branched alkanes by *Pseudomonas aeruginosa* ATCC 55925. AMB Express. doi:10.1186/2191-0855-1-9

Rodrigues LR (2015) Microbial surfactants: fundamentals and applicability in the formulation of nano sized drug delivery vectors. J Colloid Interface Sci 1:304–316

Roldan-Carrillo T, Martinez-Garcia X, Zapata-Penasco I, Castorena-Cortes G, Reyes-Avila J, Mayol-Castillo M, Olguin-Lora P (2011) Evaluation of the effect of nutrient ratios on biosurfactant production by *Serratia marcescens* using a Box-Behnken design. Colloids Surf B Biointerfaces 86(2):384–389

Rufino RD, De Luna JM, De Campos Takaki GM, Sarubbo LS (2014) Characterization and properties of the biosurfactant produced by *Candida lipolytica* UCP 0988. Electronic J Biotechnol 17:34–38

Ruhal R, Kataria R, Choudhury B (2013) Trends in bacterial trehalose metabolism and significant nodes of metabolic pathway in the direction of trehalose accumulation. Microb Biotechnol 6 (5):493–502

Saikia RR, Deka H, Goswami D, Lahkar J, Borah SN, Patowary K, Baruah P, Deka S (2014) Achieving the best yield in glycolipid biosurfactant preparation by selecting the proper carbon/nitrogen ratio. J Surfact Det 17(3):563–571

Saikia RR, Deka S, Deka M, Sarma H (2012) Optimization of environmental factors for improved production of rhamnolipid biosurfactant by *Pseudomonas aeruginosa* RS29 on glycerol. J Basic Microbiol 52:446–457

Saimmai A, Rukadee O, Onlamool T, Sobhon V, Maneerat S (2012) Isolation and functional characterization of a biosurfactant produced by a new and promising strain of *Oleomonas sagaranensis* AT18. World J Microbiol Biotechnol 28(10):2973–2986

Sana S, Bhattacharya M, Datta S, Biswas D (2015) RSM study for the production of rhamnolipid using Catla catla Fish fat. Int J Curr Microbiol App Sci 4(1):169–178

Sandrin C, Peypoux F, Michel G (1990) Co-production of surfactin and iturin A, lipopetides with surfactant and antifungal properties by *Bacillussubtilis*. Biotechnol Appl Biochem 12:370–375

Sanket KG, Yagnik BN (2013) Current trend and potential for microbial biosurfactants. Asian J Exp Biol Sci 4(1):1–8

Schmidberger A, Henkel M, Hausmann R, Schwartz T (2014) Influence of ferric iron on gene expression and rhamnolipid synthesis during batch cultivation of *Pseudomonas aeruginosa* PAO1. Appl Microbiol Biotechnol 98(15):6725–6737

Shaligram NS, Singhal RS (2010) Surfactin a review on biosynthesis, fermentation, purification and applications. Food Technol Biotechnol 48(2):119–134

Sharafi H, Abdoli M, Hajfarajollah H, Samie N, Alidoust L, Abbasi H, Fooladi J, Zahiri HS, Noghabi KA (2014) First report of a lipopeptide biosurfactant from thermophilic bacterium *Aneurinibacillus thermoaerophilus* MK01 newly isolated from municipal landfill site. Appl Biochem Biotechnol 173(5):1236–1249

Sharma D, Saharan BS, Chauhan N, Bansal A, Procha S (2014) Production and structural characterization of *Lactobacillus helveticus* derived biosurfactant. Sci W J. doi:10.1155/2014/493548

Sharma D, Singh SB (2014) Simultaneous production of biosurfactants and bacteriocins by probiotic *Lactobacillus casei* MRTL3. Int J Microbiol. doi:10.1155/2014/698713

Sierra-Garcıa IN, Alvarez CJ, Vasconcellos SP, Souza AP, Neto EV (2014) New hydrocarbon degradation pathways in the microbial metagenome from Brazilian petroleum reservoirs. PLoS ONE. doi:10.1371/journal.pone.0090087

Singh AK, Cameotra SS (2013) Efficiency of lipopeptide biosurfactants in removal of petroleum hydrocarbons and heavy metals from contaminated soil. Environ Sci Pollut Res Int 20 (10):7367–7376

Singh AK, Rautela R, Cameotra SS (2014a) Substrate dependent in vitro antifungal activity of *Bacillus* sp strain AR2. Microb Cell Fact 13:67. doi:10.1186/1475-2859-13-67

Singh BN, Rawat AK, Khan W, Naqvi AH, Singh BR (2014b) Biosynthesis of stable antioxidant ZnO nanoparticles by *Pseudomonas aeruginosa* rhamnolipids. PLoS ONE 9(9):e106937. doi:10.1371/journal.pone.0106937

Soberon-Chavez G, Maier RM (2011) Biosurfactants: a general overview. In: Soberon-Chavez G (ed) Biosurfactants. Springer, Berlin, pp 1–11

Solaiman DKY, Ashby RD, Crocker NV (2015) High-titer production and strong antimicrobial activity of sophorolipids from R*hodotorula bogoriensis*. Biotechnol Prog. doi:10.1002/btpr.2101

Sousa M, Melo VM, Rodrigues S, Santana HB, Goncalves LR (2012) Screening of biosurfactant producing *Bacillus* strains using glycerol from the biodiesel synthesis as main carbon source. Bioprocess Biosyst Eng 35(6):897–906

Sriram MI, Gayathiri S, Gnanaselvi U, Jenifer PS, Mohan Raj S, Gurunathan S (2011) Novel lipopeptide biosurfactant produced by hydrocarbon degrading and heavy metal tolerant bacterium *Escherichia fergusonii* KLU01 as a potential tool for bioremediation. Bioresour Technol 102(19):9291–9295

Stipcevic T, Knight CP, Kippin TE (2013) Stimulation of adult neural stem cells with a novel glycolipid biosurfactant. Acta Neurol Belg 113(4):501–506

Striebich RC, Smart CE, Gunasekera TS, Susan SM, Ellen MS, Brett WM, Oscar NR (2014) Characterization of the F-76 diesel and Jet-A aviation fuel hydrocarbon degradation profiles of *Pseudomonas* aeruginosa and *Marinobacter hydrocarbonoclasticus*. Int Biodeterior Biodegradation 93:33–43

Sullivan ER (1998) Molecular genetics of biosurfactant production. Curr Opin Biotechnol 9:263–269

Syldatk C, Wagner F (1987) Production of biosurfactants. In: Kosaric N, Cairns WL, Gray NCC (eds) Biosurfactants and biotechnology. Marcel Dekker, New York, pp 89–120

Thaniyavarn J, Chongchin A, Wanitsuksombut N, Thaniyavarn S, Pinphanichakarn P, Leepipatpiboon N, Morikawa M, Kanaya S (2006) Biosurfactant production by *Pseudomonas aeruginosa* A41 using palm oil as carbon source. J Gen Appl Microbiol 52:215–222

Thavasi R, Jayalakshmi S, Banat IM (2011) Application of biosurfactant produced from peanut oil cake by *Lactobacillus delbrueckii* in biodegradation of crude oil. Bioresour Technol 102:3366–3372

Valentin L, Nousiainen A, Mikkonen A (2013) Introduction to organic contaminants in soil: concepts and risks. In: Vicent J et al (eds) Emerging organic contaminants in sludges: analysis, fate and biological treatment, vol 24. Springer, Berlin, pp 1–30

Vanavil B, Perumalsamy M, Rao AS (2013) Biosurfactant production from novel air isolate *P. aeruginosa* NITT6L: screening, characterization and optimization of media. J Microbiol Biotechnol 23:1229–1243

VanBogaert INA, Ciesielska K, Devreese B, Soetaert W (2014) Sophorolipids: Microbial synthesis and application. In: Kosaric N, Sukan FV (eds) Biosurfactant production and utilization—processes, technologies, and economics, vol 159. CRC Press, London, pp 19–36

Van-Dyke MI, Couture P, Brauer M, Lee H, Trevors JT (1993) *Pseudomonas aeruginosa* UG2 rhamnolipid biosurfactants: structural characterization and their usein removing hydrophobic compounds from soil. Can J Microbiol 39(11):1071–1078

Varadavenkatesan T, Murty VR (2013) Production of a lipopeptide biosurfactant by a novel *Bacillus* sp. and its applicability to enhanced oil recovery. ISRN Microbiol. doi:10.1155/2013/621519

Varvaresou A, Iakovou K (2015) Biosurfactants in cosmetics and biopharmaceuticals. Lett Appl Microbiol. doi:10.1111/lam.12440

Velioglu Z, Ozturk Urek R (2015) Optimization of cultural conditions for biosurfactant production by *Pleurotus djamor* in solid state fermentation. J Biosci Bioeng. doi:10.1016/j.jbiosc.2015.03.007

Vollbrecht E, Rau U, Lang S (1999) Microbial conversion of vegetable oils into surface-active di-, tri- and tetrasaccharide lipids (biosurfactants) by the bacterial strain *Tsukamurella* spec. Lipid Lett 101:389–394

Wang XB, Nie Y, Tang YQ, Wu G, Wu XL (2013) n-Alkane chain length alters *Dietzia* sp. strain DQ12-45-1b biosurfactant production and cell surface activity. Appl Environ Microbiol 79(1):400–402

Wei YH, Chou CL, Chang JS (2005) Rhamnolipid production by indigenous *Pseudomonas aeruginosa* J4 originating from petrochemical wastewater. Biochem Eng J 27:146–154

Willenbacher J, Rau J, Rogalla J, Syldatk C, Hausmann R (2015) Foam-free production of surfactin via anaerobic fermentation of *Bacillus subtilis* DSM 10. AMB Express 5:21. doi:10.1186/s13568-015-0107-6

Wu JY, Yeh KL, Lu WB, Lin CL, Chang JS (2008) Rhamnolipid production with indigenous *Pseudomonas aeruginosa* EM1 isolated from oil-contaminated site. Bioresour Technol 99 (5):1157–1164

Xia W, Du Z, Cui Q, Dong H, Wang F, He P, Tang Y (2014) Biosurfactant produced by novel *Pseudomonas* sp. WJ6 with biodegradation of n-alkanes and polycyclic aromatic hydrocarbons. J Hazard Mater 276:489–498

Xia W, Zhi-Bin L, Han-Ping D, Li Y, Qing-Feng C, Yong-Qiang B (2012) Biosynthesis, characterization, and oil recovery application of biosurfactant produced by indigenous *Pseudomonas aeruginosa* WJ-1 using waste vegetable oils. Appl Biochem Biotechnol 166:1148–1166

Yin H, Qiang J, Jia Y, Ye J, Peng H, Qin H, Zhang N, He B (2009) Characteristics of biosurfactant produced by *Pseudomonas aeruginosa* S6 isolated from oil-containing wastewater. Process Biochem 44(3):302–308

Yoshida S, Koitabashi M, Nakamura J, Fukuoka T, Sakai H, Abe M, Kitamoto D, Kitamoto H (2015) Effects of biosurfactants, mannosylerythritol lipids, on the hydrophobicity of solid surfaces and infection behaviours of plant pathogenic fungi. J Appl Microbiol 119(1):215–224

Yu H, Huang GH, An CJ, Wei J (2011) Combined effects of DOM extracted from site soil/compost and biosurfactant on the sorption and desorption of PAHs in a soil-water system. J Hazard Mater 190:883–890

Yu H, Huang GH, Xiao H, Wang L, Chen W (2014) Combined effects of DOM and biosurfactant enhanced biodegradation of polycylic armotic hydrocarbons (PAHs) in soil-water systems. Environ Sci Pollut Res Int 21(17):10536–10549

Yu M, Liu Z, Zeng G, Zhong H, Liu Y, Jiang Y, Li M, He X, He Y (2015) Characteristics of mannosylerythritol lipids and their environmental potential. Carbohydr Res 407:63–72

Zhang L, Veres- Schalnat TA, Somogyi A (2012) Fatty acid cosubstrates provide β-oxidation precursors for rhamnolipid biosynthesis in *Pseudomonas aeruginosa* as evidenced by isotope tracing and gene expression assays. Appl Environ Microbiol 78:8611–8622

Zhang X, Liu X, Wang Q, Chen X, Li H, Wei J, Xu G (2014) Diesel degradation potential of endophytic bacteria isolated from Scirpus triqueter. Diesel degradation potential of endophytic bacteria isolated from Scirpus triqueter. Int Biodeterior Biodegradation 87:99–105

Zhang Y, Liu C, Dong B, Ma X, Hou L, Cao X, Wang C (2015) Anti-inflammatory activity and mechanism of surfactin in lipopolysaccharide-activated macrophages. Inflammation 38 (2):756–764

Zhao F, Shi R, Zhao J, Li G, Bai X, Han S, Zhang Y (2015) Heterologous production of *Pseudomonas aeruginosa* rhamnolipid under anaerobic conditions for microbial enhanced oil recovery. J Appl Microbiol 118(2):379–389

Biodegradation of Petroleum Hydrocarbon and Its Influence on Corrosion with Special Reference to Petroleum Industry

Aruliah Rajasekar

Abstract Crude oils vary in their physical properties, chemical composition and between fields of origin; of special importance are the relative amounts of high and low molecular weight compounds present in crude oils. Heavy oils contain a greater percentage of high molecular weight components. The low biodegradation rates of heavy oils probably reflect the resistance of the complex high molecular weight components to microbial degradation. Hydrocarbons are water insoluble substrates that often need to be emulsified prior to their degradation by microbes. Emulsification is an important phenomenon among those parameters which influence the fate of crude oil. It has been accepted that the microorganisms developing on hydrocarbons produce surface-active agents, or biosurfactants, which emulsify the substrate, enabling its transfer into cells. Biodegraded hydrocarbons can provide bio-corrosive microbial communities with an important source of nutrients. Hence, occurrence of some hydrocarbon-degrading bacteria in oil reservoir/petroleum transporting pipelines influence electrochemical corrosion reactions. These isolates able to cause corrosion of carbon steel. Microbiologically influenced corrosion is a problem commonly encountered in facilities of the oil and gas industries. New insights into microbial corrosion processes may be achieved through understanding of the interaction between biodegradation and corrosion of carbon steel in crude oil reservoir/transporting pipelines. In this chapter, the interaction between hydrocarbon-degrading bacteria on corrosion of carbon steel are elucidated. The role of biodegradation on corrosion was confirmed by gas chromatography–mass spectrometry, X-ray diffraction, electrochemical impedance spectroscopy and weight loss analyses. The potential of this consortium was evaluated for the bioremediation of oil-contaminated environments.

Keywords Biodegradation · Biocorrosion · Petroleum hydrocarbon · *Serratia marcescens* ACE2 · *Bacillus cereus* ACE4 · NMR · GCMS

A. Rajasekar (✉)
Environmental Molecular Microbiology Research Laboratory, Department of Biotechnology, Thiruvalluvar University, Serkkadu, Vellore, Tamilnadu 632115, India
e-mail: rajasekar.aruliah@gmail.com; rajasekargood@gmail.com

K. Heimann et al. (eds.), *Biodegradation and Bioconversion of Hydrocarbons*, Environmental Footprints and Eco-design of Products and Processes,
DOI 10.1007/978-981-10-0201-4_9

1 Introduction

Microbial-influenced corrosion (MIC) refers to corrosion by metabolic activity of microorganisms on a metallic surface. The microorganisms induce corrosion by altering the chemistry at the interface between the metal and the bulk fluid (Jones and Amy 2002; Little and Ray 2002). The total estimated direct cost of metallic corrosion in the US represented approximately 3.1 % of the gross domestic product in the year 2001 (13.4 billion dollars) (Koch et al. 2001). Specifically, in the oil and gas exploration and production sectors, the cost have been estimated to be about 1.4 billion dollars. While these estimates did not distinguish between abiotic and biotic corrosion phenomena, it has been suggested that microbial corrosion represents up to 20 % of the corrosion occurring in industrial systems (Ferrari et al. 1995). Corrosion is a leading cause of pipe failure and is a main component of the operating and maintenance costs of gas industry pipelines (Angostini and Young 1990; Buck et al. 1996; Farthing 1997; Graves and Sullivan 1996; Koch et al. 2001; Pope and Pope 1998). Pipeline corrosion was estimated in 1996 to cost the gas industry about $840 million/year (Buck et al. 1996) and in 2001 it was estimated that the annual cost of all forms of corrosion to the oil and gas industries was $13.4 billion, of which microbially influenced corrosion accounted for about $2 billion (Koch et al. 2001). It has been estimated that 40 % of all internal pipeline corrosion in the gas industry can be attributed to microbial corrosion (Pound 1998). Basic research to increase our understanding of the microbial species involved in microbial corrosion and their interactions with metal surfaces and with other microorganisms will be the basis for the development of new approaches for the detection, monitoring, and control of microbial corrosion. A thorough knowledge of the causes of microbially influenced corrosion and an efficient and effective means of detecting and preventing corrosion are lacking (Emde et al. 1992; Kholodenko et al. 2000; Rajasekar and Ting 2011; Strickland et al. 1996). Many studies have indicated the importance of microbial contamination of stored hydrocarbon fuels, which can lead to blocking of pipelines and filters, reduced fuel quality and corrosion of storage tanks (Gaylarde et al. 1999). The presence of water and microbial contaminants during storage of diesel oil leads to the production of metabolic products, such as acids, which can facilitate the corrosion of storage tanks (Bento et al. 2004; Bento and Gaylarde 1998). Corrosion in storage tanks occurs principally in the aqueous phase and thus in laboratory studies of the process it is fundamental to use a suitable aqueous solution. The aqueous phase must support microbial growth and have adequate electrical conductivity to allow electrochemical measurements of corrosion to be carried out. These problems of microbial contamination in stored diesel oil led us to investigate the electrochemical behavior of steel used in the construction of diesel storage tanks.

The most extensively studied microorganisms in relation to biocorrosion are the sulfate-reducing bacteria (SRB), whose participation in corrosion was evidenced decades ago (Von Wolzogen Kuhr and Van der Klugt 1934). SRB have been repeatedly detected in oil and gas-producing facilities as well as transporting and storage facilities and are most likely the cause of the biocorrosion, souring and biofouling problems that often arise at these sites (Cord-Ruwisch et al. 1988; Burger 1998). The injection of sulfate-containing seawater into the reservoirs during the secondary recovery of oil favors the proliferation of this type of bacteria. Therefore, most of the research on biocorrosion has focused on SRB; however recent studies suggest that other types of bacteria could be involved, such as metal-reducing bacteria and methanogens (Lewandowski et al. 2003; Zhang and Fang 2001; Zhang et al. 2003; Zhu et al. 2003). In the field, biocorrosion occurs when complex microbial consortia interact with metallic surfaces through the establishment of multi-species biofilms in which the different types of bacteria influence the corrosion through a cooperative global metabolism. Biocorrosion studies involving the use of natural microbial communities obtained from industrial systems are scarce. However, such studies would better address the actual problem and provide more useful information for the monitoring, control and prevention of biocorrosion in industrial facilities.

Microorganisms present in aqueous environment form biofilms on solid surfaces. Biofilms consist of populations of microorganisms and their hydrated polymeric secretions. Numerous types of organisms may exist in any particular biofilm, ranging from strictly aerobic bacteria at the water interface to anaerobic bacteria such as sulphate reducing bacteria (SRB) at the oxygen-depleted metal surface. The presence of biofilms can contribute to corrosion in three ways:

1. Physical deposition.
2. Production of corrosive by-products
3. Depolarization of the corrosion cell caused by chemical reaction.

Many of the by-products of microbial metabolism including organic acids and hydrogen sulphide which are corrosive. These materials can concentrate in the biofilm causing accelerated metal attack. Corrosion tends to be self-limited due to the build-up of corrosion reaction products. However, microbes can absorb some of these materials in their metabolism, thereby removing them from the anodic and cathodic sites. The removal of reaction products, termed depolarization, stimulates further corrosion. The surface exhibits its scattered areas of localized corrosion, unrelated to the flow pattern.

The capability of microbes to utilize hydrocarbons as sole energy and carbon resource has drawn significant interest in the last decade (Laczi et al. 2015; Rajasekar et al. 2008). The finding of reputed low molecular weight alkane metabolites in hot oilfield facilities suggests that anaerobic hydrocarbon biodegradation is intrinsic and probably involved in supporting biocorrosive biofilms (Duncan et al. 2009). The petroleum refining industry is one of the largest manufacturing industries in the world. After Western Europe and the USA, Latin America, including Mexico, has the highest refining capacity of the rest of the world. Huge amounts are spent each year on capital equipment, modernization and maintenance, including prevention and treatment of microbial contamination. The major microbial problem in the industry is contamination of stored products, which can lead to loss of product quality, formation of sludge and deterioration of pipework and storage tanks, both in the refinery and at the end-user. Reports of such contamination have increased substantially in recent years (Hill 1987; Hill and Hill 1993), probably due to increasing demand for diesel fuel and high quality gasolines and jet fuel (Batts and Fathomi 1991; Gaylarde et al. 1999). One of the enhancing factors of deterioration of crude oil and its products is the occurrence of microbial corrosion of pipelines and tanks caused by complex action of various bacteria inside the biofilm structure formed on the surface of the metals (Costerton and Lashen 1984; Muthukumar et al. 2003). It is whispered that the bacteria involved in a biological cycle of sulphur, exclusively SRB, play an ultimate role in biocorrosion (Yemashova et al. 2007). As a result, the quality of crude oil detoriated over the time due to generation of hydrogen sulphide by SRB. Although, the MIC is common in any metals and metal alloys, it is of major concern for petroleum industries. In petroleum transportation and storage systems, the hydrocarbon acts as an excellent food source for a wide variety of microorganisms for example *Acinetobacter* sp., *Pseudomonas* sp., *Flavobacterium* sp., *Mycococcus* sp., *Geobacillus* sp., *Micrococcus* sp., *Aeromonas* sp., and *Bacillus* sp. (Rahman et al. 2002; Olliver and Magot 2005;). Water also subsists in diverse solution with hydrocarbon together with other elements including carbon sulphur, nitrogen, and phosphorus that are needed to support microbial life (Coetser and Cloete 2005).

In the production, transport, and processing of oil and gas, a contact of metals with water and soil, where microorganisms and algae are almost always present, brings about biocorrosion involving such biogenic products as hydrogen sulfide (localized corrosion attacks, such as pitting, and intensified general corrosion) and carbon dioxide (carbonate deposits). Bacterial slime and the increased viscosity of water promotes sludging, which results in an increased pressure in producing and injection of oil wells, power over consumption, rock leaching, a change in the water–salt composition of sub oil and stratal waters, and decomposition of process reagents. When adsorbed at rocks, cell biomass intensifies both evolution of H_2S and CO_2 and release of organic acids, thus deteriorating the pool permeability. Finely divided ferrous sulfide and the biomass of dead bacteria fills up the face zone of injection wells, reducing the productivity by 30–40 %.

Obuekwe and Westlake (1987) reported the occurrence of bacteria in the Pembina crude oil pipeline system and their potential role in the corrosion process. The crude oil stream from the Pembina fields of North Central Alberta, Canada, contained a relatively high detectable load of bacteria. The oil and produced water contained aerobic and anaerobic microorganisms capable of producing sulphides from sulphates and sulphite, and ferrous ions from ferric compounds. The ability to produce sulphur and iron (Fe(II)) in solution is considered very important in the corrosion phenomenon in the pipeline system. Apart from SO_4 reducers (*Desulfovibrio* sp.) the organisms found in the crude oil system and capable of generating corrosive environments were mainly members of the genus *Pseudomonas*.

It may seem surprising to look at the percentage of contribution of MIC towards total corrosion failures. Pipeline corrosion was estimated in 1996 to cost the gas industry about $840 million/year (Buck et al. 1996), and in 2001 it was estimated that the annual cost of all forms of corrosion to the oil and gas industries was $13.4 billion, of which microbially influenced corrosion accounted for about $2 billion (Koch et al. 2001). According to a current inspection, damage due to corrosion in the United States is estimated at $276 billion (Little and Lee 2007). The yearly cost of corrosion in the USA oil and gas industry is over $27 billion (Table 1); leading some to estimate the global annual corrosion cost of the oil and gas industry as exceeding ~$60 billion (Papavinasam 2013). Similar surveys in United Kingdom, Australia, Germany and Japan estimated that the cost due to corrosion loss up to 1–5 % of the gross national product. It has been estimated that 40 % of all internal pipeline corrosion in the gas industry can be attributed to microbial corrosion (Pound 1998; Muthukumar 2014).

Microbial diversity associated biodegradation of the hydrocarbons and bio-hazards through their activities is a well established phenomenon. Among these recalcitrant compounds, polycyclic aromatic hydrocarbons (PAHs) represent a unique class of petroleum hydrocarbons due to their pyrogenic nature and the complexity of the assemblages in which they occur (Irish and Richardson 1981; Kanaly and Harayama 2000). Although many genera of microorganisms have the ability to degrade these recalcitrant compounds and use them as a source of carbon or energy, such phenomena are not commonly encountered in enteric bacteria. The enteric bacteria in the family enterobacteriaceae are mainly regarded as inhabitants of animal guts (Diaz et al. 2001) where *Leclercia adecarboxylata* is noticed in oily-sludge contaminated soil. In this article an enteric bacterium, *Serratia marcescens*, in a cross-country pipeline in North India is reported. Microbial corrosion studies involving the use of natural individual species obtained from industrial systems are scarce. However, such studies would better address the actual problem and increase the understanding of the microbial species involved in the degradation of diesel as well as their role in microbial corrosion and their interactions with the metal surfaces which will be the basis for the development of new approaches for the detection, monitoring and control of microbial corrosion in industrial facilities. Hence, a focus of chemolithotrophic bacteria in elucidating corrosion is essential for petroleum product pipeline. In the present investigation the role of *Serratia marcescens* ACE2 and *Bacillus cereus* ACE4 identified from the corrosion products of a diesel pipeline, on biodegradation of diesel and its influence on the corrosion of API 5LX steel has been studied.

Table 1 Percentage of biodegradation efficiency of diesel degradation by *S. marcescens*

Retention time (time)	Compounds	Control	Strain ACE2	Biodegradation efficiency BE (%)	Strain ACE4	
		Relative abundance (RA)	Relative abundance (RA)		Relative abundance (RA)	Biodegradation efficiency BE (%)
2	C 10	21	0	100	0	100
3	C 11	38	0	100	0	100
5.1	C 12	61	20	67.21	0	100
6	C 13	80	36	55	0	100
7.1	C 14	83	38	54	0	100
8.2	C 15	97	64	34	100	100
11.8	C 16	100	45	55	0	100
12.6	C 17	99	46	53	0	100
14.5	C 18	95	0	100	0	100
15.1	C 19	78	38	51	0	100
16.5	C 20	38	24	27	0	100
3.7	Napthalene derivatives Compound-I	18	0	100	0	100
4	Napthalene derivatives Compound-II	50	48	4	0	100
5.8	Branched alkanes Compound-I	39	0	100	37	5
9.5	Branched alkanes Compound-I	100	39	61	45	65
16	Phytane Compound-I	45	0	100	35	22
13.1	pristane Compound-I	98	43	56	40	59
18.1	pristane Compound-II	14	0	100	12	14
11.5	Farnithane Compound-I	98	38	61	35	64
7.6	Farnithane Compound-I	30	0	100	28	7
BE (total percentage)				69		77

2 Materials and Methods

2.1 Culture Condition and Media Composition

Two strains are used throughout this study are *Serratia marcescens* ACE2 and *Bacillus cereus* ACE4 were isolated from a corrosion product at the diesel transporting pipelines in a northwestern region of India and identified as described earlier (Rajasekar et al. 2010). The 16S rDNA gene analysis and the nucleotide sequence data of ACE2 and ACE4 were deposited in GenBank under accession numbers DQ092416 and AY912105. Bacterial cultures ACE4 and ACE2 were pre-cultured overnight in 250 mL Erlenmeyer flasks containing 100 mL of Bushnell Hass medium (BH) consisting of magnesium sulphate 0.20 g/L; calcium chloride 0.02 g/L; monopotassium phosphate 1 g/L; dipotassium phosphate 1 g/L; ammonium nitrate 1 g/L; ferric chloride 0.05 g/L; plus diesel (10 g/L), pH was adjusted to 7 and cultures were incubated at 30 °C in a rotary shaker at 150 rpm until turbid growth was observed. These cultures were used for further studies.

2.2 Utilization of Diesel by Serratia Marcescens ACE2 and Bacillus Cereus ACE4

The bacterial strains ACE2 and ACE4 were grown in batch cultures in 500 mL Erlenmeyer flasks containing 300 mL of Bushnell Hass broth supplemented with 1 g of diesel fuel as a sole carbon source. Diesel samples were sterilized by filtering through a Millipore 0.45 μm pore size membrane filter. The flasks were inoculated with 2 mL of inoculam of 10^8 CFU/mL of ACE2 and ACE4 and incubated aerobically at 37 °C on a rotary shaker operated at 200 rpm for 30 days. An un-inoculated control flask was incubated in parallel to monitor abiotic losses of the diesel substrate. All experiments including un-inoculated controls were performed in duplicate.

2.3 Diesel Degradation Analyses

At the end of the 30 days of incubation period, the residual diesel in the entire flask was extracted first with an equal volume of dichloromethane from the un-inoculated and inoculated systems. Evaporation of solvent was carried out in a water bath at 40 °C. The resultant diesel solution (1 μL) was dissolved in dichloromethane and analyzed by gas chromatography/mass spectrometry (GC–MS) and Nuclear Magnetic Resonance spectroscopy (NMR).

2.3.1 Gas Chromatography Mass Spectrum Analysis (GC–MS)

The GC–MS (Thermo Finnigan) used was equipped with a RTX-5 capillary column (30 m long × 0.25 mm internal diameter) and highly pure nitrogen constituted the carrier gas. Oven temperature was between 80 and 250 °C, ramping at 10 °C/min. To determine the extent of degradation, various hydrocarbon components of diesel were quantified by interpreting the areas of individual peaks and expressed as percentage of degradation relative to the amount of the corresponding peak remaining in the appropriate abiotic control samples. The chromatographic ratios C_{15}/farnesane, C_{17}/pristane and C_{18}/phytane were calculated from the peak heights. The degradation of diesel as a whole was expressed as the percentage of diesel degraded in relation to the amount of the remaining fractions in the appropriate abiotic control samples. The biodegradation efficiency (BE), based on the decrease in the total concentration of hydrocarbons, was evaluated by using the following expression (Eq. 1) as described by Luigi Michaud et al. (2004),

$$\mathrm{BE}\,(\%) = 100 - (A_{\mathrm{s}} \times 100/A_{\mathrm{ac}}) \tag{1}$$

where A_s = total area of peaks in each sample, A_{ac} = total area of peaks in the appropriate abiotic control, BE (%) = efficiency of biodegradation.

2.3.2 Nuclear Magnetic Resonance Spectroscopy (NMR)

NMR analysis (Bruker, 300 MHz) was used to detect the protons in diesel hydrocarbon. The sample of diesel was dissolved using deuterated chloroform as a solvent. Tetra Methyl Silane (TMS) was used as a reference standard.

2.4 Aryl Hydrocarbon Hydroxylase Assay

Aryl hydrocarbon hydroxylase (AHH) was measured in bacterial samples as described by Dehnen et al. (1973). All assays were carried out in the presence of NADPH (reduced β-nicotinamide adenine dinucleotide phosphate) and measured fluorometrically (excitation 460 nm, emission 517 nm) under yellow light. One milliliter of assay volume contained 100 mM triethanolamine-HCl (pH 7.25), 4 mM NADPH, 60 μM benzo[a]pyrene, and 100 μL of microsomal homogenate, which was incubated for 10 min. Arbitrary fluorescence units were converted into picomoles of benzo[a]pyrenephenols formed using intercalibration between quinine sulfate and 3-hydroxybenzo[a]pyrene. The metabolite 3-hydroxybenzo-[a]pyrene was obtained from the NCI Chemical Carcinogen Repository, USA.

2.5 *Corrosion Studies*

2.5.1 Weight Loss Method

Steel API 5LX (C-0.29 max, S-0.05 max, P-0.04 max, Mn-1.25 max, balance Fe) coupons of size 2.5 × 2.5 × 0.5 cm were mechanically polished to mirror finish and then degreased using trichloroethylene. In the present study, 500 mL of diesel with 2 % water containing 120 ppm chloride was used as the control system, while 500 mL diesel with 2 % of water containing 120 ppm chloride inoculated with 2 mL of ACE2 inoculum (about 10^8 CFU/mL) was used as the experimental system I and 500 mL diesel with 2 % of water containing 120 ppm chloride inoculated with 2 mL of ACE4 inoculum (about 10^8 CFU/mL) as the experimental system II. After 7 days, the coupons were removed, pickled and washed with water and dried with an air drier. Final weights of the six coupons in each system were taken and the average corrosion rates were calculated. The standard deviation for each system is also presented.

2.5.2 Electrochemical Studies

For electrochemical studies, a mixture of diesel and water (containing 120 ppm chloride ion) in the ratio, 2:1 was made (De Schiapparelli and de Meybaum 1980). The API 5LX steel coupons were embedded in araldite with an exposed area of 1 cm^2 as a working electrode. In the present study, 75 mL of 1 % BH broth (containing 120 ppm chloride) and 150 mL of diesel was used as the control system, while 75 mL of 1 % BH broth (containing 120 ppm chloride) and 150 mL of diesel inoculated with 2 mL of inoculum of ACE2 and ACE4 about 10^6 CFU/mL was used as the experimental system I and II respectively. The diesel/water mixtures were stirred vigorously for the periods of 48 and 120 h to simulate the field condition. After the 4th and 10th day, the coupons and electrolytes were removed and potentio-dynamic polarizations and impedance studies were carried out (De Schiapparelli and de Meybaum 1980). Polarization and impedance were taken after 30 min of immersion using Solatron Electrochemical Analyzer (Model 1280B). The polarization was carried out using Corware software from a cathodic potential of −0.2 V to an anodic potential of +0.2 V with respect to the corrosion potential at a sweep rate of 120 mV/min. A coupon of 1 cm^2 API 5LX as working electrode, a standard calomel electrode (SCE) and a platinum wire as counter electrode were employed for polarization study. The tafel polarization curves were obtained by scanning from the open circuit potential towards 200 mV anodically and cathodically. The impedance measurements were carried out using AC signals of 10 mV amplitude for the frequency spectrum from 100 kHz to 0.01 Hz.

2.6 *Surface Analysis*

2.6.1 X-Ray Diffractometer Analysis

A computer controlled X-ray diffraction technique (XRD), JEOL Model JDX – 8030 was used to scan the corrosion products between 10^0 and $85^0 - 2\theta$ with copper K α radiation (Ni filter) at a rating of 40 kV, 20 mA. The dried corrosion products were collected and crushed into a fine powder and used for XRD analysis for determining the nature of oxides present in the corrosion product.

2.6.2 Scanning Electron Microscope

The surface morphological characteristics of the control and experimental coupons API 5LX were observed under scanning electron microscope (SEM) (Hitachi model S-3000H) at magnification ranging from 50 to 200× operated at accelerating voltage of 25 kV. To examine biofilms on metal surfaces, coupon size 1 × 1 cm were exposed to diesel/water mixtures for the formation of biofilm. After seven days, the coupons were removed and immersed in 2–4 % cacodylate-buffered gludaraldehyde solution for 4 h under refrigerated condition. After 1 h, the coupons were subjected to a series of distilled water washes. Distilled water rinses were followed by a graded series of water/acetone washes (100:0, 75:25, 50:50, 25:75, 0:100 v/v water/acetone). Removal of acetone was accomplished through a series of graded washes of acetone—xylene. Air-dried, coated with gold sputter using lon sputter Jeol model JFC 1100. Coupons were examined under SEM.

3 Results and Discussions

Microbiologically influenced corrosion is a well documented phenomenon in corrosion, which causes deleterious effect on petroleum product pipeline, storage tanks and various industries. A diverse group of bacteria and fungi has been reported in hydrocarbon degradation (Calvo et al. 2002; Richard and Vogel 1999; Frederick and Starkey 1948; Gerdes et al. 2005). In oil pipelines, water can also stratify at the bottom of the line, if the velocity is less than that required to drain water and sweep it through the pipeline system. Liquids (hydrocarbon) stratify along the bottom of the pipe, with water forming a separate layer beneath the liquids where microbial hydrocarbon degradation occurs at the interface (Rajasekar et al. 2005). Hence, the role of bacteria on oil degradation and corrosion is an important area in petroleum product pipelines. However there are only a few reports available on involvement of individual bacterial species on diesel degradation and corrosion. This is the first study that enlightens the role of the manganese oxidizer *Serratia marcescens* ACE2 and *Bacillus cereus* ACE4 on biodegradation and corrosion process in a tropical country pipeline. ACE2 is a Gram-negative, rod-shaped, and non-spore-forming bacterium that exists in dirt and it will grow at 20–40 °C. ACE4 is a Gram-positive, rod-shaped spore-forming bacterium usually found in soil and oil-contaminated sites and grows at 20–40 °C. Previous microbiological studies have concluded that SRB play a major role in microbially influenced corrosion (Hamilton 1985; Voordouw et al. 1992; Fardeaua et al. 2003; Bethencourt et al. 2006). Phylogenetic characterization and environmental scanning microscopy analysis of corrosive bacterial consortium in sour gas pipeline, however, revealed a low abundance of SRB in sour gas pipelines (Zhu et al. 2003; Jan-Roblero et al. 2004). These studies demonstrate that SRB need not be present in abundance in all microbial communities responsible for microbially influenced corrosion in the petroleum industry. The bacterial strain ACE4 was isolated from diesel transporting pipeline where SRB could not detectable in the pipeline. It can be explained that flow velocity may create uniform distribution of oxygen, which may suppress the distribution of SRB (Zhu et al. 2003; Maruthamuthu et al. 2005).

The ability of the enteric bacterium, *Serratia marcescens* to degrade hydrocarbon appears to be an unexpected finding since this feature has always been associated with typical soil bacteria. Besides, enteric bacteria are mainly regarded as an inhabitant of animal guts (Holt et al. 1994). Present understanding of role of hydrocarbon degradation on corrosion by enteric bacterium is clearly in its infancy. This study provides new insights into the influence of degradation of diesel fuel on corrosion by enteric bacteria. Such studies can help in adopting efficient and predictable bioremediation strategies. However, only a handful of reports address utilization of aromatic compounds by *Enterobacteria*, particularly those of the genera *Klebsiella*, *Escherichia*, and *Hafnia* (Grant 1967; Ijah 1998; Diaz et al. 2001). Although there are several reports of bioremediation of high-molecular-weight polyaromatic hydrocarbons (PAHs), research pertaining to biodegradation of these substances by enteric bacteria has been relatively rare (Diaz et al. 2001; Kanaly and Harayama 2000; Rajasekar and Ting 2010). Sarma et al. (2004) discovered a new enteric bacterium *Leclercia adecarboxylata* PS 4040 in oil sludge-contaminated soil and reported the degradation of PAH at the Digboi Oil Refinery in Northeastern India.

Recently, Wongsa et al. (2004) identified *S marcescens* which has the ability to degrade a broad spectrum of hydrocarbon present in crude oil-contaminated soil. Muthukumar et al. (2003) reported that *Brucella* sp. and *Gallionella* sp. could degrade diesel in a transporting pipeline in Northwest India, while Rajasekar et al. in 2005 detected bacterial genera *Pseudomonas* sp., *Bacillus* sp., *Gallionella* sp., *Siderocapsa* sp., *Thiobacillus* sp., *Thiospira* sp., *Sulfolobus* sp. and *Vibrio* sp. in a naphtha pipeline, and reported on interactions between heterotrophs and chemolithotrophs in naphtha-transporting pipelines in Southwest India. Hence, an investigation on the role of facultative anaerobic species and manganese oxidizer *Serratia marcescens* ACE2 and *Bacillus cereus* ACE4 on corrosion process is worthwhile to study in the diesel pipeline in the absence of SRB.

3.1 Biodegradation Studies and Chemical Characterization

Biodegradation of diesel hydrocarbon by *Bacillus cereus* ACE4 and *Serratia marcescens* ACE2 were confirmed by GC–MS and NMR analysis. No evaporation of diesel fuel from the screw capped flasks was observed throughout the experiments. From GC–MS analysis, it was observed that the control system consisted of *n*-alkanes (C_{10}–C_{20}), branched alkanes, naphthalene derivatives, substituted naphthalenes and iso preniod alkanes (pristane, phytane) (Fig. 1a). The inoculated system of *Serratia marcescens* ACE2 degraded almost all *n*-alkanes (C_{10}–C_{20}) and many of the branched alkanes found in diesel (Fig. 1b). Inoculated system of *Bacillus cereus* ACE4 degraded the *n*-alkanes (C_{10}–C_{20}) (Fig. 1c) and many of the branched alkanes found in diesel where C_{10} to C_{21}, *n*-alkanes (numbers designate the number of C atoms); Br, branched alkanes; Na, substituted naphthalenes; Pr, pristane; Ph, phytane. The alkanes, naphthalene, phytane, and pristane peaks were identified by comparison of their retention times and mass spectra with authentic standards.

Branched alkanes and naphthalene derivatives could be found in the degraded products, which is a common feature of many other alkane-degrading microorganisms (Van Beilen et al. 1994; Whyte et al. 1998; Wongsa et al. 2004).

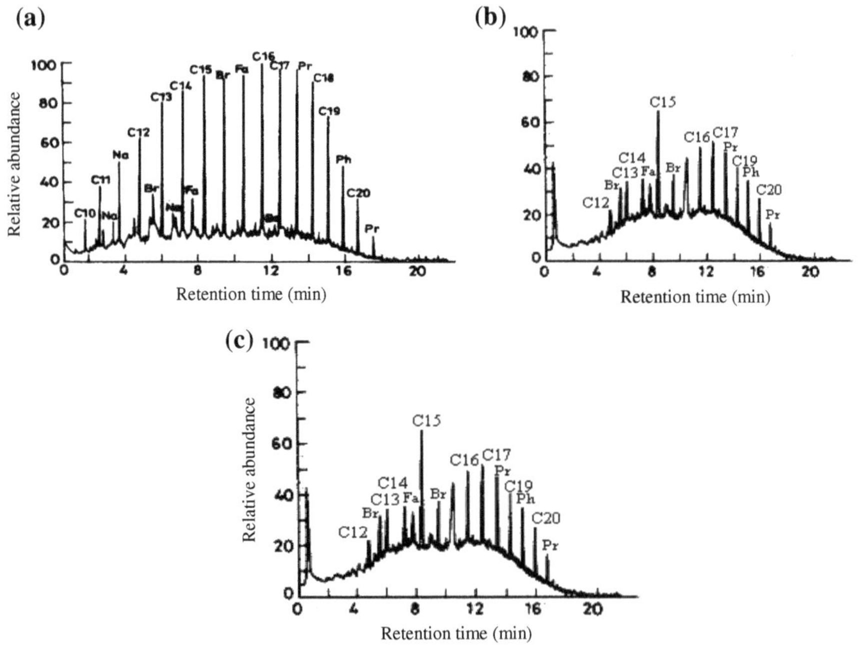

Fig. 1 **a** GC–MS profiles of abiotic control (uninoculated). **b** GC–MS of diesel inoculated with *S. marcescens* ACE2. **c** GC–MS of diesel inoculated with *B. cereus*ACE4

Compared to ACE2 (Fig. 1b). The biodegradation efficiency of diesel hydrocarbon by the strains ACE2 and ACE4 are presented in the Table 1. Bacteria ACE2 and ACE4 showed 69 and 77 % biodegradation efficiency. The GC–MS analysis suggested that ACE2 and ACE4 degrades all the hydrocarbon present in diesel (C_{10} and C_{22}). Moderately degraded 4–5 δ indicates the presence of (–O–CH_2–) after the degradation by bacterial strains. Aliphatic proton (0–3 chemical shifts (δ) peaks and benzene peaks [6–7 chemical shifts (δ)] disappeared because of biodegradation. It reveals inclusion of oxygen with carbon atoms and it can be explained that aliphatic hydrogen (–CH_2–CH_2–) was consumed by degrading bacteria and that the carbon–hydrogen bond was cleaved into carbon–oxygen bond (–O–CH_2–).

From this it is evident that aliphatic protons (CH_2–CH_2) are completely broken by bacterial species, and converted as (–O–CH_2) group. Further, it is certain that the strain is involved in the utilization of both aliphatic and aromatic hydrocarbons present in the diesel. The ACE2 and ACE4 degrades both aliphatic and aromatic hydrocarbon present in the diesel. It was reported that benzene is converted to catechol (an orthohydroxy phenol) and fission occurs between the carbon atoms being two hydroxyl groups to formcis-cis-muconic acid which is then converted into 3-oxoadipic acid via the action of a catechol 1,2-oxygenase (Traxler and Flannery 1968). During degradation *n*-alkanes were converted into (–O–CH_2–) which was identified in NMR studies (Fig. 2b, c). The conversion was due to hydrogen consumption and the very rapid addition of oxygen in the place of hydrogen by bacterial metabolic activity (Muthukumar et al. 2003; Rajasekar et al. 2005). Overall, the bacterium is capable of utilizing a broad range of alkanes and aliphatic components present in diesel (Rajasekar et al. 2007a, b, c, d, e).

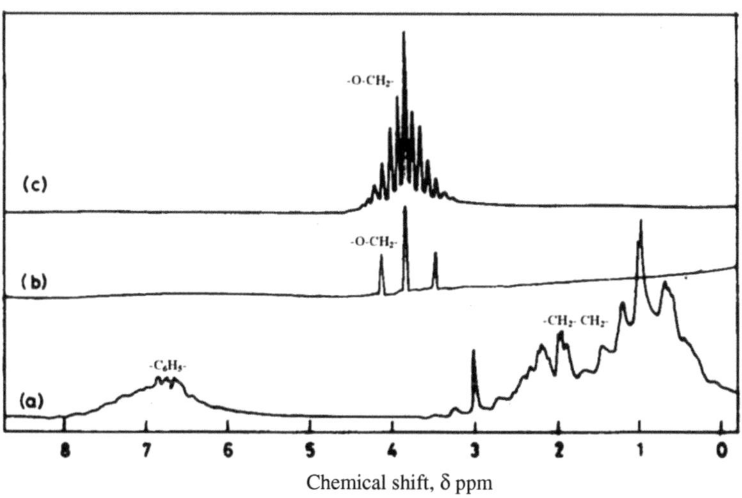

Fig. 2 NMR Spectrum of diesel and diesel inoculated with ACE2 and ACE4. **a** NMR spectrum of diesel. **b** NMR spectrum of diesel inoculated with ACE. **c** NMR spectrum of diesel inoculated with ACE4

3.1.1 Aryl Hydrocarbon Hydroxylase Assay

The activities of AHH were measured during biodegradation of diesel. In the stage of biodegradation, the activity of AHH in ACE2 was measured as 11.38 pmol/min. After 20 days of bacterial inoculation in diesel, the quantity of AHH decreased significantly to below the detection limit (BDL, <0.001 pmol/min). Similarly, in the case of ACE4, the activity of AHH was 15.91 pmol/min, which decreased to BDL after 20 days. This observation indicates that at the initial stage of biodegradation the inoculated bacteria may produce a large quantity or detectable levels of AHH enzyme (Jan et al. 2005). The *Bacillus* species-mediated conversion of PAH with 3–5 rings (acenaphthene, fluoranthene, pyrene, benzo[e]pyrene) was demonstrated earlier (Feitkenhauer et al. 2003). Aitke et al. (1998) also observed mineralization of benzo(a)pyrene (one of the PAHs) by different types of bacterial strains. Since diesel is the major organic nutrient source for bacteria, they consume the carbon from diesel with the help of hydrocarbon metabolizing enzymes such as AHH. After degradation, the availability of simple molecules in diesel or metabolites of diesel/hydrocarbons may inhibit AHH activity and act as substrate inhibitor.

3.2 Corrosion Studies

3.2.1 Weight Loss Method

The corrosion rates of API 5LX steel in the presence and absence of bacteria are presented in Table 2. The corrosion rate in the absence of bacteria was 0.1616 mm/year. However, in the presence of ACE2, the corrosion rate was higher 0.4237 mm/year while in ACE4 inoculated system it was 0.2047 mm/year.

3.2.2 Electrochemical Studies

S. Marcescens ACE2

Figure 3 shows the polarization curve for API 5LX steel in the diesel-water system with and without the bacterium ACE2 on the 4th and 10th days. It can be seen that

Table 2 Corrosion rates of carbon steel API 5LX

System	Weight loss (mg)	Corrosion rate (mm/year)
500 mL oil + 2 %water (120 ppm chloride)	45 ± 2	0.1616
500 mL oil + 2 %water (1 % BH and 120 ppm chloride) + ACE2 culture	118 ± 1	0.4237
500 mL diesel + 2 % water (120 ppm chloride) + ACE4 culture	57.5 ± 1	0.2047

on the 4th day, the corrosion current for the control system was 7.26×10^{-6} A/cm^2 while it was 1.08×10^{-5} A/cm^2 in presence of the bacterium (Table 3). The corrosion potential for the control system was −607 mV while it was −708 mV in the presence of the bacterium at 4th day. This indicates that the potential of carbon steel is shifted by ACE2 towards the cathodic side. It can also be seen from the curves on the 4th day that the anodic current is higher at −550 mV whereas the cathodic current is lower as compared to the control at −750 mV the increasing current is due to the anodic dissolution.

E_{corr} for the control system at the 10th day was −651 mV and in presence of bacteria the E_{corr} potential was −602 mV. On the 10th day, the corrosion current for the control and experimental systems were 1.36×10^{-6} A/cm^2 and 2.94×10^{-6} A/cm^2, respectively. The nature of the curves indicates that the anodic current is lower as compared to that of the control. Besides, insignificant difference in the cathodic currents between bacteria-inoculated and un-inoculated systems can be noticed in all the potentials. On the 4th day, the increasing of anodic current may be due to utility of electrons from of ferric ion which will be converted to ferric oxide by ACE2. The reduction in the cathodic current may be attributed to consumption of oxygen by ACE2 which reduces the oxygen reduction process. On the 10th day, the products of degraded components along with nutrients (1 % BH) may adsorb on the metal surface and reduce the anodic current. At the same time, the converted ferric ions may be reduced to ferrous and supply electrons to bacterium and encourage the cathodic reduction process (Angerer et al. 1992). Hence, it can be concluded that the formation of organic metallic complexes may adsorb on the metal surface and reduce the anodic current. Busalmen et al. (2002) suggested the extracellular production of catalase enzyme (catalase positive) indicates the production of hydrogen peroxide, which encourages

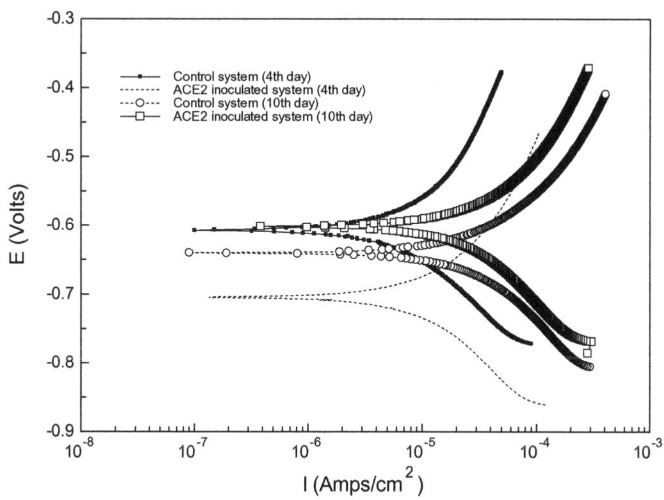

Fig. 3 Polarizationcurves for API 5LX in diesel water system in presence/absence of ACE2 at different periods

Table 3 Polarization study for API 5LX in presence and absence of *S. marcescens* ACE2

Systems	Immersion periods (day)	E_{corr} (mV)	b_a (mV/decade)	b_c (mV/decade)	I_{corr} (A/cm^2)
Control system 75 mL of 1 % BH broth (containing 120 ppm chloride) and 150 mL of diesel	4	−607	0.210	0.148	7.26×10^{-6}
	10	−651	0.113	0.073	1.36×10^{-6}
Experimental system 75 mL of 1 % BH broth (containing 120 ppm chloride) and 150 mL of diesel and inoculated with 2 mL of inoculum ACE2 about 10^6 CFU/ml	4	−708	0.189	0.155	1.08×10^{-5}
	10	−602	0.188	0.168	2.94×10^{-6}

an autocatalytic mechanism by which the oxygen reduction current increases. Data obtained from the impedance measurements were analyzed using the Z view programme, in order to find an equivalent circuit and to give physical meaning to the corrosion process observed during the experimental work. The equivalent circuit shown below was used to adjust the impedance curve obtained for the steel exposed to presence/absence of the bacterium ACE2. The circuit includes the following elements. The obtained data from the equivalent circuit fit are tabulated in Table 4. A resistor (R_s), represents (Eq. 3) the solution resistance, a parallel combination of a resistor (R_{ct}) and a capacitance (Cdl). Appearance of bacterial metabolites with a catalytic effect on the oxygen reduction in the presence of bacteria. Based on the present study, it can be explained that the bacterial values of R_s and R_{ct} were derived from impedance measurements and are presented in Table 4 and Fig. 4. The R_s values for the all the systems are in the range 138–265 Ω cm^2. The R_{ct} value for the control system on the 4th day was 1.57 kΩ cm^2, which was lower as compared to that for the bacterial inoculated system (4.75 kΩ cm^2). On the 10th day, the R_{ct} values for the control and bacterial inoculated systems were 0.46 and 0.60 kΩ cm^2, respectively. The nature of the curves on the 4th day suggests an activation control reaction both in the presence and absence of the bacterial system (Fig. 4b). In both the systems, two capacitive loops can be noticed, the first one being small. The first capacitive loop appears well defined on 10th day with a higher frequency range (Fig. 4d). The high-frequency part represents the formation of the adsorbed film (Ashassi and Nabavi 2002; Rajasekar et al. 2007a, b) whereas the low-frequency data points are associated with the Faradaic processes occurring on the bare metal through defects and pores in the adsorbed diesel components on the metal surface.

Table 4 Impedance parameters for API 5LX in presence/absence of *S. marcescens* ACE2 and *B. cereus* ACE4

Systems	Immersion periods (day)	Solution resistance R_s (Ω)	Charge transfer resistance R_{ct} (k Ω cm^2)
Control system 75 mL of 1 % BH broth (containing 120 ppm chloride) and 150 mL of diesel	4	138	9.3
	10	229	0.46
Experimental system 75 mL of 1 % BH broth (containing 120 ppm chloride) and 150 mL of diesel and inoculated with 2 mL of inoculum ACE2 about 10^6 CFU/mL	4	238	1.57
	10	265	0.60
75 mL of 1 % BH broth (containing 120 ppm chloride) and 150 mL of diesel and inoculated with 2 mL of inoculum ACE4 about 10^6 CFU/mL	4	300	2.8
	10	498	6.9

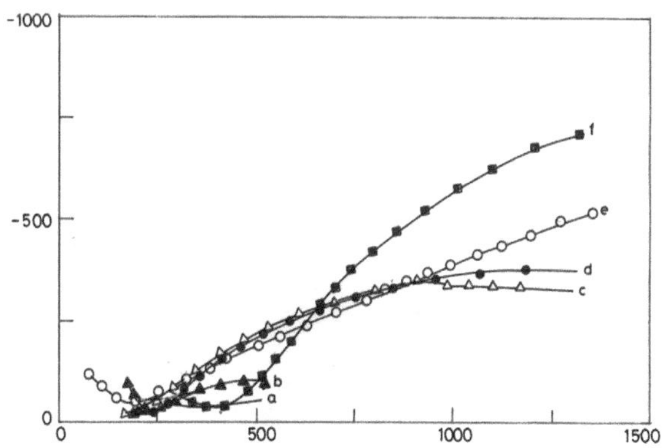

Fig. 4 Impedance curves for API 5LX in diesel water system in presence and absence of ACE2 at different periods. **a** Control—uninoculated system (4th day). **b** ACE2 inoculated system (4th day). **c** ACE4 inoculated system (4th day). **d** ACE2 inoculated system (10th day). **e** Control—uninoculated system (10th day). **f** ACE4 inoculated system (10th day)

B. Cereus ACE4

Figure 5 shows the polarization curve for steel API 5LX in diesel-water system in presence/absence of bacteria on 4th and 10th day. On 4th day, the corrosion current for control system is 4.025×10^{-5} A/cm^2 while in presence of bacteria the current is 2.62×10^{-5} A/cm^2 (Table 5). The corrosion potential for control system is

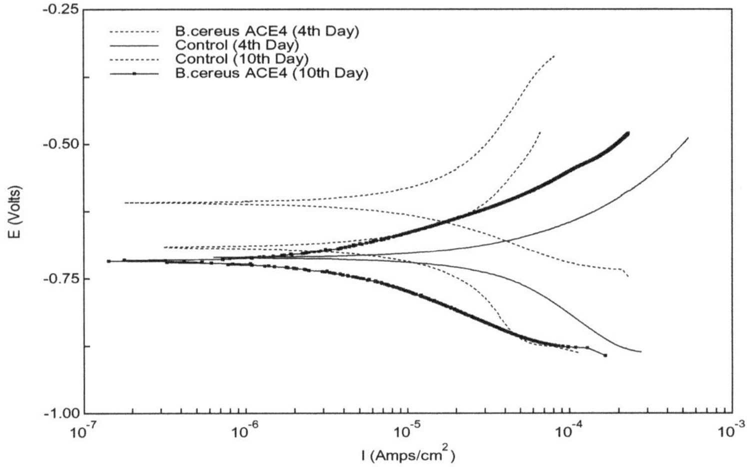

Fig. 5 Polarization curves for API 5LX in diesel water system in presence/absence of ACE4 at different periods

−711 mV versus SCE and in presence of bacteria the potential value is −679 mV versus SCE at 4th day. On the 10th day, the corrosion current for control and experimental system is 1.085×10^{-5} and 3.666×10^{-6} A/cm^2 respectively. The nature of the curves for the 4th day, both anodic and cathodic curves shifted to left side, which indicates that the bacterium ACE4 suppresses both anodic and cathodic reactions.

It may be due to the production of metabolic activity of ACE4 by phosphate based compounds and oxygen consumption in the electrolyte. On 10th day, the

Table 5 Polarization study for API 5LX in presence/absence of *B. cereus* ACE4

Systems	Immersion periods (day)	E_{corr} (mV)	b_a (mV/decade)	b_c (mV/decade)	I_{corr} (A/cm^2)
Control system 75 ml of 1 % BH broth (containing 120 ppm chloride) and 150 ml of diesel	4	−711	158	206	4.025×10^{-5}
	10	−601	283	94	1.085×10^{-5}
Experimental system 75 ml of 1 % BH broth (containing 120 ppm chloride) and 150 ml of diesel and inoculated with 2 ml of inoculum ACE4 about 10^6 CFU/ml	4	−879	330	278	2.62×10^{-5}
	10	−717	106	121	3.666×10^{-6}

potential goes to negative side from −601 to −717 mV versus SCE. In control, slight passivation could be noticed. At the same time the anodic curve shifted to right side and enhances the anodic current on 10th day where the Tafel corrosion current is lower in bacterial system. It can be assumed that though i_{corr} does not show any significant variation on 10th day the formation of pit on the metal sample may be the reason for the higher anodic current.

Impedance parameters for API 5LX in presence of ACE4 is also presented in Table 4 and Fig. 5c, f. In presence of ACE4, the resistance of the API 5LX was 2.8 kΩ cm^2 at the 4th day (Fig. 5c) and on 7th day the resistance was 6.9 kΩ cm^2 (Fig. 5f). The higher resistance can be explained on the basis that, since ACE4g-positive bacteria and spore formers, they enhance the resistance on API 5LX. GC–MS data analysis reveals that ACE4 is a high potent diesel degrader which enhances the resistance by adsorption of degraded diesel products. It is also possible, it may be temporary passivation which may accelerate corrosion with time. Bento et al. (2003) also noticed higher passivation in fungal inoculated system at diesel–water interface.

3.3 Surface Analysis

Figure 6a–c present the details of XRD data corresponding to the phases present in the corrosion product sample collected from control and experimental systems. α-Iron oxide hydroxide (α-FeO(OH), ferrous hydroxide ($Fe(OH)_2$) and ferrous chloride ($FeCl_2$) were observed in the control system (Fig. 6a). Very intense peaks of α-Ferric oxide (Fe_2O_3), Iron oxide (FeO) and manganese dioxide (MnO_2) were noticed in the ACE2 inoculated experimental system (Fig. 6b). XRD results reveal the presence of ferric oxides (Fe_2O_3) and manganese oxides (λ-MnO_2) indicating the role of ACE2 on manganese/iron deposition during the formation of corrosion product and accelerate the microbial corrosion directly on the pipeline. XRD results (9c) reveal the presence of ferric oxides (Fe_2O_3) and manganese oxides (λ-MnO_2) indicating the role of *Bacillus cereus* ACE4 on manganese/iron deposition during the formation of corrosion product and accelerate the microbial corrosion directly on the pipeline.

API 5LX steel samples were exposed to the bacterial cultures of ACE2 and ACE4 examined under SEM after the corrosion studies. Due to the bacterial attack, the coupons were covered by a biofilm, which was thick as seen in Fig. 7a–c at different magnifications. This behavior was observed in most of the experiments, causing a pronounced effect on corrosion damage occurring at the metal surface. The surface morphological characteristics of the control and experimental steel API 5LX coupons were observed under scanning electron microscope. Figure 7a–d show the SEM micrograph of the API 5LX steel coupons after 7 days of exposure in the bacteria inoculated system (ACE2 and ACE4) and un-inoculated system after removing the corrosion products and biofilm. Uniform corrosion was noticed in control system (Fig. 8a) whereas in experimental system severe pitting attacks were

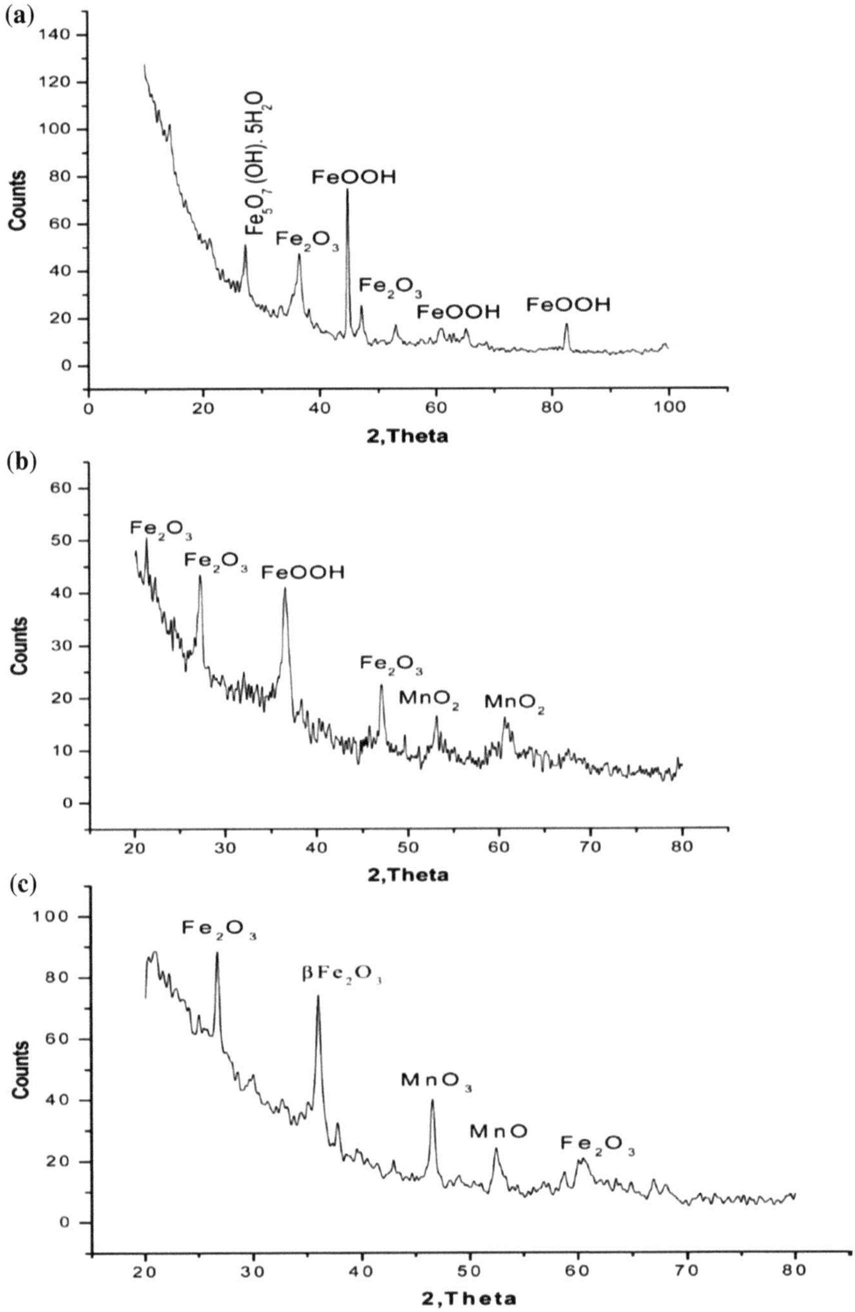

Fig. 6 XRD pattern of corrosion products collected from various systems: **a** un-inoculated system; **b** experimental system, inoculated with *S. marcescens* ACE2; **c** experimental system, inoculated with *B. cereus* ACE4

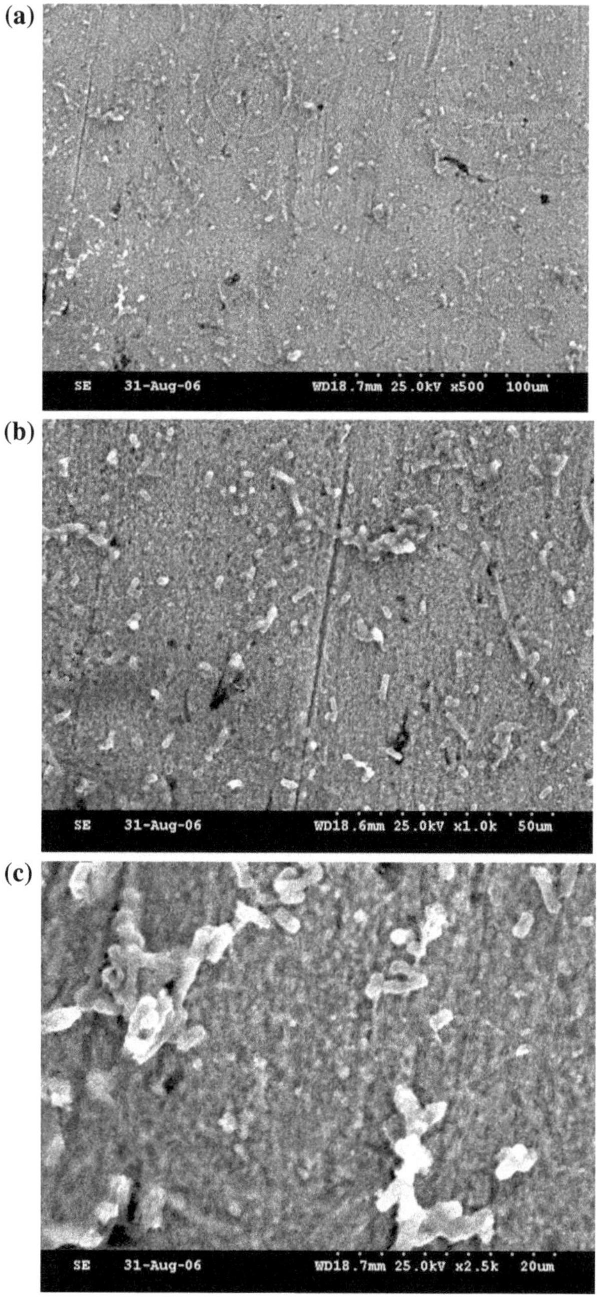

Fig. 7 **a** SEM micrographs of biofilm (ACE2 and ACE4) developed on API 5LX steel (500X). **b** Dense colonies of bacillary cells (2.5 μm; ×1000). **c** Closed view of biofilm (1.5 μm; ×2500)

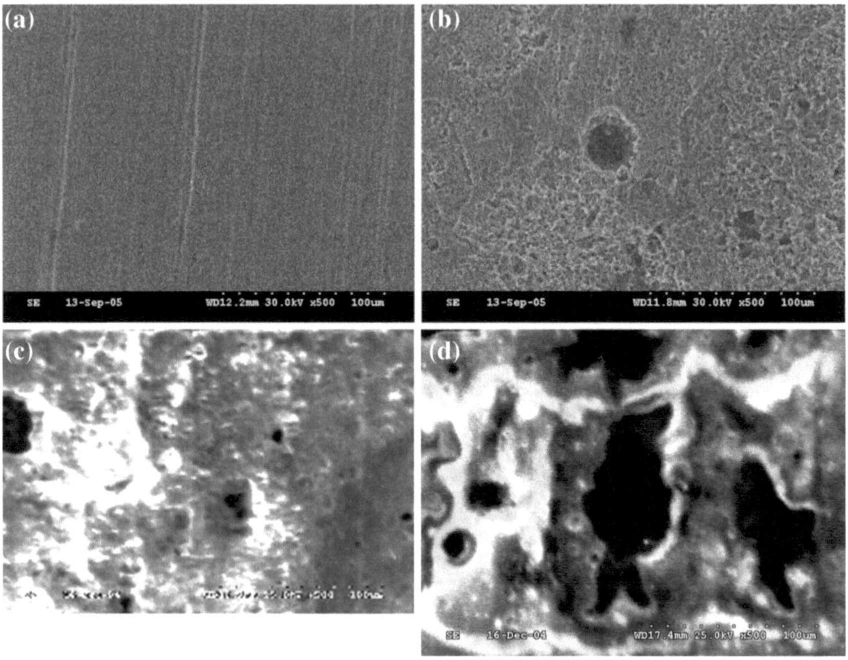

Fig. 8 **a** Scanning electron microscopy (SEM) of API 5LX (control system). **b** SEM of steel API 5LX, showing localized attack, following exposure to ACE2. **c** SEM of steel API 5LX. **d** Closer view of the pit on steel showing localized attack, following exposure to ACE4

observed over the surface of steel (Fig. 8b). Figure 8c shows the magnification of the pit and the size of the pit was about 200 μm in diameters. The SEM study shows that the ACE2 and ACE4 were able to cause pitting type of corrosion in diesel-water interface.

In the same pipeline in Northwest India, *Brucella* sp. and *Gallionella* sp. were reported by Muthukumar et al. (2003). *Brucella* sp. was reported to be a major aromatic degrader where *Gallionella* sp. was reported to be an aliphatic degrader. Besides, *Gallionella* sp. produced more (–O–CH_2–) group due to very rapid addition of oxygen. The observation noticed in the presence of *S. marcescens* supports the findings of Muthukumar et al. (2003). In the present study, *S. marcescens* breaks both the aromatic and aliphatic hydrocarbons as energy source. The addition of oxygen can also be noticed because of the presence of catalase enzyme. Hence, it can be concluded that *S. marcescens* encourages the formation of ferric oxide. Zhu et al. (2003) reported the presence of *Eschercia coli* (AF511430) in a gas pipeline and its association with microbiologically influenced corrosion. Jobson (1975) also reported that intermediate hydrocarbon degradation products make energy sources available for the physiological activities of the corrosion bacterium *Desulfovibrio* sp. (Obuekwe and Westlake 1987). This supply of utilizable hydrocarbon degradation products explains why corrosion is intense in the

Pembiana oil pipeline. Toluene and ethyl benzene are used as sources of carbon and energy by microbes, whereas ethyl benzene is degraded by mono-oxygenase enzyme (Prenafeta-Boldu et al. 2002). Furthermore, *Rhodococcus rhodochrons* S-2 produces extracellular polysaccharides that help to live in aromatic fraction (Iwabuchi et al. 2002). From the present study it can be concluded that *S. marcescens* ACE2 degrades both aliphatic and aromatic hydrocarbons and encourages oxygen addition in aliphatic compounds. It is possible that the converted compounds accelerate the formation of ferric oxide. Subsequently, the same microbe encourages the corrosion process by forming Fe_2O_3.

3.4 Corrosion Mechanism by Serratia Marcescens ACE2

Serratia marcescens ACE2 is a facultative anaerobe and biochemical characteristics indicate the presence of catalase and cytochrome oxidase. This bacterium was isolated from petroleum product pipeline. Since ACE2 has a peroxidase enzyme, which produces hydrogen peroxide for the degradation of petroleum products. Moreover, it produces catalase that overcomes the toxic nature of hydrogen peroxide, breaking peroxide into water and oxygen (Busalmen et al. 2002). During the respiratory process, oxygen is consumed by ACE2 and converted into water (Eq. 2), wherein H^+ is utilized from the degraded product and electrons are supplied by cytochrome oxidase enzyme.

$$1/2O_2 + 2H^+ + 2e^- \rightleftarrows H_2O \tag{2}$$

It can be also assumed that ACE2 favours the Fenton reaction (Touati 2000) by reducing ferric iron, leading to the production of hydroxyl radicals, which can damage any biological macromolecules (Eq. 3).

$$Fe^{2+} + H_2O_2 \rightarrow Fe^{3+} + OH + OH^- \tag{3}$$

It is inferred that the formation of Fe^{3+} combines with OH ions and degraded products and form as iron organic as end product. Since hydrogen and carbon are consumed by ACE2 from diesel, (–O–CH_2–) is formed. The oxygen from peroxide and H^+ molecules from the degraded products combine with Fe^{++} to produce Fe^{+++}. Since this bacteria live in low pH environments, (Maruthamuthu et al. 2005), Fe^{+++} formation is encouraged by peroxide production and degradation according to the Eq. 4.

$$Fe^{2+} + 1/4O_2 + H^+ \rightarrow Fe^{3+} + 1/2H_2O \tag{4}$$

ACE2 converts ferric to ferric oxides by inclusion of oxygen from the degraded products and finally (–CH_2–CH_2–) is formed in the degradation pathway. Finally the ferric organic complex may be formed as corrosion product on the metal surface. In the petroleum product pipeline *Serratia* is a rare species, involved in the corrosion reaction individually without any help from other bacteria.

3.5 Corrosion Mechanism by B. Cereus *ACE4*

Weight loss study indicates the bacterial corrosion on the metal surface. The electrochemical study carried out on the 4th day indicates the bacterium ACE4 suppress both anodic and cathodic reaction whereas on 10th day, ACE4 accelerates the anodic reaction and reduces the cathodic reaction (Fig. 5). It can be explained that due to the adsorption of degraded products with Fe^{++} (organometallic components) on the metal surface, there is a reduction in current flow as such it does not show significant difference in i_{corr} values (Bento et al. 2004). Hence, it can be concluded that this strain ACE4 is also capable of producing oxides of iron and manganese (Nealson et al. 1989; David and Michael 2003). The enzymatic processes are classified by three groups of organisms: (1) Those which oxidize dissolved Mn^{2+} and (2) organisms which oxidize Mn^{2+} pre-bound to certain solids. Both groups utilize oxygen as the terminal electron acceptor and certain organisms may derive useful energy from this reaction. Group (3) organisms oxidize dissolved Mn^{2+} by the metabolic H_2O_2 via catalase. All manganese oxidizing bacteria known so far are aerobes and the oxidation of Mn^{2+} is often closely related to the oxidation of Fe^{2+}. The most manganese oxidizing bacteria form manganese oxides in the cell envelope or completely outside the cell, we may speculate about the possibility of a direct coupling of a Mn redox enzyme to a metal substrate. If this is the case, we could expect anodic potential shifts of passive metals without massive (observable) formation of MnO_2. A similar situation arises if we assume that the manganese oxide, produced by cells on the surface of a metal is reduced immediately by corrosion currents when it is formed. This could mean large corrosion rates for non-passivating metals such as carbon steel (Linhardt 1997).

Since ferric/manganese has high affinity to oxygen, it takes oxygen from ($-O-CH_2-$) and encourages the formation of ferric/manganese oxides and accelerates the corrosion and again ($-CH_2-CH_2-$) was formed. Since electrons are needed continuously for ACE4 for its activity, it converts the Fe^{2+} to Fe^{3+}, Mn^{2+} to $Mn^{3+}$and forms ferric oxides (Fe_2O_3) and manganese oxides (MnO_2) by continuous addition of oxygen. This bacterium is capable of oxidizing manganous ions to manganic ions with concomitant deposition of manganese dioxide and dense accumulation on metal surface and promotes corrosion. Jones and Amy (2002) suggests that the insoluble product is a mixture of ferric oxide (FeO_3) and ferric hydroxide ($Fe(OH)_3$), with an approximate formula is FeO(OH)n. The ferric ions are not precipitated completely, especially in the acidic crevice regions. Ferric ions in solution serve as highly oxidizing species and tend to accelerate corrosion. Moreover, in waters containing chloride ions, iron-oxidizing bacteria may be directly involved in the production of ferric chloride which is an extremely corrosive substance that can

Fig. 9 Proposed model for biodegradation and corrosion process by *S. marcescens* ACE2 and *B. cereus* ACE4 in a tropical country pipeline

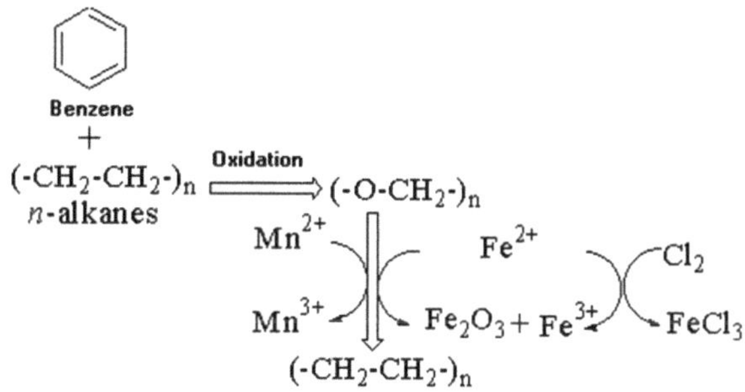

concentrate under nodules (Borenstein and Lindsay 1988; Tiller 1982). Chloride ions can migrate into a crevice location by neutralizing the increased charge via anodic dissolution and then combine with the oxidation of ferrous and manganous ions to ferric and manganic ions by ACE4 and it is observed in the corrosion products (Fig. 6c).

In the present study, though complex is formed on the metal surface, due to the degradation of diesel at the interface act as humic substances resulting in the high supply of electron to the metal surface and accelerates the corrosion. It can be inferred that the degraded diesel hydrocarbon coordinated with Fe^{2+} through the –C–O–C– and carbonyl oxygen resulting in the formation of Fe^{2+} hydrocarbon complex on the metal surface. The corrosion damage observed on the steel surface exposed to the isolate ACE4 was more severe pitting attack on the steel (Fig. 8c, d) than the damage observed under control system (Fig. 5a). It is observed that the cavities created by the bacterium, bioleached the tunnels through the iron/manganese oxides deposit as a result of microbiologically influenced corrosion (Jeffery and Melchers 2003).

Based on the overall result we put forward a model (Fig. 9). It can be concluded that the degraded organic compounds in diesel accelerate the formation of ferric oxide, subsequently the ACE2 and ACE4 encourages the corrosion process by formation of Fe_2O_3. Degraded carbon (diesel) acts as a good nutrient for bacteria, which increases the proliferation of bacteria on the steel and determine the nature of degradation and corrosion.

4 Conclusion

MIC causes hundreds of millions of dollars damage to the production, transport and storage of petroleum products. In summary, this study provided a first look on the role of *Serratia marcescens* ACE2 and *Bacillus cereus* ACE4 in biodegradation of diesel and in the corrosion process of a tropical country pipeline. It can be concluded that *Bacillus cereus* ACE4 and *Serratia marcescens* ACE2 is the most proficient diesel degrader among the culturable organisms isolated from corrosion product collected from diesel transporting pipeline. It is a major aromatic and aliphatic degrader that breaks the benzene ring and aliphatic (–O–CH_2–) chain allowing inclusion of oxygen and accelerating the corrosion by ferric oxide formation in diesel environment. The presence of manganese and ferric oxides in the corrosion products indicated that *Serratia marcescens* ACE2 and *Bacillus cereus* ACE4 is capable of converting the ferric and manganese on metal into oxides and accelerates severe pitting attack on the surface of steel API 5LX. This investigation clearly indicates the possibility of breakdown of diesel by *Bacillus cereus*ACE4at the stagnant areas of the pipelines.

Corrosion studies carried out by weight loss technique and electrochemical study. XRD identified the nature of corrosion product. While inoculating the bacteria (initial state), the bacteria produce large quantity of aryl hydrocarbon hydroxylase. In the later stage, the quantity of the enzyme is low. Since diesel is the major organic nutrient source for bacteria, it cannot consume the carbon from diesel directly. Hence, it secretes 15.91 pmol/min of aryl hydroxylase enzyme for breaking of the diesel for their metabolic activity. After degradation, the availability of simple molecule in diesel does not encourage the secretion of aryl hydroxylase. Total degradation of hydrocarbon was estimated and concluded that the activity of ACE4 on degradation of diesel is vigorous than ACE2. There is no significant difference in weight loss between two species and subsequently, it accelerates pitting corrosion which was examined by SEM. It can be concluded that the degraded organic compounds in diesel accelerate the formation of ferric oxide, subsequently the bacteria ACE2 and ACE4 encourages the corrosion process by formation of Fe_2O_3. Degraded carbon (diesel) acts as a good nutrient for bacteria, which increases the proliferation of bacteria on the steel and determine the nature of degradation and corrosion. Even though, these bacteria could be useful in the bioremediation of diesel polluted habitat, their presence in diesel pipeline and transportation facilities would lead to the reduction in the quality of diesel and in turn economic loss.

References

Aitke MD, Stringfellow WT, Nagel RD, Kazunga C, Chen SH (1998) Characteristics of phenanthrene-degrading bacteria isolated from soils contaminated with polycyclic aromatic hydrocarbons. Can J Microbiol 44:743–752

Angerer A, Klupp B, Braun V (1992) Iron transport systems of *Serratia marcescens*. J Bacteriol 174(4):1378–1387

Angostini RA, Young RD (1990) A case history: investigation of microbially influenced corrosion in a wet Texas water flood. In: Proceedings of the NACE annual conference, corrosion/ 90, paper 119, NACE International, Houston, Texas 1–14

Ashassi SH, Nabavi A (2002) Polarization and impedance methods in corrosion inhibition study of carbon steel by amines in petroleum-water mixtures. Electrochem Acta 47:2239–2244

Batts BD, Fathomi AZ (1991) A literature review on fuel stability studies with particular emphasis on diesel oil. Energy Fuels 5(1):2–21

Bento FM, Gaylarde CC (1998) Biofilms formed on coated or uncoated metal surfaces into diesel/water systems. In: Proceedings of symposium aquatic microbiology (SAM), 1998, Manchester, IMS, London, p 64

Bento FM, Camargo FAO, Okeke B, Frankenberger W (2003) Bioremediation approaches will be soil contaminated with diesel oil. Braz J Microbiol 34:65–68

Bento FM, Englert GE, Gaylarde CC, Muller IL (2004) Influence of aqueous phase on electrochemical biocorrosion tests in diesel/water systems. Mater Corros 55(8):577–585

Bethencourt M, Botana FJ, Maria JC, González L, Mariano MB, Jose MS (2006) Proteccion frente a la corrosion de LA AA7050 en medios con cloruros: sistemas basados en sales de cerio. Perspectiva de la investigacion sobre materiales en Espana en el siglo XXI

Borenstein SW, Lindsay PB (1988) Microbiologically influenced corrosion failures of austenitic stainless steel analysis. Mater Perform 27(8):62

Buck E, Maddux GC, Sullivan RL (1996) Intern corrosion cost impact study—United States natural gas exploration and production industry. GRI-96/0056 document no 96-1466. Gas Research Institute, Des Plaines

Burger ED (1998) Method for inhibiting microbiologically influenced corrosion. US Patent 5,753,180

Busalmen JP, Vazquez M, de Sanchez SR (2002) New evidence on the catalse mechanism of microbial corrosion. Electrochem Acta 47:1857–1865

Calvo C, Martínez-Checa F, Toledo FL, Porcel J, Quesada E (2002) Characteristics of bioemulsifiers synthesized in crude oil media by *Halomonas eurihalina* and their effectiveness in the isolation of bacteria able to grow in the presence of hydrocarbons. Appl Microbiol Biotechnol 60(3):347–351

Coetser SE, Cloete TE (2005) Biofouling and biocorrosion in industrial water systems. Crit Rev Microbiol 31:213–232

Cord-Ruwisch R, Hans-Jiirgen S, Ralf C (1988) The capacity of hydrogenotrophic anaerobic bacteria to compete for traces of hydrogen depends on the redox potential of the terminal electron acceptor. Arch Microbiol 149:350–357

Costerton JW, Lashen ES (1984) Influence of biofilm on efficacy of biocides on corrosion-causing bacteria. Mater Perform 23:13–18

David GK, Michael EM (2003) Emerging themes in manganese transport, biochemistry and pathogenesis in bacteria. FEMS Microbiol Rev 27(2–3):263–290

De Schiapparelli ER, de Meybaum BR (1980) Microbial contamination and corrosion of aircraft integral fuel storage tanks—evaluation risk control. Mater Perform 19(10):47

Dehnen W, Tomingas R, Ross J (1973) A modified method for assay of benzo(a)pyrene hydroxylase. Anal Biochem 53:373–383

Diaz E, Fernandez A, Prieto MA, Garcia JL (2001) Bioremediation of aromatic compounds by *Escherichia coli*. Microbiol Mol Biol Rev 65:523–569

Duncan KE, Gieg LM, Parisi VA, Tanner RS, Tringe SG, Bristow J, Suflita JM (2009) Biocorrosive thermophilic microbial communities in Alaskan north slope oil facilities. Environ Sci Technol 43:7977–7984

Emde KME, Smith DW, Facey R (1992) Initial investigation of microbially influenced corrosion (MIC) in a low temperature water distribution system. Water Res 26:169–175

Fardeaua L, Fernandezb Florina R, Jean-Luc C, Pierre T, Jean-Louis G, Bernard O (2003) *Desulfovibrio capillatus* sp. nov., a novel sulfate-reducing bacterium isolated from an oil field separator located in the Gulf of Mexico. Anaerobe 9:97–103

Farthing S (1997) Company combats MIC with aggressive control program. Pipe Line Gas Ind 80 (10):43

Feitkenhauer H, Muller R, Markl H (2003) Degradation of polycyclic aromatic hydrocarbons and long chain alkanes at 60–70 degrees C by *Thermus* and *Bacillus* spp. Biodegradation 14:367–372

Ferrari MD, De Mele FL, Videla HA (eds) (1995) Manual practico de biocorrosión y biofouling para la industria programa iberoamericano de ciencia y tecnología para el desarrollo, Red Tematica XVc. Biocorr p8

Frederick LR, Starkey RL (1948) Bacterial oxidation of sulphur in pipe sealing mixtures. J Am Water Works Assoc 40:729–736

Gaylarde CC, Bento FM, Kelley J (1999) Microbial contamination of stored hydrocarbon fuels and its control: a mini-review. Rev Microbiol 30:01–10

Gerdes B, Brinkmeyer R, Dieckmann G, Helmke E (2005) Influence of crude oil on changes of bacterial communities in Arctic sea-ice. FEMS Microbiol Ecol 53(1):129–139
Grant DJW (1967) Kinetic aspects of the growth of *Klebsiella aerogenes* with some benzenoid carbon sources. J Gen Microbiol 46:213–224
Graves JW, Sullivan EH (1996) Internal corrosion in gas gathering system and transmission lines. Mater Prot 5:33
Hamilton WA (1985) Sulphate-reducing bacteria and anaerobic corrosion. Annu Rev Microbiol 39:195–217
Hill EC (1987) Fuels. In: Hill EC, Shennan J, Watkinson R (eds) Microbial problems in the offshore oil industry. Wiley, New York, pp 219–230
Hill EC, Hill GC (1993) Microbiological problems in distillate fuels. Trans Inst Mar Eng 104:119
Holt JG, Kreig NR, Sneath PHA, Stanely JT (1994) In: Williams ST (ed) Bergey's manual of determinative bacteriology. Williams and Wilkins Publishers, Maryland
Ijah UJJ (1998) Studies on relative capability of bacterial and yeast isolates from trophical soil degrading crude oil. Waste Manag 18:293–299
Irish GE, Richardson RL (1981) ASTM STP 751, 76
Iwabuchi N, Sunairi M, Urai M, Itoh C, Anzai H, Nakajima M, Harayama S (2002) Extracellular polysaccharides of *Rhodococcus rhodochrous* S-2 stimulate the degradation of aromatic components in crude oil by indigenous marine bacteria. Appl Environ Microbiol 68:2337–2343
Jan B, Van Beilen E, Funhoff G (2005) The expanding the alkane oxygenase fool box: new enzymes and applications. Curr Opin Biotechol 16:308–314
Jan-Roblero J, Romero JM, Amaya M, Le Borgne S (2004) Phylogenetic characterization of a corrosive consortium isolated from a sour gas pipeline. Appl Microbiol Biotehnol 64:862–867
Jeffery R, Melchers E (2003) Bacteriological influence in the development of iron sulphide species in marine immersion environments. Corros Sci 45(4):693–714
Jobson AM (1975) Ph.D. thesis, University of Alberta, Edmonton
Jones DA, Amy PS (2002) A thermodynamic interpretation of microbiologically influenced corrosion. Corrosion 58:638–645
Kanaly RA, Harayama S (2000) Bio-degradation of high-molecular-weight polycycic aromatic hydrocarbons by bacteria. J Bacteriol 182(8):2059–2067
Kholodenko VP, Jigletsova SK, Chugnov VA, Rodin VB, Kobelev VS, Karpov SV (2000) Chemico microbiological diagnostics of stress corrosion cracking of trunk pipelines. Appl Biochem Microbiol 36(6):594–601
Koch GH, Brongers MPH, Thompson NG, Virmani YP, Payer JH (2001) Corrosion costs and preservative strategies in the United States FHWA-RD-01-156 (on-line), Federal Highway Administration, Washington, DC. http://corrosioncost.com
Laczi K, Kis A, Horvath B, Maroti G, Hegedus B, Perei K, Rakhely G (2015) Metabolic responses of *Rhodococcus erythropolis* PR4 grown on diesel oil and various hydrocarbons. Appl Microbiol Biotechnol. doi:10.1007/s00253-015-6936-z
Lewandowski Z, Eugene Cloete T, Dexter SC, Dickinson WH, Kikuchi Y, Little B, Mansfeld F, Rossmoore HS, Videla HA (2003) MIC issues: commentary from the corrosion 2002 MIC panel discussion. CORROSION/2003, paper no. 03560, NACE International, Houston
Linhardt P (1997) Corrosion of metals in natural waters influenced by manganese oxidizing microorganisms. Biodegradation 8:201–210
Little, BJ, Lee, JS (2007) In: Revie RW (ed) Wiley series in corrosion; microbiologically influenced corrosion. John Wiley & Sons, Inc., Hoboken, New Jersey
Little B, Ray R (2002) A perspective on corrosion inhibition by biofilms. Corrosion 58:424–428
Maruthamuthu S, Mohanan S, Rajasekar A, Muthukumar N, Ponmarippan S, Subramanian P, Palaniswamy N (2005) Role of corrosion inhibitors on bacterial corrosion in petroleum product pipeline. Ind J Chem Tech 12(5):567–575
Michaud Luigi, Giudice Angelina Lo, Saitta Marcello, De Domenico Maria, Bruni V (2004) The biodegradation efficiency on diesel oil by two psychrotrophic Antarctic marine bacteria during a two-month-long experiment. Mar Res Bull 49:405–409

Muthukumar N (2014) Petroleum products transporting pipeline corrosion—a review: the role of colloidal systems in environmental protection, 1st edn. M Fanun, Elsevier, pp 528–571
Muthukumar N, Rajasekar A, Maruthamuthu S, Mohanan S, Ponmarriapan S, Palaniswamy N (2003) Microbiologically influenced corrosion in petroleum product pipelines—a review. Ind J Exp Biol 41:1012–1022
Nealson KH, Tebo BM, Rosson RA (1989) Occurrence and mechanisms of microbial oxidation of manganese. Adv Appl Microbiol 33:279–318
Obuekwe CO, Westlake DWS (1987) Occurence of bacteria in Pembina oil pipeline system and their role in corrosion process. Appl Microbiol Biotechnol 26:389–393
Olliver B, Magot M (2005) Petroleum microbiology, 1st edn. ASM Press, Washington
Papavinasam S (2013) Corrosion control in the oil and gas industry. Gulf Professional Publishing is an imprint of Elsevier
Pope DH, Pope RM (1998) Guide for the monitoring and treatment of microbiologically induced corrosion in the natural gas industry. GRI report GRI-96/0488. Gas Research Institute, Des Plaines, IL
Pound BG (1998) Gap analysis of the pipeline research committee international (PRCI)/Gas Research Institute (GRI) research program on internal corrosion. GRI contract 6008. Topical report SF26363.000/AOTO/1198/BP02. Gas Research Institute, Des Plaines, IL
Prenafeta-Boldu FX, Vervoort J, Gvnestijn JTC (2002) Substrate interactions during the bio-degradation of benzene, toluene, ethylbenzene and xylene (BTEX) hydrocarbons by the fungus *Cladophialophora* sp. strain T1. Appl Environ Microbiol 68:2660–2665
Rahman PKSM, Rahman T, Lakshmanaperumalsamy P, Banat IM (2002) Occurrence of crude oil degrading bacteria in gasoline and diesel station soils. J Basic Microbiol 42:284–291
Rajasekar A, Ting YP (2010) Microbial corrosion of aluminum 2024 aeronautical alloy by hydrocarbon degrading bacteria *Bacillus cereus* ACE4 and *Serratia marcescens* ACE2. Ind Eng Chem Res 49:6054–6061
Rajasekar A, Ting YP (2011) Inhibition of biocorrosion of aluminum 2024 aeronautical alloy by conductive ladder polymer poly(o-phenylenediamine). Ind Eng Chem Res 50:2040–2046
Rajasekar A, Maruthamuthu S, Muthukumar N, Mohanan S, Subramanian P, Palaniswamy N (2005) Bacterial degradation of naphtha and its influence on corrosion. Corros Sci 47:257–271
Rajasekar A, Maruthamuthu S, Palaniswamy N, Rajendran A (2007a) Role of *Serratia marcescens* on diesel degradation and its influence on corrosion. J Ind Microbiol Biotechnol 34(9):589–598
Rajasekar A, Ponmariappan S, Maruthamuthu S, Palaniswamy N (2007b) Bacterial degradation and corrosion of naphtha in transporting pipeline. Curr Microbiol 55:374–381
Rajasekar A, Ganesh Babu T, Karutha Pandian S, Maruthamuthu S, Palaniswamy N, Rajendran A (2007c) Biodegradation and corrosion behaviour of *Bacillus cereus* ACE4 in diesel transporting pipeline. Corros Sci 49:2694–2710
Rajasekar A, Ganesh Babu T, Maruthamuthu S, Karutha Pandian S, Mohanan S, Palaniswamy N (2007d) Biodegradation and corrosion behavior of *SerratiamarcescensACE2* isolated from Indian-diesel transporting pipeline. World J Microbiol Biotechnol 23:1065–1074
Rajasekar A, Maruthamuthu S, Palaniswamy N, Rajendran A (2007e) Biodegradation of corrosion inhibitors and their influence on petroleum product pipeline. Microbiol Res 162:355–368
Rajasekar A, Maruthamuthu S, Ting YP (2008) Electrochemical behavior of *Serratia marcescens* ACE2 on carbon steel API5-LX-60 in organic-aqueous phase. Ind Eng Chem Res 47:6925–6932
Rajasekar A, Anandkumar B, Maruthamuthu S, Ting YP, Rahman PKSM (2010) Characterization of corrosive bacterial consortia isolated from petroleum product transporting pipelines. Appl Microbiol Biotechnol 85:1175–1188
Richard JY, Vogel TM (1999) Characterization of a soil bacterial consortium capable of degrading diesel fuel. Int Biodeterior Biodegradation 44(2–3):93–100
Sarma PM, Bhattacharya D, Krishnan S, Lal Banwari (2004) Degradation of polycyclic aromatic hydrocarbons by a newly discovered enteric bacterium, *Leclercia adecarboxylata*. Appl Environ Microbiol 70:3163–3166

Strickland KLN, Fortnum RT, Du Bose BW (1996) A case history of microbiologically influence of corrosion in the Lost Hills oilfield, Kern country, California. In: Proceedings of the NACE annual conference, corrosion 96, paper 297, NACE International, Houston, TX, pp 1–11
Tiller AK (1982) Aspects of microbial corrosion in corrosion processes (RN Purkiss, ed) Appl Sci Pub NY, pp 115–159
Touati D (2000) Iron and oxidative stress in bacteria. Arch Biochem Biophy 373(11):1–6
Traxler RW, Flannery WL (1968) Mechanism of hydrocarbon degradation. In: Walters HA, Elphick J (eds) Biodeterioration of materials, microbiological and allied aspects. Elsevier publishing Co. Ltd, Amsterdam, pp 44–53
Van Beilen JB, Wubbolts MG, Witholt WE (1994) Genetic of alkane oxidation by *Pseudomonas oleovorans*. Biodegradation 5:161–174
Von Wolzogen Kuhr CAH, Van der Klugt IS (1934) The graphitization of cast iron as an electrochemical process in anaerobic solid. Water 18:147
Voordouw G, VoordouwJK Jack TR, Foght J, Fedorak PM, Westlake WS (1992) Identification of distinct communities of sulfate reducing bacteria in oil fields by reverse sample genome probing. Appl Environ Microbiol 58:3542–3552
Whyte LG, Jalal H, Edward Z, Luc B, William EI, Charles WG (1998) Biodegradation of variable-chainlength alkanes at low temperatures by a Psychrotrophic *Rhodococcus* sp. Appl Environ Microbiol 64(7):2578–2584
Wongsa P, Tanaka M, Ueno A, Hasanuzzaman M, Yumoto Okuya I (2004) Isolation and characterization of novel strains of *Pseudomonas aeruginosa* and *Serratia marcescens* possessing high efficiency to degrade gasoline, kerosene, diesel oil, and lubricating oil. Curr Microbiol 49(6):415–422
Yemashova NA, Murygina VP, Zhukov DV, Zakharyantz AA, Gladchenko MA, Appanna V, Kalyuzhnyi SV (2007) Biodeterioration of crude oil and oil derived products: a review. Rev Environ Sci Biotechnol 6:315–337
Zhang T, Fang HHP (2001) Phylogenetic diversity of a SRB-rich marine biofilm. Appl Microbiol Biotechnol 57:437–440
Zhang T, Fang HHP, Ko BCB (2003) Methanogen population in a marine biofilm corrosive to mild steel. Appl Microbiol Biotechnol 63(1):101–106
Zhu XY, Lubeck J, Kilbane JJ (2003) Characterization of microbial communities in gas industry pipelines. Appl Environ Microbiol 69(9):5354–5363

HC-0C-03: Biological Treatments to Improve the Quality of Heavy Crude Oils

Eduardo J. Gudiña and José A. Teixeira

Abstract Petroleum or crude oil is a crucial source of energy and one of the main factors driving the World's economy. The progressive depletion of high-quality crudes in the last decades will make necessary the exploitation of unconventional low-quality heavy and extra-heavy crudes to meet future energy demands. However, their exploitation requires the application of special techniques in order to facilitate their recovery, transportation and refining. Chemical and thermal methods commonly employed (e.g. use of gases, polymers or solvents; hydraulic fracturing; in situ combustion) are expensive and environmentally hazardous. The application of biological treatments to reduce the viscosity and density of unconventional oils can be a cheaper and environmentally friendly alternative or a complementary technology. The bioconversion of crude oil is a process where heavy oil fractions are converted into lighter ones due to the action of microorganisms or enzymes, resulting in an enrichment in lighter hydrocarbons. However, it is necessary to select microorganisms and enzymes with the ability of degrading preferentially the heavy oil fractions (long chain alkanes, aromatics, resins and asphaltenes). As a result, the oil viscosity is reduced (usually from 10^3–10^2 cP up to 10 cP) and its mobility is improved, which contributes to increase oil recovery efficiency and, at the same time, increases the quality of the oil. This chapter will review the latest advances in the use of biological treatments to reduce the viscosity and improve the quality of heavy crude oils.

Keywords Heavy crude oil · Bioconversion · Asphaltenes · Microbial enhanced oil recovery · API gravity

E.J. Gudiña (✉) · J.A. Teixeira
CEB – Centre of Biological Engineering,
University of Minho, 4710-057 Braga, Portugal
e-mail: egudina@deb.uminho.pt

J.A. Teixeira
e-mail: jateixeira@deb.uminho.pt

K. Heimann et al. (eds.), *Biodegradation and Bioconversion of Hydrocarbons*,
Environmental Footprints and Eco-design of Products and Processes,
DOI 10.1007/978-981-10-0201-4_10

1 Introduction

The steady worldwide population increase, rapid improvements in living standards of developing countries and the robust growth of new economies have led to an increasing energy demand in the last years (Mirchi et al. 2012; Sen 2008). Nowadays, fossil fuels remain the largest source of energy worldwide, representing between 80 and 90 % of the current global energy production, with crude oil and gas contributing to about 60 % (Graus et al. 2011; Strubinger et al. 2015). Furthermore, it is estimated that, in the next 20 years, despite the efforts to develop renewable energy resources such as biofuels, solar energy and wind energy, at least 80 % of the World's energy requirements will come from petroleum, natural gas and coal (IEA 2008; Mirchi et al. 2012; Shibulal et al. 2014). As a result, the demand for crude oil to produce refined fuels and petrochemicals is increasing. In contrast, the conventional light crude oil reserves are declining at an average of 4.5 % annually (Höök et al. 2009). According to the International Energy Agency, unconventional oil reserves (including heavy and extra-heavy crude oils, bitumen, oil sands and shale oil) represent between 50 and 70 % of the recoverable oil resources of the World (IEA 2008). These vast oil reserves will play an important role in securing the global energy supply in the next decades, replacing the loss of conventional crudes (He et al. 2015).

Historically, the demand for heavy and extra-heavy oils has been marginal, due to their high viscosity, high density and complex composition when compared with conventional light oils. Their physico-chemical properties represent serious operational problems in their production and downstream processing; as a result, they cannot be recovered, transported and refined using conventional methods, requiring the application of special and costly technologies, which increases their price and can result in severe environmental impacts (Harner et al. 2011; Madden and Morawski 2011). However, the progressive depletion of high quality light crudes has led to an increase in the production of unconventional oils, which is expected to continue rising in the next decades. Their incorporation to the energy markets presents important challenges that require the development of sustainable technologies to recover and upgrade these emerging sources of energy (León and Kumar 2005; Shibulal et al. 2014; Strubinger et al. 2015).

2 Properties of Unconventional Crude Oils

Petroleum or crude oil is a naturally occurring heterogeneous and complex mixture of organic compounds, mainly water-insoluble aliphatic and aromatic hydrocarbons of various molecular weights. Although there is not a strict definition for unconventional crude oils, the main properties that distinguish them when compared with conventional light oils are their high viscosity and density. Unconventional oils are characterized by a heavy molecular composition, containing low percentages of low molecular weight compounds. They also contain high levels of heteroatoms (sulfur, nitrogen and oxygen) and heavy metals (nickel, iron, copper and vanadium) when compared with light oils (León and Kumar 2005; Castro and Vázquez 2009; Madden and Morawski 2011; Shibulal et al. 2014; He et al. 2015).

The American Petroleum Institute (API) gravity, a measure of how heavy or light a petroleum liquid is compared to water, is a parameter commonly used to classify oils according to their density. The API gravity is calculated as follows:

$$^{\circ}API = \frac{141.5}{\rho} - 131.5 \quad (1)$$

where ρ is the density of the oil. Although the API gravity is a dimensionless parameter, it is often expressed in degrees. According to their API gravity, crude oils are classified as light (API gravity > 31.1°), medium (API gravity between 22.3° and 31.1°), heavy (API gravity between 10.0° and 22.3°) or extra-heavy (API gravity < 10.0°). This means that extra-heavy oils have a density greater than water (more than 1000 kg/m^3) (Castro and Vázquez 2009).

The carbon/hydrogen (C/H) mass ratio is another parameter commonly used to express the quality of crude oils. Unconventional oils are in general characterized by high C/H mass ratios (above 6.5 and sometimes even up to 11) due to a high concentration of aromatic compounds (Speight 2014).

The complex nature of unconventional oils makes the study of their molecular composition difficult and expensive. As in the case of conventional oils, SARA fractionation has been widely used to characterize them. This methodology separates crude oil into four main fractions (Saturates, Aromatics, Resins and Asphaltenes) according to their different polarity and solubility.

Saturates are light components of crude oil, and consist mainly of non-polar alkanes with linear or branched chains, and aliphatic cyclic paraffins. Aromatics are species with one or more aromatic rings linked with aliphatic chains, and normally contain embedded heteroatoms. Aromatics have a similar molecular weight as saturates (450–550 g/mol), but a higher C/H mass ratio. Resins are operationally defined as the fraction of crude oils that is soluble in heptane and pentane, but insoluble in liquid propane. Despite the lack of well characterized molecular structures, resins are mainly polar, poly-nuclear molecules composed of aromatic rings, aliphatic side chains and heteroatoms. Their molecular weights vary from 700 to 950 g/mol (Miller et al. 1998; Groenzin and Mullins 2000; Castro and Vázquez 2009; He et al. 2015).

Asphaltenes are the heaviest and more polar fraction of crude oil; they are soluble in aromatic solvents (such as benzene and toluene) but insoluble in *n*-alkanes. The chemical constituents of asphaltenes are similar to those of resins but with a higher molecular weight, and contain higher amounts of heteroatoms and heavy metals. The molecular structure of asphaltene molecules has been an enigma for several decades due to their complexity and their tendency to form aggregates, which presents difficulties in studying them. Their molecular weights, for instance, were reported to spread over a wide range, from less than 10^3 to 10^9 g/mol, which

Table 1 API gravity and composition of light, medium, heavy and extra-heavy crude oils

	Crude oil			
	Light[a]	Medium[b]	Heavy[c]	Extra-heavy[d]
Origin	Turkey	Mexico	Mexico	Venezuela
API gravity	31.5	27.1	11.6	8.1
Composition				
Saturates (%)	41.6	43.1	10.8	17.0
Aromatics (%)	46.0	29.9	42.3	33.1
Resins (%)	8.3	18.2	23.1	35.6
Asphaltenes (%)	4.0	8.7	23.7	14.2
Sulfur (%)	0.4	2.5	4.8	2.6

[a]Kok and Gul (2013)
[b]Sanchez-Minero et al. (2013)
[c]Castorena-Cortés et al. (2012)
[d]Strubinger et al. (2015)

can be explained by the formation of complex aggregates (Miller et al. 1998; Groenzin and Mullins 2000; Desando and Ripmeester 2002; He et al. 2015). There still exists uncertainty regarding if asphaltene molecules consist of one single polycyclic aromatic hydrocarbon (PAH) or several cross-linked PAHs, known as the island and archipelago model of asphaltenes, respectively. Nowadays, the most widely accepted model refers the asphaltene molecules as containing one single, moderately large fused aromatic hydrocarbon ring linked to peripheral alkyl groups (Pineda-Flores et al. 2004; León and Kumar 2005; He et al. 2015).

The SARA content of unconventional oils from different reservoirs around the World varies considerably. However, in general, these oils contain low levels of saturates and high percentages of aromatics, resins and asphaltenes. The higher the content of asphaltenes and resins in the oil, the heavier and more viscous the oil is. Furthermore, the asphaltenic fraction and its strong intermolecular interactions, which form a continuous structure throughout the oil, are thought to be largely responsible for the adverse properties of unconventional oils, such as their high viscosity and their high propensity to form emulsions, polymers, and coke (León and Kumar 2005; Shibulal et al. 2014; Speight 2014).

Table 1 provides comparative details of the API gravity and the composition of heavy and extra-heavy crude oils when compared with light and medium oils.

3 Distribution of Unconventional Crude Oils

There are vast deposits of unconventional crude oils located in more than 30 countries around the World. Their estimated reserves are more than 6 trillion barrels of oil, accounting for about 70 % of total World oil reserves (Lavania et al. 2012; Shibulal et al. 2014; He et al. 2015). The origin of these deposits is the microbial biodegradation of light oils over millions of years, which modifies the molecular composition and physical properties of crude oil: the percentage of low molecular weight compounds, saturates and aromatic fractions is reduced, whereas the concentration of resins, asphaltenes and sulfur is increased, leading to high oil densities and viscosities and to a decrease in the quality of the oil (Head et al. 2003; Harner et al. 2011).

The largest extra-heavy crude oil reservoir in the World is the Orinoco Oil Belt, located in the north of the Orinoco River in Venezuela. In October 2009, the United States Geological Survey estimated that it contained 513 billion barrels of recoverable oil, making this area the World's first recoverable oil deposit, ahead of Saudi Arabia, which estimated oil reserves are 240 billion barrels (León and Kumar 2005; Shibulal et al. 2014). Other important extra-heavy oil reservoirs are the Athabasca Oil Sands (Alberta, Canada), which oil reserves considered to be technically recoverable are estimated to be 280–300 billion barrels (Harner et al. 2011; Madden and Morawski 2011; Shibulal et al. 2014). In Mexico, it is estimated that 61.9 % of the oil reserves correspond to heavy or extra-heavy crudes (Castorena-Cortés et al. 2012). In China, heavy crude oil reserves are also significant (1.64 billion tons), and account for 20 % of the Chinese oil reserves (Zhang and Xiang 2010).

Table 2 summarizes the viscosity and API gravity of some heavy and extra-heavy crude oils obtained from different reservoirs.

4 Difficulties Associated with the Use of Unconventional Oils

The extraction, transportation and processing of unconventional crude oils are more difficult when compared with light and medium oils, and require the application of special and expensive technologies.

Table 2 Viscosity and API gravity of heavy and extra-heavy crude oils from different reservoirs (Rana et al. 2007; Harner et al. 2011; Shibulal et al. 2014; He et al. 2015)

Location	Viscosity (cP)	API gravity (°)
Athabasca Oil Sands (Canada)	10^4–10^6	5.8–12.0
Cold Lake (Canada)	–	6.8
West Sak field (Alaska, USA)	20–90	17.0–21.0
Ugnu field (Alaska, USA)	200–10^4	8.0–12.0
Orinoco Oil Belt (Venezuela)	10^3–10^4	8.5–10.0
Cantarell Field (Mexico)	–	7.9
Utah Oil Sands (USA)	>106	8.0–14.0
Lianhua oil field (China)	400–5600	–
Bikaner-Nagaur basin (India)	10^4	–
Safaniya oil field (Saudi Arabia)	–	13.0

Unconventional oils are often trapped in highly heterogeneous, difficult to extract deposits, and are usually tightly bound to the porous spaces of the reservoir matrix. Furthermore, their high density and viscosity (due to the presence of high concentrations of heavy constituents such as resins and asphaltenes), makes their flow extremely difficult, if not impossible, and can cause the blockage of crude oil extraction equipment. Therefore, the recovery of unconventional crude oils is a challenging and expensive process (Harner et al. 2011; He et al. 2015).

There are two main methods that are used to recover unconventional crude oils. The first one includes non-thermal techniques, where water, gases, chemicals or mechanical methods are used to release the oil from the substrate. They include water flooding, gas injection, alkaline flooding, polymer flooding, solvent injection, hydraulic fracturing or surface mining, among others. The second approach is thermal production, which includes the injection of hot fluids or steam into the reservoir, as well as in situ combustion. These techniques increase the temperature of the oil, allowing its flow to the production wells through the use of mechanical pumps or gas lift (Sen 2008).

The transportation of unconventional crude oils presents also many operational difficulties. Pipelining is the most suitable and common way for transportation of crude oils and their derived products continuously and economically. However, transportation of heavy and extra-heavy crude oils through pipelines is difficult due to their low mobility, the deposition of heavy oil components on the pipeline wall surfaces and the tendency of asphaltenes to flocculate, which slows the oil flow and causes partial plugging in transport pipes. Usually, the pipelining of heavy and extra-heavy oils over significant distances requires the addition of solvents or diluents in order to reduce their viscosity and increase their flow rate (Shibulal et al. 2014; Martínez-Palou et al. 2011).

Furthermore, these unconventional oils must be rigorously treated in order to convert them into upgraded crudes before they can be refined to produce fuels and petrochemicals. Heavy and extra-heavy crudes contain large quantities of non-distillable hydrocarbons. The presence of high concentrations of asphaltenes and heavy metals increases their refining costs, decreasing the economic benefit when compared with conventional light and medium oils. Upgrading can increase the quantity and the quality of distillates obtained, reducing the corresponding vacuum residue (Le Borgne and Quintero 2003; Garcia-Arellano et al. 2004; Pineda-Flores et al. 2004; León et al. 2007; Rana et al. 2007).

5 Biological Upgrading of Heavy Crudes

The expected increase in the production of heavy and extra-heavy crude oils has led oil companies to develop new technologies in order to improve their properties, as well as to facilitate their recovery and transportation. Chemical and thermal methods commonly used are expensive and troublesome, which together with the increasing environmental restrictions resulted in an increased interest in the application of biological treatments to reduce the viscosity and density of unconventional crude oils, as a cheaper and environmentally friendly alternative or as a complementary technology (León and Kumar 2005; Rana et al. 2007; Harner et al. 2011).

For instance, solvents are usually employed to reduce the viscosity of heavy crude oils and help in their recovery and transportation. These chemical compounds could be replaced by biosurfactants, surface-active compounds synthesized by microorganisms, which can perform the same function and, due to their biological origin, are less toxic and more easily biodegradable, thus implying a greater environmental compatibility (Gudiña et al. 2013).

Special attention has been paid to the biological upgrading of heavy and extra-heavy crude oils, the goal of which is to improve their value and make them easier to produce, transport and process. The reactions that could improve the properties of heavy crude oils include cleavage of large aliphatic chains, aromatic rings, heterocycles, resins and asphaltene molecules into smaller compounds to reduce the average molecular weight of crude oil and, at the same time, modify their interactions with other molecules, resulting in the rearrangement of aggregates to liberate the smaller trapped molecules. As a result, the viscosity of the oil is reduced and its flow properties are improved, which enhances the mobility of crude oil through the reservoirs and the pipelines (Premuzic and Lin 1999; Premuzic et al. 1999; Tavassoli et al. 2012).

The degradation of crude oil by microorganisms is a sequential process in which *n*-alkanes are generally degraded first, followed by *iso*-alkanes, cycloalkanes, aromatic compounds (with 1–3 rings) and finally polyaromatics. However, the typical pattern of degradation can vary in different microorganisms, as well as with the type and composition of the oil (Lavania et al. 2012).

In recent years, different bacterial strains (e.g. *Bacillus* and *Petrotoga* species) isolated from oil reservoirs or oil-polluted places with the ability of degrading preferentially the large alkyl chains of different paraffinic mixtures and heavy oils have been described (Wang et al. 2006; Das and Mukherjee 2007; She et al. 2011; Gudiña et al. 2012, 2013; Sugai et al. 2014). These microorganisms increase the relative weight fraction of short *n*-alkanes in crude oil, reducing its viscosity.

As it was previously stated, the asphaltenic fraction is the main component responsible for the high viscosity of heavy and extra-heavy crude oils. Microbial degradation of asphaltenes is an environmentally sound technology to improve the properties of those unconventional oils. However, asphaltenes are the most recalcitrant fraction of crude oil; their high hydrophobicity limits their mass transfer in aqueous reactions and due to their high molecular weight and complex molecular structure, they are usually the least preferred component of crude oil to be degraded by most microorganisms (Pineda-Flores et al. 2004; He et al. 2015). For a long time, there was no clear evidence that asphaltene molecules could be degraded by microorganisms, and it was considered that those compounds were recalcitrant to microbial attack. However, nowadays there is evidence for degradation of asphaltenes by different microorganisms, which can be useful for upgrading heavy oils.

The first reports on biodegradation of asphaltenes by different consortia of microorganisms did not provide sufficient information to claim that the asphaltenic fraction supported bacterial growth or that changes in the asphaltene content occurred. In those assays, asphaltenes were usually determined gravimetrically after precipitation with *n*-alkanes, and the changes reported could be attributed to the disruption of the asphaltenic matrix (for instance due to the production of biosurfactants), which resulted in the release of entrapped hydrocarbons. Therefore, most of the asphaltene losses observed during microbial activity could be considered to be the result of abiotic processes (Rontani et al. 1985; Lacotte et al. 1996; Premuzic et al. 1999). The application of more precise technologies, such as Fourier Transformed Infrared Spectroscopy (FT-IR) to compare the chemical structure of asphaltenes before and after treatment, or spectroscopy techniques (e.g. Nuclear Magnetic Resonance) to study the biodegradation of asphaltenes, allowed a more precise study of their biodegradation (Desando and Ripmeester 2002; Tavassoli et al. 2012; Jahromi et al. 2014).

Several bacterial strains with the ability of using asphaltenes as the sole carbon and energy source have been reported. Most of these bacteria were isolated from heavy and extra-heavy crude oil samples. They include Gram-negative (*Pseudomonas*, *Acinetobacter*, *Alcaligenes*, *Flavimonas*, *Flavobacterium*) and Gram-positive bacteria (*Corynebacterium*, *Bacillus*, *Brevibacillus*, *Staphylococcus*). Bacterial consortia are more effective in bioconversion of asphaltenes than pure cultures, because the effective transformation of the different constituents of the asphaltenic fraction requires the synergistic action of different microorganisms with broad substrate specificity (Pendrys 1989; Pineda-Flores et al. 2004; Tavassoli et al. 2012).

The thermophilic bacterial strain *Thermus* sp. TH-2, isolated from an oil reservoir, degraded aromatic hydrocarbons, resins and asphaltenes from crude oil, reducing the oil viscosity between 10 and 55 % (Hao et al. 2004). *Serratia liquefasciens* and *Bacillus* sp. strains isolated from hydrocarbon-polluted soils degraded 50 % of the asphaltenic fraction of crude oil after 168 h of treatment (Rojas-Avelizapa et al. 2002). Also microbial strains isolated from oil-polluted soil samples of an Iranian oil reservoir (*Pseudomonas* sp., *Bacillus licheniformis*, *Bacillus lentus*, *Bacillus cereus* and *Bacillus firmus*) degraded between 35 and 48 % of asphaltenes after incubation at different temperatures (28–40 °C) for 2 months, using pure or mixed cultures (Tavassoli et al. 2012).

Some microorganisms degraded preferentially the aromatic and asphaltenic fractions of crude oil. The anaerobic and thermophilic microbial consortium TERIB#90 (formed by strains of the genus *Moorella*) isolated from an oil reservoir in India degraded preferentially the asphaltenic fraction of crude oil (up to 56 % of degradation), reducing the crude oil viscosity by 98 % after 30 days of treatment at 70 °C (Lavania et al. 2015). The bacterial strain *Garciaella petrolearia* TERIG02, isolated from a sea buried oil pipeline on the western coast of India, also exhibited the ability of degrading preferentially asphaltenes. In assays performed using Jodhpur heavy crude oil (with a viscosity of 2637 cP and an API gravity of 10°) alone or supplemented with molasses as an additional carbon source, the oil viscosity was reduced 37 and 42 %, respectively, after 30 days of incubation at 50 °C under anaerobic conditions. The strain exhibited preference towards degradation of aromatic and asphaltenic fractions (45 and 55 % of degradation, respectively) when compared with saturates (25 % of degradation) (Lavania et al. 2012).

Jahromi et al. (2014) reported the degradation of asphaltenes by bacterial strains isolated from oil contaminated soils and sludge, including *Pseudomonas aeruginosa*, *Pseudomonas fluorescens*, *Citrobacter amalonaticus*, *Enterobacter cloacae*, *Staphylococcus hominis*, *B. cereus* and *Lysinibacillus fusiformis*. Using different consortia of microorganisms, between 21 and 51 % of asphaltenes were degraded. FT-IR analysis of the asphaltenic fraction showed that, before the treatment, the main functional groups present in asphaltenes were alkyne, aldehyde and alkene groups, whereas after the treatment, only the aldehyde groups remained, due to their higher resistance to the microbial attack (Jahromi et al. 2014).

In contrast, other authors reported on the difficulty of improving the properties of extra-heavy crude oils through biological treatments. Strubinger et al. (2015) studied the effect of a microbial consortium (*Ochrobactrum intermedium*, *Comamones testosteroni* and *Pseudomonas putida*) on the properties of two extra-heavy crude oils from the Orinoco Oil Belt (Venezuela), with API gravities between 8° and 11°. Different culture media, incubation times (up to 45 days) and temperatures were tested, using oil-in-water emulsified systems. After microbial treatments, increases in the initial boiling point and consumption of the light fractions (including *n*-alkanes) of the crude oils were observed. The API gravity, as well as the concentration of heteroatoms and asphaltenes remained unchanged in both crude oils for all conditions tested, and no significant degradation of heavy oil components was observed (Strubinger et al. 2015).

Besides bacteria, yeasts and filamentous fungi with the ability of upgrading heavy and extra-heavy crude oils have also been reported. The fungi *Emericella nidulans*, *Graphium putredinis*, *Eupenicillium javanicum* and *Aspergillus flavipes* degraded up to 28 % of resins and 40 % of asphaltenes from crude oil (Oudot et al. 1993). Also the filamentous fungi *Paecilomyces variotii* and *Fusarium decemcellulare*, and the yeasts *Candida palmioleophila* and *Pichia guilliermondii* degraded between 10 and 15 % of resins and asphaltenes (Chaillan et al. 2004).

Naranjo et al. (2007) isolated several filamentous fungi from crude oil polluted soils that can use extra-heavy crude oil and PAHs as the sole carbon and energy source. The isolates were identified as belonging to the genera *Fusarium*, *Penicillium*, *Trichoderma*, *Aspergillus*, *Neosartorya*, *Pseudallescheria*, *Cladosporium*, *Pestalotiopsis*, *Phoma* and *Paecillomyces*. They synthesized extracellular oxidative enzymes (belonging to the lignin-degrading enzyme systems), suggesting a relationship between those enzymes and the bioconversion of extra-heavy crude oils. Lignin is a highly complex, stable and irregular polymer, which is structurally similar to resins and asphaltene molecules found in crude oil. The lignin-degrading enzyme systems, which exhibit broad substrate specificity, include different oxido-reductases, hydroxylases and ligninolytic peroxidases (such as laccases, lignin peroxidases or manganese peroxidases), which catalyze the oxidation of diverse organic and inorganic compounds. The results obtained by these authors suggest that the bioconversion of heavy crude oils requires the synergistic action of different enzymes with broad substrate specificity. These enzymes, through the partial oxidation of resins and asphaltene molecules, reduced the aromaticity of heavy crude oils, as well as their viscosity and density, but without the loss of carbon atoms, keeping the calorific value of the oil, which makes them of great interest for application in upgrading heavy crude oils (Naranjo et al. 2007).

A *Neosartorya fischeri* strain isolated from a natural asphalt lake of Sucre State (Venezuela), was able of growing using asphaltenes as the sole carbon and energy source, degrading up to 13 % of them. Laccase activity was detected in the medium, suggesting its possible involvement in their degradation (Uribe-Álvarez et al. 2011).

A co-culture of the fungi *Pestalotiopsis* sp. NG007 and *Polyporus* sp. S133 exhibited a high degradation of the asphaltenic fraction of crude oil. 81 % of resins and 79 % of asphaltenes were degraded after 30 days of incubation at 25 °C. A synergistic effect, which enhanced the production of degradative enzymes (catechol dioxygenases, laccase, manganese peroxidase and lignin peroxidase) in the co-culture when compared with the pure cultures, was observed. Also in biodegradation assays performed with the crude enzymes extracted from fungal cultures, the mixture of the enzymes produced by both fungi was more effective in degrading the asphaltenic fraction when compared with the crude enzymes produced by each fungus (Yanto and Tachibana 2014).

Although there are a number of reports on bioconversion of asphaltenes by different microorganisms, none of them described the exact mechanisms involved or the final products obtained. Asphaltenes are probably depolymerized by rupture at active sites containing heteroatoms, allowing the liberation of small entrapped molecules. However, more research is needed to elucidate the metabolic pathways responsible for these transformations at molecular level to further improve the activity of these microorganisms.

Despite some promising results achieved in bioconversion of heavy oil fractions using different microorganisms, a limiting factor is that long incubation times are required, in most cases about 15–60 days or more. The bioconversion of crude oil's heaviest components (such as asphaltenes) using whole cells is difficult, due to their low solubility and low mass transfer rates in aqueous systems (where the microbial activity usually takes place), and the difficulty of their transportation across the cell membrane, which results in low transformation rates. Therefore, it is desirable the development of biotechnological processes that can upgrade heavy and extra-heavy oils in a one phase, non-aqueous system. Enzymes require less water than microorganisms to perform their catalytic activity. In this case, crude oil itself can act as a reaction matrix, thus avoiding or minimizing the addition of water, and consequently increasing the solubility and the bioavailability of hydrophobic substrates. The ability of enzymes of working in the absence of water or with very low amounts of water represents an advantage when compared with the use of microorganisms for crude oil upgrading.

There exists experimental evidence that different enzymes are able to modify asphaltene molecules. Naranjo-Briceño et al. (2012) studied the effect of laccase produced by the fungus *Pestalotiopsis palmarum* BM-04 on extra-heavy crude oil, and observed its ability to oxidize asphaltene molecules. Also chloroperoxidase (a protein with high peroxidase activity) produced by the fungus *Caldariomyces fumago* was able to transform asphaltenes, with a higher efficiency in systems containing organic solvents when compared with aqueous systems (Fedorak et al. 1993).

Garcia-Arellano et al. (2004) modified a cytochrome C in order to increase its activity, its substrate range and its stability at high temperatures. The ability of the enzyme to modify the asphaltene molecules of heavy crude oils was evaluated. The enzyme was active in organic solvent systems with low water content (5 %) and catalyzed the oxidation of sulfur and carbon atoms in the petroporphyrin rich-fraction of asphaltenes (Garcia-Arellano et al. 2004).

Zhang et al. (2015) studied the applicability of fungal extracellular oxidative enzymes to reduce the viscosity of crude oil. Extracellular enzymes were obtained from six cultures of *Aspergillus* spp. strains isolated from oil-contaminated soil samples. All the crude enzyme preparations degraded between 13 and 35 % of resins and 24–34 % of asphaltenes from crude oil after 4 days of treatment at 40 °C. As a result, the oil viscosity was reduced between 40 and 59 % (Zhang et al. 2015).

5.1 Biological Upgrading: In Situ Versus Ex Situ

The application of microorganisms or enzymes to improve the physical and chemical properties of unconventional crude oils can be performed in situ (process occurring in the reservoir) or ex situ (after oil recovery).

In situ treatments offer several advantages over surface upgrading technologies. Bioconversion of heavy crude oils into lighter ones in the oil reservoirs can contribute to recover higher amounts of entrapped heavy oil, avoiding the problems related to their extraction and transportation. It constitutes a low cost and eco-friendly alternative to the traditional chemical or thermal processes employed to facilitate the recovery of these unconventional crudes.

Regarding the use of microorganisms, for application in situ, they should be adapted to the harsh environments of oil reservoirs (e.g. lack of oxygen, high pressure, presence of hydrocarbons and toxic compounds). Extremophile microorganisms or isolates obtained from oil reservoirs with the ability of upgrading heavy oils (as some of those described previously in this chapter) would be useful for this application (Gudiña et al. 2012, 2013; Lavania et al. 2012, 2015; Sugai et al. 2014). However, the in situ upgrading exhibits some problems: difficulty in controlling the microbial growth, variability between different reservoirs, etc.

As in the case of microorganisms, enzymes from extremophile microorganisms are more suitable as they are more adapted to the oil reservoir conditions. It has been demonstrated that enzymatic preparations can be directly injected into the oil reservoirs, resulting in the degradation of heavy oil fractions and reducing the oil viscosity, exhibiting a great potential for enhancing oil recovery (Zhang et al. 2015).

The main advantages of ex situ upgrading include a more precise control of the processes and less severe conditions.

6 Conclusions

The increasing energy demand in the upcoming years will make necessary the exploitation of unconventional crude oil reserves, including heavy and extra-heavy crude oils. However, their exploitation requires the application of special techniques in order to facilitate their recovery, transportation and refining. The chemical and thermal methods commonly used are expensive and environmentally hazardous, which resulted in an increased interest in the application of biological treatments to reduce the viscosity and density of unconventional crude oils, as a cheaper and environmentally friendly alternative or as a complementary technology. The bio-conversion of crude oil is a process where heavy oil fractions are converted into lighter ones due to the action of microorganisms, resulting in an enrichment in lighter hydrocarbons, and consequently, a reduction in oil viscosity and an increase in its mobility, which contributes to increase oil recovery and, at the same time, the quality of the oil. The available research shows that biological upgrading of crude oil, at least to some extent, is a technically feasible process that may be at least a feasible complementary process, if not a replacement for conventional upgrading techniques. However, its application at an industrial level has not been achieved yet, and more research will be necessary to implement this promising technology.

Acknowledgments The authors acknowledge the financial support from the Portuguese Foundation for Science and Technology (FCT) and the European Community fund FEDER, through Program COMPETE, under the scope of the Projects FCOMP-01-0124-FEDER-007025 (PTDC/AMB/68393/2006), PEst-OE/EQB/LA0023/2013, PEST-C/FIS/UI607/2013, RECI/BBB-EBI/0179/2012 (FCOMP-01-0124-FEDER-027462), the strategic funding of UID/BIO/04469/2013 unit, and the Projects “BioEnv - Biotechnology and Bioengineering for a sustainable world” and “Matepro–Optimizing Materials and Processes”. NORTE-07-0124-FEDER-000048, co-funded by the Programa Operacional Regional do Norte (ON. 2—O Novo Norte), QREN, FEDER.

References

Castorena-Cortés G, Roldán-Carrillo T, Reyes-Avila J, Zapata-Peñasco I, Mayol-Castillo M, Olguín-Lora P (2012) Coreflood assay using extremophile microorganisms for recovery of heavy oil in Mexican oil fields. J Biosci Bioeng 114:440–445

Castro LV, Vázquez F (2009) Fractionation and characterization of Mexican crude oils. Energy Fuels 23:1603–1609

Chaillan F, Fleche A, Bury E, Phantavong Y, Grimont P, Saliot A, Oudot J (2004) Identification and biodegradation potential of tropical aerobic hydrocarbon degrading microorganisms. Res Microbiol 155:587–595

Das K, Mukherjee AK (2007) Crude petroleum-oil biodegradation efficiency of *Bacillus subtilis* and *Pseudomonas aeruginosa* strains isolated from a petroleum-oil contaminated soil from North-East India. Bioresour Technol 98:1339–1345

Desando MA, Ripmeester JA (2002) Chemical derivatization of Athabasca oil sand asphaltene for analysis of hydroxyl and carboxyl groups via nuclear magnetic resonance spectroscopy. Fuel 81:1305–1319

Fedorak PM, Semple KM, Vazquez-Duhalt R, Westlake DWS (1993) Chloroperoxidase-mediated modifications of petroporphyrins and asphaltenes. Enzyme Microb Technol 15:429–437

Garcia-Arellano H, Buenrostro-Gonzalez E, Vazquez-Duhalt R (2004) Biocatalytic transformation of petroporphyrins by chemical modified cytochrome C. Biotechnol Bioeng 85:790–798

Graus W, Roglieri M, Jaworski P, Alberio L, Worrell E (2011) The promise of carbon capture and storage: evaluating the capture-readiness of new EU fossil fuel power plants. Clim Policy 11:789–812

Groenzin H, Mullins O (2000) Molecular size and structure of asphaltenes from various sources. Energy Fuels 14:667–684

Gudiña EJ, Pereira JFB, Rodrigues LR, Coutinho JAP, Teixeira JA (2012) Isolation and study of microorganisms from oil samples for application in microbial enhanced oil recovery. Int Biodeter Biodegr 68:56–64

Gudiña EJ, Pereira JFB, Costa R, Coutinho JAP, Teixeira JA, Rodrigues LR (2013) Biosurfactant-producing and oil-degrading *Bacillus subtilis* strains enhance oil recovery in laboratory sand-pack columns. J Hazard Mater 261:106–113

Hao R, Lu A, Zeng Y (2004) Effect on crude oil by thermophilic bacterium. J Petrol Sci Eng 43:247–258

Harner NK, Richardson TL, Thompson KA, Best RJ, Best AS, Trevors JT (2011) Microbial processes in the Athabasca Oil Sands and their potential applications in microbial enhanced oil recovery. J Ind Micorbiol Biotechnol 38:1761–1775

He L, Lin F, Li X, Sui H, Xu Z (2015) Interfacial sciences in unconventional petroleum production: from fundamentals to applications. Chem Soc Rev. doi:10.1039/C5CS00102A

Head IM, Jones DM, Larter SR (2003) Biological activity in the deep subsurface and the origin of heavy oil. Nature 426:344–352

Höök M, Hirsch R, Aleklett K (2009) Giant oil field decline rates and their influence on world oil production. Energy Policy 37:2262–2272

International Energy Agency (IEA). World Energy Outlook (2008) http://www.worldenergyoutlook.org/media/weowebsite/2008-1994/WEO2008.pdf. Accessed on 1 June 2015

Jahromi H, Fazaelipoor MH, Ayatollahi Sh, Niazi A (2014) Asphaltenes biodegradation under shaking and static conditions. Fuel 117:230–235

Kok MV, Gul KG (2013) Thermal characteristics and kinetics of crude oils and SARA fractions. Thermochim Acta 569:66–70

Lacotte DJ, Mille G, Acquaviva M, Bertrand JC (1996) Arabian light 150 asphaltene biotransformation with *n*-alkanes as co-substrates. Chemosphere 32:1755–1761

Lavania M, Cheema S, Sarma PM, Mandal AK, Lal B (2012) Biodegradation of asphalt by *Garciaella petrolearia* TERIG02 for viscosity reduction of heavy oil. Biodegradation 23: 15–24

Lavania M, Cheema S, Lal B (2015) Potential of viscosity reducing thermophillic anaerobic bacterial consortium TERIB#90 in upgrading heavy oil. Fuel 144:349–357

Le Borgne S, Quintero R (2003) Biotechnological processes for the refining of petroleum. Fuel Process Technol 81:155–169

León V, Kumar M (2005) Biological upgradation of heavy crude oil. Biotechnol Bioprocess Eng 10:471–481

León V, Cordova J, Munoz S, De Sisto A, Naranjo L (2007) Process for the upgrading of heavy crude oil, extra-heavy crude oil or bitumens through the addition of a biocatalyst. US patent no US2007/0231870 A1
Madden P, Morawski J (2011) The future of the Canadian oil sands: engineering and project management advances. Energy Environ 22:579–596
Martínez-Palou R, Mosqueira ML, Zapata-Rendón B, Mar-Juárez E, Bernal-Huicochea C, Clavel-López JC, Aburto J (2011) Transportation of heavy and extra-heavy crude oil by pipeline: a review. J Petrol Sci Eng 75:274–282
Miller JT, Fisher RB, Thiyagarajan P, Winans RE, Hunt JE (1998) Subfractionation and characterization of Mayan asphaltene. Energy Fuels 12:1290–1298
Mirchi A, Hadian S, Madani K, Rouhani OM, Rouhani AM (2012) World energy balance outlook and OPEC production capacity: implications for global oil security. Energies 5:2626–2651
Naranjo L, Urbina H, de Sisto A, Leon V (2007) Isolation of autochthonous non-white rot fungi with potential for enzymatic upgrading of Venezuelan extra-heavy crude oil. Biocatal Biotransform 25:341–349
Naranjo-Briceño L, Pernía B, Guerra M, Demey JR, De Sisto A, Inojosa Y, González M, Fusella E, Freites M, Yegres F (2012) Potential role of oxidative exoenzymes of the extremophilic fungus *Pestalotiopsis palmarum* BM-04 in biotransformation of extra-heavy crude oil. Microb Biotechnol 6:720–730
Oudot JP, Dupont J, Haloui S, Roquebert MF (1993) Biodegradation potential of hydrocarbon-degrading fungi in tropical soil. Soil Biol Biochem 25:1167–1173
Pendrys JP (1989) Biodegradation of asphalt cement-20 by aerobic bacteria. Appl Environ Microbiol 55:1357–1362
Pineda-Flores G, Boll-Arguello G, Lira-Galeana C, Mesta-Howard AM (2004) A microbial consortium isolated from a crude oil sample that uses asphaltenes as a carbon and energy source. Biodegradation 15:145–151
Premuzic ET, Lin MS (1999) Induced biochemical conversions of heavy crude oils. J Petrol Sci Eng 22:171–180
Premuzic ET, Lin MS, Bohenek M, Zhou WM (1999) Bioconversion reactions in asphaltenes and heavy crude oils. Energy Fuels 13:297–304
Rana MS, Sámano V, Ancheyta J, Diaz JAI (2007) A review of recent advances on process technologies for upgrading of heavy oils and residua. Fuel 86:1216–1231
Rojas-Avelizapa NG, Cervantes-González E, Cruz-Camarillo R, Rojas Avelizapa LI (2002) Degradation of aromatic and asphaltenic fractions by *Serratia liquefasciens* and *Bacillus* sp. Bull Environ Contam Toxicol 69:835–842
Rontani JF, Bosser-Joulak F, Rambeloarisoa E, Bertrand JC, Faure GR (1985) Analytical study of asphalt crude oil and asphaltenes biodegradation. Chemosphere 14:1413–1422
Sanchez-Minero F, Ancheyta J, Silva-Oliver G, Flores-Valle S (2013) Predicting SARA composition of crude oil by means of NMR. Fuel 110:318–321
Sen R (2008) Biotechnology in petroleum recovery: the microbial EOR. Prog Energy Combust 34:714–724
She YH, Zhang F, Xia JJ, Kong SQ, Wang ZL, Shu FC, Hu JM (2011) Investigation of biosurfactant-producing indigenous microorganisms that enhance residue oil recovery in an oil reservoir after polymer flooding. Appl Biochem Biotechnol 163:223–234
Shibulal B, Al-Bahry SN, Al-Wahaibi YM, Elshafie AE, Al-Bemani AS, Joshi SJ (2014) Microbial enhanced heavy oil recovery by the aid of inhabitant spore-forming bacteria: an insight review. Sci World J. Article ID 309159
Speight JG (2014) The chemistry and technology of petroleum. CRC Press, Boca Raton
Strubinger A, Ehrmann U, León V, DeSisto A, González M (2015) Changes in Venezuelan Orinoco belt crude after different biotechnological approaches. J Petrol Sci Eng 127:421–432
Sugai Y, Komatsu K, Sasaki K, Mogensen K, Bennetzen MV (2014) Microbial-induced oil viscosity reduction by selective degradation of long-chain alkanes. In: SPE-171850-MS. Proceedings of the Abu Dhabi international petroleum exhibition and conference, 10–13 November 2014, Abu Dhabi, UAE

Tavassoli T, Mousavi SM, Shojaosadati SA, Salehizadeh H (2012) Asphaltene biodegradation using microorganisms isolated from oil samples. Fuel 93:142–148

Uribe-Álvarez C, Ayala M, Perezgasga L, Naranjo L, Urbina H, Vázquez-Duhalt R (2011) First evidence of mineralization of petroleum asphaltenes by a strain of *Neosartorya fischeri*. Microb Biotechnol 4:663–672

Wang L, Tang Y, Wang S, Liu RL, Liu MZ, Zhang Y, Liang FL, Feng L (2006) Isolation and characterization of a novel thermophilic *Bacillus* strain degrading long-chain *n*-alkanes. Extremophiles 10:347–356

Yanto DHY, Tachibana S (2014) Potential of fungal co-culturing for accelerated biodegradation of petroleum hydrocarbons in soil. J Hazard Mater 278:454–463

Zhang X, Xiang T (2010) Review of Microbial Enhanced Oil Recovery technology and development in China. Int J Pet Sci Technol 4:61–80

Zhang JH, Xue QH, Gao H, Ma X, Wang P (2015) Degradation of crude oil by fungal enzyme preparations from *Aspergillus* spp. for potential use in enhanced oil recovery. J Chem Technol Biotechnol. doi:10.1002/jctb.4650

HC-0C-07: Isolation and Characterisation of Crude Oil Degrading Microorganisms from Petrochemical Wastewater

Chooi Yit Heng, Madihah Md. Salleh, Adibah Yahya, Zaharah Ibrahim and Huszalina Hussin

Abstract Bacteria with the abilities to degrade crude oil were isolated from soil, activated sludge and biological treatment lagoon of the local petrochemical industries. For the biodegradation process, n-alkanes, of varying carbon chain length, C_{16}–C_{38}, were used. Out of the 12 cultures of bacteria isolated, 3 of the best oil degraders were partially identified via biochemical tests; 2 of which were *Acinetobactor* spp while another one belonged to *Proteus* sp. Degradation of the n-alkanes in crude oil was monitored under agitated and non-agitated condition using gas chromatography technique. Generally, non-agitated cultures showed higher degradation rates. One of the *Acinetobacter* sp. showed the highest degradation rate, in which 80–100 % of the alkanes (C_{16}–C_{38}) in crude oil was degraded without any addition of organic nitrogen and phosphorus. It is of interest to highlight another of the *Acinetobacter* sp. which showed the ability to degrade longer chain alkanes more rapidly than shorter ones; C_{36} and C_{38} were fully degraded in 2 days. Only one bacterium, *Proteus* sp showed increased rates of degradation under agitated condition.

Keywords Crude oil degrading bacteria · Degradation rates · Petrochemical wastewater · n-Alkanes · C_{16}–C_{38} · Gas chromatography

C.Y. Heng · M.Md. Salleh (✉) · A. Yahya · Z. Ibrahim · H. Hussin
Sustainability Resources Research Alliance, Faculty of Biosciences and Biomedical Engineering, Universiti Teknologi Malaysia, 81310 Skudai, Johor, Malaysia
e-mail: madihah@fbb.utm.my

A. Yahya
e-mail: adibah@fbb.utm.my

Z. Ibrahim
e-mail: zaharah@fbb.utm.my

H. Hussin
e-mail: huszalina@fbb.utm.my

K. Heimann et al. (eds.), *Biodegradation and Bioconversion of Hydrocarbons*, Environmental Footprints and Eco-design of Products and Processes, DOI 10.1007/978-981-10-0201-4_11

1 Introduction

The biodegradation of petroleum or its alteration by bacteria has long been discovered and it is a widespread phenomenon in nature (Leahy and Colwell 1990). The interest in the research on hydrocarbon-utilizing microorganisms never ceased since the 1940s when the interactions of microorganisms and petroleum were first reported (Zobell 1946). The research in petroleum microbiology kept evolving through the years, where each decade is marked by intense interest in a different aspect of this field (Brito et al. 2015; Shibulal et al. 2014; Voordouw 2011; Atlas 1981, 1995).

Although interactions of microorganisms and petroleum has been intensively researched for decades in many advanced countries, such as United States and United Kingdom (Venosa et al. 2001), it is a relatively a new field of research in countries such as Malaysia (Othman et al. 2011). For the oil-degrading microbes to be applied in local environments, it is essential to isolate microorganisms that are adapted to the local climatic condition prior to the optimisation process. Comparatively it is more expensive to import the technology from other countries than to tailor-make our own technology for our own needs.

Crude oils in general can vary greatly in composition, viscosity, density, and flammability consisting of organic compounds ranging from methane to extremely heavy hydrocarbon molecules with up to 80 carbon atoms (Leahy and Colwell 1990). Among them, alkanes constitute the major fraction. Alkanes can constitute up to 50 % of crude oil, depending on the oil source, but are also produced by many living organisms such as plants, green algae, bacteria or animals. As alkanes are polar molecules that are chemically very inert (Labinger and Bercaw 2002), their metabolism by microorganisms poses challenges related to their low water solubility, their tendency to accumulate in cell membranes, and the energy needed to activate the molecule. However, several microorganisms, both aerobic and anaerobic, can use diverse alkanes as their source of carbon and energy. Several reviews have covered different aspects of the physiology, enzymes and pathways responsible for the degradation of alkanes (Watkinson and Morgan 1990; Ashraf et al. 1994; van Beilen et al. 2003; van Hamme et al. 2003; Coon 2005; van Beilen and Funhoff 2007; Wentzel et al. 2007). The chemical composition of crude oils may vary between regions and even within the same geologic formations (Head et al. 2006; Harayama et al. 2004, Margerin et al. 2003).

In view of the diverse microbial metabolism and adaptation mechanisms, the success of biodegradation of crude oil is very much dependent on the selection of microbes that are adapted to the local climatic condition. Hence this study focused on the isolation of local bacterial species able to degrade crude oil. These bacteria were partially characterised via biochemical tests and the profiles of n-alkane degradation by the selected bacteria are presented.

2 Materials and Methods

2.1 Chemicals Reagents

All chemicals reagents used for medium preparation, crude oil extraction and various biochemical tests were of analytical grade, except solvents used for gas chromatography analyses which were of HPLC-grade.

2.2 Crude Oil Samples

The crude oil samples were obtained from local petrochemical industries. The crude oil was autoclaved at 121 °C for 15 min prior to use.

2.3 Source of Microorganisms

Twelve bacteria that were isolated and taken from various sampling points within the vicinity of the petrochemical industries and their ability to biodegrade crude oil was investigated in this research. Bacterial sources were activated sludge, soil sludge farm, influent flotation tank, biological treatment lagoon effluent, aeration basin wastewater and treatment lagoon. These samples were serially diluted and inoculated into enrichment media (Sect. 2.5) and were incubated at 37 °C, 250 rpm in a shaking incubator overnight or until the culture medium became turbid. The bacteria were then isolated using the dilution streak method until pure colonies were obtained.

2.4 Subculturing and Storage of Microorganisms

All isolated microorganisms were subcultured directly from the stock culture on slant agar and plated on fresh nutrient agar (NA). Overnight cultures of the 12 isolates were transferred to nutrient agar slants complemented with 0.5 mL paraffin oil for storage at 5 °C. Bead cultures were also prepared for selected strains for long-term storage. Overnight cultures were centrifuged at 5000 rpm at 4 °C for 30 min and the cell pellet obtained was resuspended in nutrient broth. The cell suspension was then transferred to the bead culture system and stored at −70 °C.

2.5 Enrichment Medium

The bacteria were grown in an enrichment medium based on basal salts medium consisting of the inorganic salts 0.5 g/L KH_2PO_4, 1.0 g/L NH_4Cl, 2.0 g/L Na_2SO_4, 2.0 g/L KNO_3, 0.001 g/L $CaCl_2 \cdot 6H_2O$, 1.0 g/L $MgSO_4 \cdot 7H_2O$ and $FeSO_4$ as a trace element. The salts were dissolved in a total volume of 1000 mL distilled water and autoclaved at 121 °C for 15 min. Crude oil, 2–3 % (v/v) was added to the medium in addition to 5 mM of glucose. The glucose solution was prepared as a 50 mM stock solution and was autoclaved separately under the same condition.

2.6 Experimental Design

The experiments were carried out under both agitated and non-agitated condition in 250 mL flasks to determine which condition was more favourable for the microbial degradation of crude oil. One loop-full of stock cultures were first inoculated into the enrichment medium and incubated overnight at 37 °C. Then, 10 % (v/v) of the overnight cultures were transferred to the same enrichment medium and cultured under agitation at 250 rpm and non-agitated conditions. Control experiments in the absence of the microorganisms were also carried out. Growth was monitored throughout the process. Samples were collected after 2 and 6 days of cultivation and solvent extraction of the organic compounds (Sect. 2.8) prior to gas chromatographic analysis was performed.

2.6.1 Degradation of Crude Oil Under Agitated Condition

The inoculum was prepared by inoculating a loop-full of fresh overnight cultures plated on nutrient agar into 10 mL of the basal salts medium in a 25 mL universal bottle supplemented with 5 mM glucose and 2 % (v/v) of crude oil. The cultures were incubated at 37 °C in a shaking incubator at 250 rpm for 24 h prior to inoculation.

10 % (v/v) of inoculum was added to 250 mL Erlenmeyer flask containing 87 mL of basal salts medium supplemented with 5 mM glucose and 3 % (v/v) crude oil to a final culture volume of 100 mL. Cultures of each isolate were incubated in a shaking incubator at 30 °C, 250 rpm for 6 days.

2.6.2 Degradation of Crude Oil Under Non-agitated Condition

An inoculum of the overnight suspension culture (1 mL; 4 % (v/v)) was transferred into a universal bottle containing 23.5 mL basal salts medium supplemented with 5 mM glucose and 2 % (v/v) of crude oil. This was to minimize air in the head space. The cultures were incubated at 37 °C without any agitation for 6 days.

The universal bottles used for the non-agitated culture were sealed using the rubber stopper. They were positioned upside down to ensure the seal was fitted properly. A hole was inserted in the lid of the universal bottles using the syringe needle for sampling. All universal bottles were autoclaved before use.

2.7 Growth Determination

Growth of the cultures was monitored using a UV–Vis spectrophotometer at 550 nm.

2.8 Crude Oil Extraction

Samples of the agitated and non-agitated cultures were collected at regular time intervals after inoculation with the isolated bacterial strains. Residual crude oil was extracted by adding approximately 2 mL of n-pentane to the sample. The mixture was shaken briefly and filtered using a Whatman silicon-treated filter paper that held the aqueous portion while allowing hydrophobic compounds to pass through. The filtrate was collected in a 1.7 mL sample bottle and was vacuum-evaporated to dryness using a BUCHI B-169 vacuum system. The residual crude oil was recovered by using 1 mL of toluene and stored at 0 °C prior to gas chromatography (GC) analysis.

2.9 Gas Chromatography

The separation and determination of n-saturated alkanes was achieved by GC using a Hewlet-Packard (HP) 6890 gas chromatograph equipped with a flame ionization detector and a 25 m (0.2 mm inner diameter) Ultra-1 capillary column containing cross-linked methyl siloxane. The operating temperature of the detector was 320 °C and that of the injector was 300 °C. The column temperature was set at 50 °C for the first minute, then increased to 310 °C at a rate of 6 °C/min and the final temperature was maintained for 5 min.

Alkanes varying in length from C_{16} to C_{38} (Alltech Associates Inc.) were used as standards. The un-inoculated control and the hydrocarbons standards with chain length varying from *CWC38* were first analysed by gas chromatography and results obtained were compared. The resultant chromatograms showed the presence of all the standard compounds in the crude oil extracted from the un-inoculated control with retention times between 20 and 50 min. Peaks with differences in retention time of ±0.5 min compared to the standards were assumed as similar compounds or having the same carbon number.

Degradation was calculated by difference in percentage area of the peaks between un-inoculated control and experimental culture extracts.

$$\%\text{Degradation} = \frac{\text{Peak Area of Un inoculated Control} - \text{Peak Area of Experimental Extracts}}{\text{Peak Area of Un-inoculated Control}} \times 100$$

2.10 *Biochemical Tests*

Strains selected on the basis of best oil-degradation were subjected to further biochemical characterization. The bacteria were partially identified according to the Bergey's Manual of Bacteriology.

3 Results and Discussion

3.1 *Screening for Potential Crude Oil Degrading Microorganisms*

Mineralization or complete biodegradation of a hydrocarbon compound can occur under either aerobic or non aerobic conditions using agitated and non-agitated system. As we aimed to identify potential crude oil degrading microorganisms, both systems were used for the screening process.

Twelve bacteria isolated from two petrochemical industries were screened for their abilities to degrade crude oil under both agitated and non-agitated conditions (Table 1). Bacteria showing high biodegradation rates of crude oil, under agitated and non-agitated culture was selected for further work.

The outcomes on performance of non-agitated and agitated systems for the biodegradation of crude oil are presented in Table 2. Non-agitated biodegradation showed better degradation rates compared to the agitated cultures. All 12 isolates showed abilities to degrade C_{16}–C_{38} within 2–6 days under non-agitated conditions.

Some isolates were able to degrade long chain carbon completely after day 2 and 6 of incubation in non-agitated systems. Isolate BTL (+g)-Cr7 was able to degrade long chain C_{30}–C_{36} completely after 2 days of incubation at 37 °C. Another isolate showing ability to degrade C_{34}–C_{38} was RETL-Cr3. Isolate STL-Cr7 and STL (-g)-Cr8 consistently degraded C_{16}–C_{38} up to 100 % after 6 days in non-agitated system. Strain STL-(-g)-Cr8 had the highest biodegradation rates achieving 79.7 % biodegradation after 2 days of incubation. Although RETL-Cr3 and STL-Cr7 also achieved 80 % of biodegradation after 6 days of incubation, the initial biodegradation rate was slower compared to STL-(-g)-Cr8. Hence STL-(-g)-Cr8 was selected as the best crude oil-degrader among the 12 isolates and were partially

Table 1 Sampling source of 12 microorganisms for biodegradation of crude oil

No.	Isolate code	Sampling source
1.	RAS-CrI	Activated sludge
2.	BSSF-Crl	Soil-sludge farm
3.	RFTA-02	FTA influent
4.	MFTA-Wl	FTA influent
5.	RSSF-Crl	Soil-sludge farm
6.	BAS-Crl	Activated sludge
7.	RETL-Cr3	Biological treatment lagoon effluent
8.	MAB-CRl	Aeration basin wastewater
9.	RETL-Crl	Biological treatment lagoon effluent
10.	BTL(+g)-Cr7	Treatment lagoon
11.	STL-Cr7	Treatment lagoon
12.	STL(-g)-Cr8	Treatment lagoon

characterised via biochemical tests. Alongside STL-(-g)-Cr8, bacteria coded as BTL-(+g)-Cr7 were also selected because of its ability to utilize longer chain alkanes.

Most of the petroleum biodegradation studies show that agitation significantly increased biodegradation of crude oils (Foster 1962; Gibson and Subramaniam 1984; Watkinson 1980; Britton 1984). Conversely, this situation was not observed in this study. Screening of potential degrading microorganisms using agitated system showed contradicting results compared with non-agitated system. Most of the agitated cultures showed poor performance for the degradation of C_{16}–C_{38} as compared to cultures in non-agitated condition. The only bacterium that showed high biodegradation rates for short chain alkanes under agitated condition was RSSF-Cr1. None of the isolates were able to degrade long chain alkanes by more than 50 % after 6 days of incubation. This might be due to the origin of the microorganisms. Almost all of the microorganisms used in this research were from petroleum wastes, which were covered by oily substances preventing the diffusion of oxygen. The high degree of aeration provided might be toxic to these microorganisms (Madigan et al. 1997). Another other possibility is that agitation subjected the isolates to high shear stress (Morgan and Watkinson 1990). Some of the specific enzymes responsible for catalysing degradation of crude oil components are very sensitive to oxygen. While enzymes are synthesised inside the cell, their secretion is important for the initiation of the degradation of extracellular macromolecules, which must be broken down into smaller subunits prior to uptake (Rojo 2009; Ajisebutu et al. 2001; Maier et al. 1999).

The reduction of the crude oil in the cultures was established for most isolates after a few days of incubation in the agitated system (Fig. 1a, b). Some problems occurred prevented sampling processing of a few bacteria, RAS-Crl, BSSF-Crl, RFTA-02 and BAS-Crl. Formation of slimy and foam-like oil clumps was observed after a few hours to 2 days of cultivation (Fig. 1a, b), making it was difficult to obtain homogenous aliquots of the cultures. The oil clumps remained in clusters or

Table 2 Performance of 12 isolated bacteria for crude oil biodegradation after 6 days under non-agitated (NA) and agitated (A) conditions

	Percentage of Crude Oil Component Biodegradation after 6 days in incubation at 37 °C in agitated (A) and non-agitated (NA) Systems																						
Carbon	C16		C18		C20		C22		C24		C26		C28		C30		C32		C34		C36		C38	
Strain Code	NA	A	NA	A	NA	A	NA	A	NA	A	NA	A	NA	A	NA	A	NA	A	NA	A	NA	A	NA	A
RAS-Crl	62.2	nd	59.2	nd	57.6	nd	51.4	nd	45.9	nd	47.4	nd	46.8	nd	51.6	nd	51.7	nd	45.1	nd	28.1	nd	41.0	nd
BSSF-Crl	79.5	nd	76.5	nd	74.1	nd	69.6	nd	66.2	nd	66.4	nd	66.2	nd	68.1	nd	67.3	nd	61.6	nd	48.9	nd	58.2	nd
RFTA-02	65.4	nd	64.5	nd	64.9	nd	65.3	nd	67.2	nd	68.8	nd	69.9	nd	71.6	nd	67.7	nd	61.6	nd	57.0	nd	65.4	nd
MFTA-W1	33.4	53.9	35.1	19.1	35.7	25.9	27.1	22.6	23.7	25.0	26.0	31.6	28.5	32.7	31.6	32.1	32.4	16.9	21.0	27.3	19.1	32.9	0	37.4
RSSF-Crl	67.8	35.7	65.5	45.1	63.9	23.7	60.3	23.8	59.1	23.4	62.2	29.6	63.0	0	64.3	4.35	67.5	0	61.8	12.4	50.0	0	62.7	18.5
BAS-Crl	0	nd	0	nd	0	nd	0	nd	2.62	nd	4.84	nd	6.84	nd	6.50	nd	0.77	nd	0	nd	0	nd	0	nd
RETL-Cr3	82.8	nd	80.7	nd	80.4	nd	80.7	nd	81.8	nd	82.9	nd	83.3	nd	83.9	nd	80.7	nd	100	nd	100	nd	100	nd
MAB-Crl	21.3	36.9	21.7	0	23.2	23.1	16.2	16.2	16.9	33.7	19.1	26.6	23.6	18.3	22.6	29.3	23.0	7.62	8.59	9.60	0	25.6	0	12.1
RETL-Cr1	78.3	55.1	76.9	26.7	75.1	52.0	70.0	29.4	66.6	23.7	67.1	15.4	64.9	25.2	67.5	31.6	67.2	0	56.7	27.4	37.9	8.43	55.0	12.1
BTL(+g)-Cr7	45.8	49.3	45.3	21.3	46.4	39.4	49.5	38.9	65.4	15.8	74.8	29.8	82.8	21.3	100	22.0	100	8.22	100	7.91	100	10.6	100	59.0
STL-Cr7	86.4	36.3	84.0	45.9	82.5	40.1	80.9	14.6	80.2	18.8	81.7	21.7	79.1	23.5	83.1	38.3	83.5	0	82.3	0	74.4	20.2	100	44.6
STL(-g)-Cr8	85.5	28.6	82.7	8.7	82.2	41.1	77.4	5.34	78.1	20.8	78.9	19.9	64.1	37.6	78.6	21.1	77.1	12.7	69.2	34.5	100	44.4	100	40.6

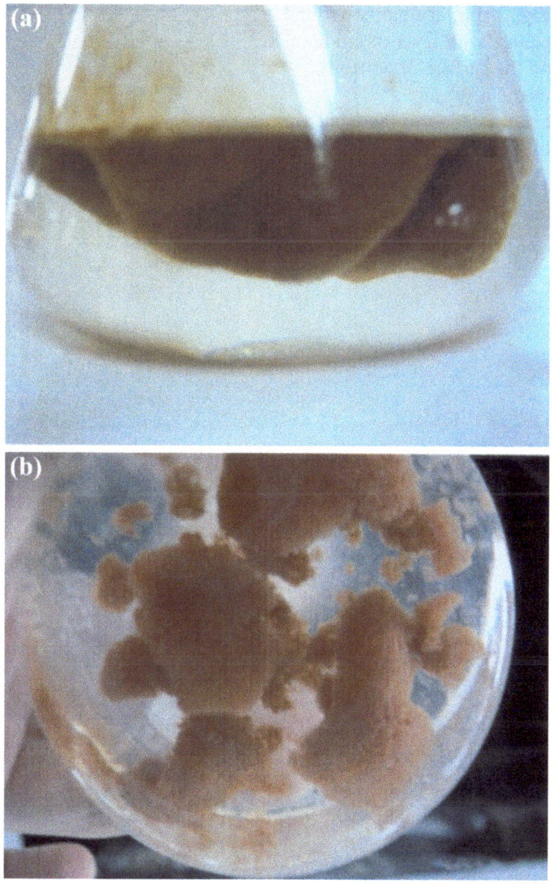

Fig. 1 Oil clumps observed in BAS-Crl culture from side view (**a**) for BSSF-Crl culture from the bottom view of the flask (**b**)

aggregates and smaller clusters could not be obtained despite of vigorous shaking of the flasks.

Besides the formation of oil clumps by a few strains, most of the cultures showed signs of emulsification (Fig. 2a, b). This emulsification of the crude oil into tiny oil droplets could be observed within hours of shaking at 250 rpm for RSSF-Crl, RETL-Crl, STL-Cr7, STL-(-g)-Cr8 and MFTA-Wl. This observation is very common in most petroleum biodegradation and biosurfactant studies (Shibulal et al. 2014; Bryant 1987; Cooper 1982). The results provided the evidence for the production of biosurfactants by these microorganisms. Biosurfactants and extra-cellular enzymes are important agents in the degradation of aliphatic macro-molecules (Robert and Philip 2012; Maier et al. 1999).

Complete biodegradation or mineralization involves the oxidation of the parent compound to form carbon dioxide and water, a process that provides both carbon and energy for growth and reproduction of cells (Rojo 2009; Maier et al. 1999,). Low degradation in all agitated culture might be due to lack of internal or extra-cellular enzymes. Lack of appropriate biodegrading enzymes is one of the common

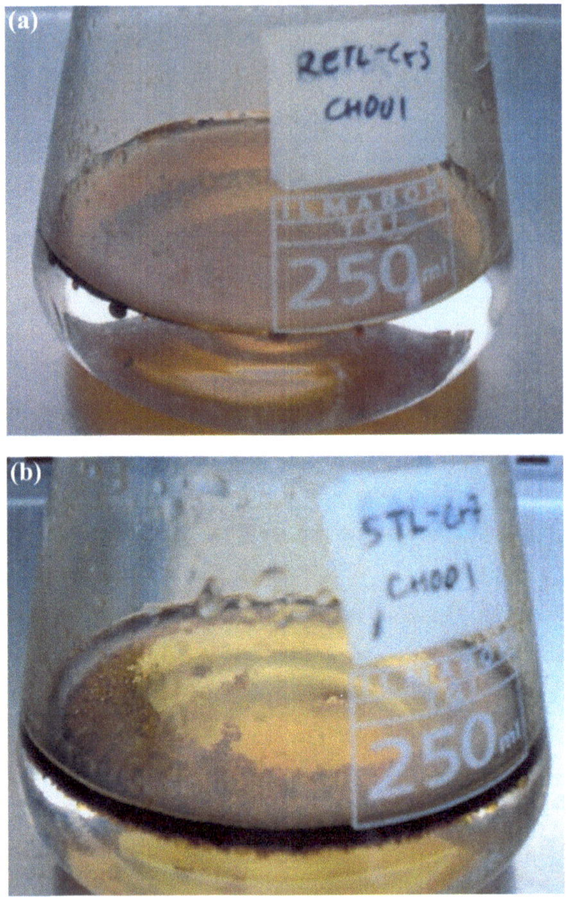

Fig. 2 Initial condition of the culture with layer of crude oil on the surface of aqueous medium (**a**) and emulsification of the crude oil by STL-Cr7 after 2 days (**b**)

reasons for persistence of organic contaminants, particularly those with unusual chemical structures that existing enzymes do not recognize.

3.2 *Characterization of Selected Bacteria*

The 3 strains that were selected for biochemical testing were STL-(-g)-Cr8, RSSF-Cr1 and BTL-(+g)-Cr7. Twelve different biochemical tests had to be carried out to characterize the 3 selected isolates, and the results are shown in Table 3.

All three isolates were Gram-negative. RSSF-Crl was rod shape, while STL-(-g)-Cr8 and BTL-(+g)-Cr7 were cocci, with fresh cultures, however, appearing to be short rods. Coccobacilli are often short rods becoming spherical in stationary phase of growth. Based on the results of the biochemical tests and their cellular morphologies, RSSF-Crl was identified as *Proteus* sp., while STL-(-g)-Cr8 and BTL-(+g)-Cr7 were identified as the genus *Acinetobacter*.

Table 3 Biochemical tests on the three selected isolates

Biochemical tests	STL-(-g)-Cr8	RSSF-Crl	BTL-(+g)-Cr7
Gram-stain	Gram negative	Gram negative	Gram negative
O–F test	F+	F+	F+
Spore	–	–	–
Motility	+	–	–
Catalase	+	+	+
Oxidase	–	–	–
H_2S	–	–	–
Gelatinase	–	–	–
Nitrate reductase	+	+	+
Citrase	+	+	+
Urease	+	+	–
Lactose fermentation	+	–	–

Strain STL-(-g)-Cr8 and BTL-(+g)-Cr7 were rod shaped *Acinetobacter* bacteria with 0.9–1.6 μm in diameter and 1.5–2.5 μm in length becoming spherical in stationary phase of growth [4]. They commonly occur in pairs and also in chains of variable length. Cells do not form spores. Cells are Gram-negative. Swimming motility does not occur, but the cells display "twitching motility" when observed under the microscope, presumably because of the presence of polar fimbriae (Baltimore 1994). They grow well on non-selective media such as nutrient agar. They are oxidase negative and catalase positive. Most bacteria grow well in defined media (as in this case was basal salts medium) containing a single carbon and energy source; they use ammonium or nitrate salts as the source of nitrogen and display no growth factor requirement (Baltimore 1994). Species of this genus occur naturally in soil, water and sewage (Baltimore 1994). In this study, both STL-(-g)-Cr8 and BTL-(+g)-Cr7 were from a treatment lagoon at a petroleum refinery. *Acinetobacter* sp. is also one of the genera commonly reported to be encountered in hydrocarbon-contaminating soils (Kampher and Steiof 1991; Scott and Finerty 1976).

RSSF-CrI bacteria were straight rods with 0.4–0.8 μm in diameter and 1–3 μm in length (Baltimore 1994). The bacterium was motile when observed by light microscope, which is one of the characteristics of the genus *Proteus*. The bacteria spread in a uniform film over moist surfaces of nutrient media solidified with agar. *Proteus* bacteria have both a respiratory and a fermentative type of metabolism (Baltimore 1994). The *Proteus* sp. isolated was oxidase-negative and catalase-positive, and was capable of reducing nitrates. Species of this genus commonly occur in the intestines of humans and a wide variety of animals; they also occur in manure, soil and polluted waters (Baltimore 1994). As for RSSF-Crl, it originated from a soil-sludge farm. *Proteus* sp. has rarely been used in studies as an oil-degrading strain (Shaw et al. 1985; Kalish et al. 1964).

3.3 Biodegradation of Crude Oil by Selected Bacteria

This section described the biodegradation potential of the selected bacteria in greater details. STL-(-g)-Cr8 showed the highest biodegradation rate among all isolates; RSSF-Crl, was the best n-alkanes degrader under agitated conditions, and BTL-(+g)-Cr7 exhibited ability to degrade higher molecular weight n-alkanes.

Figure 3a–c shows the crude oil biodegradation by STL-(-g)-Cr8 *(Acinetobacter* sp.) under non-agitated condition. The chromatograms of the control (a) and the partially degraded crude oil by STL-(-g)-Cr8 (b) were used as comparison. In general, there was a significant decrease in all hydrocarbon peaks after day-2 (Fig. 3b) and day-6 (Fig. 3c) of cultivation as compared to the control (Fig. 3a). The short chain alkanes below C_{10} were not detected in the chromatogram because they are generally lost by evaporation or by dissolution in water (Brown 1987). The performance of STL-(-g)-Cr8 in non-agitated system is presented in Table 4. The percentage biodegradation of each standard compound (C_{16}–C_{38} alkanes) after days 2 and 6 of incubation were more than 30 % degraded on day 2. Biodegradation of alkanes below C_{34} was achieved above 80 % and C_{34}–C_{38} alkanes were fully degraded after days 6. Published reports show biodegradation of up to 90.6 % on day 10 by *Oscillatoria salina* (Raghukumar et al. 2001). However, the results could not be compared directly since the parameters and conditions were different. Further optimization of the conditions and parameters might improve the biodegradation process.

The change of the peaks of respective alkanes is presented in Fig. 4. The shorter length alkanes decrease more rapidly as compared to the longer ones. This finding is in a good agreement with most published research (Brown 1987; Blasig et al. 1988). It is commonly known that the greater the chain length and amount of branching, the more resistant compounds are to microbial attack. It is possible that shorter alkanes are metabolised preferentially, since less energy is required in breaking down simple compounds (Blasig et al. 1988; Maier et al. 1999). The rate of biodegradation of crude oil by strain STL-(-g)-Cr8 was calculated as 0.65–1.15 % h^{-1}. Figure 5 shows the profile of crude oil degradation by BTL-(+g)-Cr7. BTL-(+g)-Cr7 showed a preference for metabolising longer chain n-alkanes. C_{36} and C_{38} compounds were first fully degraded within 2 days while by C_{30}–C_{34} were fully degraded after 6 days. The degradation rate varied from 1.05 to 2.08 % h^{-1} which is 2 times higher than for STL-(-g)-Cr8.

However, all other compounds were degraded less than 70 % after days 6 of incubation, except for C_{26} and C_{28} in which 74.75 and 82.84 % were degraded, respectively. The profiles of crude oil biodegradation observed were the opposite to other published data (Brown 1987). This may be explained by the conditions of the original habitat from which the strain was isolated which are dominated by long chain alkanes. Results suggest that degradation of shorter alkanes could be enhanced using mixed cultures, that is, combining potential oil-degrading strains, which will complement each other's role.

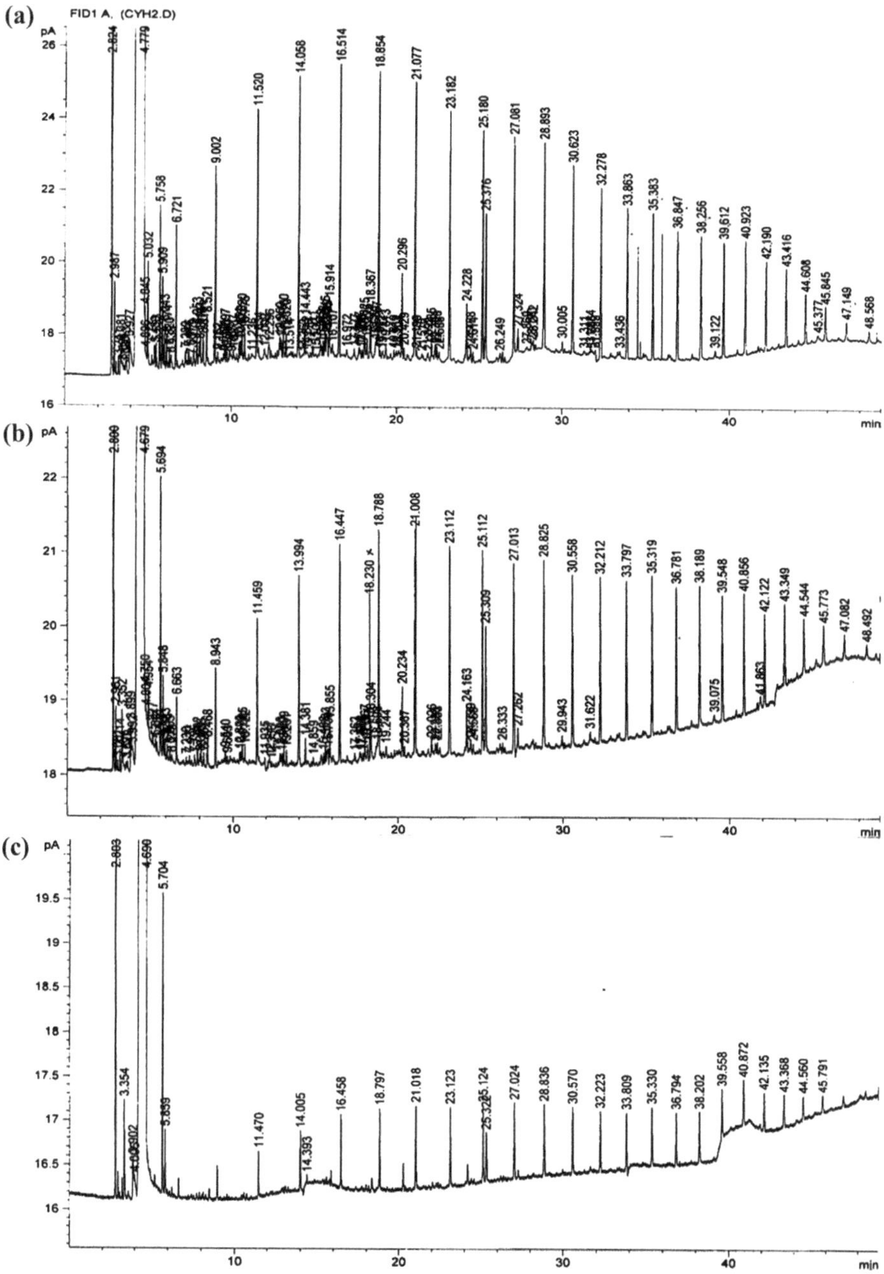

Fig. 3 Chromatogram of biodegradation of n-alkanes in crude oil by STL-(-g)-Cr8 under non-agitated system. **a** Un-inoculted control, **b** after 2 days of incubation and **c** after 6 days of incubation

Table 4 Partial biodegradation of n-alkanes in crude oil by STL-(-g)-Cr8 under non- agitated system

N-alkanes	% Biodegradation	
	Day-2	Day-6
C_{16}	55.28	82.79
C_{18}	54.17	80.77
C_{20}	48.15	80.36
C_{22}	50.43	80.66
C_{24}	36.08	81.82
C_{26}	37.29	82.90
C_{28}	38.10	83.27
C_{30}	41.12	83.88
C_{32}	42.72	80.74
C_{34}	34.85	100.00
C_{36}	31.46	100.00
C_{38}	38.96	100.00

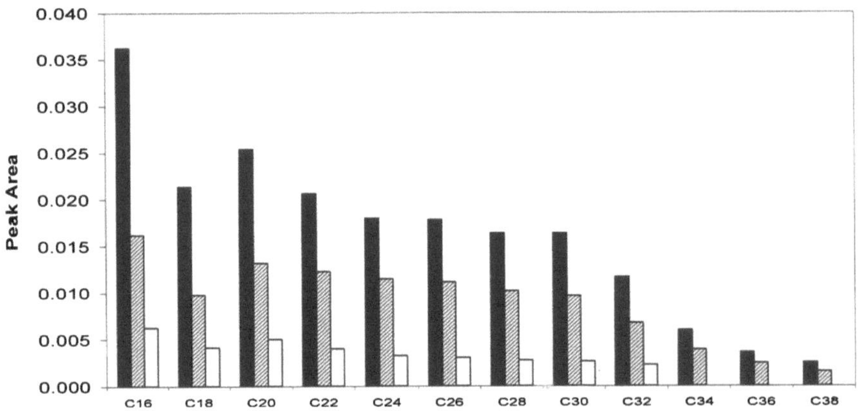

Fig. 4 Biodegradation of n-alkanes in crude oil by STL-(-g)-Cr8 under non-agitated conditions. *Shaded bar* represents the initial peaks at day 0; *striped bar* represents the peaks on day-2 and *white bar* on day-6

The potential biodegradation of crude oil by RSSF-Crl (*Proteus* sp.) under agitated condition was studied. Generally, agitated conditions showed no major improvement to the biodegradation process. Only the bacteria coded as RSSF-Crl performed better under agitated condition. During the biodegradation process under non-agitated condition, RSSF-Crl degraded C_{20} alkanes at 0.93 % h^{-1}, while the rate increased to 1.30 % h^{-1} when agitation was provided, which means the rate of biodegradation had increased by 39.7 %. Other studies on certain bacteria had shown a higher increase in biodegradation rate, ranging from 50–60 % when agitation was provided (Blasig et al. 1988; Okoh et al. 2001; Raghukumar et al. 2001). Figure 6 shows the comparison of biodegradation of crude oil by RSSF-Crl under non-agitated and agitated conditions. All compounds showed increase in

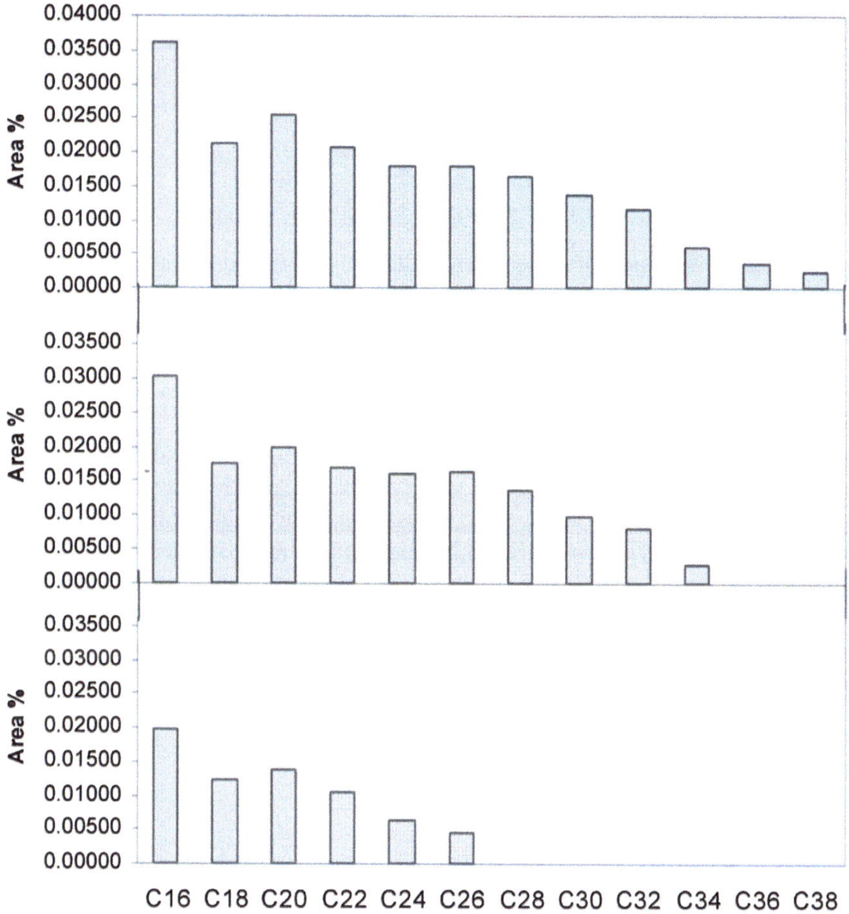

Fig. 5 Biodegradation of n-alkanes in crude oil by BTL-Cr7 in agitated culture. **a** Un-inoculated control, **b** after day 2 and **c** after day 6 of incubation

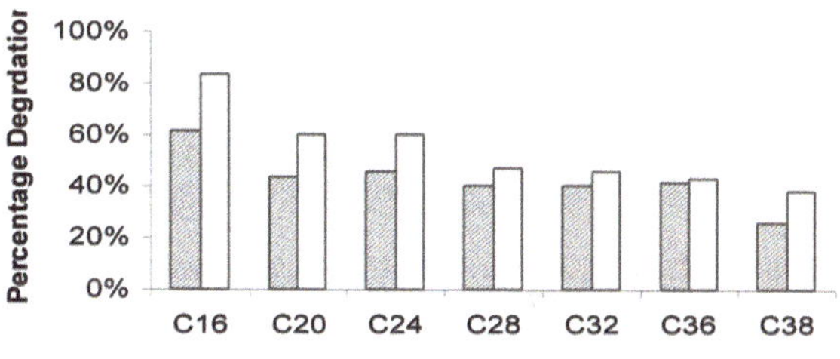

Fig. 6 Comparison of the biodegradation of crude oil by RSSF-Crl under non-agitated (*striped bar*) and agitated conditions (*white bar*)

biodegradation under agitated condition. The differences in the biodegradation on shorter chain alkanes were more significant than for the longer ones, implying aeration was required for the better degradation of shorter chain alkanes.

4 Conclusions

In this study, 12 cultures of bacteria with potential ability to degrade n-alkanes were successfully isolated. Three of the best n-alkanes degraders were partially identified via biochemical tests; two were *Acinetobactor* spp while the other belonged to *Proteus* sp. STL-(-g)-Cr8, identified as *Acinetobacter* sp. yielded the highest biodegradation rate, in which 80–100 % of the standard alkanes were degraded within 6 days at varying rates, ranging from 0.65 to 1.51 % h^{-1}. It was able to degrade most of the n-alkanes (C_{16}–C_{38}) present in crude oil without any addition of organic nitrogen and phosphorus source. Overall, the biodegradation rate of non-agitated cultures was nearly 39.7 % higher than in agitated cultures. The only bacteria that showed improvement in biodegradation under agitated condition was RSSF-Crl *(Proteus* sp.). RSSF-Crl biodegradation rate on C_{20} was higher than non-agitated cultures. BTL-(+g)-Cr7, an *Acinetobacter* sp. had an ability to degrade longer chain alkanes more rapidly than shorter ones. C_{36} and C_{38} were fully degraded within 2 days at a rate varying from 1.05 to 2.08 % h^{-1}.

References

Ajisebutu S, Babalola G, Trejo-Hernandez MR (2001) Potential of *Burkholderia cepacia* RQ1 in the biodegradation of heavy crude oil. Int Microbiol 4:83–87

Al-Wasify RS, Hamed SR (2014) Bacterial biodegradation of crude oil using local isolates. Int J Bacteriol 1–8

Ashraf W, Mihdhir A, Murrell JC (1994) Bacterialoxidation of propane. FEMS Microbiol Lett 122:1–6

Atlas RM (1981) Microbial degradation of petroleum hydrocarbons: an environmental perspective. Microbiol Rev 45(1):180–209

Atlas RM (1995) Bioremediation of petroleum pollutants. Int Biodeterior Biodegradation 35(1–3):317–327

Baltimore M (1994) Bergey's manual of determinative bacteriology, 9th edn. Williams & Wilkins, Baltimore

Blasig R, Mauersberger S, Riege P, Schunck W-H, Jockisch W, Franke P, Muller H-G (1988) Degradation of long-chain n-alkanes by the yeast *Candida maltosa* II. Osidation of n-alkanes and intermediate using microsomal membrane fractions. Appl Microbiol Biotechnol 28:589–597

Brito EM, De la Cruz Barron M, Caretta CA, Goni-Urriza M, Andrade LH, Cuevas-Rodriguez G, Malm O, Torres JP, Simon M, Guyoneaud R (2015) Impact of hydrocarbon, PCBs and heavy metals on bacterial communities in Lerma River, Salamanca, Mexico: investigation of hydrocarbon degradation potential. Sci Total Environ 521–522:1–10

Britton LN (1984) Microbial degradation of aliphatic hydrocarbons. In: Gibson DT (ed) Microbial degradation of organic compounds. Marcel Dekker, New York, pp 89–129

Brown LR (1987) Oil degrading microorganisms. Chem Eng Prog October 35–40

Bryant RS (1987) Potential uses of microorganisms in petroleum recovery technology. Proc Okla Acad Sci 67:97–104

Coon MJ (2005) Omega oxygenases: nonheme-iron enzymes and P450 cytochromes. Biochem Biophys Res Commun 338:378–385

Cooper DG (1982) Biosurfactant and enhanced oil recovery. Proc Int Conf Microbial Enhanced Oil Recovery. Afton, Oklahoma, pp 112–113

Das N, Chandran P (2011) Microbial degradation of petroleum hydrocarbon contaminants: an overview–SAGE Hindawi. Biotechnol Res Int 1–13

Foster JW (1962) Hydrocarbons as substrate for microorganisms. Anton van Leuwenhooke Microbiol Serol 28:241–274

Gibson DT, Subramaniam V (1984) Microbial degradation of aromatic hydrocarbons. In: Gibson ET (ed) Microbial degradation of organic compounds. Marcel Dekker, Inc., New York, pp 181–252

Harayama S, Hara A, Baik A, Shutsubo K, Misawa N, Smits THM, van Beilen JB (2004) Cloning and functional analysis of AlkB genes in *Alcanivorax borkumensis* SK2. Environ Microbiol 6:191–197

Head IM, Jones DM, Rolly WF (2006) Marine microorganism make a meal of oil. Nat Rev Microbiol 4:173–182

Kalish PJ, Stewart JA, Rogers WF, Bennett EO (1964) The effect of bacteria on sandstone permeability. J Petrol Technol 16(7):805–814

Kampher P, Steiof M (1991) Microbiological characterization of a fuel-oil contaminated site including numerical identification of heterotrophic water and soil bacteria. Microbial Ecol 21:227–251

Labinger JA, Bercaw JE (2002) Understanding and exploiting C–H bond activation. Nature 417:507–514

Leahy JG, Colwell RR (1990) Microbial degradation of hydrocarbons in the environment. Microbiol Rev 54(3):305–315

Madigan MT, Martinko JM, Parker J (1997) Brock biology of microorganisms. Prentice Hall, Englewood Cliffs, pp 473–531

Maier RM, Pepper IL, Gerba CP (1999) Environmental microbiology. Academic Press, London, pp 365–402

Morgan P, Watkinson RJ (1990) Physiology of aliphatic hydrocarbon-degrading microorganisms. In: Ratledge C (ed) Physiology of biodegradative microorganisms. Kluwer, Dordrecht, pp 79–92

Okoh, Ajisebutu S, Babalola G, Trejo-Hernandez M (2001) Potential of *Burkholderia cepacia* RQ1 in the biodegradation of heavy crude oil. Int Microbiol 4(2):83–87

Othman N, Irwan JM, Hussain N, Abdul-Talib S (2011) Bioremediation a potential approach for soil contaminated with polycyclic aromatic hydrocarbon: an overview. Int J Sustain Construct Eng Technol 2(2):48–53

Raghukumar C, Vipparty V, David JJ, Chandramohan D (2001) Degradation of crude oil by marine cyanobacteria. Appl Microbiol Biotechnol 57:433–436

Robert JW, Philip M (2012) Physiology of aliphatic hydrocarbon-degrading microorganisms. In: Colin R (ed) Physiology of biodegradative microorganism, 2nd edn. Springer, Heidelberg, pp 79–92

Rojo F (2009) Minireview: degradation of alkanes by bacteria. Environ Microbiol 11(10):2477–2490

Scott CCL, Finerty WR (1976) Characterization of intra cytoplasmic hydrocarbon inclusions from the hydrocarbon-oxidizing *Acinobacter sp.* HO1-N. J Bacteriol 127:481–489

Shaw JC, Bramhill B, Wardlaw NC, Costerton JW (1985) Bacterial fouling in a model core system. Appl Environ Microbiol 49(3):693–701

Shibulal B, Al-Bahry SN, Al-Wahaibi YM, Elshafie AE, Al-Bemani AS, Joshi SJ (2014) Microbial enhanced heavy oil recovery by the aid of inhabitant spore-forming bacteria: an insight review. Sci World J 1–12

van Beilen JB, Funhoff EG (2007) Alkane hydroxylases involved in microbial alkane degradation. Appl Microbiol Biotechnol 74:13–21

van Beilen JB, Li Z, Duetz WA, Smits THM, Witholt B (2003) Diversity of alkane hydroxylase systems in the environment. Oil Gas Sci Technol 58:427–440

van Hamme JD, Singh A, Ward OP (2003) Recent advances in petroleum microbiology. Microbiol Mol Biol Rev 67:503–549

Venosa AD, Zhu X, Suidan MT, Lee K (2001) Guidelines for the bioremediation of marine shorelines and freshwater wetlands. U.S. Environmental Protection Agency, National Risk Management Research Laboratory Cincinnati, Ohio

Voordouw G (2011) Production-related petroleum microbiology: progress and prospects. Curr Opin Biotechnol 22:1–5

Watkinson RJ (1980) Interaction of microorganisms with hydrocarbons. In: Harrison DEF, Higgins IJ, Watkinson RJ (eds) Hydrocarbons in biotechnology. Heyden, London, pp 1–24

Watkinson RJ, Morgan P (1990) Physiology of aliphatic hydrocarbon-degrading microorganisms. Biodegradation 1:79–92

Wentzel A, Ellingsen TE, Kotlar HK, Zotchev SB, Throne-Holst M (2007) Bacterial metabolism of long-chain n-alkanes. Appl Microbiol Biotechnol 76:1209–1221

Zobell CE (1946) Action of microorganisms on hydrocarbons. Bacteriological Rev 10:1–49

MIX
Papier aus verantwortungsvollen Quellen
Paper from responsible sources
FSC® C105338

If you have any concerns about our products,
you can contact us on
ProductSafety@springernature.com

In case Publisher is established outside the EU,
the EU authorized representative is:
Springer Nature Customer Service Center GmbH
Europaplatz 3, 69115 Heidelberg, Germany

Printed by Libri Plureos GmbH
in Hamburg, Germany